W9-AGU-339

LOGICOMIX

LOGICOMIX

APOSTOLOS DOXIADIS
CHRISTOS H. PAPADIMITRIOU

ART
ALECOS PAPADATOS

COLOR
ANNIE DI DONNA

BLOOMSBURY

NEW YORK · BERLIN · LONDON

Published by Bloomsbury USA, New York

All papers used by Bloomsbury USA are natural, recyclable products made from
wood grown in well-managed forests. The manufacturing processes conform to
the environmental regulations of the country of origin.

LIBRARY OF CONGRESS CATALOGING-IN-PUBLICATION DATA HAS BEEN APPLIED FOR.

ISBN-10 1-59691-452-1
ISBN-13 978-1-59691-452-0

First U.S. Edition 2009

1 3 5 7 9 10 8 6 4 2

Printed and bound in Singapore by Tien Wah Press

CONCEPT & STORY
Apostolos Doxiadis
Christos H. Papadimitriou

SCRIPT
Apostolos Doxiadis

CHARACTER DESIGN & DRAWINGS
Alecos Papadatos

COLOR
Annie Di Donna

INKING
Dimitris Karatzaferis
Thodoris Paraskevas

VISUAL RESEARCH & LETTERING
Anne Bardy

To our children,
Eirene, Emma, Isabel, Io,
Kimon, Konstantinos,
Tatiana, Yorgos

Ὑμὲς δ' ἔσεσθε πολλῷ κάρρονες.

OVERTURE

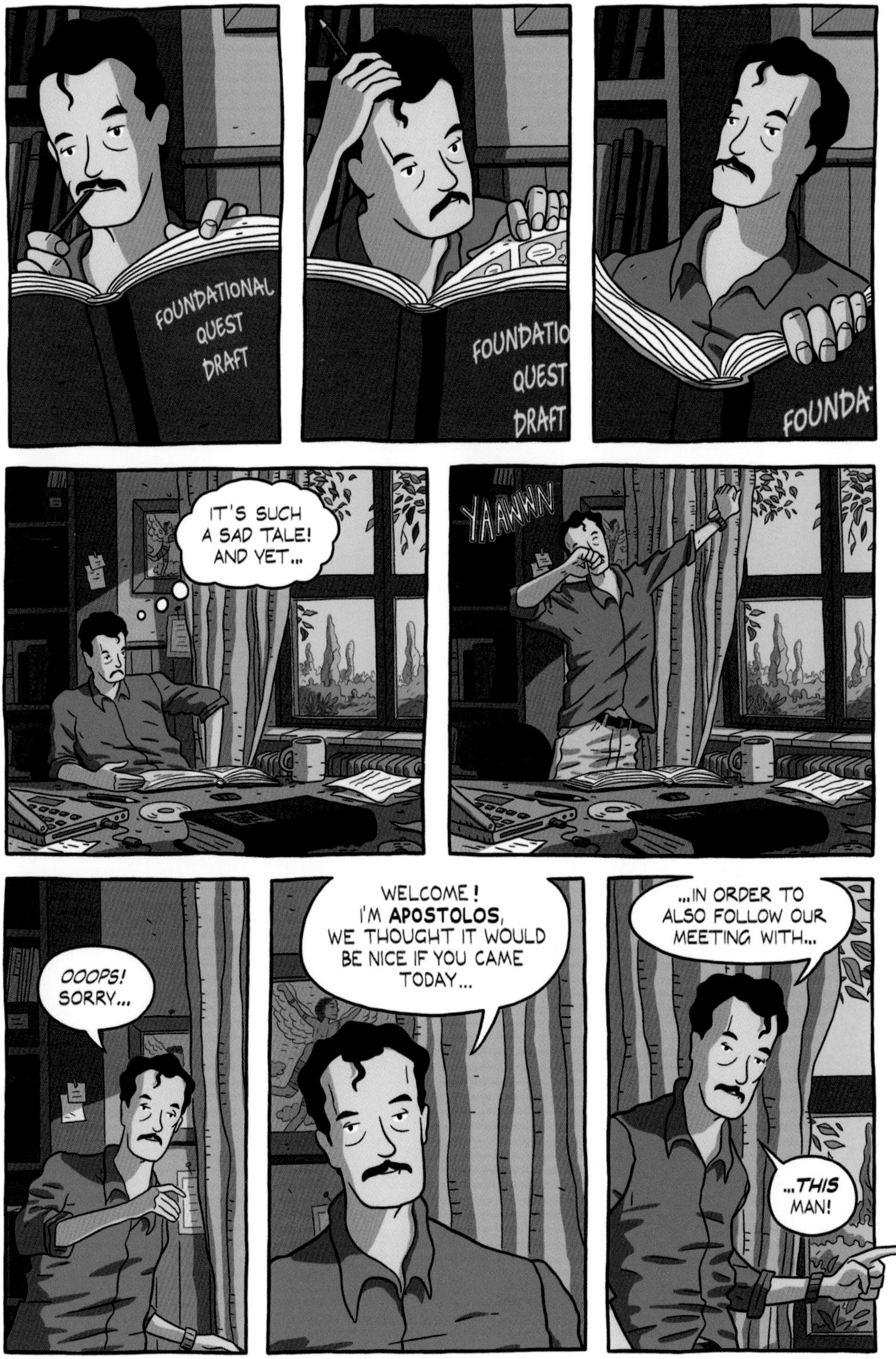
FOUNDATIONAL QUEST DRAFT
FOUNDATIO QUEST DRAFT
FOUNDA-
IT'S SUCH A SAD TALE! AND YET...
YAAWWN
OOOPS! SORRY...
WELCOME! I'M APOSTOLOS, WE THOUGHT IT WOULD BE NICE IF YOU CAME TODAY...
...IN ORDER TO ALSO FOLLOW OUR MEETING WITH...
...THIS MAN!

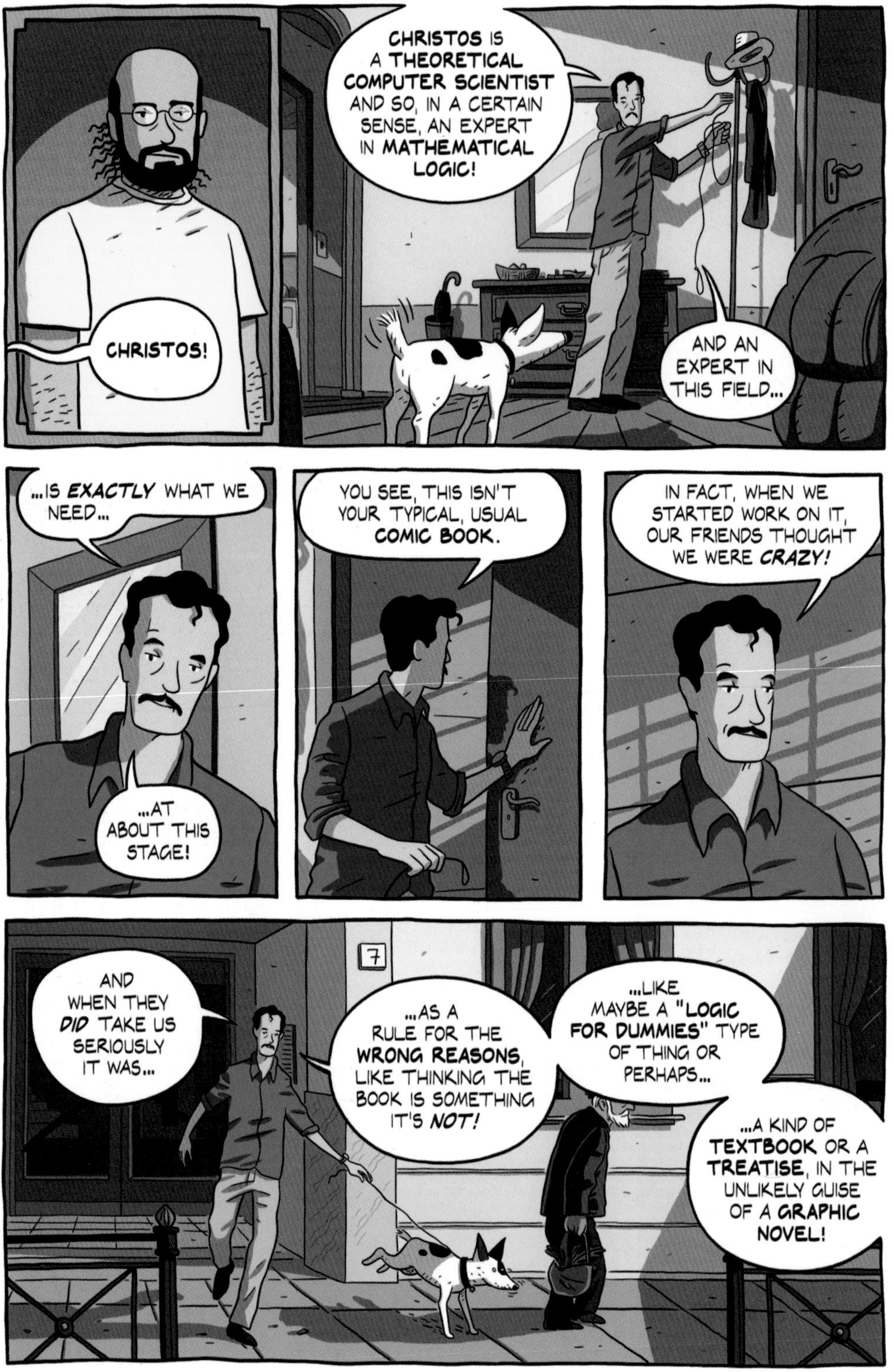
CHRISTOS!
CHRISTOS IS A THEORETICAL COMPUTER SCIENTIST AND SO, IN A CERTAIN SENSE, AN EXPERT IN MATHEMATICAL LOGIC!
AND AN EXPERT IN THIS FIELD...
...IS EXACTLY WHAT WE NEED...
...AT ABOUT THIS STAGE!
YOU SEE, THIS ISN'T YOUR TYPICAL, USUAL COMIC BOOK.
IN FACT, WHEN WE STARTED WORK ON IT, OUR FRIENDS THOUGHT WE WERE CRAZY!
7
AND WHEN THEY DID TAKE US SERIOUSLY IT WAS...
...AS A RULE FOR THE WRONG REASONS, LIKE THINKING THE BOOK IS SOMETHING IT'S NOT!
...LIKE MAYBE A "LOGIC FOR DUMMIES" TYPE OF THING OR PERHAPS...
...A KIND OF TEXTBOOK OR A TREATISE, IN THE UNLIKELY GUISE OF A GRAPHIC NOVEL!

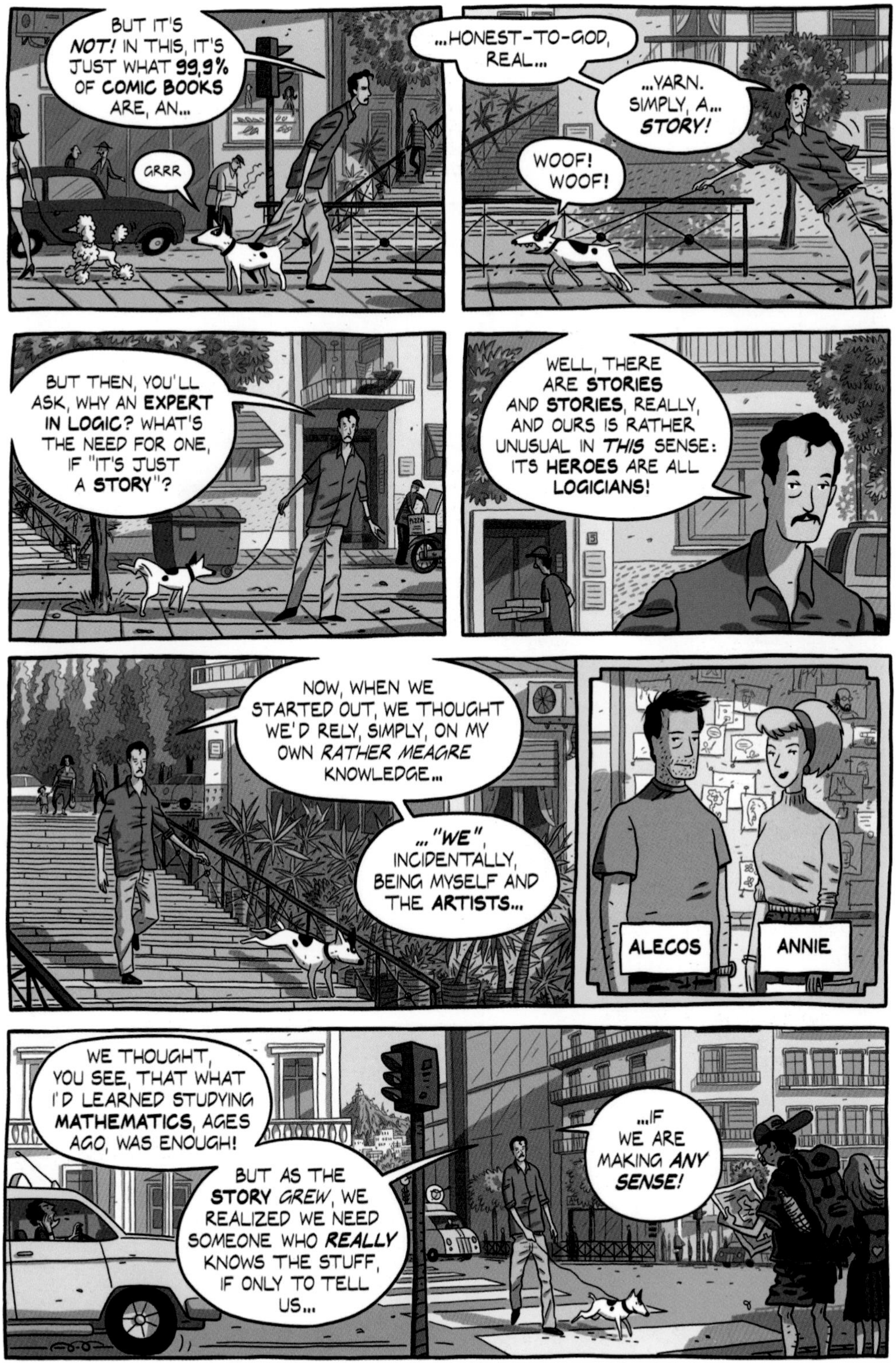
BUT IT'S NOT! IN THIS, IT'S JUST WHAT 99,9% OF COMIC BOOKS ARE, AN...
GRRR
...HONEST-TO-GOD, REAL...
...YARN. SIMPLY, A... STORY!
WOOF! WOOF!
BUT THEN, YOU'LL ASK, WHY AN EXPERT IN LOGIC? WHAT'S THE NEED FOR ONE, IF "IT'S JUST A STORY"?
WELL, THERE ARE STORIES AND STORIES, REALLY, AND OURS IS RATHER UNUSUAL IN THIS SENSE: ITS HEROES ARE ALL LOGICIANS!
NOW, WHEN WE STARTED OUT, WE THOUGHT WE'D RELY, SIMPLY, ON MY OWN RATHER MEAGRE KNOWLEDGE...
..."WE", INCIDENTALLY, BEING MYSELF AND THE ARTISTS...
ALECOS
ANNIE
WE THOUGHT, YOU SEE, THAT WHAT I'D LEARNED STUDYING MATHEMATICS, AGES AGO, WAS ENOUGH!
BUT AS THE STORY GREW, WE REALIZED WE NEED SOMEONE WHO REALLY KNOWS THE STUFF, IF ONLY TO TELL US...
...IF WE ARE MAKING ANY SENSE!

BUT ENOUGH SAID IN THE WAY OF A PREAMBLE. LET'S NOW GO, AT LAST, AND MEET CHRISTOS!

AH, THERE HE IS!

GOOD TO SEE YOU, MAN!

BUT WE BETTER...

...NOT TELL HIM...

...WE'RE "RECORDING LIVE", AS IT WERE. OK?

...KEEP HIS STYLE MORE NATURAL!

COME AGAIN?

I MUST SAY, I'M SO GLAD YOU CAME!

SO AM I! IT'S NOT EVERY DAY THAT I'M ASKED TO CONTRIBUTE TO A COMIC BOOK ON...

...THE "QUEST FOR THE FOUNDATIONS OF MATHEMATICS"!

I HOPE YOU ENJOY THE BOOK!

OH, THANKS. I SURE HOPE SO TOO...

BUT TELL ME MORE! WHAT'S YOUR LINE OF ATTACK ON THE "QUEST"?

MM, LET'S SAY IT'S RATHER OBLIQUE!

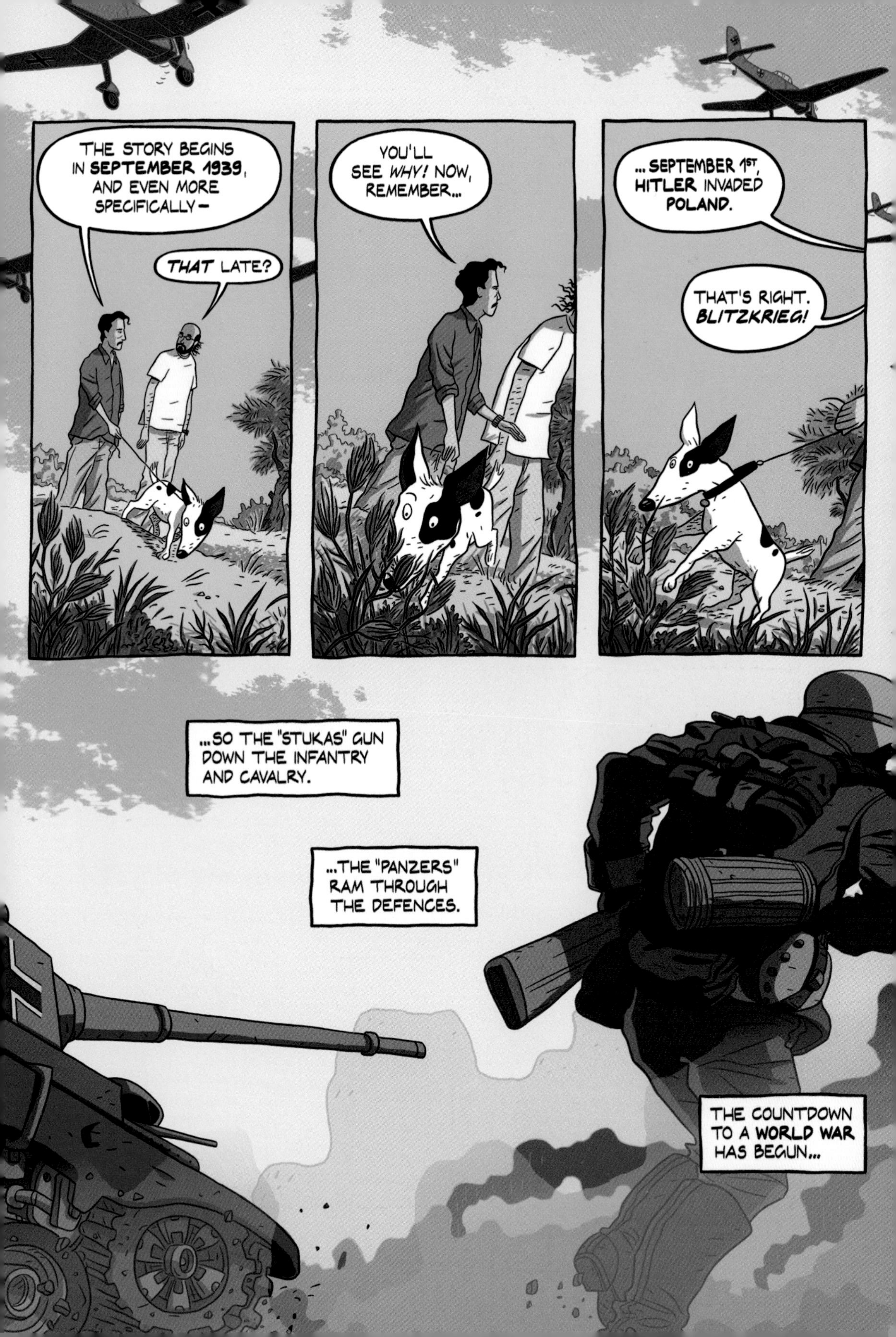
THE STORY BEGINS IN **SEPTEMBER 1939**, AND EVEN MORE SPECIFICALLY—
THAT LATE?
YOU'LL SEE *WHY!* NOW, REMEMBER...
... **SEPTEMBER 1ST**, **HITLER** INVADED **POLAND**.
THAT'S RIGHT. ***BLITZKRIEG!***
...SO THE "STUKAS" GUN DOWN THE INFANTRY AND CAVALRY.
...THE "PANZERS" RAM THROUGH THE DEFENCES.
THE COUNTDOWN TO A **WORLD WAR** HAS BEGUN...

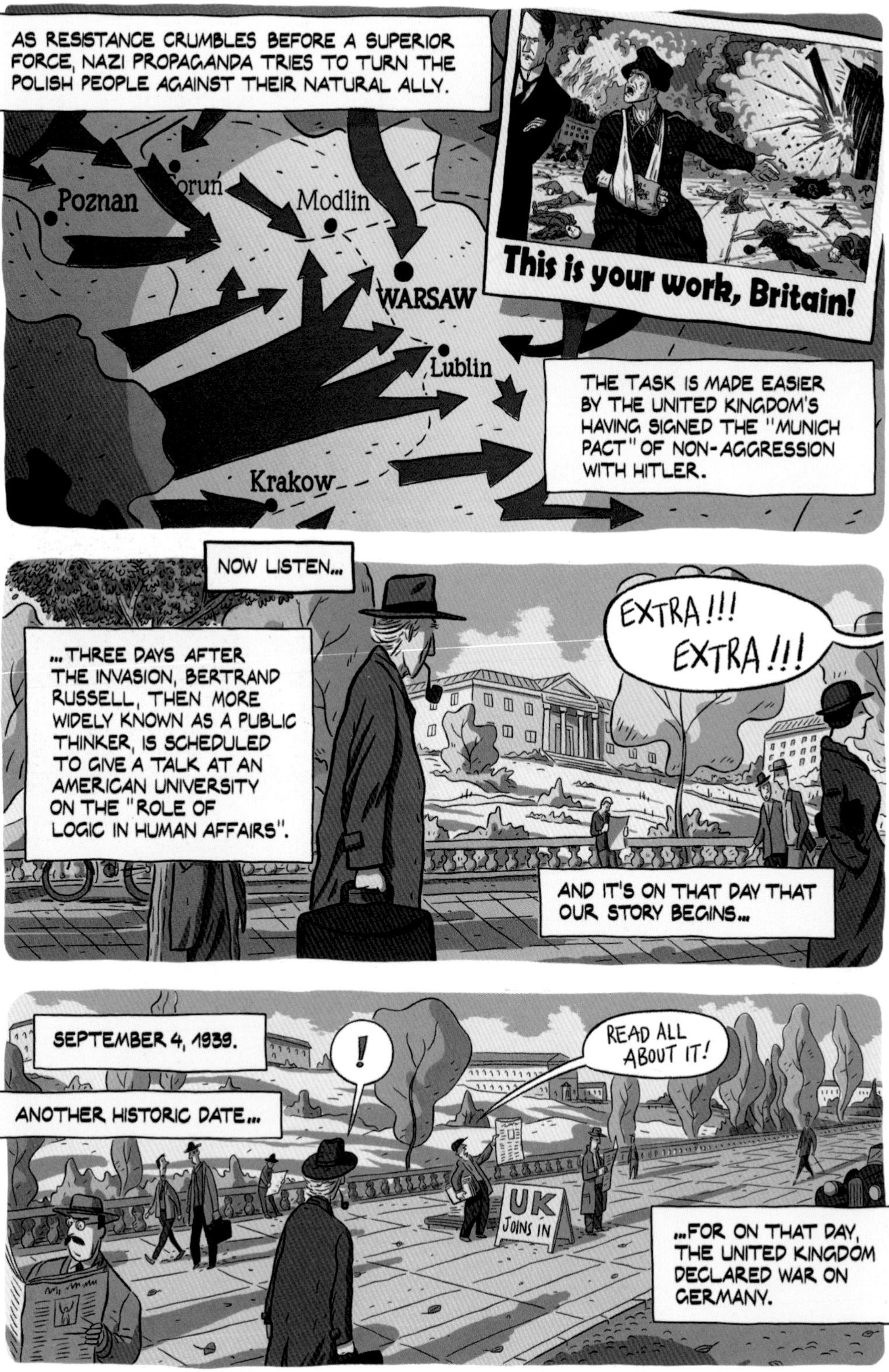

AS RESISTANCE CRUMBLES BEFORE A SUPERIOR FORCE, NAZI PROPAGANDA TRIES TO TURN THE POLISH PEOPLE AGAINST THEIR NATURAL ALLY.
Poznan
Modlin
WARSAW
Lublin
Krakow
This is your work, Britain!
THE TASK IS MADE EASIER BY THE UNITED KINGDOM'S HAVING SIGNED THE "MUNICH PACT" OF NON-AGGRESSION WITH HITLER.
NOW LISTEN...
...THREE DAYS AFTER THE INVASION, BERTRAND RUSSELL, THEN MORE WIDELY KNOWN AS A PUBLIC THINKER, IS SCHEDULED TO GIVE A TALK AT AN AMERICAN UNIVERSITY ON THE "ROLE OF LOGIC IN HUMAN AFFAIRS".
EXTRA!!! EXTRA!!!
AND IT'S ON THAT DAY THAT OUR STORY BEGINS...
SEPTEMBER 4, 1939.
ANOTHER HISTORIC DATE...
!
READ ALL ABOUT IT!
UK JOINS IN
...FOR ON THAT DAY, THE UNITED KINGDOM DECLARED WAR ON GERMANY.

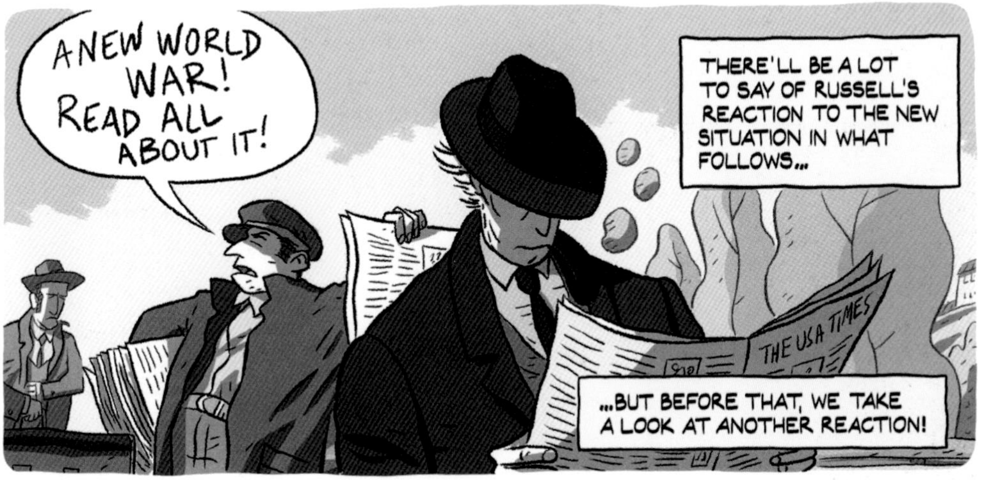
A NEW WORLD WAR! READ ALL ABOUT IT!
THERE'LL BE A LOT TO SAY OF RUSSELL'S REACTION TO THE NEW SITUATION IN WHAT FOLLOWS...
THE USA TIMES
...BUT BEFORE THAT, WE TAKE A LOOK AT ANOTHER REACTION!

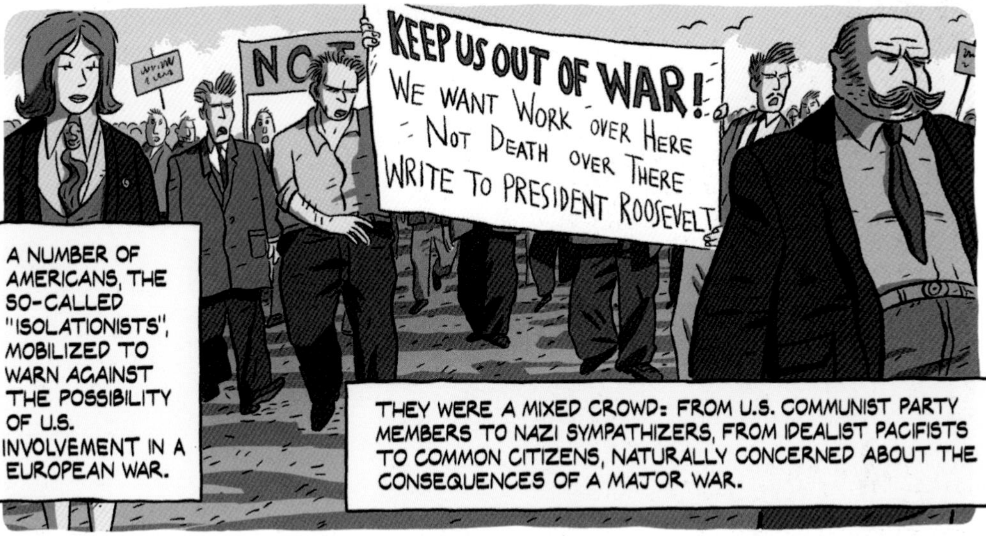
KEEP US OUT OF WAR! WE WANT WORK OVER HERE NOT DEATH OVER THERE WRITE TO PRESIDENT ROOSEVELT
A NUMBER OF AMERICANS, THE SO-CALLED "ISOLATIONISTS", MOBILIZED TO WARN AGAINST THE POSSIBILITY OF U.S. INVOLVEMENT IN A EUROPEAN WAR.
THEY WERE A MIXED CROWD: FROM U.S. COMMUNIST PARTY MEMBERS TO NAZI SYMPATHIZERS, FROM IDEALIST PACIFISTS TO COMMON CITIZENS, NATURALLY CONCERNED ABOUT THE CONSEQUENCES OF A MAJOR WAR.

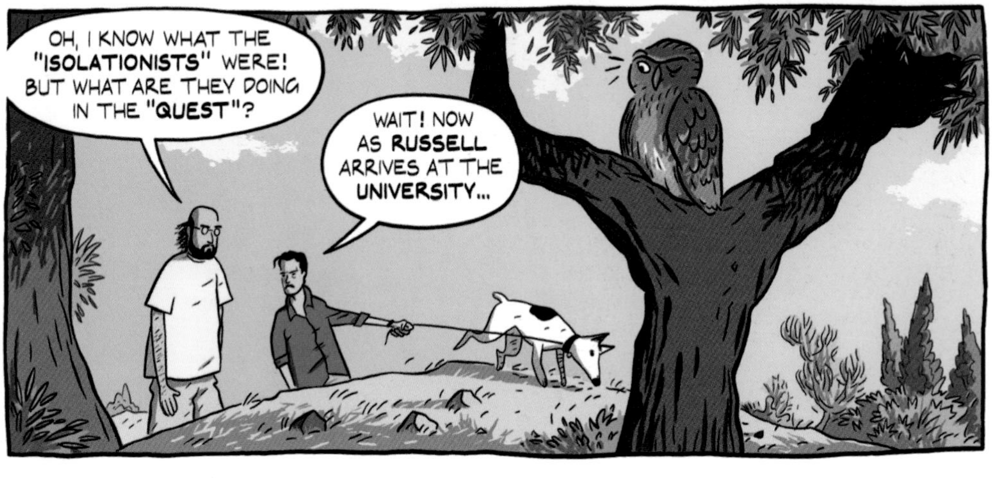
OH, I KNOW WHAT THE "ISOLATIONISTS" WERE! BUT WHAT ARE THEY DOING IN THE "QUEST"?
WAIT! NOW AS RUSSELL ARRIVES AT THE UNIVERSITY...

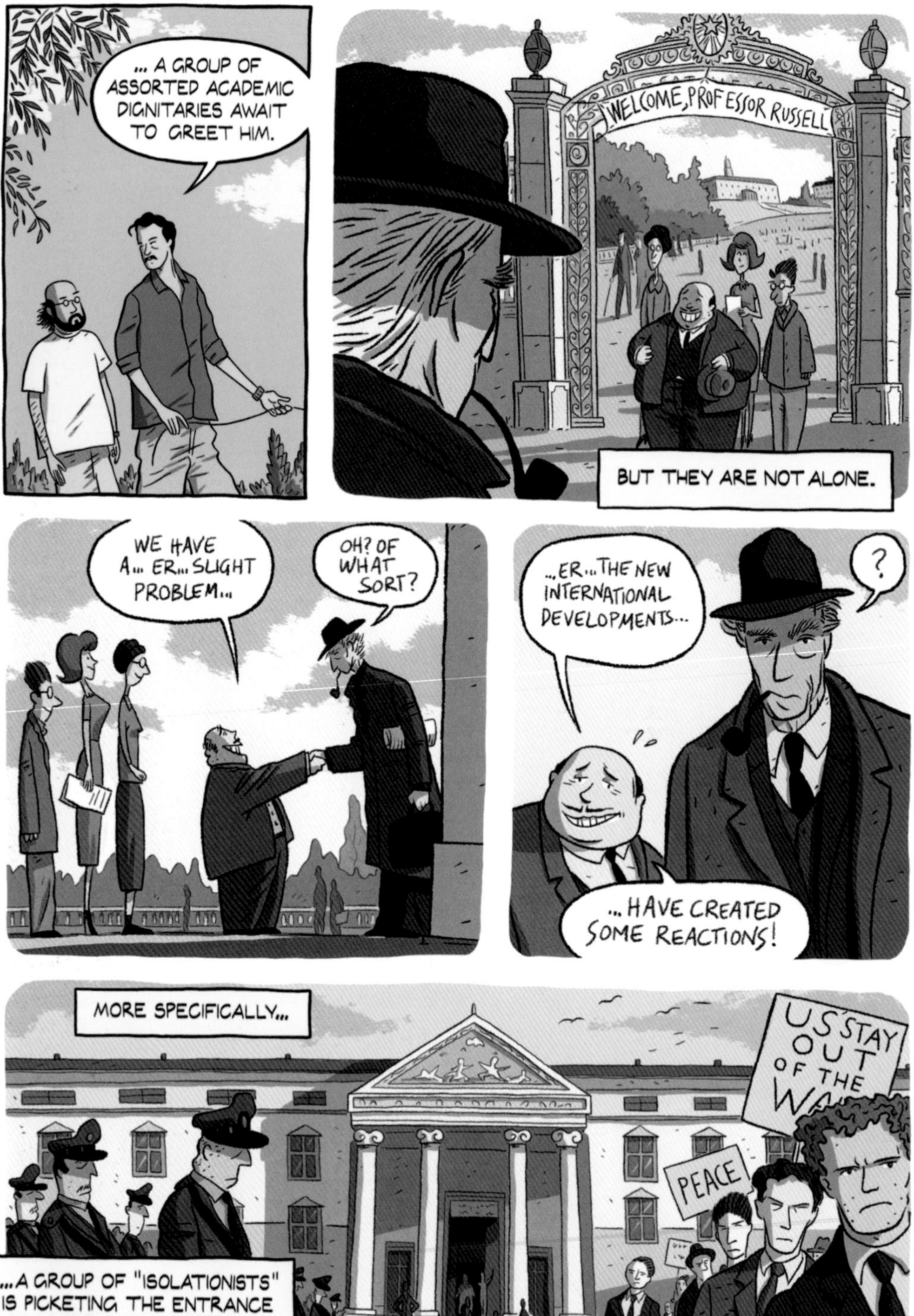
... A GROUP OF ASSORTED ACADEMIC DIGNITARIES AWAIT TO GREET HIM.
WELCOME, PROFESSOR RUSSELL
BUT THEY ARE NOT ALONE.
WE HAVE A... ER... SLIGHT PROBLEM...
OH? OF WHAT SORT?
...ER... THE NEW INTERNATIONAL DEVELOPMENTS...
?
... HAVE CREATED SOME REACTIONS!
MORE SPECIFICALLY...
US STAY OUT OF THE
PEACE
...A GROUP OF "ISOLATIONISTS" IS PICKETING THE ENTRANCE TO THE BUILDING WHERE HE WILL BE GIVING HIS TALK!
THEY HAVE A VERY SPECIFIC REQUEST OF THE SPEAKER...

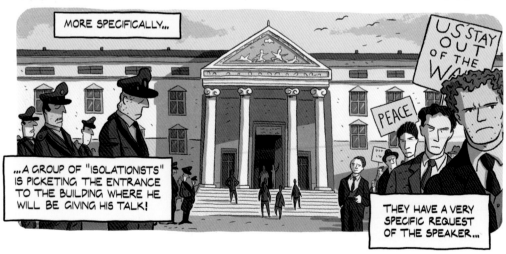

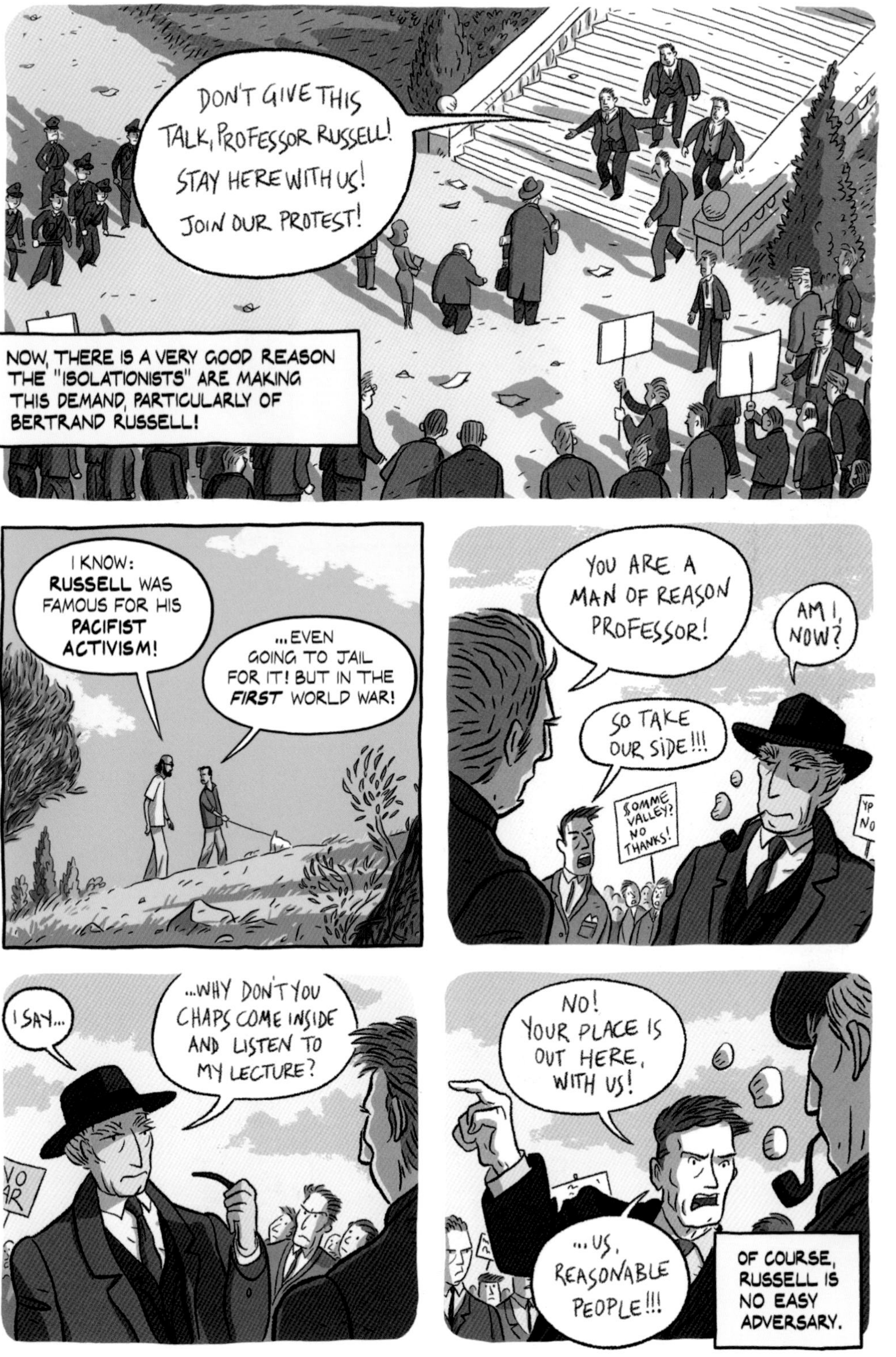
DON'T GIVE THIS TALK, PROFESSOR RUSSELL! STAY HERE WITH US! JOIN OUR PROTEST!
NOW, THERE IS A VERY GOOD REASON THE "ISOLATIONISTS" ARE MAKING THIS DEMAND, PARTICULARLY OF BERTRAND RUSSELL!
I KNOW: RUSSELL WAS FAMOUS FOR HIS PACIFIST ACTIVISM!
...EVEN GOING TO JAIL FOR IT! BUT IN THE FIRST WORLD WAR!
YOU ARE A MAN OF REASON PROFESSOR!
AM I, NOW?
SO TAKE OUR SIDE!!!
SOMME VALLEY? NO THANKS!
I SAY...
...WHY DON'T YOU CHAPS COME INSIDE AND LISTEN TO MY LECTURE?
NO! YOUR PLACE IS OUT HERE, WITH US!
...US, REASONABLE PEOPLE!!!
OF COURSE, RUSSELL IS NO EASY ADVERSARY.

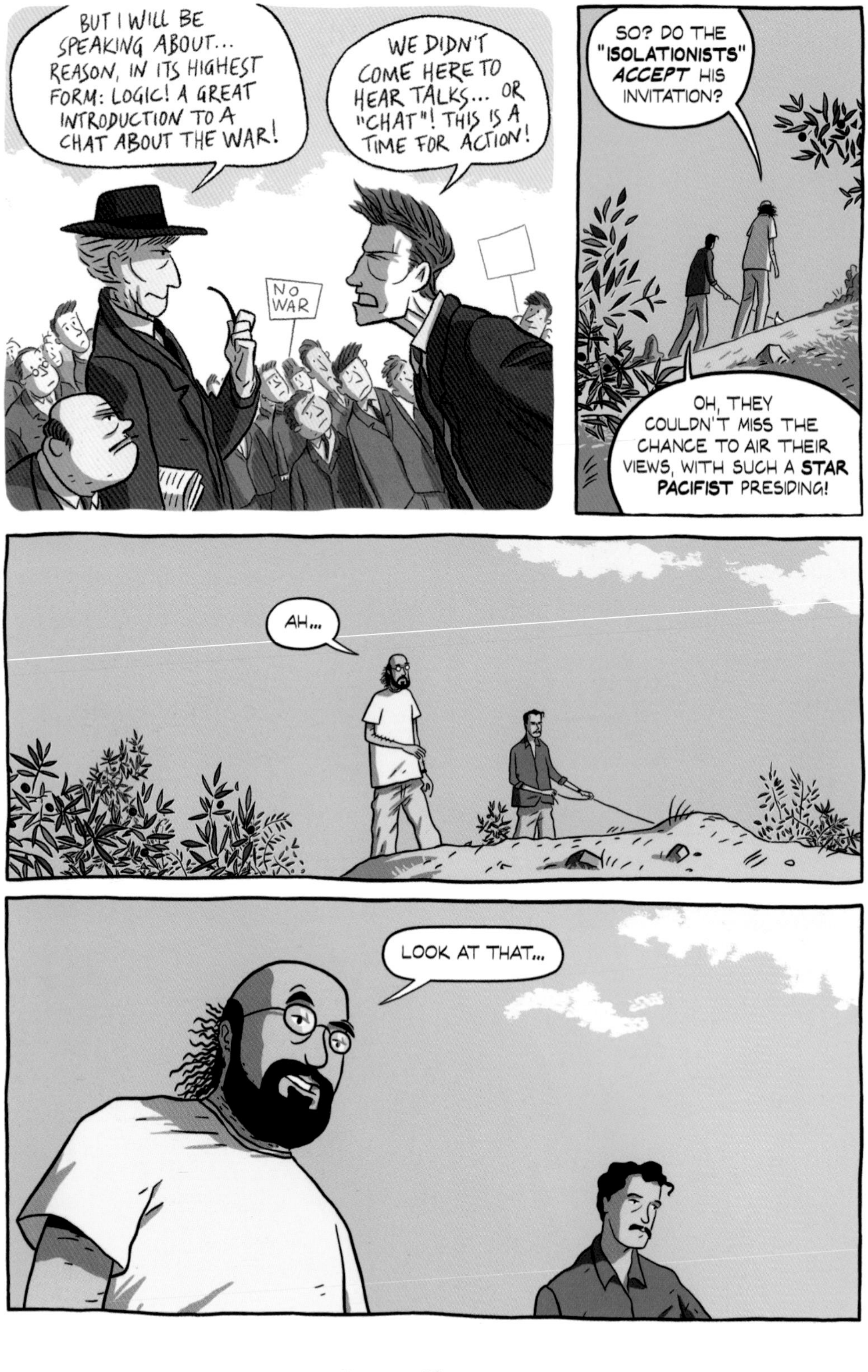
BUT I WILL BE SPEAKING ABOUT... REASON, IN ITS HIGHEST FORM: LOGIC! A GREAT INTRODUCTION TO A CHAT ABOUT THE WAR!
WE DIDN'T COME HERE TO HEAR TALKS... OR "CHAT"! THIS IS A TIME FOR ACTION!
NO WAR
SO? DO THE "ISOLATIONISTS" ACCEPT HIS INVITATION?
OH, THEY COULDN'T MISS THE CHANCE TO AIR THEIR VIEWS, WITH SUCH A STAR PACIFIST PRESIDING!
AH...
LOOK AT THAT...

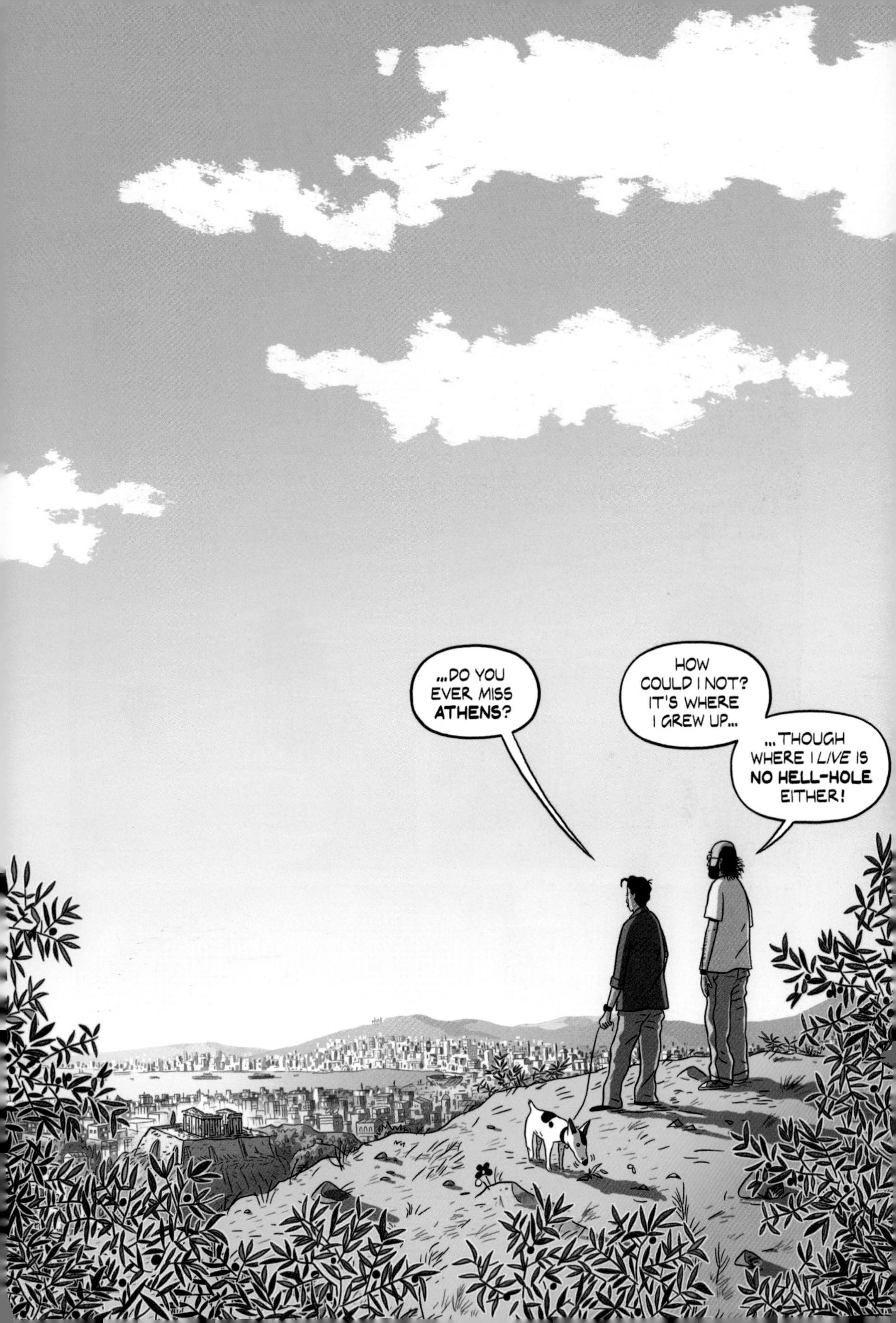
...DO YOU EVER MISS ATHENS?
HOW COULD I NOT? IT'S WHERE I GREW UP...
...THOUGH WHERE I LIVE IS NO HELL-HOLE EITHER!

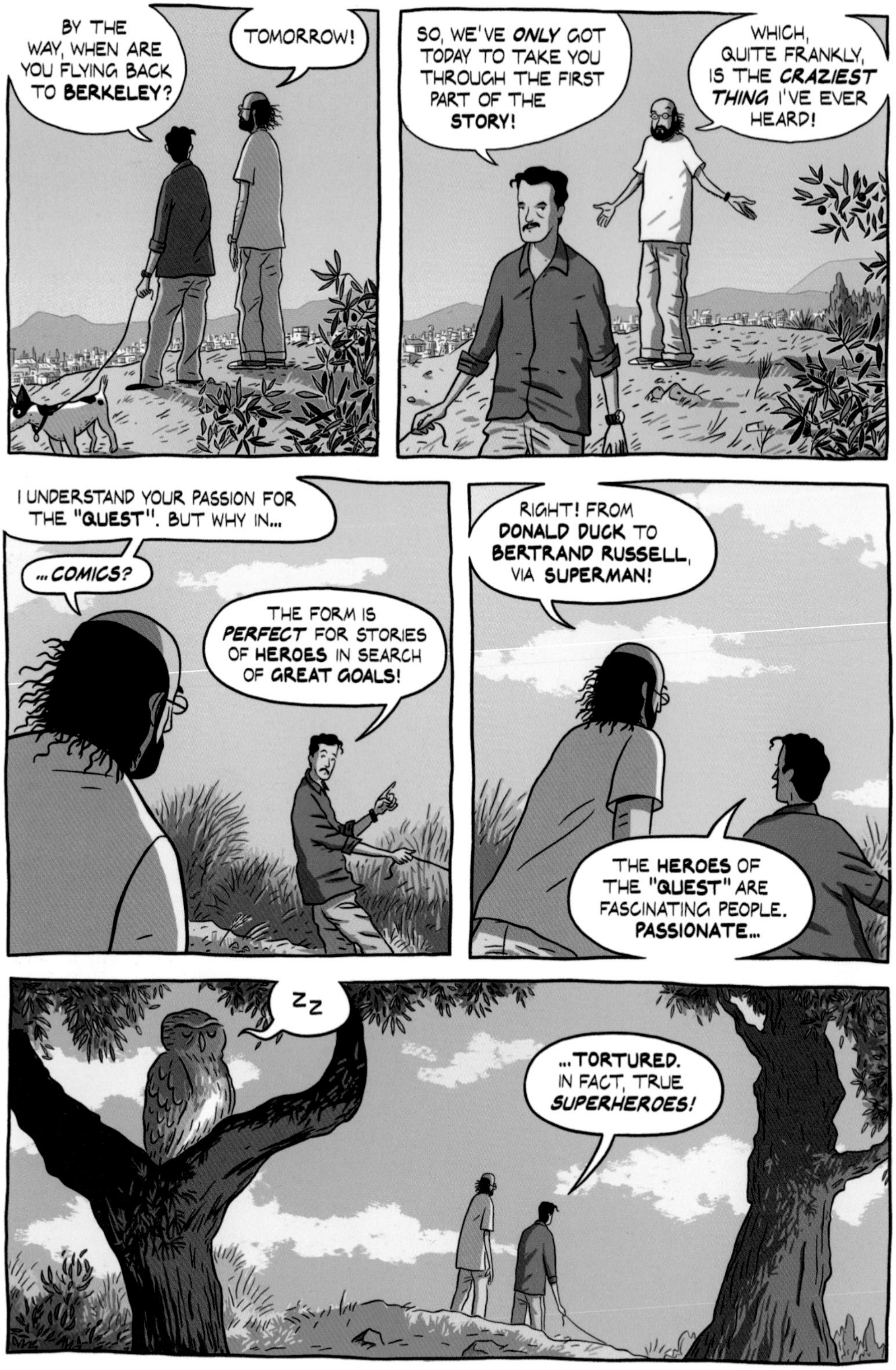
BY THE WAY, WHEN ARE YOU FLYING BACK TO BERKELEY?
TOMORROW!
SO, WE'VE ONLY GOT TODAY TO TAKE YOU THROUGH THE FIRST PART OF THE STORY!
WHICH, QUITE FRANKLY, IS THE CRAZIEST THING I'VE EVER HEARD!
I UNDERSTAND YOUR PASSION FOR THE "QUEST". BUT WHY IN...
...COMICS?
THE FORM IS PERFECT FOR STORIES OF HEROES IN SEARCH OF GREAT GOALS!
RIGHT! FROM DONALD DUCK TO BERTRAND RUSSELL, VIA SUPERMAN!
THE HEROES OF THE "QUEST" ARE FASCINATING PEOPLE. PASSIONATE...
ZZ
...TORTURED. IN FACT, TRUE SUPERHEROES!

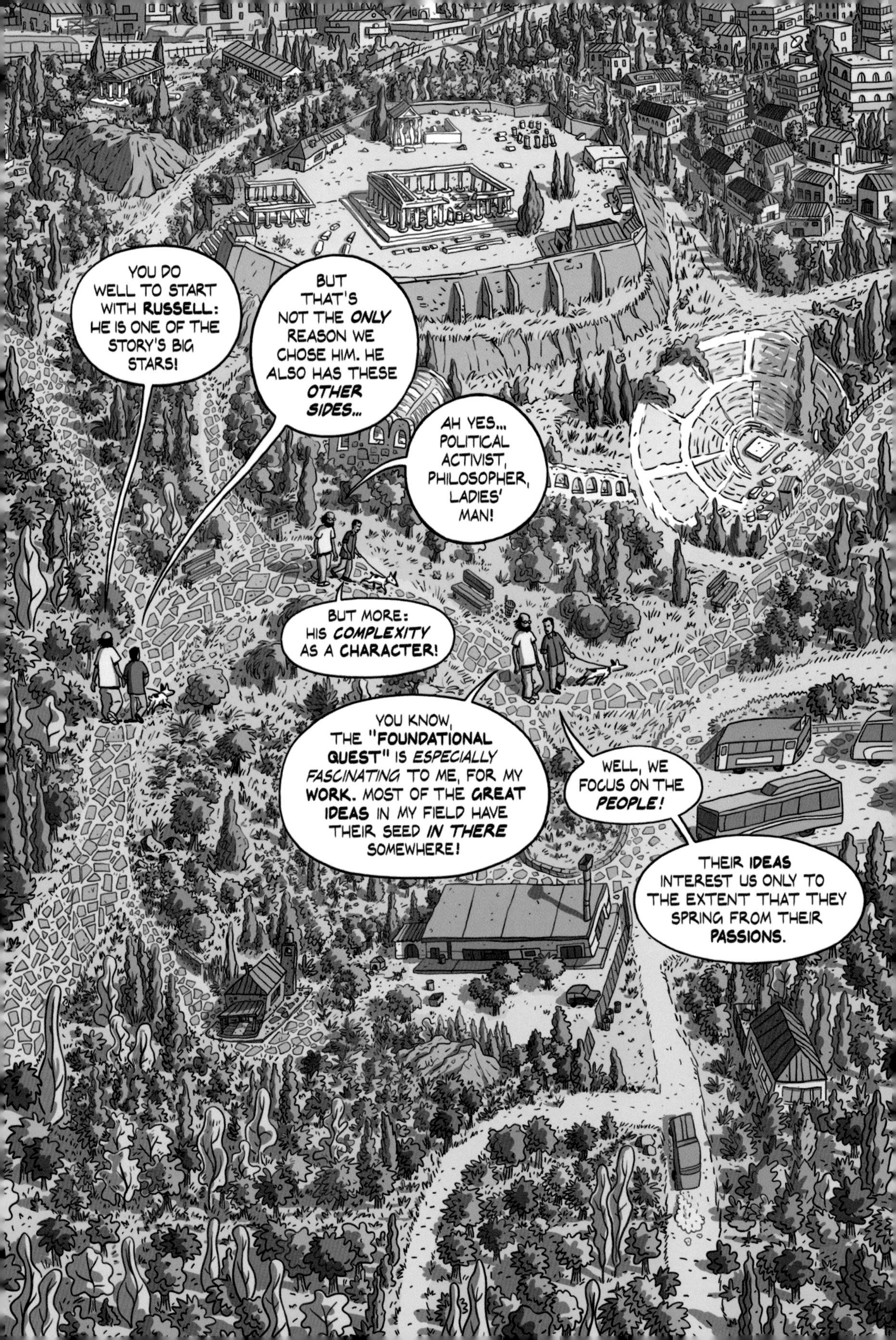
YOU DO WELL TO START WITH **RUSSELL**: HE IS ONE OF THE STORY'S BIG STARS!
BUT THAT'S NOT THE ***ONLY*** REASON WE CHOSE HIM. HE ALSO HAS THESE ***OTHER SIDES***...
AH YES... POLITICAL ACTIVIST, PHILOSOPHER, LADIES' MAN!
BUT MORE: HIS ***COMPLEXITY*** AS A **CHARACTER**!
YOU KNOW, THE "**FOUNDATIONAL QUEST**" IS *ESPECIALLY FASCINATING* TO ME, FOR MY **WORK**. MOST OF THE **GREAT IDEAS** IN MY FIELD HAVE THEIR SEED ***IN THERE*** SOMEWHERE!
WELL, WE FOCUS ON THE ***PEOPLE***!
THEIR **IDEAS** INTEREST US ONLY TO THE EXTENT THAT THEY SPRING FROM THEIR **PASSIONS**.

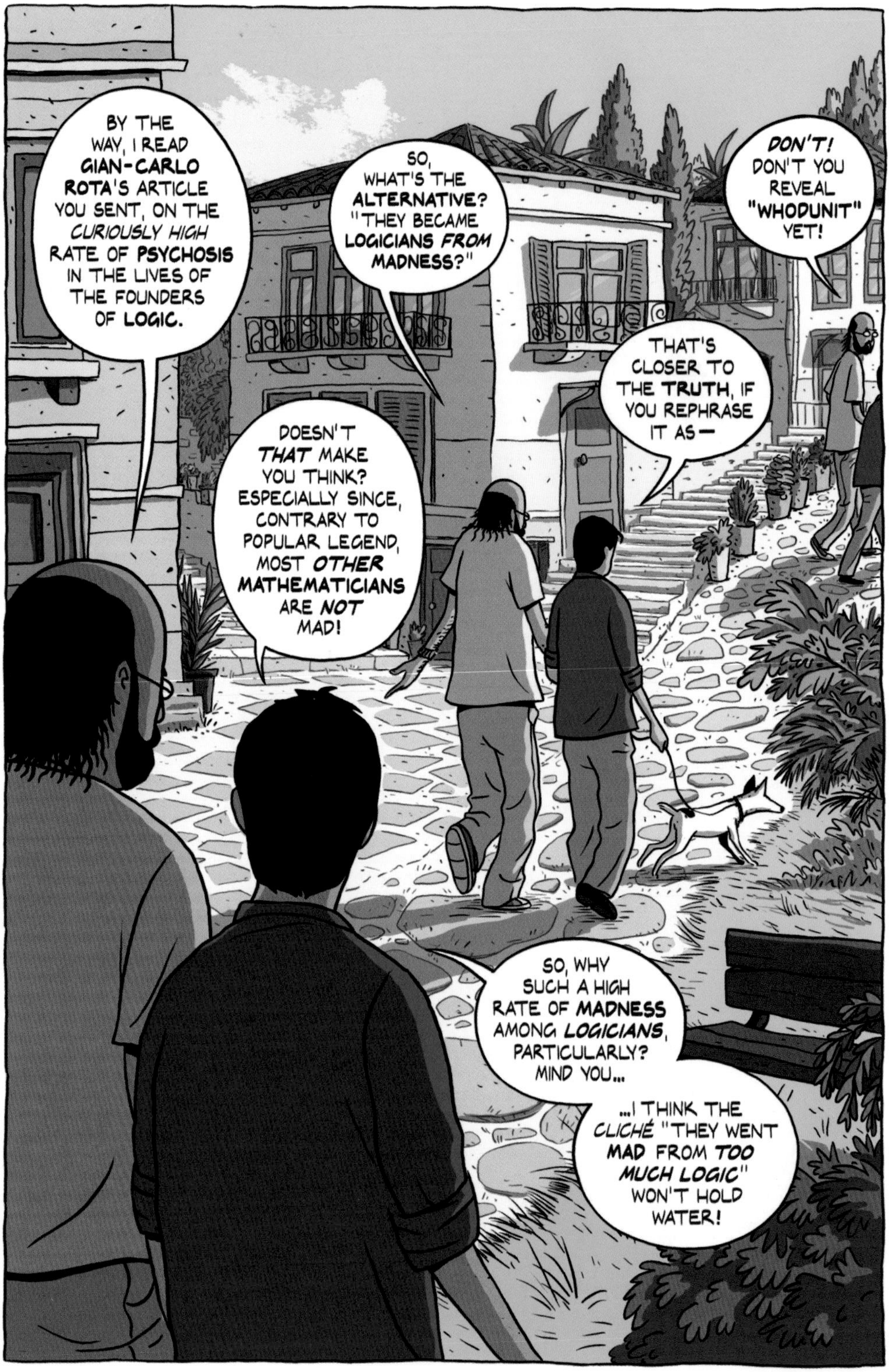
BY THE WAY, I READ GIAN-CARLO ROTA'S ARTICLE YOU SENT, ON THE CURIOUSLY HIGH RATE OF PSYCHOSIS IN THE LIVES OF THE FOUNDERS OF LOGIC.
DOESN'T THAT MAKE YOU THINK? ESPECIALLY SINCE, CONTRARY TO POPULAR LEGEND, MOST OTHER MATHEMATICIANS ARE NOT MAD!
SO, WHY SUCH A HIGH RATE OF MADNESS AMONG LOGICIANS, PARTICULARLY? MIND YOU...
...I THINK THE CLICHÉ "THEY WENT MAD FROM TOO MUCH LOGIC" WON'T HOLD WATER!
SO, WHAT'S THE ALTERNATIVE? "THEY BECAME LOGICIANS FROM MADNESS?"
THAT'S CLOSER TO THE TRUTH, IF YOU REPHRASE IT AS—
DON'T! DON'T YOU REVEAL "WHODUNIT" YET!

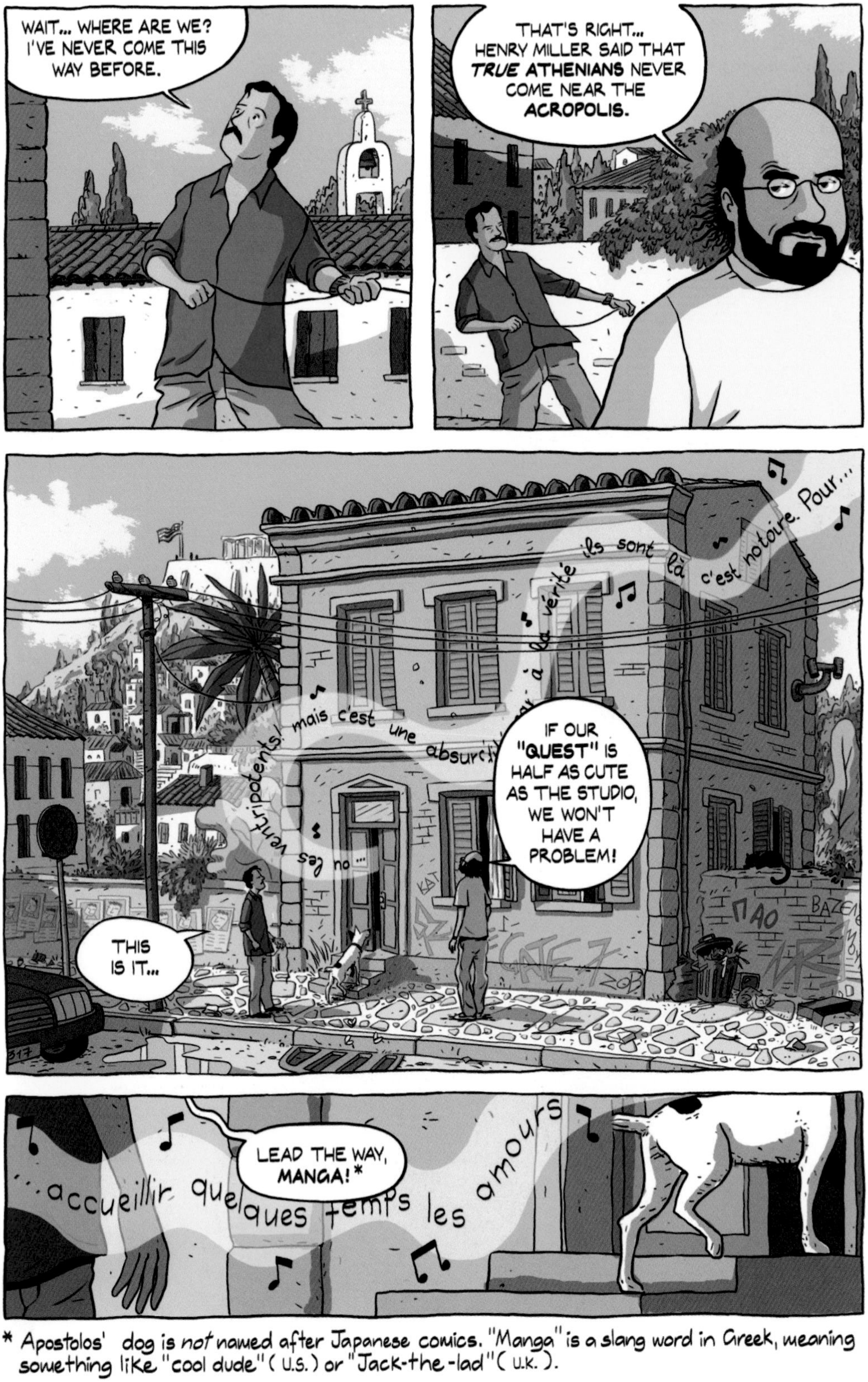
WAIT... WHERE ARE WE? I'VE NEVER COME THIS WAY BEFORE.
THAT'S RIGHT... HENRY MILLER SAID THAT *TRUE* **ATHENIANS** NEVER COME NEAR THE **ACROPOLIS.**
no... les ventripotents! mais c'est une absurd
à la vérité ils sont là c'est notoire. Pour...
IF OUR **"QUEST"** IS HALF AS CUTE AS THE STUDIO, WE WON'T HAVE A PROBLEM!
THIS IS IT...
...accueillir quelques temps les amours
LEAD THE WAY, **MANGA!***

* Apostolos' dog is *not* named after Japanese comics. "Manga" is a slang word in Greek, meaning something like "cool dude" (U.S.) or "Jack-the-lad" (U.K.).

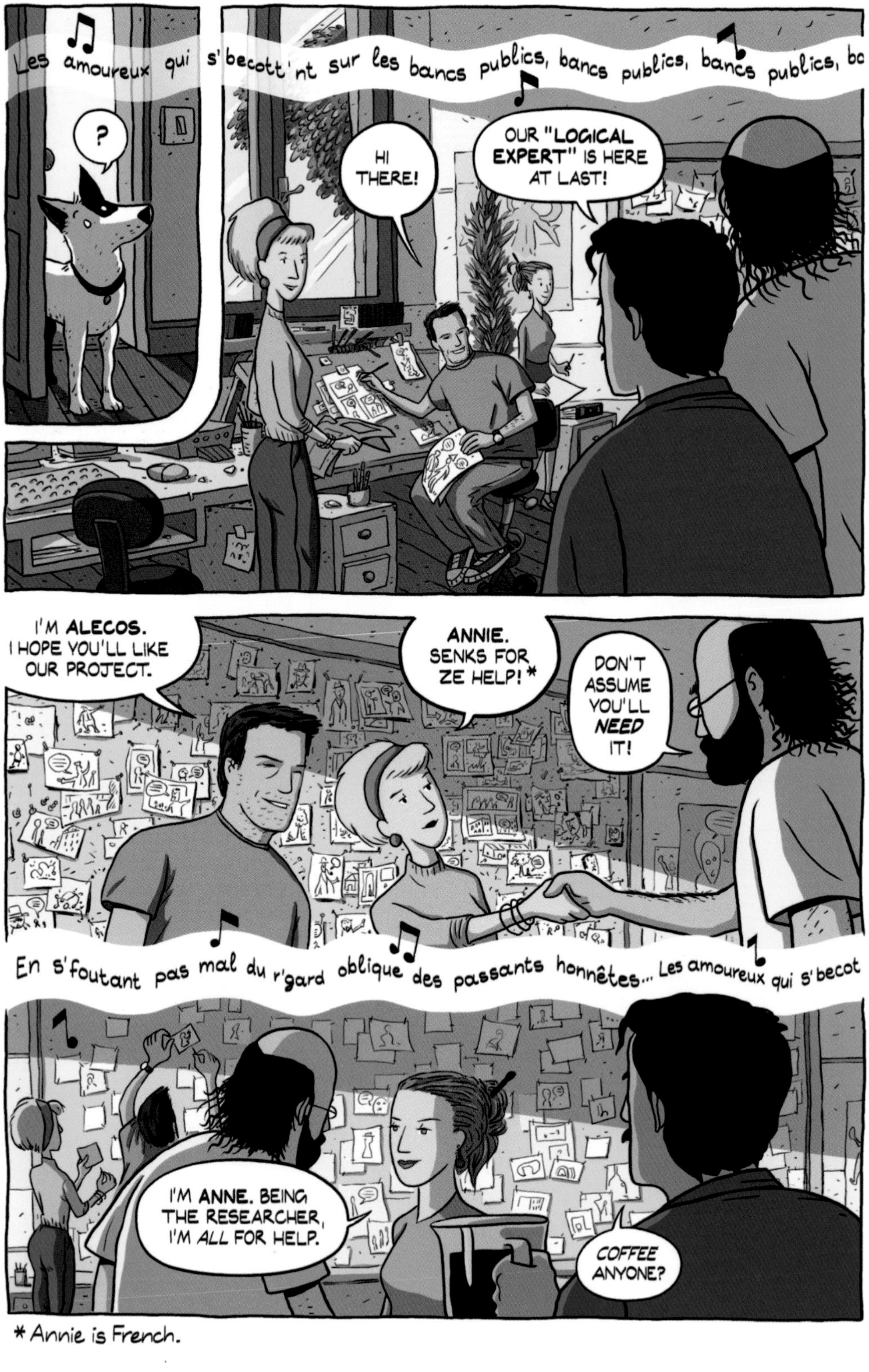
Les amoureux qui s'becott'nt sur les bancs publics, bancs publics, bancs publics, ba
?
HI THERE!
OUR "LOGICAL EXPERT" IS HERE AT LAST!
I'M ALECOS. I HOPE YOU'LL LIKE OUR PROJECT.
ANNIE. SENKS FOR ZE HELP! *
DON'T ASSUME YOU'LL NEED IT!
En s'foutant pas mal du r'gard oblique des passants honnêtes... Les amoureux qui s'becot
I'M ANNE. BEING THE RESEARCHER, I'M ALL FOR HELP.
COFFEE ANYONE?

* Annie is French.

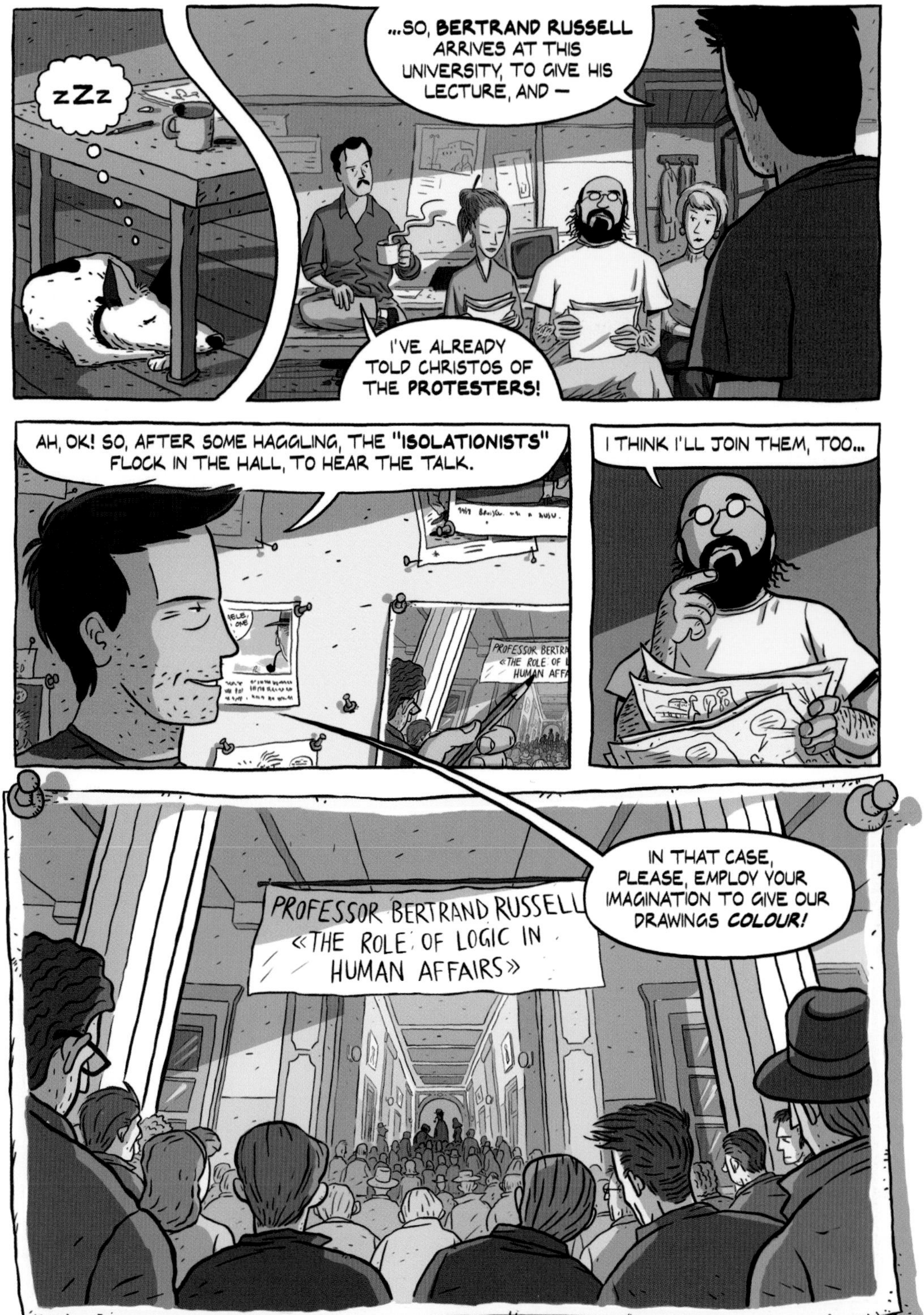
zZz
...SO, BERTRAND RUSSELL ARRIVES AT THIS UNIVERSITY, TO GIVE HIS LECTURE, AND —
I'VE ALREADY TOLD CHRISTOS OF THE PROTESTERS!
AH, OK! SO, AFTER SOME HAGGLING, THE "ISOLATIONISTS" FLOCK IN THE HALL, TO HEAR THE TALK.
I THINK I'LL JOIN THEM, TOO...
PROFESSOR BERTRAND RUSSELL «THE ROLE OF LOGIC IN HUMAN AFFAIRS»
IN THAT CASE, PLEASE, EMPLOY YOUR IMAGINATION TO GIVE OUR DRAWINGS COLOUR!

1. PEMBROKE LODGE

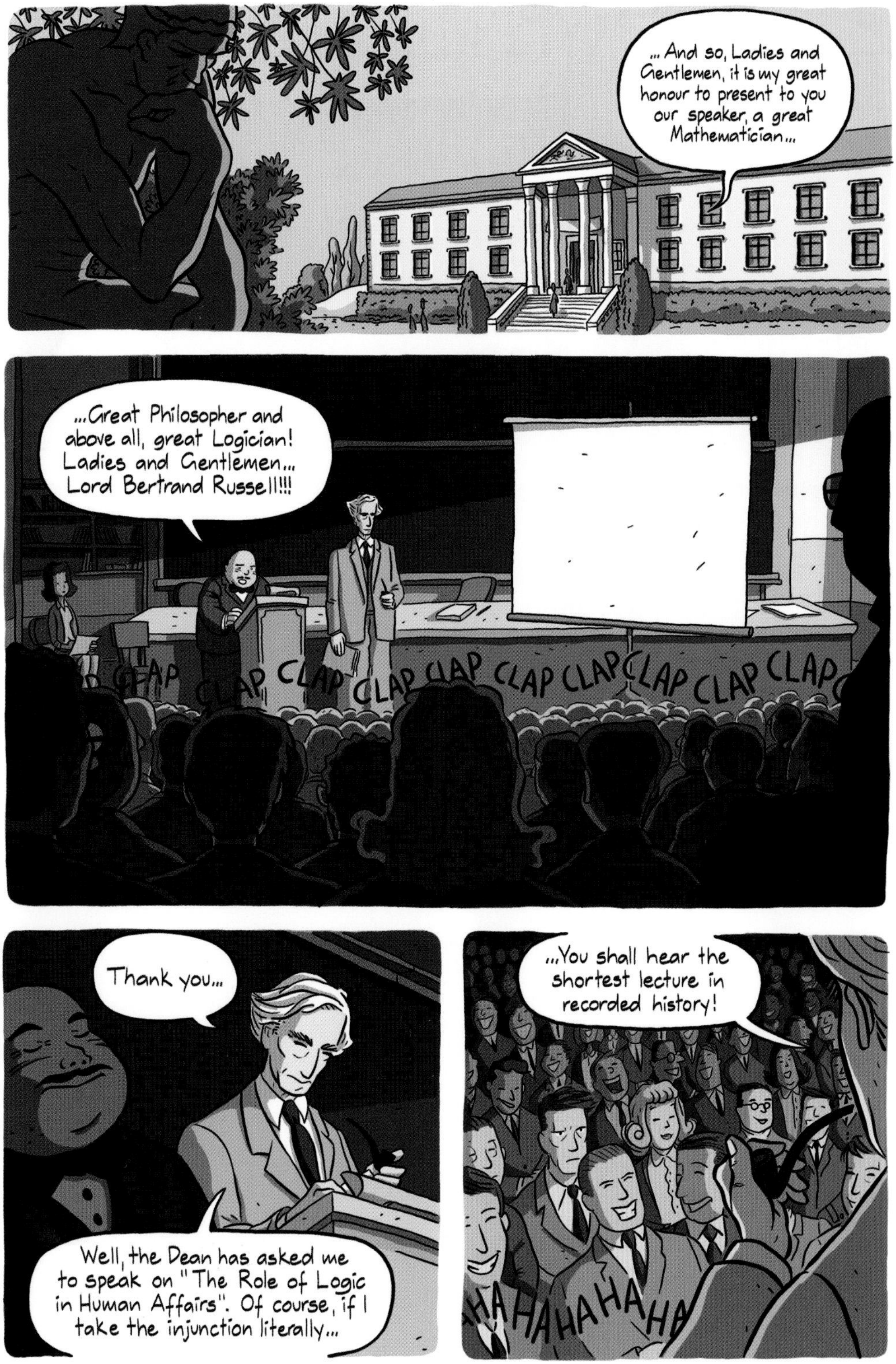
... And so, Ladies and Gentlemen, it is my great honour to present to you our speaker, a great Mathematician...
...Great Philosopher and above all, great Logician! Ladies and Gentlemen... Lord Bertrand Russell!!!
CLAP CLAP CLAP CLAP CLAP CLAP CLAP CLAP CLAP
Thank you...
Well, the Dean has asked me to speak on "The Role of Logic in Human Affairs". Of course, if I take the injunction literally...
...You shall hear the shortest lecture in recorded history!
HAHAHAHA HA

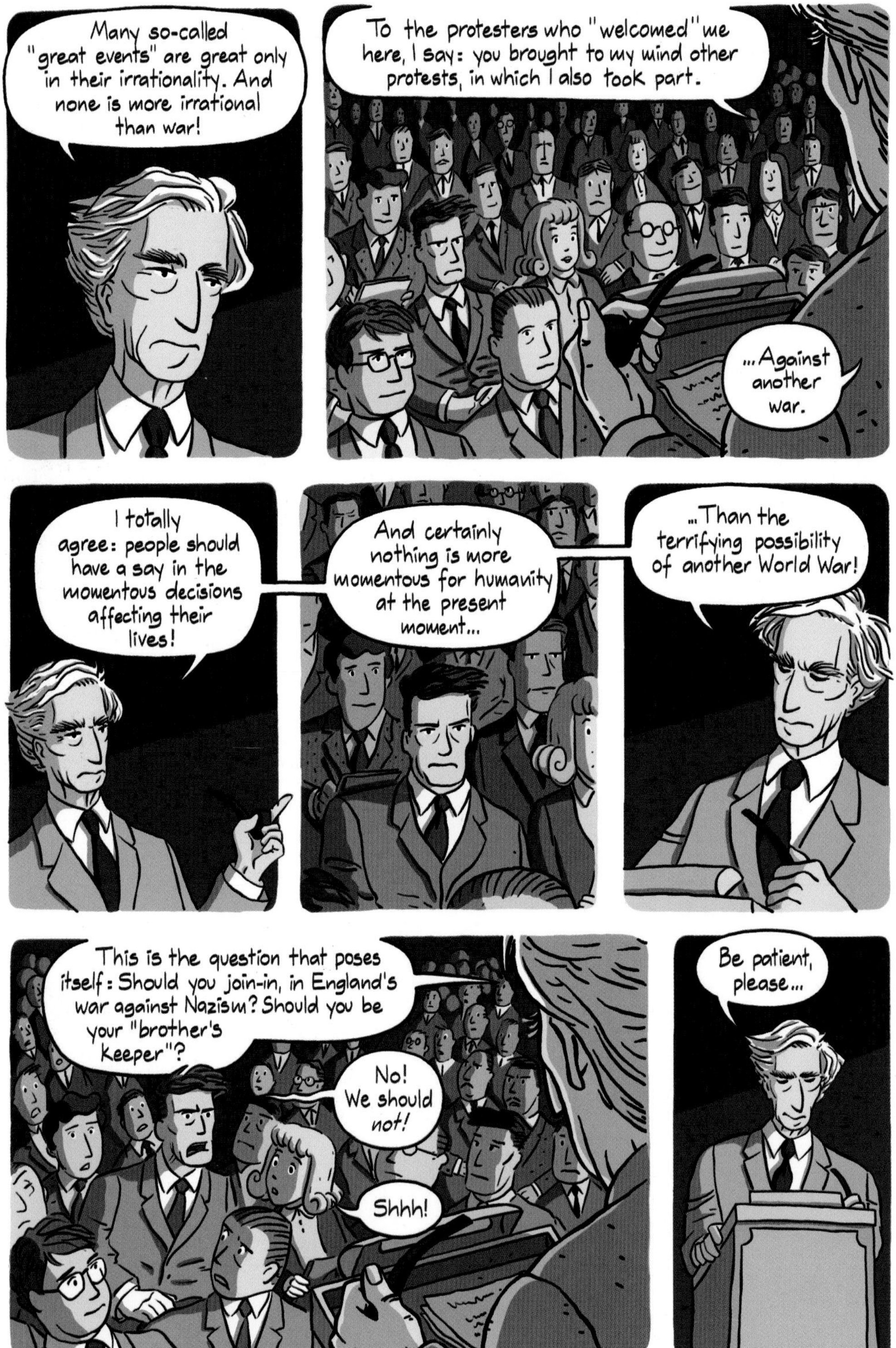
Many so-called "great events" are great only in their irrationality. And none is more irrational than war!
To the protesters who "welcomed" me here, I say: you brought to my mind other protests, in which I also took part.
...Against another war.
I totally agree: people should have a say in the momentous decisions affecting their lives!
And certainly nothing is more momentous for humanity at the present moment...
...Than the terrifying possibility of another World War!
This is the question that poses itself: Should you join-in, in England's war against Nazism? Should you be your "brother's keeper"?
No! We should not!
Shhh!
Be patient, please...

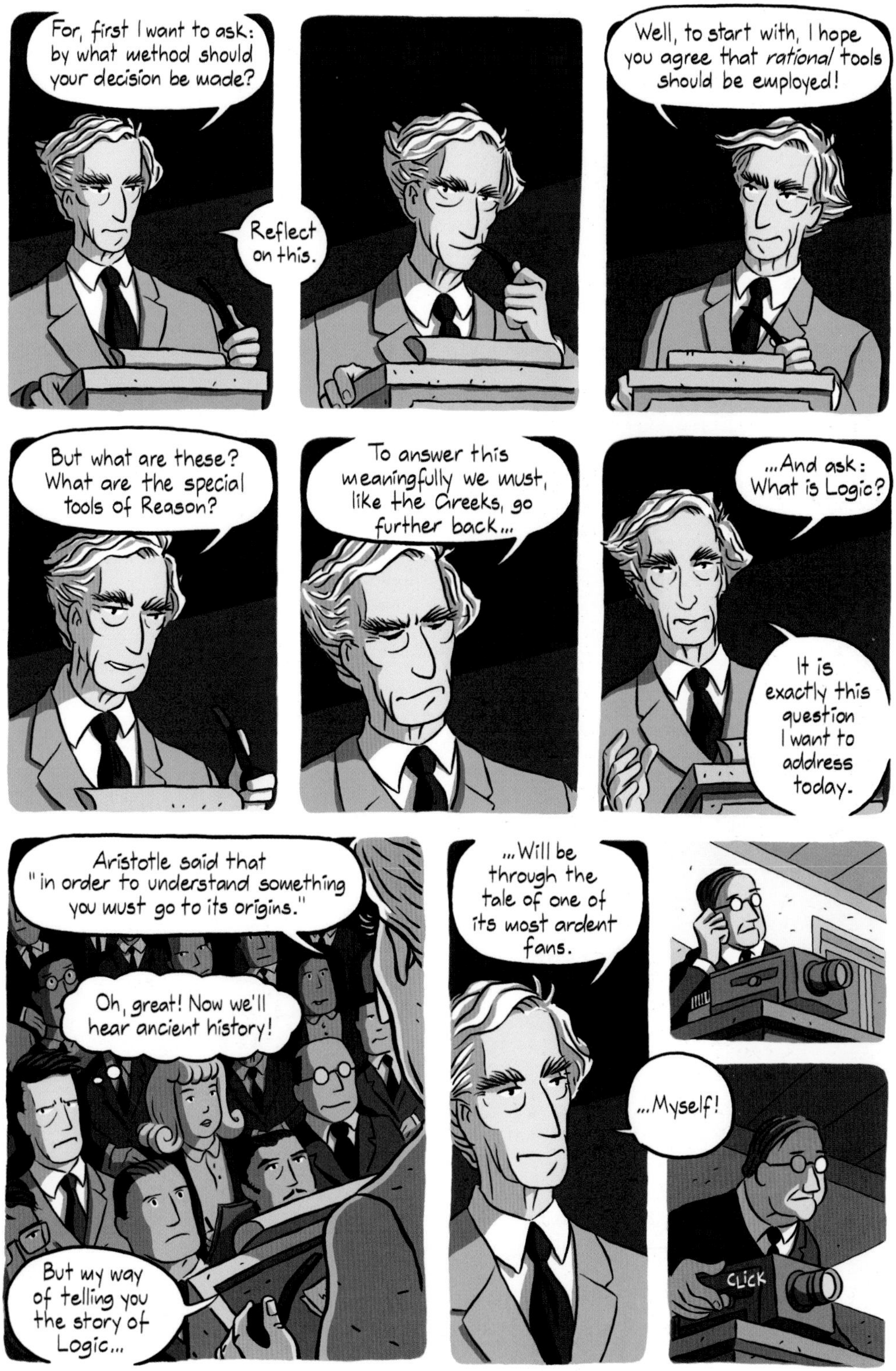

For, first I want to ask: by what method should your decision be made?
Reflect on this.
Well, to start with, I hope you agree that *rational* tools should be employed!
But what are these? What are the special tools of Reason?
To answer this meaningfully we must, like the Greeks, go further back...
...And ask: What is Logic?
It is exactly this question I want to address today.
Aristotle said that "in order to understand something you must go to its origins."
Oh, great! Now we'll hear ancient history!
But my way of telling you the story of Logic...
...Will be through the tale of one of its most ardent fans.
...Myself!
CLICK

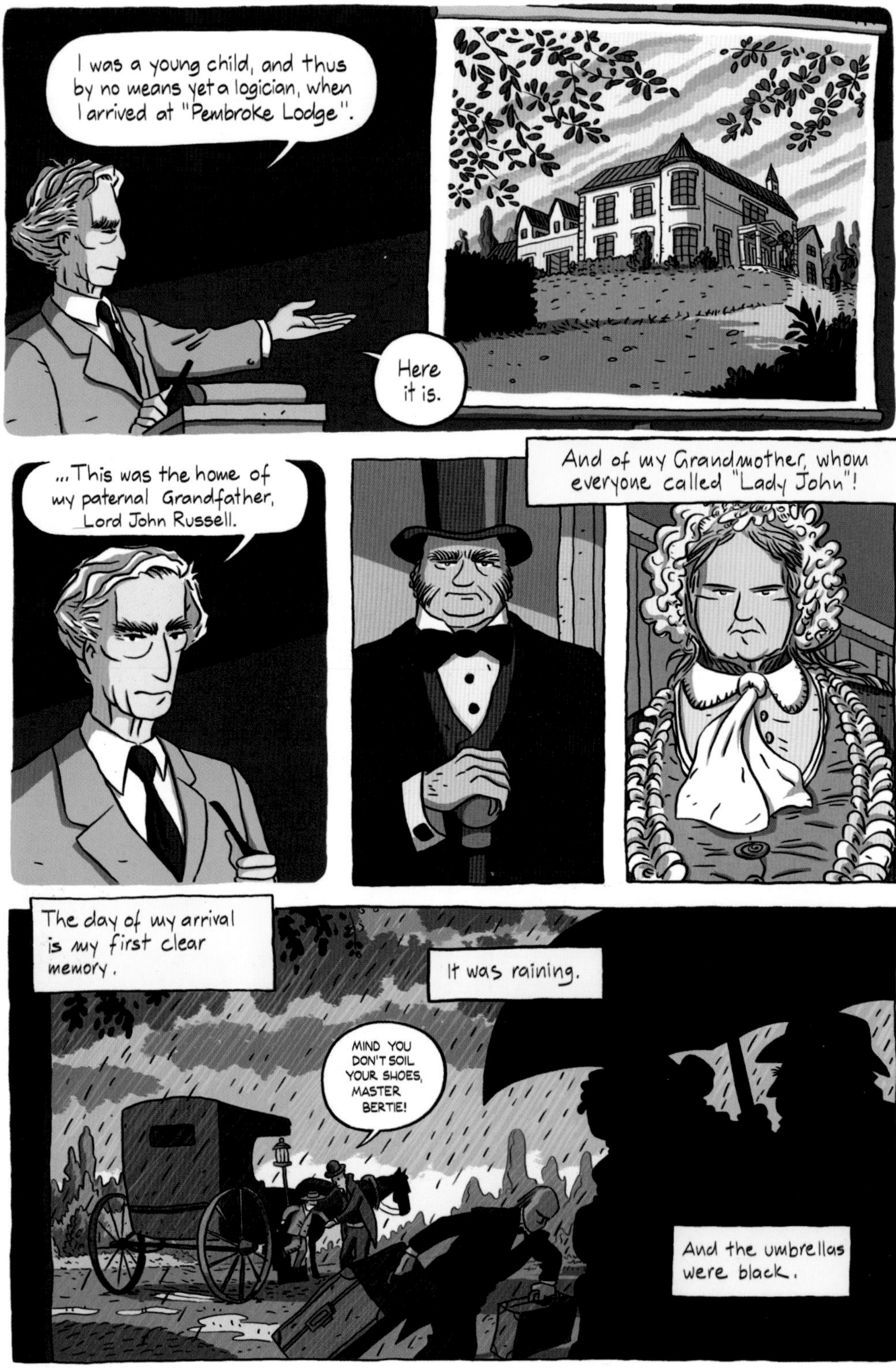
I was a young child, and thus by no means yet a logician, when I arrived at "Pembroke Lodge".
Here it is.
...This was the home of my paternal Grandfather, Lord John Russell.
And of my Grandmother, whom everyone called "Lady John"!
The day of my arrival is my first clear memory.
It was raining.
MIND YOU DON'T SOIL YOUR SHOES, MASTER BERTIE!
And the umbrellas were black.

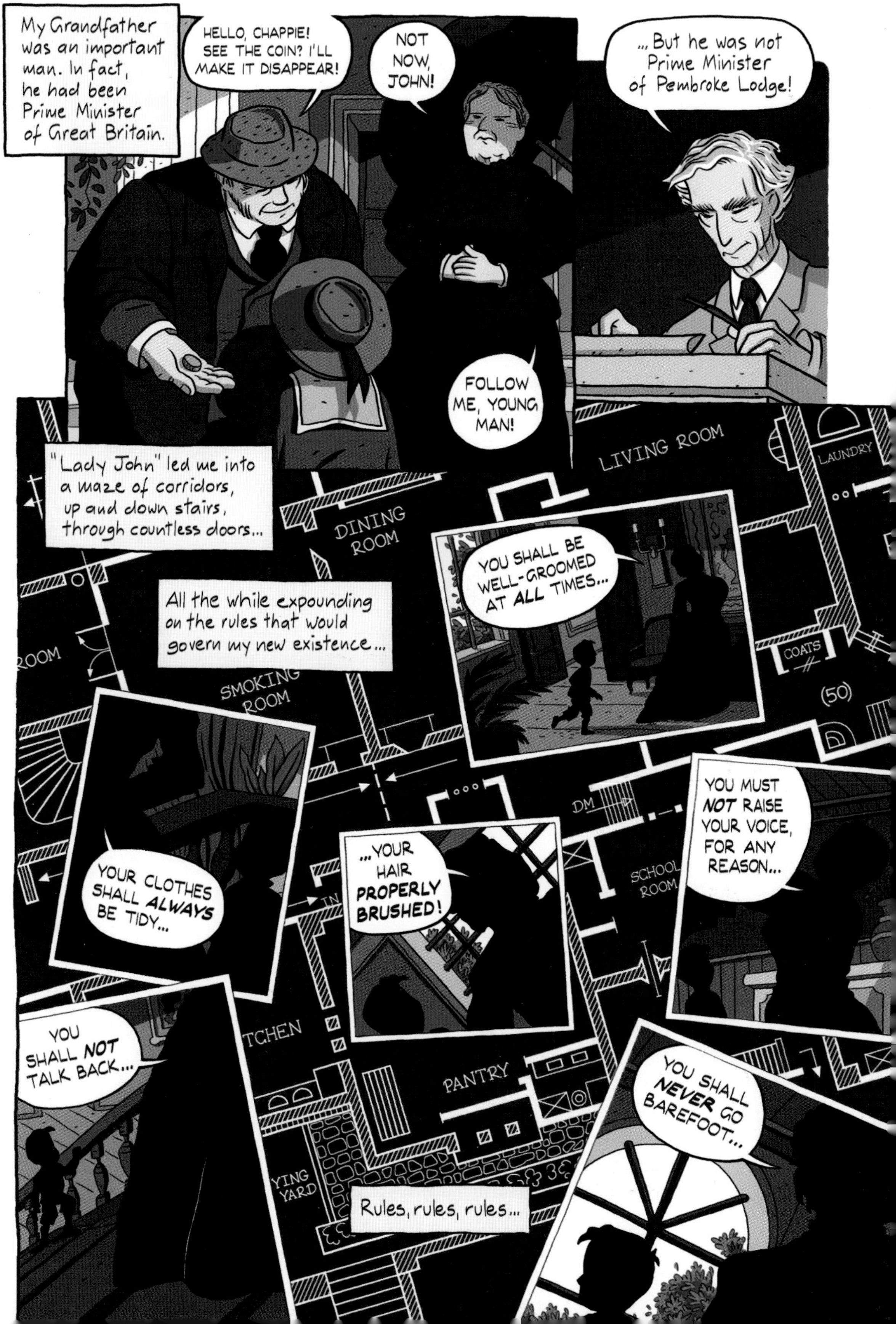

My Grandfather was an important man. In fact, he had been Prime Minister of Great Britain.
HELLO, CHAPPIE! SEE THE COIN? I'LL MAKE IT DISAPPEAR!
NOT NOW, JOHN!
FOLLOW ME, YOUNG MAN!
...But he was not Prime Minister of Pembroke Lodge!
"Lady John" led me into a maze of corridors, up and down stairs, through countless doors...
DINING ROOM
LIVING ROOM
LAUNDRY
All the while expounding on the rules that would govern my new existence...
YOU SHALL BE WELL-GROOMED AT ALL TIMES...
ROOM
SMOKING ROOM
COATS
(50)
YOUR CLOTHES SHALL ALWAYS BE TIDY...
...YOUR HAIR PROPERLY BRUSHED!
DM
SCHOOL ROOM
IN
YOU MUST NOT RAISE YOUR VOICE, FOR ANY REASON...
YOU SHALL NOT TALK BACK...
TCHEN
PANTRY
YING YARD
Rules, rules, rules...
YOU SHALL NEVER GO BAREFOOT...

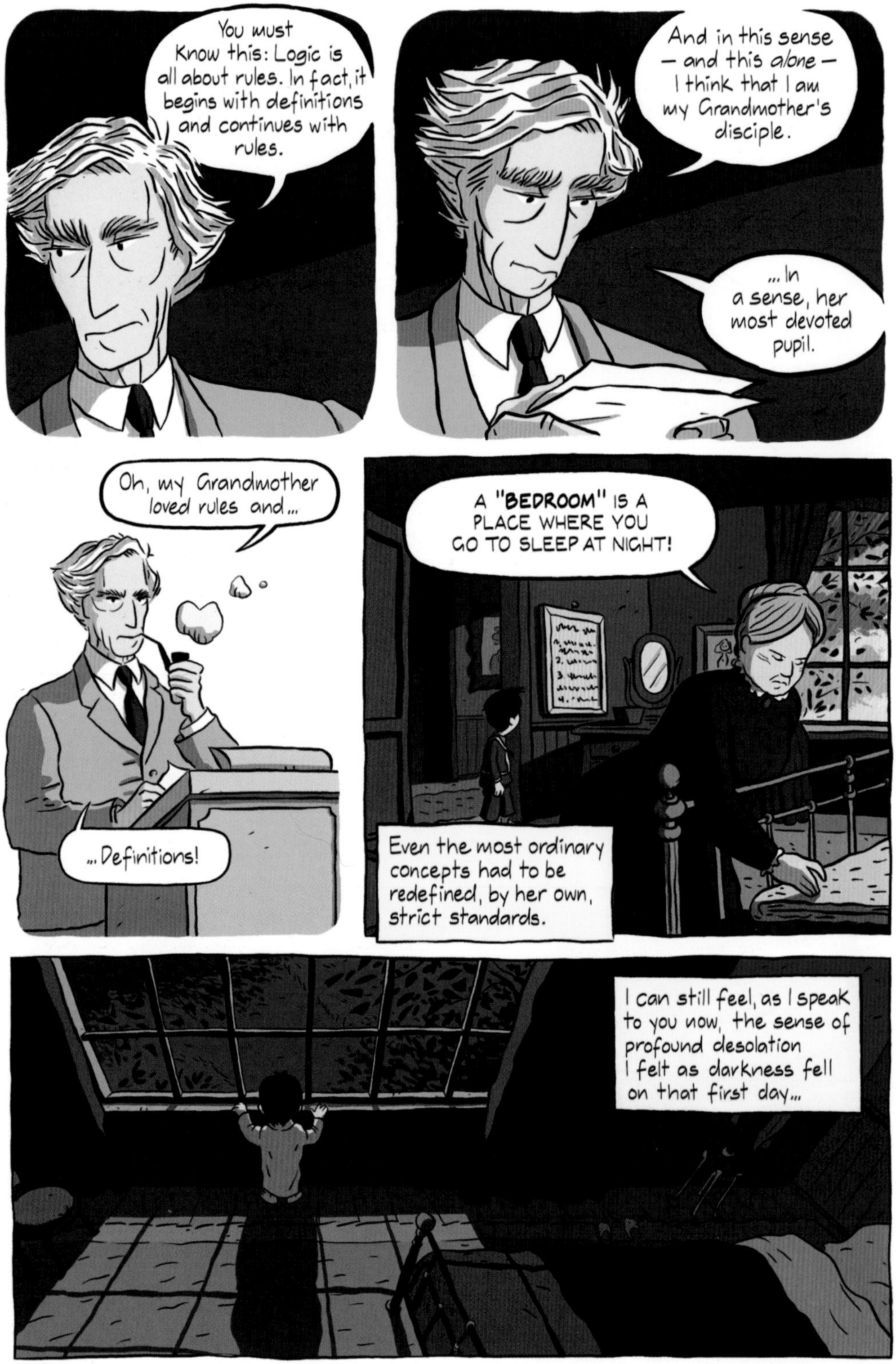
You must Know this: Logic is all about rules. In fact, it begins with definitions and continues with rules.
And in this sense — and this *alone* — I think that I am my Grandmother's disciple.
...In a sense, her most devoted pupil.
Oh, my Grandmother loved rules and...
...Definitions!
A **"BEDROOM"** IS A PLACE WHERE YOU GO TO SLEEP AT NIGHT!
Even the most ordinary concepts had to be redefined, by her own, strict standards.
I can still feel, as I speak to you now, the sense of profound desolation I felt as darkness fell on that first day...

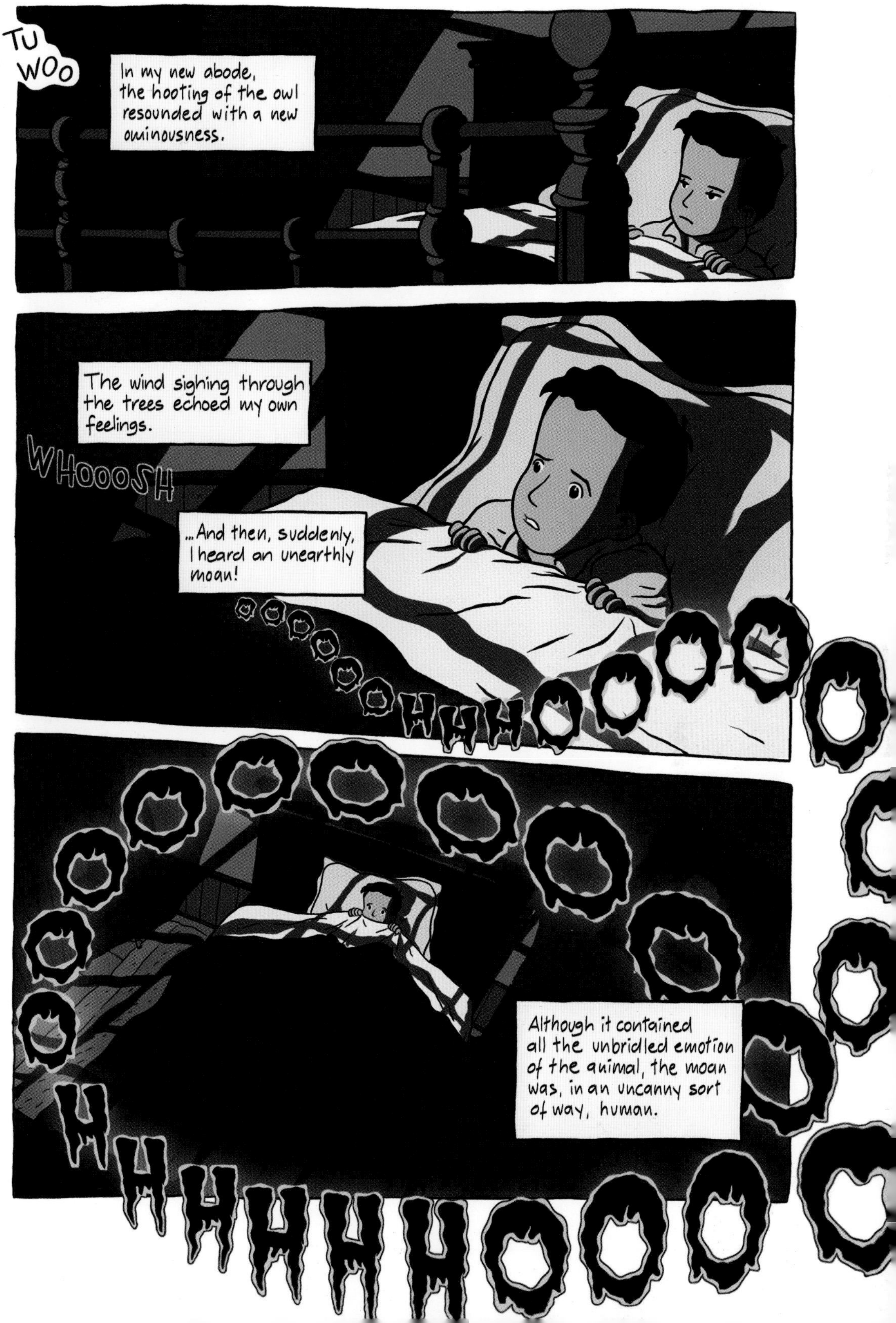
TU WOO
In my new abode, the hooting of the owl resounded with a new ominousness.
The wind sighing through the trees echoed my own feelings.
WHOOOSH
...And then, suddenly, I heard an unearthly moan!
oooooooHHHOOOOOOO
OOOOOOOOOOO
Although it contained all the unbridled emotion of the animal, the moan was, in an uncanny sort of way, human.
HHHHHHOOOOO

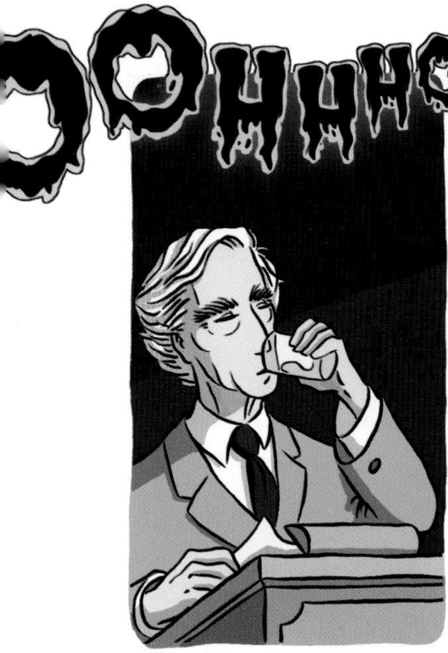
OOHHHOOoooo
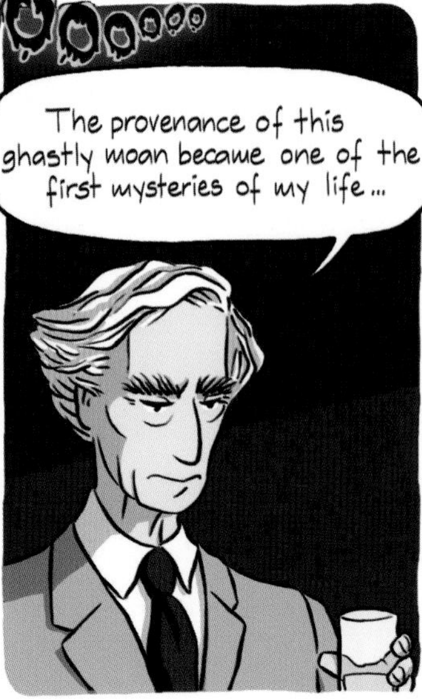
The provenance of this ghastly moan became one of the first mysteries of my life...
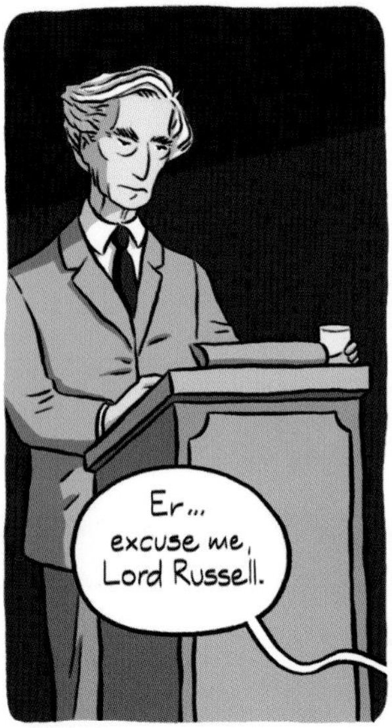
Er... excuse me, Lord Russell.
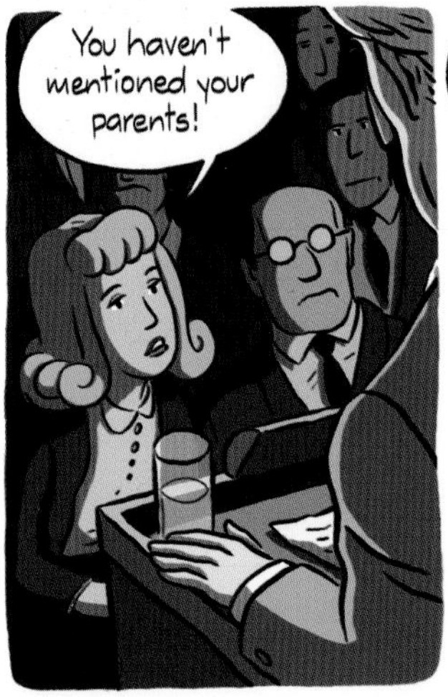
You haven't mentioned your parents!
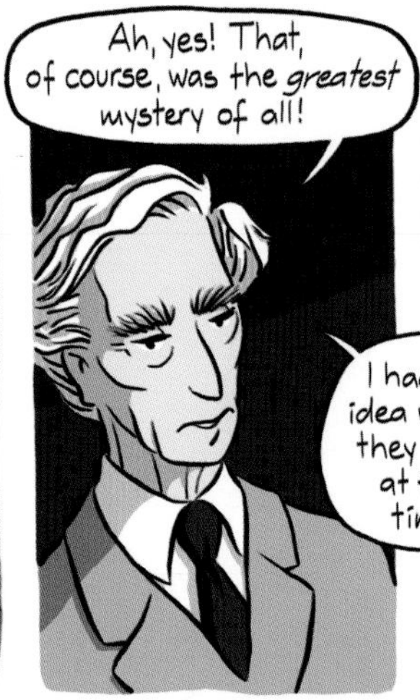
Ah, yes! That, of course, was the *greatest* mystery of all!
I had no idea where they were, at this time.
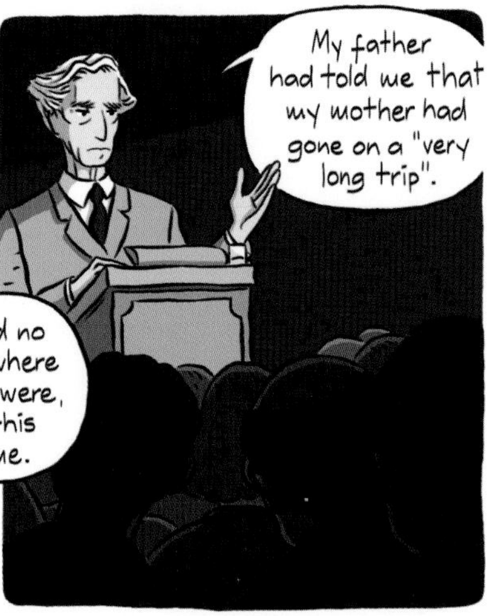
My father had told me that my mother had gone on a "very long trip".
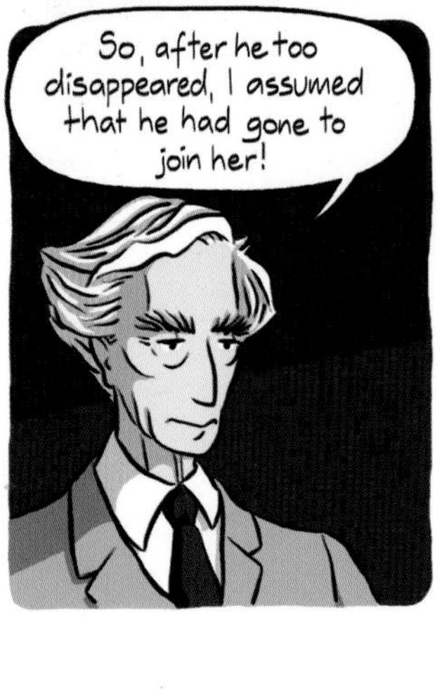
So, after he too disappeared, I assumed that he had gone to join her!
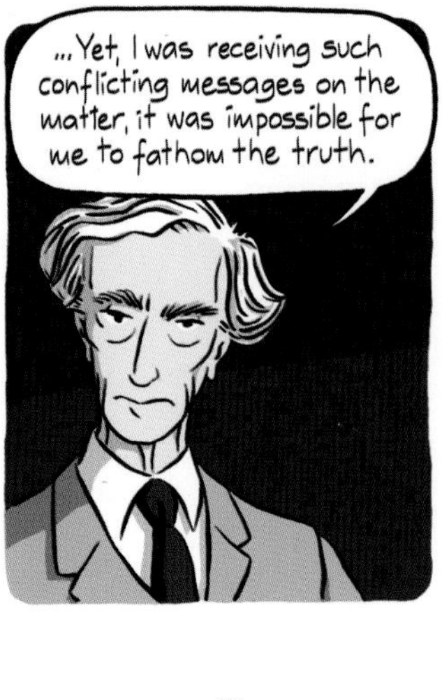
...Yet, I was receiving such conflicting messages on the matter, it was impossible for me to fathom the truth.
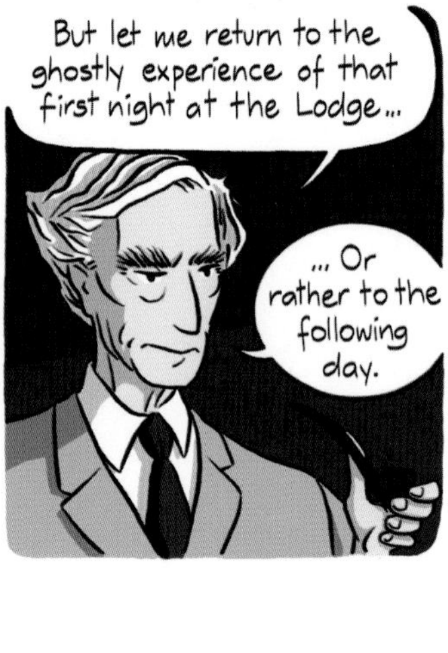
But let me return to the ghostly experience of that first night at the Lodge...
...Or rather to the following day.

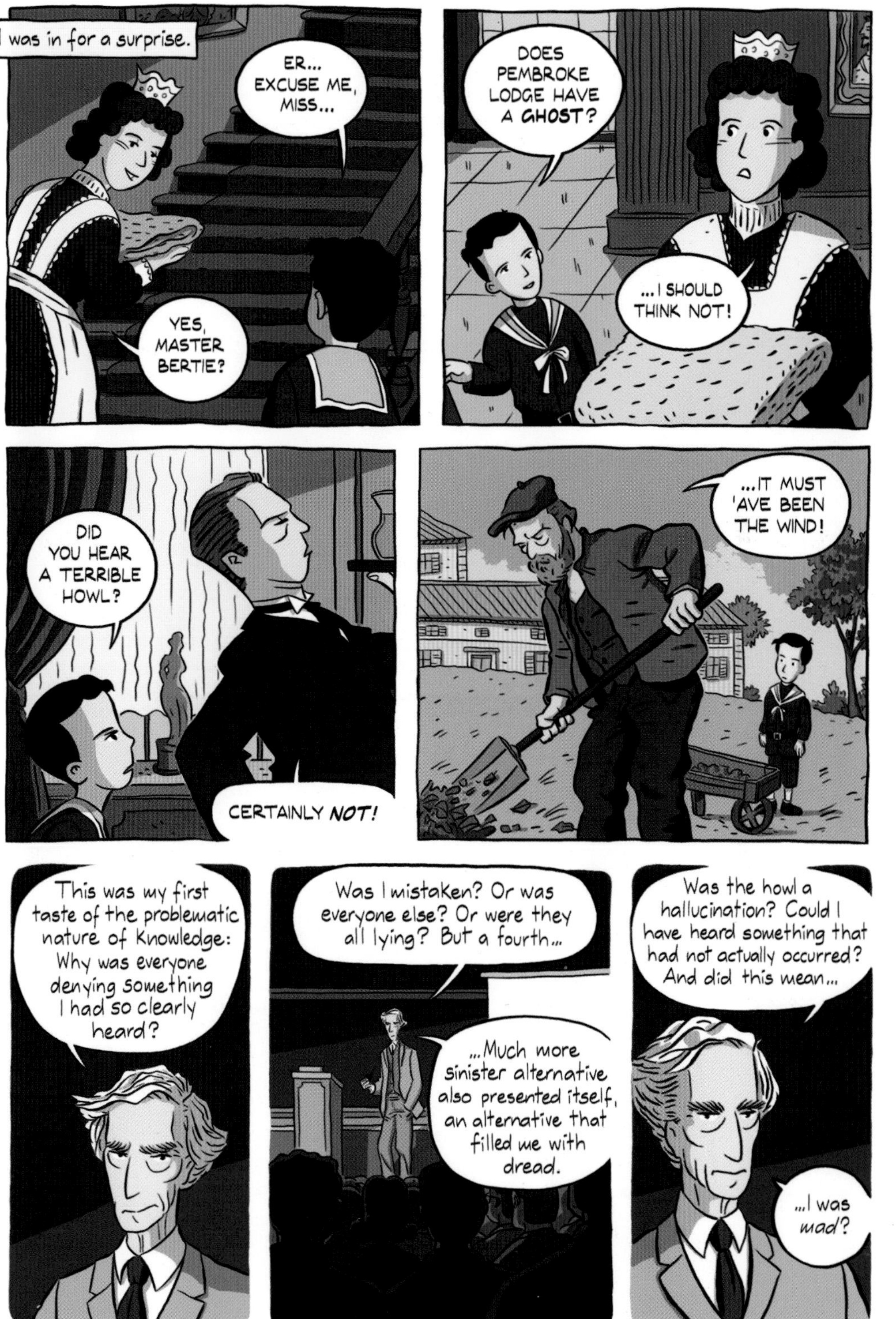
I was in for a surprise.
ER... EXCUSE ME, MISS...
YES, MASTER BERTIE?
DOES PEMBROKE LODGE HAVE A **GHOST**?
...I SHOULD THINK NOT!
DID YOU HEAR A TERRIBLE HOWL?
CERTAINLY ***NOT!***
...IT MUST 'AVE BEEN THE WIND!
This was my first taste of the problematic nature of Knowledge: Why was everyone denying something I had so clearly heard?
Was I mistaken? Or was everyone else? Or were they all lying? But a fourth...
...Much more sinister alternative also presented itself, an alternative that filled me with dread.
Was the howl a hallucination? Could I have heard something that had not actually occurred? And did this mean...
...I was *mad*?

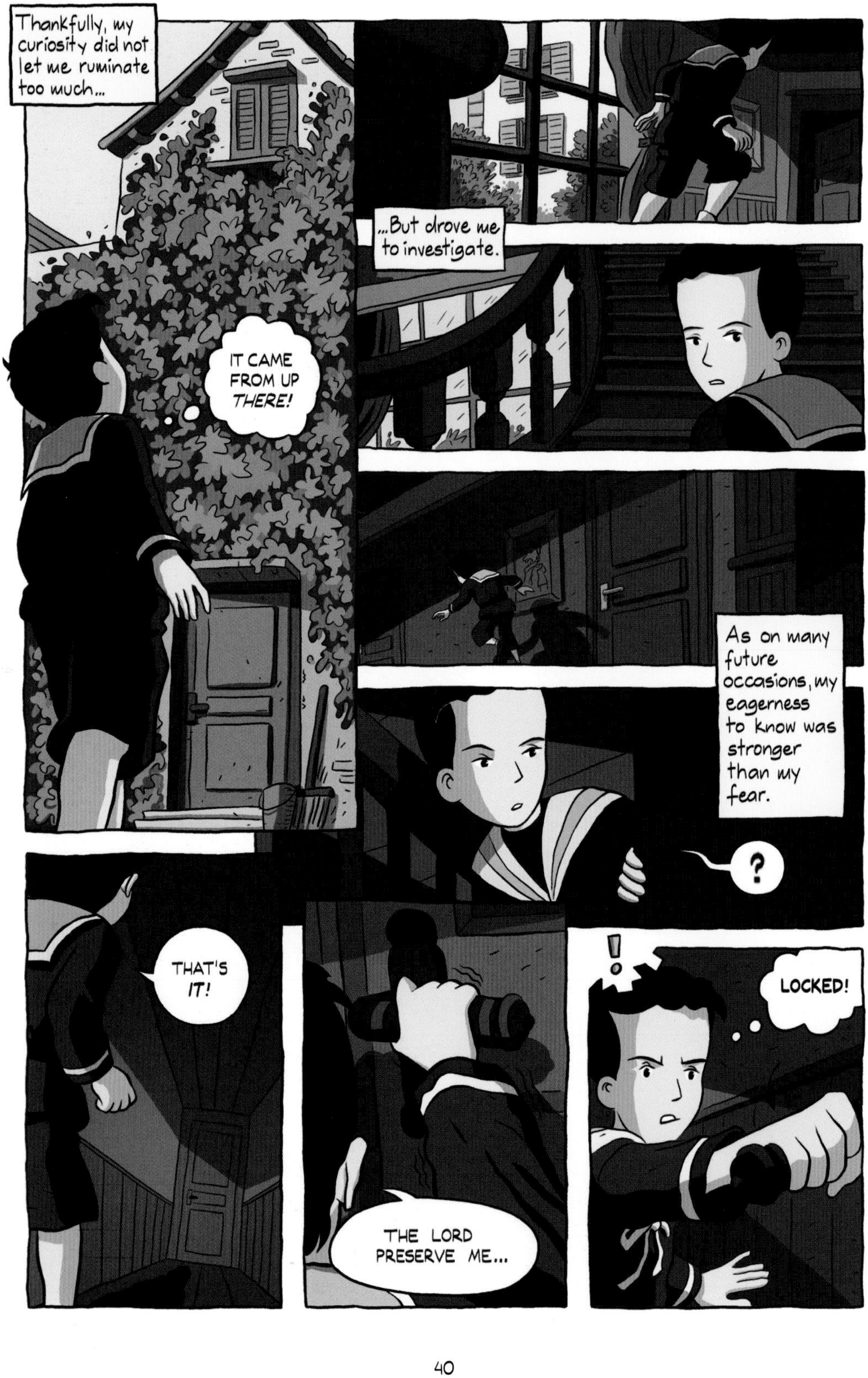
Thankfully, my curiosity did not let me ruminate too much...
IT CAME FROM UP *THERE!*
...But drove me to investigate.
As on many future occasions, my eagerness to know was stronger than my fear.
?
THAT'S *IT!*
THE LORD PRESERVE ME...
!
LOCKED!

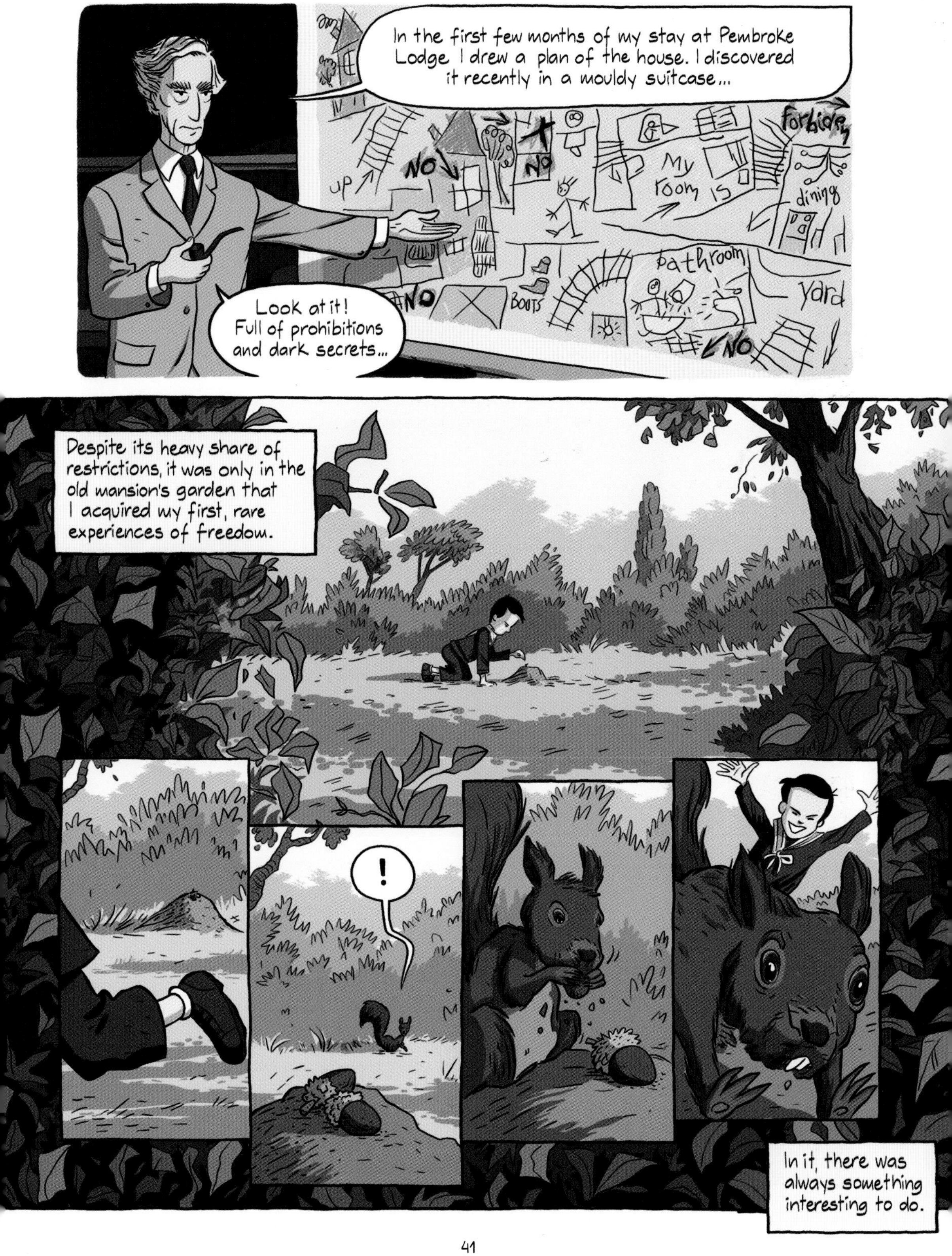
In the first few months of my stay at Pembroke Lodge I drew a plan of the house. I discovered it recently in a mouldy suitcase...
forbiden
NO
NO
up
My room is
dining
bathroom
NO
BOOTS
yard
NO
Look at it! Full of prohibitions and dark secrets...
Despite its heavy share of restrictions, it was only in the old mansion's garden that I acquired my first, rare experiences of freedom.
!
In it, there was always something interesting to do.

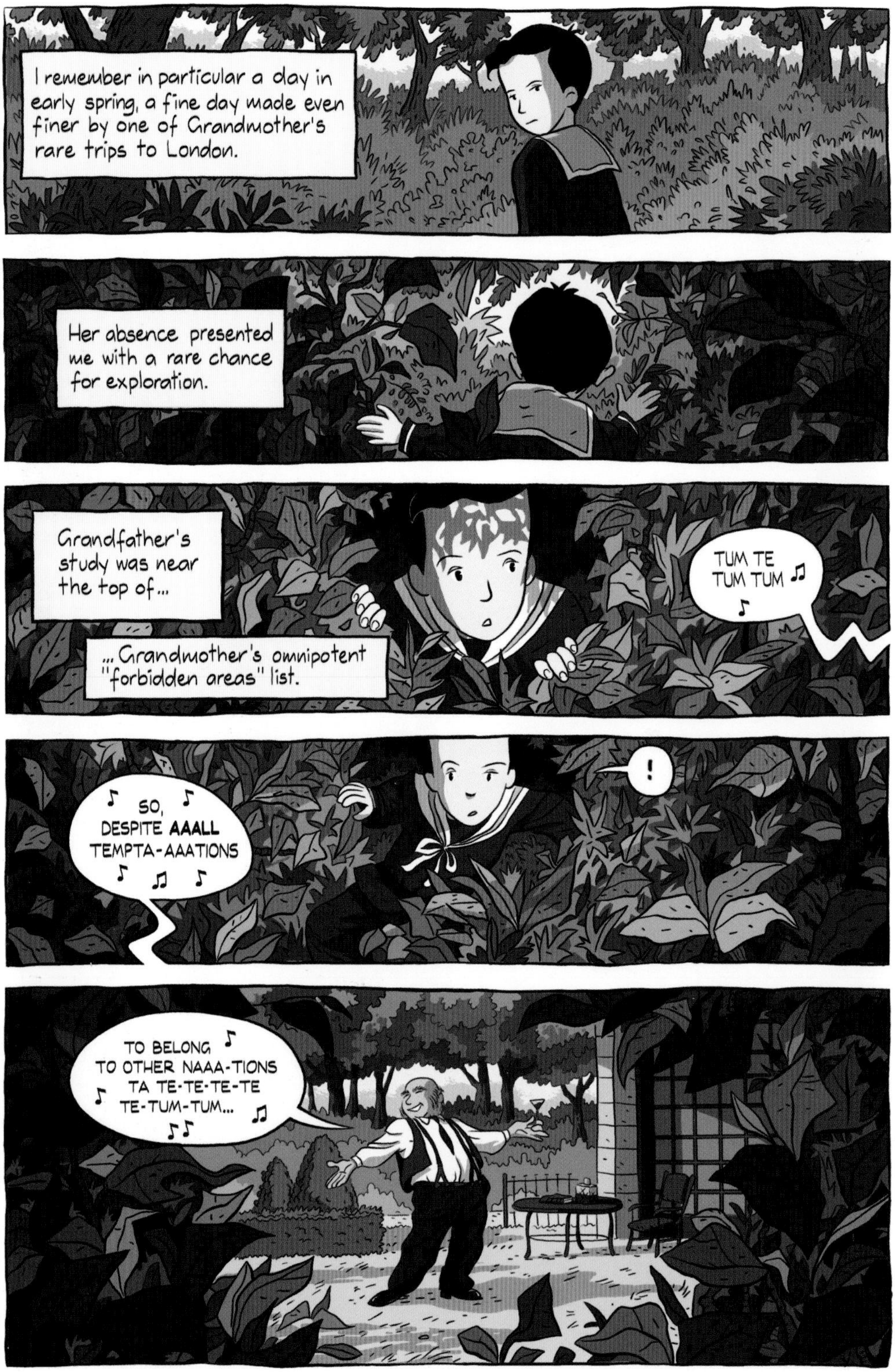
I remember in particular a day in early spring, a fine day made even finer by one of Grandmother's rare trips to London.
Her absence presented me with a rare chance for exploration.
Grandfather's study was near the top of...
...Grandmother's omnipotent "forbidden areas" list.
TUM TE TUM TUM
!
SO, DESPITE **AAALL** TEMPTA-AATIONS
TO BELONG TO OTHER NAAA-TIONS TA TE-TE-TE-TE TE-TUM-TUM...

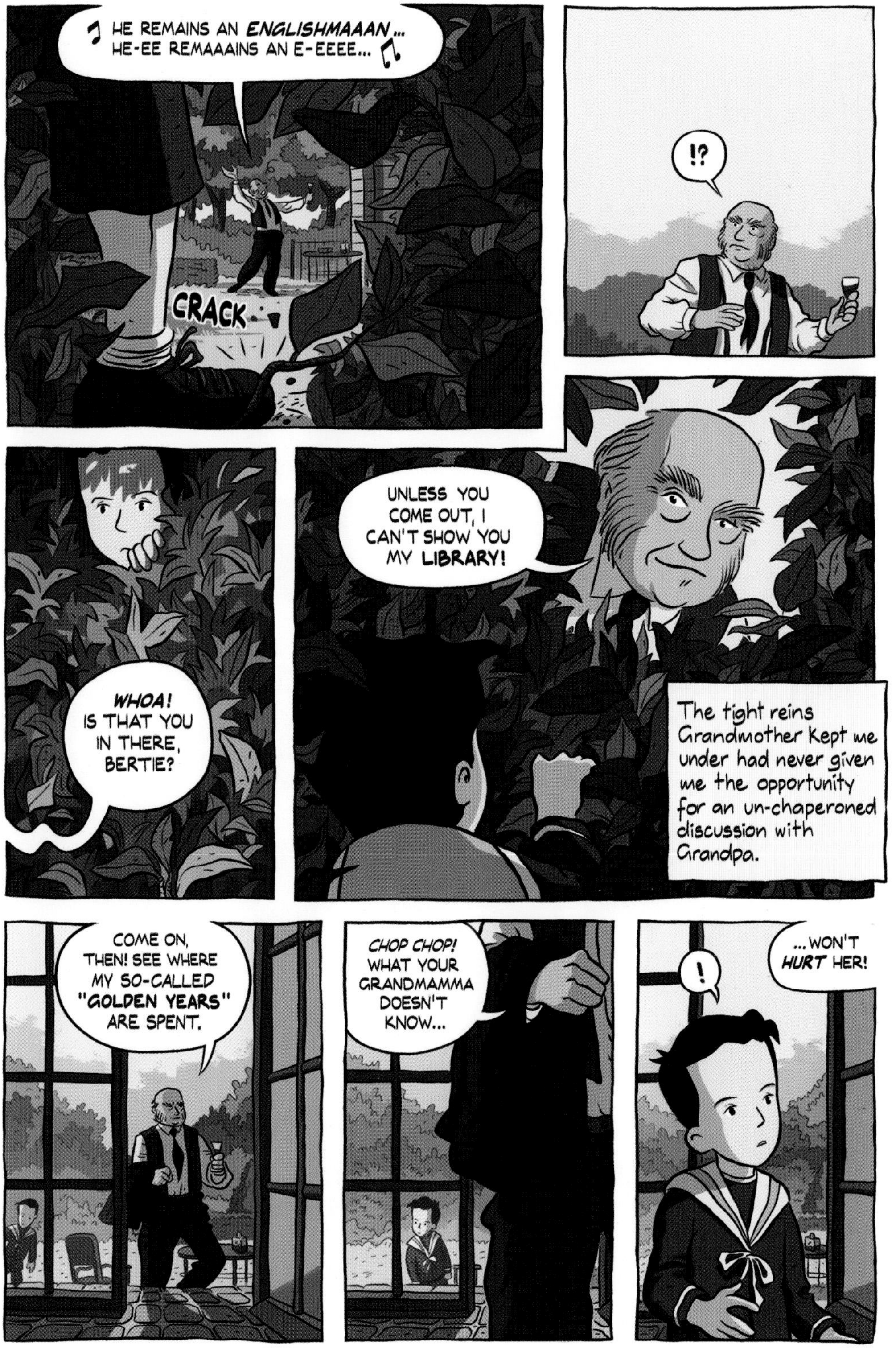
HE REMAINS AN ENGLISHMAAAN... HE-EE REMAAAINS AN E-EEEE...
CRACK
!?
WHOA! IS THAT YOU IN THERE, BERTIE?
UNLESS YOU COME OUT, I CAN'T SHOW YOU MY LIBRARY!
The tight reins Grandmother kept me under had never given me the opportunity for an un-chaperoned discussion with Grandpa.
COME ON, THEN! SEE WHERE MY SO-CALLED "GOLDEN YEARS" ARE SPENT.
CHOP CHOP! WHAT YOUR GRANDMAMMA DOESN'T KNOW...
!
...WON'T HURT HER!

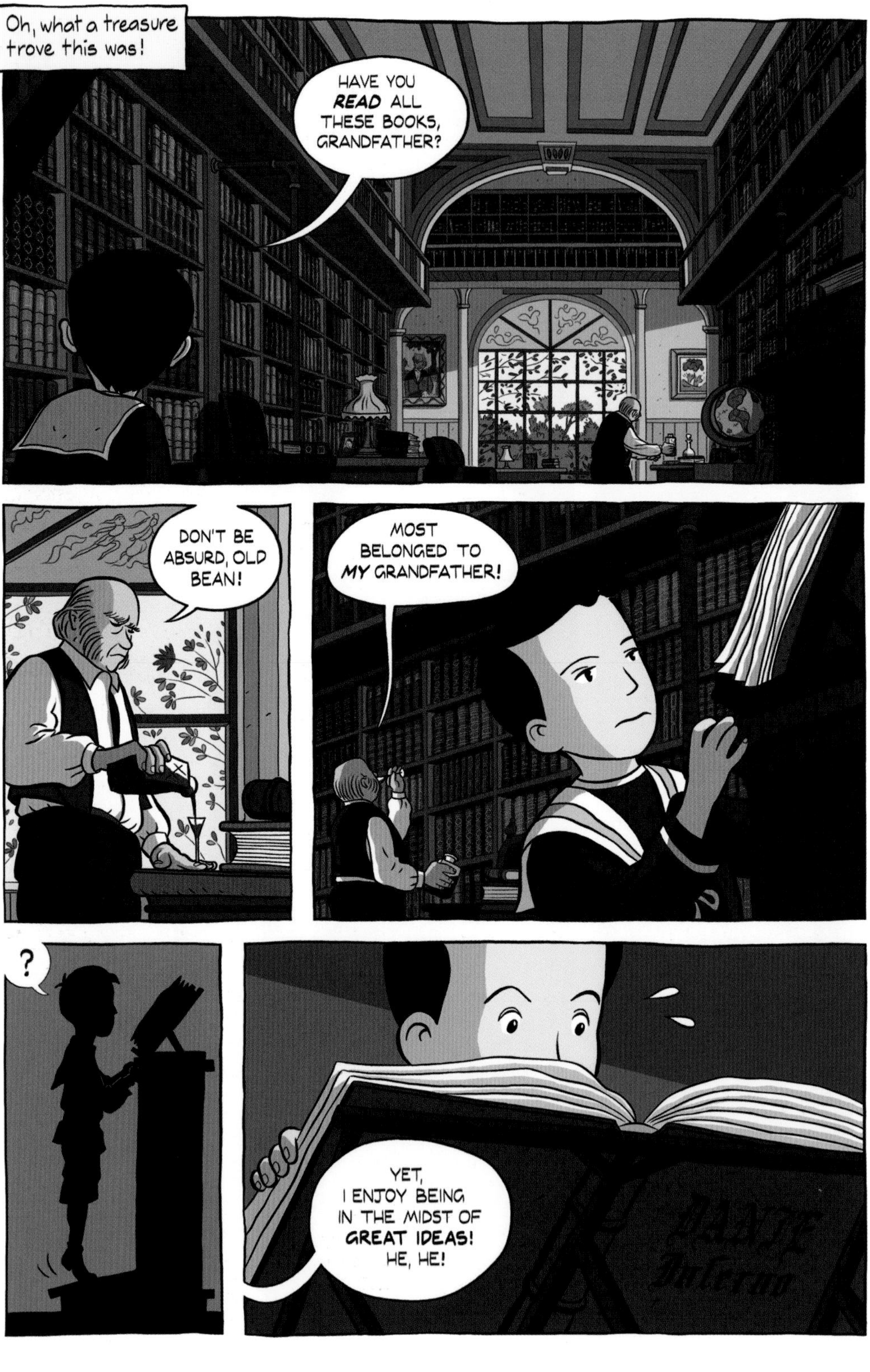
Oh, what a treasure trove this was!
HAVE YOU **READ** ALL THESE BOOKS, GRANDFATHER?
DON'T BE ABSURD, OLD BEAN!
MOST BELONGED TO ***MY*** GRANDFATHER!
?
YET, I ENJOY BEING IN THE MIDST OF **GREAT IDEAS!** HE, HE!

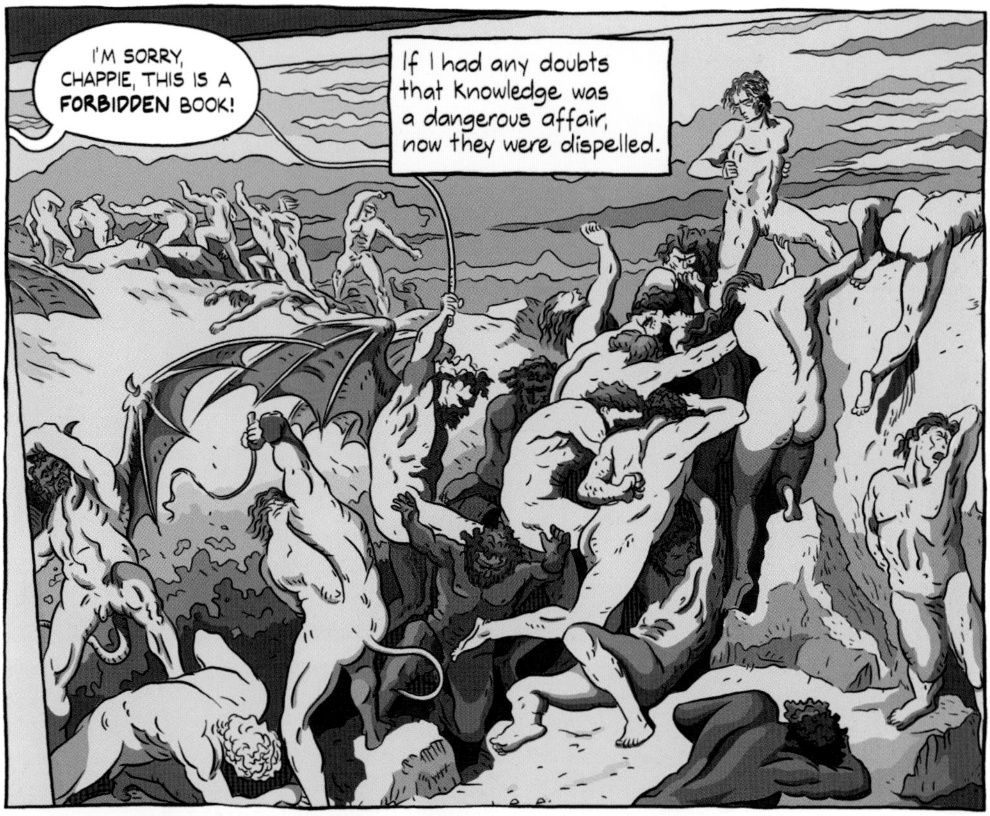
I'M SORRY, CHAPPIE, THIS IS A **FORBIDDEN** BOOK!
If I had any doubts that knowledge was a dangerous affair, now they were dispelled.

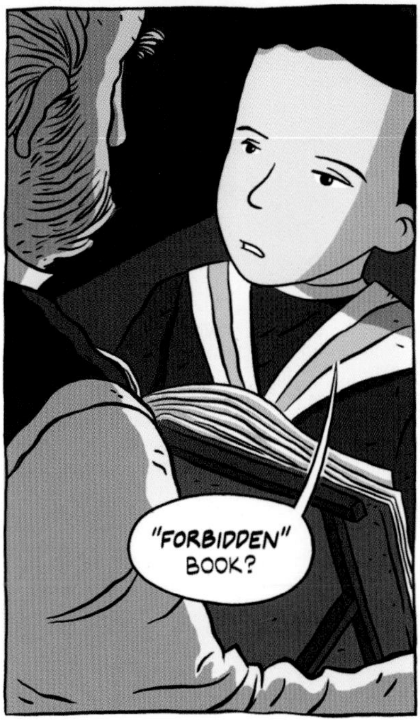
"**FORBIDDEN**" BOOK?

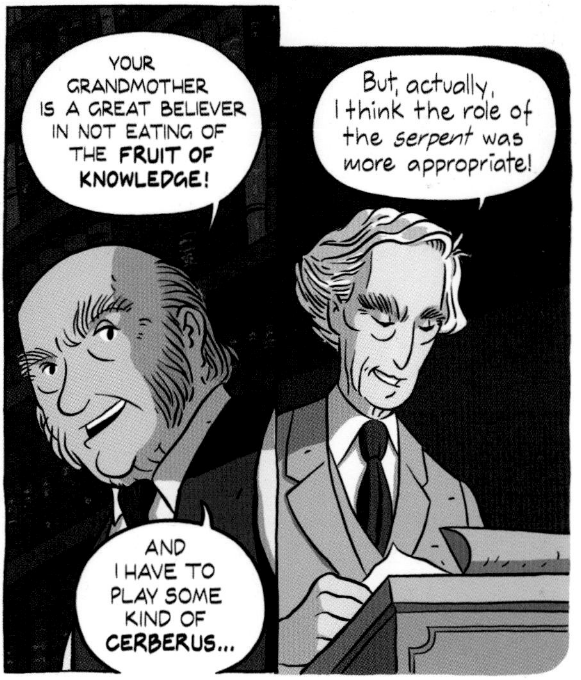
YOUR GRANDMOTHER IS A GREAT BELIEVER IN NOT EATING OF THE **FRUIT OF KNOWLEDGE!**
AND I HAVE TO PLAY SOME KIND OF **CERBERUS...**
But, actually, I think the role of the *serpent* was more appropriate!

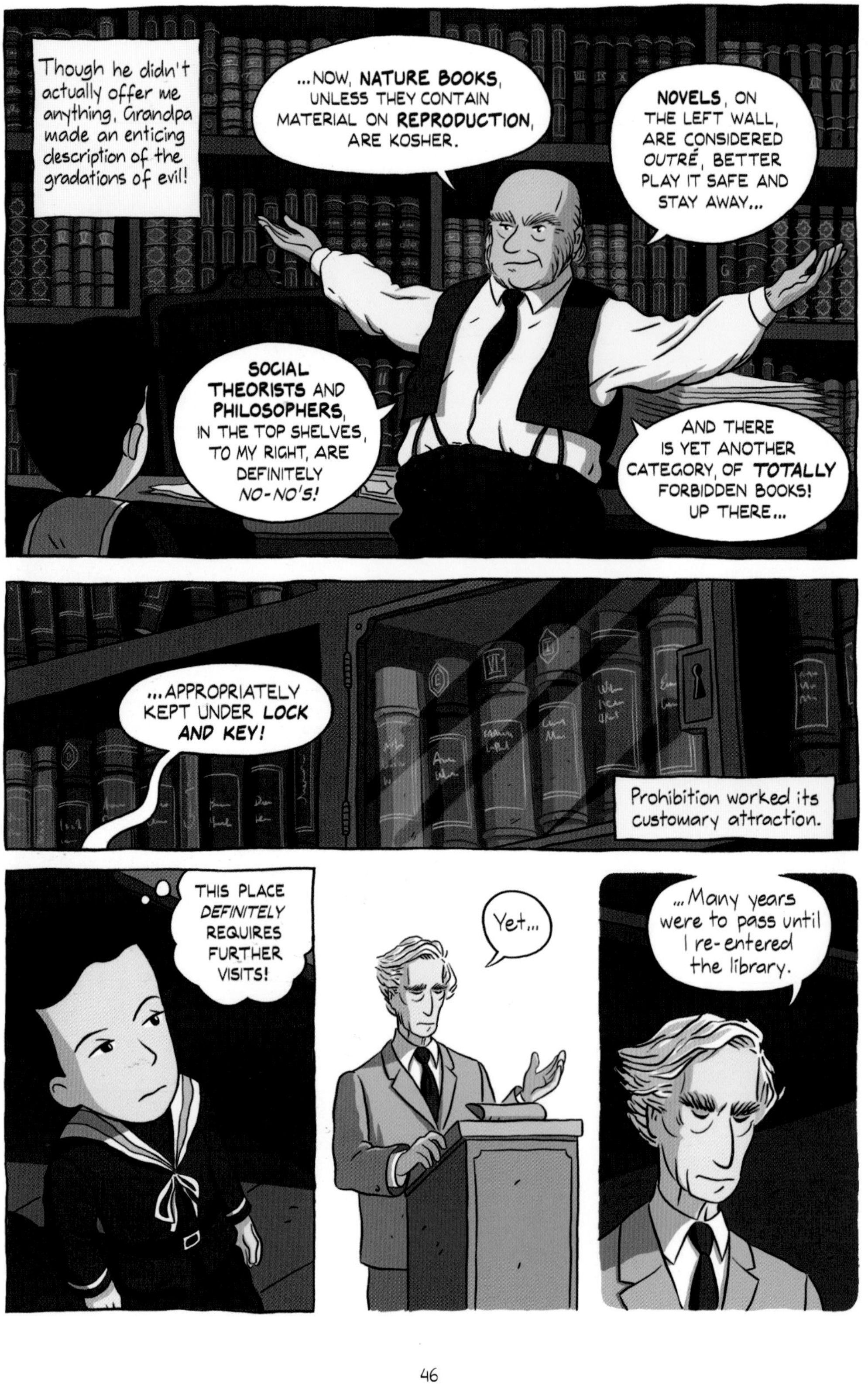

Though he didn't actually offer me anything, Grandpa made an enticing description of the gradations of evil!
...NOW, NATURE BOOKS, UNLESS THEY CONTAIN MATERIAL ON REPRODUCTION, ARE KOSHER.
NOVELS, ON THE LEFT WALL, ARE CONSIDERED OUTRÉ, BETTER PLAY IT SAFE AND STAY AWAY...
SOCIAL THEORISTS AND PHILOSOPHERS, IN THE TOP SHELVES, TO MY RIGHT, ARE DEFINITELY NO-NO'S!
AND THERE IS YET ANOTHER CATEGORY, OF TOTALLY FORBIDDEN BOOKS! UP THERE...
...APPROPRIATELY KEPT UNDER LOCK AND KEY!
Prohibition worked its customary attraction.
THIS PLACE DEFINITELY REQUIRES FURTHER VISITS!
Yet...
...Many years were to pass until I re-entered the library.

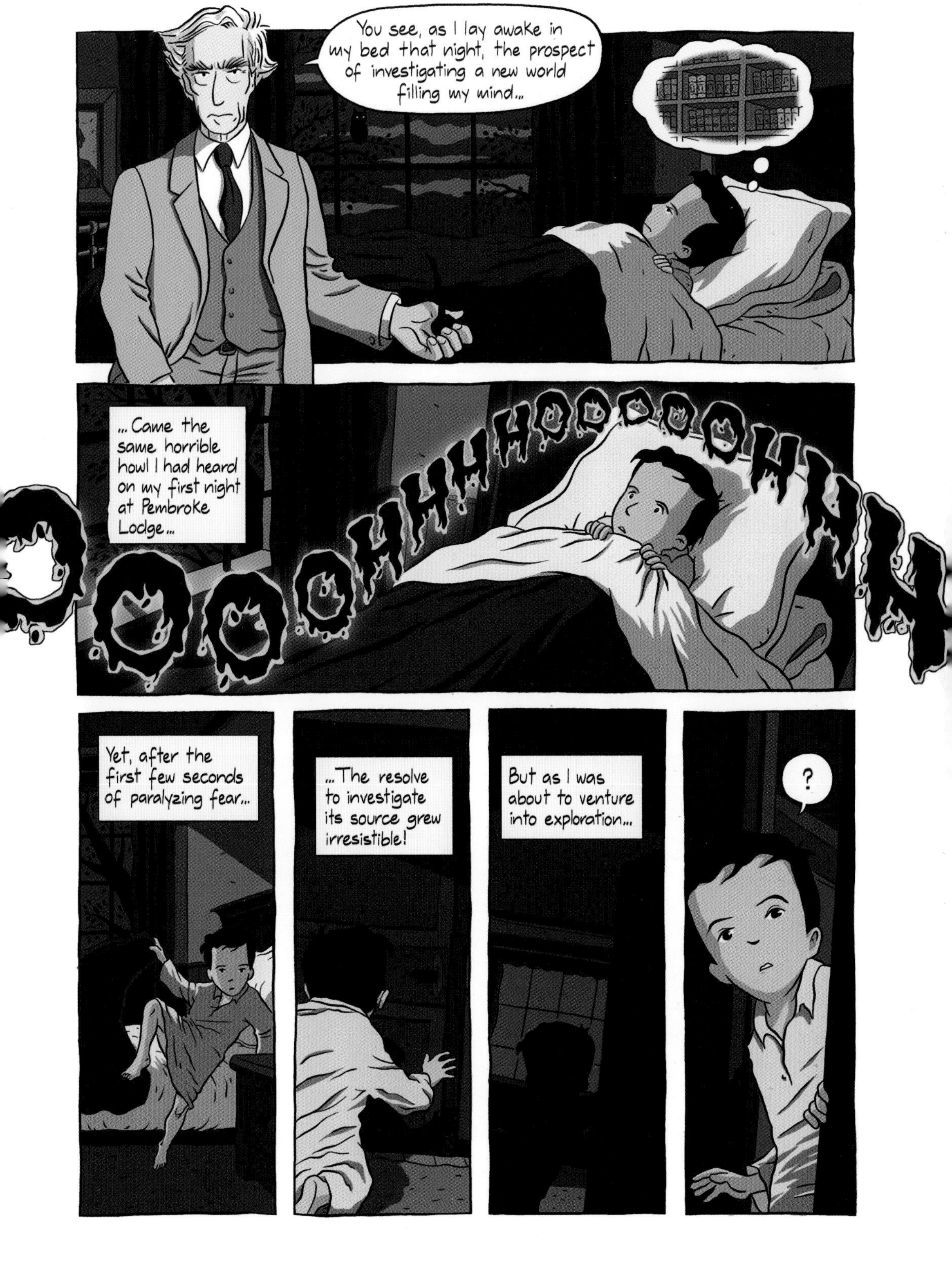
You see, as I lay awake in my bed that night, the prospect of investigating a new world filling my mind...
...Came the same horrible howl I had heard on my first night at Pembroke Lodge...
OOOOHHHHHHOOOOOOHHHH
Yet, after the first few seconds of paralyzing fear...
...The resolve to investigate its source grew irresistible!
But as I was about to venture into exploration...
?

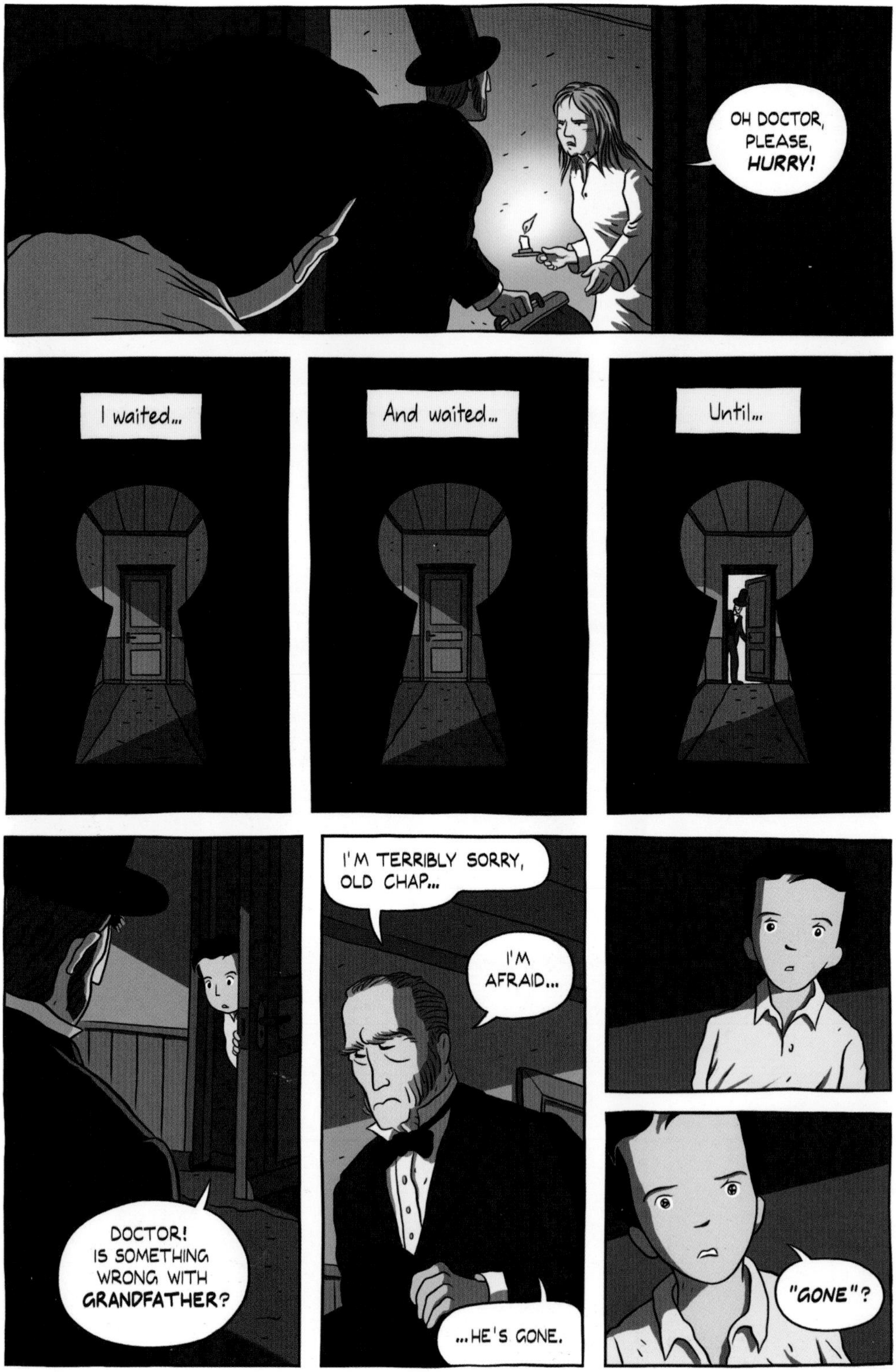
OH DOCTOR, PLEASE, **HURRY!**
I waited...
And waited...
Until...
DOCTOR! IS SOMETHING WRONG WITH **GRANDFATHER?**
I'M TERRIBLY SORRY, OLD CHAP...
I'M AFRAID...
...HE'S GONE.
"**GONE**"?

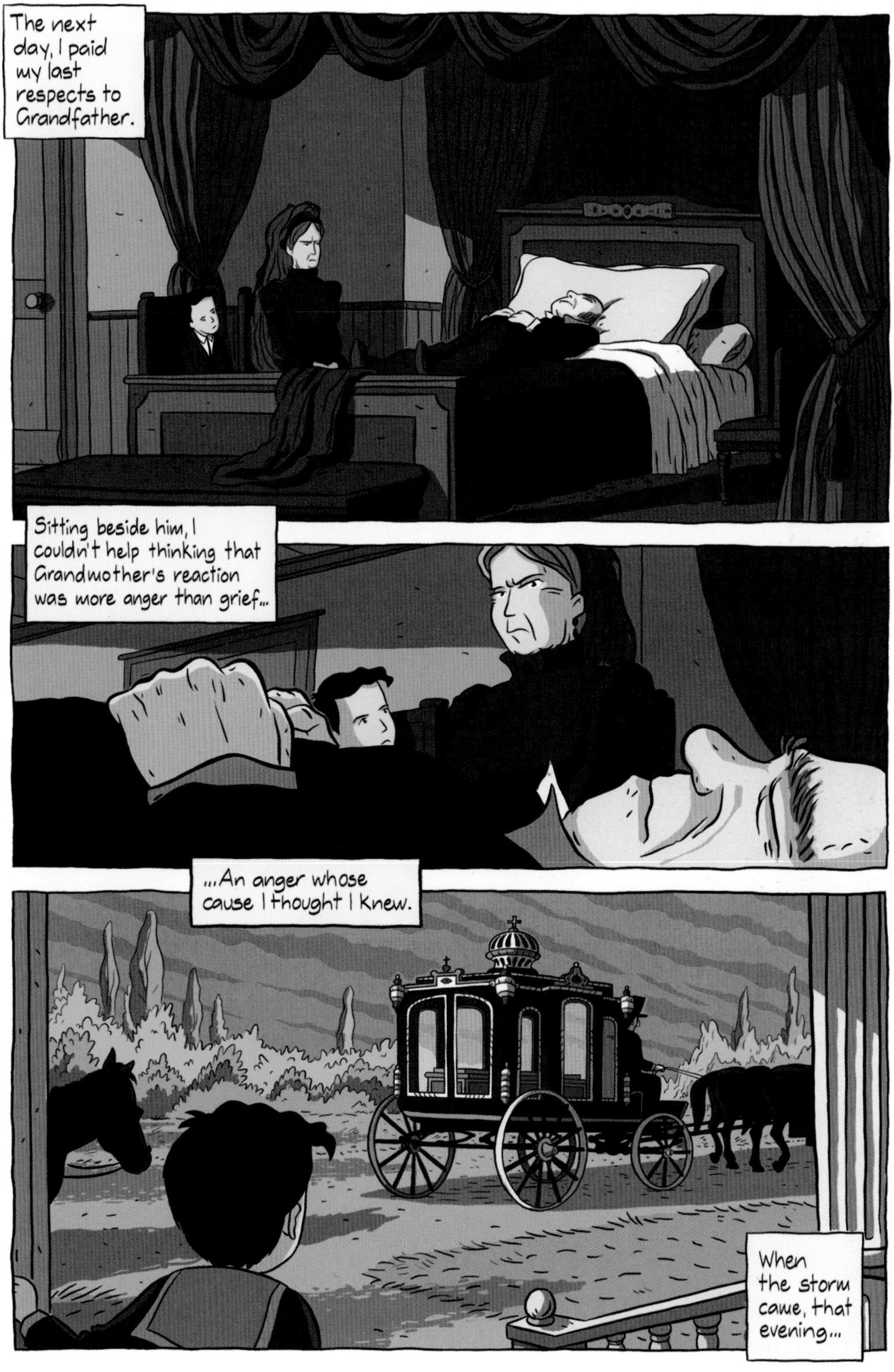
The next day, I paid my last respects to Grandfather.
Sitting beside him, I couldn't help thinking that Grandmother's reaction was more anger than grief...
...An anger whose cause I thought I knew.
When the storm came, that evening...

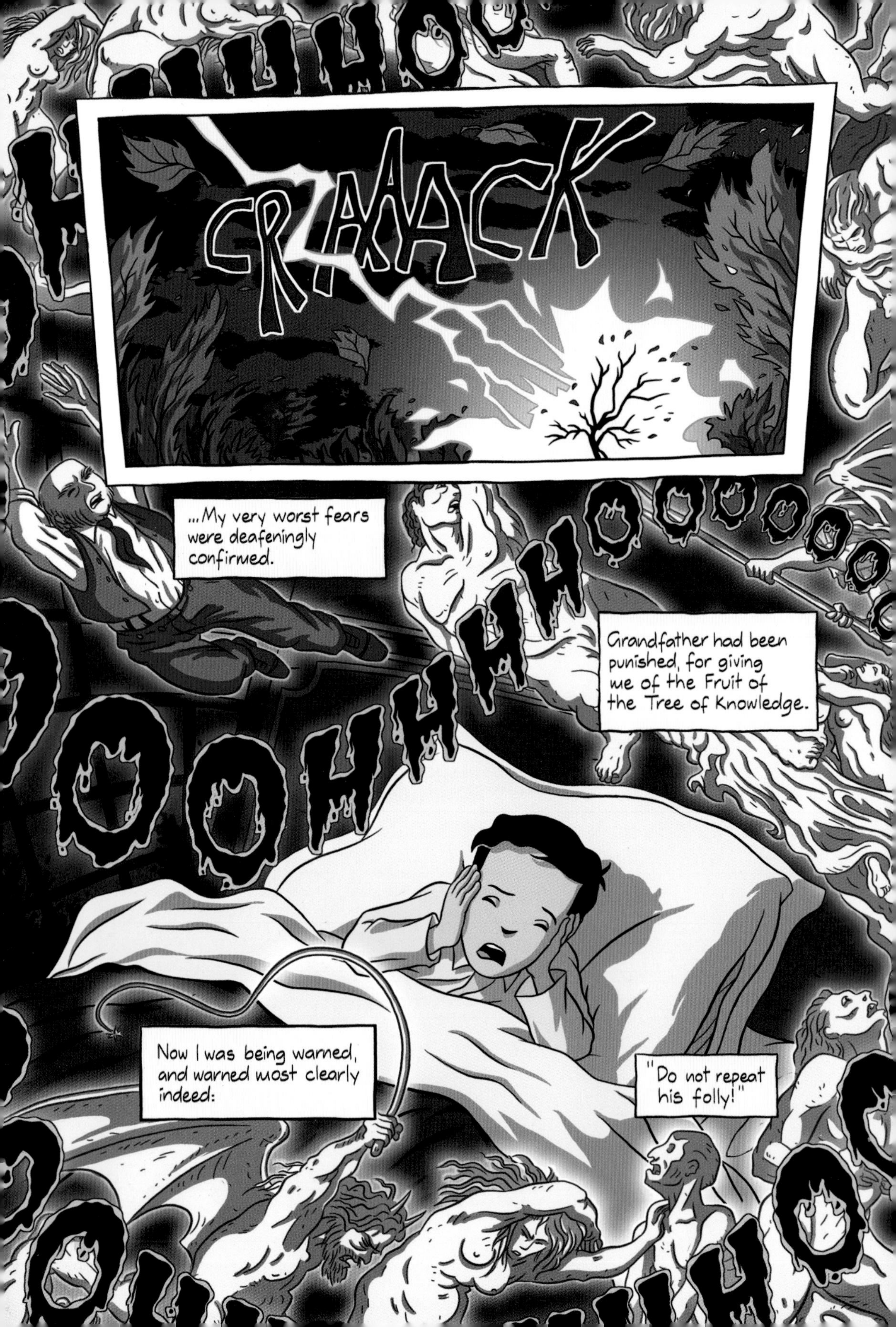
CRAAAACK
...My very worst fears were deafeningly confirmed.
Grandfather had been punished, for giving me of the Fruit of the Tree of Knowledge.
Now I was being warned, and warned most clearly indeed:
"Do not repeat his folly!"

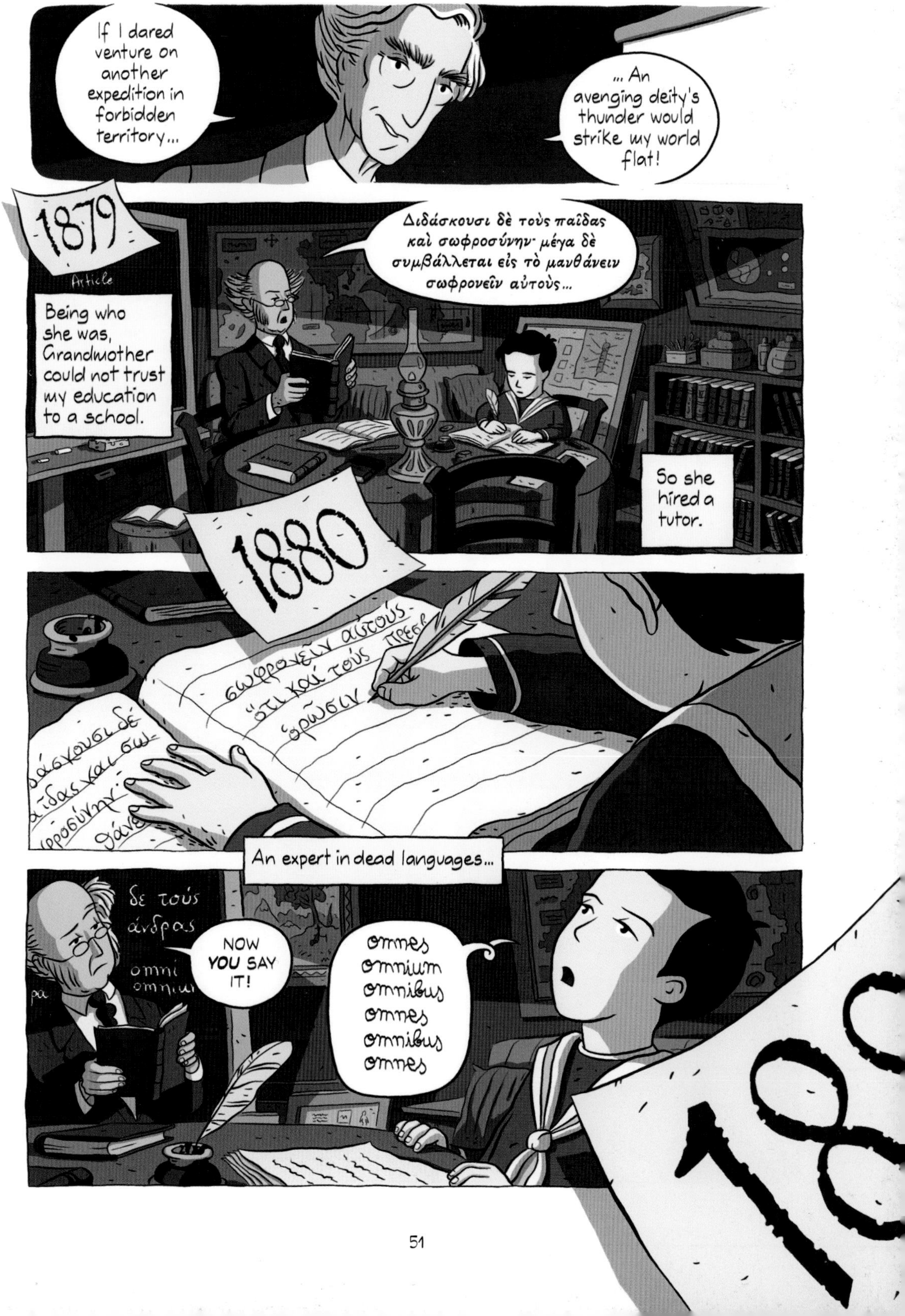
If I dared venture on another expedition in forbidden territory...
... An avenging deity's thunder would strike my world flat!
1879
Being who she was, Grandmother could not trust my education to a school.
Διδάσκουσι δὲ τοὺς παῖδας καὶ σωφροσύνην· μέγα δὲ συμβάλλεται εἰς τὸ μανθάνειν σωφρονεῖν αὐτούς...
So she hired a tutor.
1880
σωφρονεῖν αὐτούς
ὅτι καὶ τούς
ὁρῶσιν
An expert in dead languages...
δε τούς ἄνδρας
NOW YOU SAY IT!
omnes
omnium
omnibus
omnes
omnibus
omnes

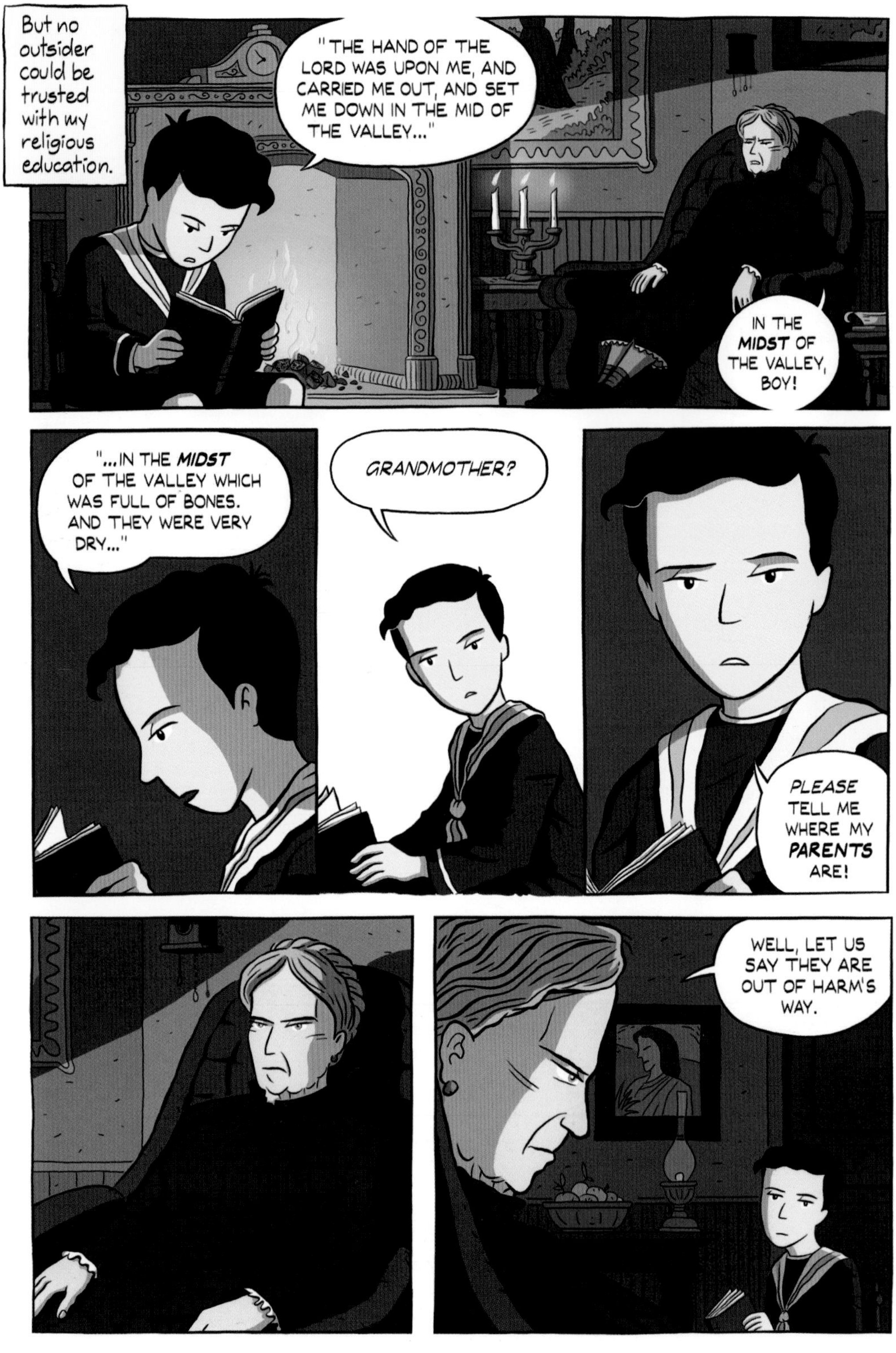
But no outsider could be trusted with my religious education.
"THE HAND OF THE LORD WAS UPON ME, AND CARRIED ME OUT, AND SET ME DOWN IN THE MID OF THE VALLEY..."
IN THE **MIDST** OF THE VALLEY, BOY!
"...IN THE **MIDST** OF THE VALLEY WHICH WAS FULL OF BONES. AND THEY WERE VERY DRY..."
GRANDMOTHER?
PLEASE TELL ME WHERE MY **PARENTS** ARE!
WELL, LET US SAY THEY ARE OUT OF HARM'S WAY.

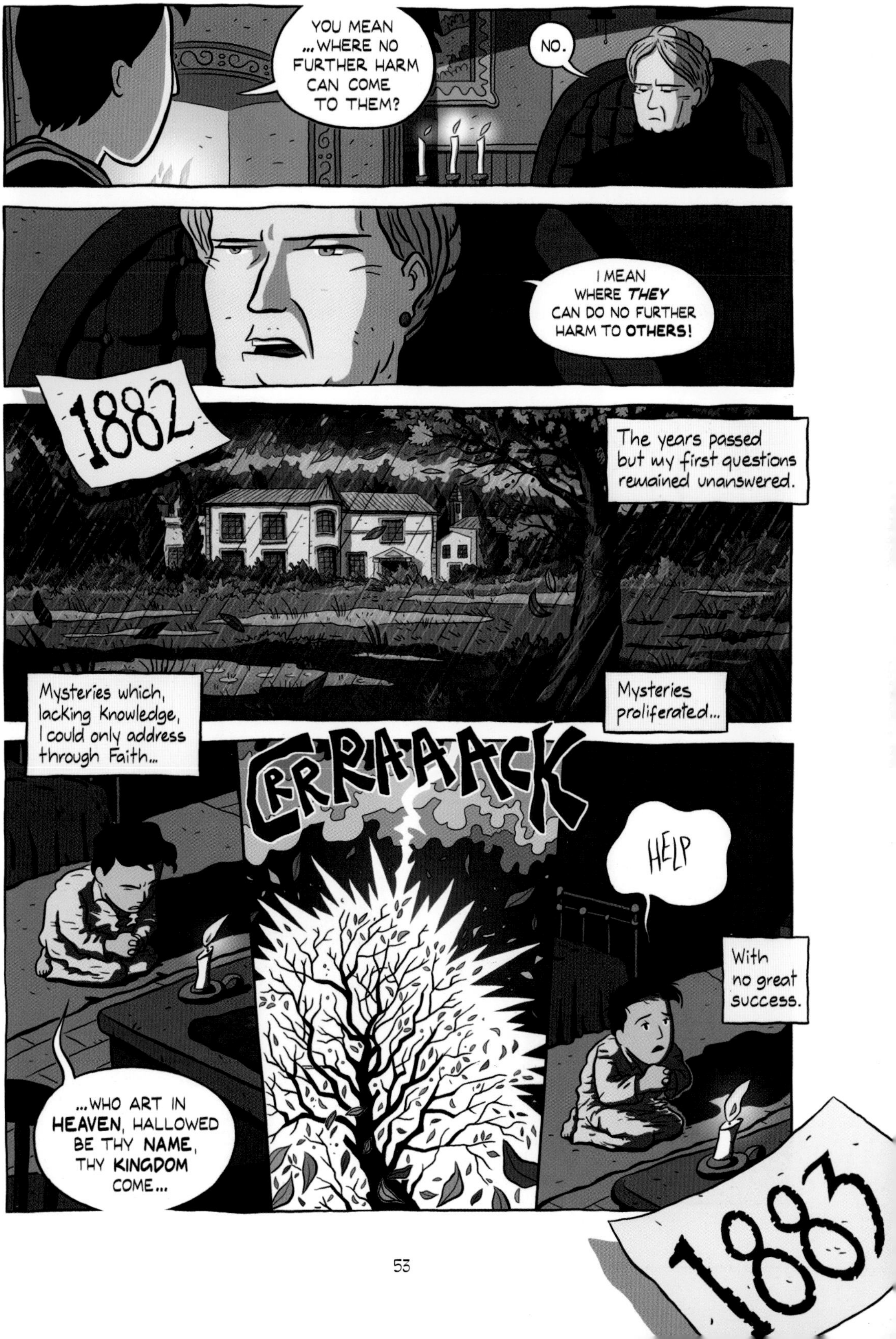

YOU MEAN ...WHERE NO FURTHER HARM CAN COME TO THEM?
NO.
I MEAN WHERE THEY CAN DO NO FURTHER HARM TO OTHERS!
1882
The years passed but my first questions remained unanswered.
Mysteries which, lacking knowledge, I could only address through Faith...
Mysteries proliferated...
CRRRAAACK
HELP
With no great success.
...WHO ART IN HEAVEN, HALLOWED BE THY NAME, THY KINGDOM COME...
1883

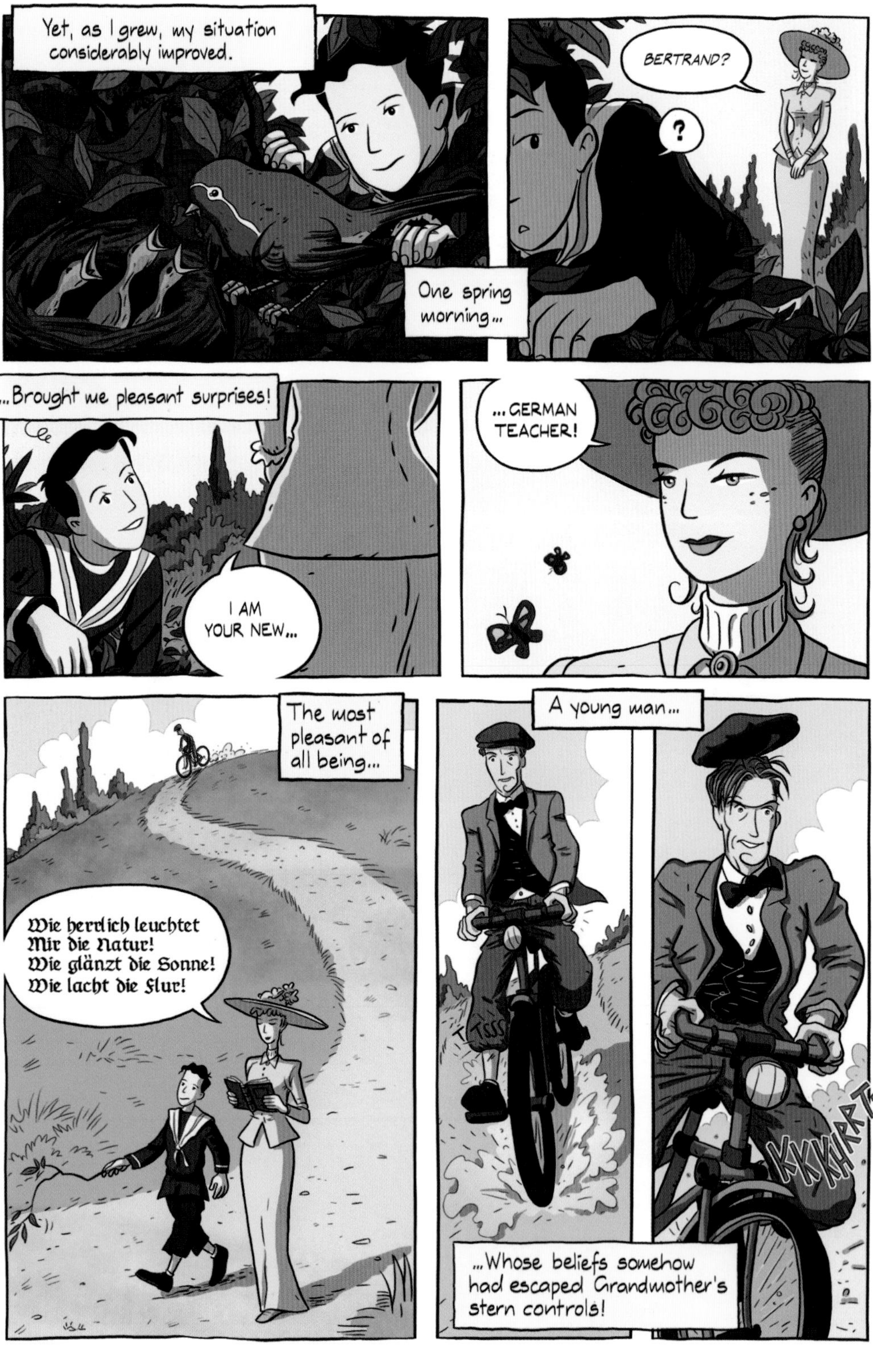
Yet, as I grew, my situation considerably improved.
One spring morning...
BERTRAND?
?
...Brought me pleasant surprises!
I AM YOUR NEW...
...GERMAN TEACHER!
The most pleasant of all being...
Wie herrlich leuchtet
Mir die Natur!
Wie glänzt die Sonne!
Wie lacht die Flur!
A young man...
TSSS
KKKHRRTT
...Whose beliefs somehow had escaped Grandmother's stern controls!

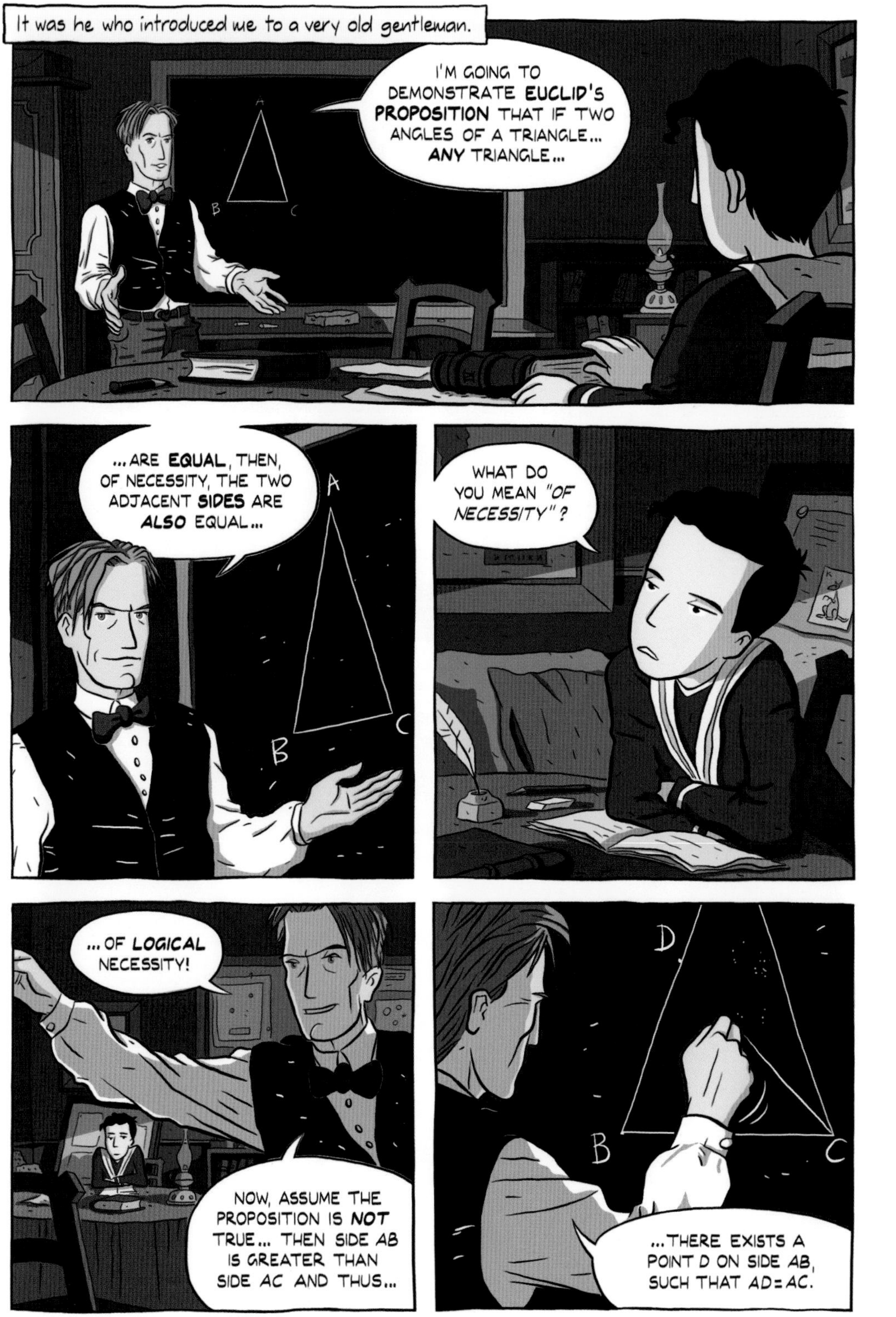
It was he who introduced me to a very old gentleman.
I'M GOING TO DEMONSTRATE **EUCLID**'S **PROPOSITION** THAT IF TWO ANGLES OF A TRIANGLE... **ANY** TRIANGLE...
...ARE **EQUAL**, THEN, OF NECESSITY, THE TWO ADJACENT **SIDES** ARE **ALSO** EQUAL...
WHAT DO YOU MEAN *"OF NECESSITY"*?
...OF **LOGICAL** NECESSITY!
NOW, ASSUME THE PROPOSITION IS **NOT** TRUE... THEN SIDE *AB* IS GREATER THAN SIDE *AC* AND THUS...
...THERE EXISTS A POINT *D* ON SIDE *AB*, SUCH THAT *AD*=*AC*.

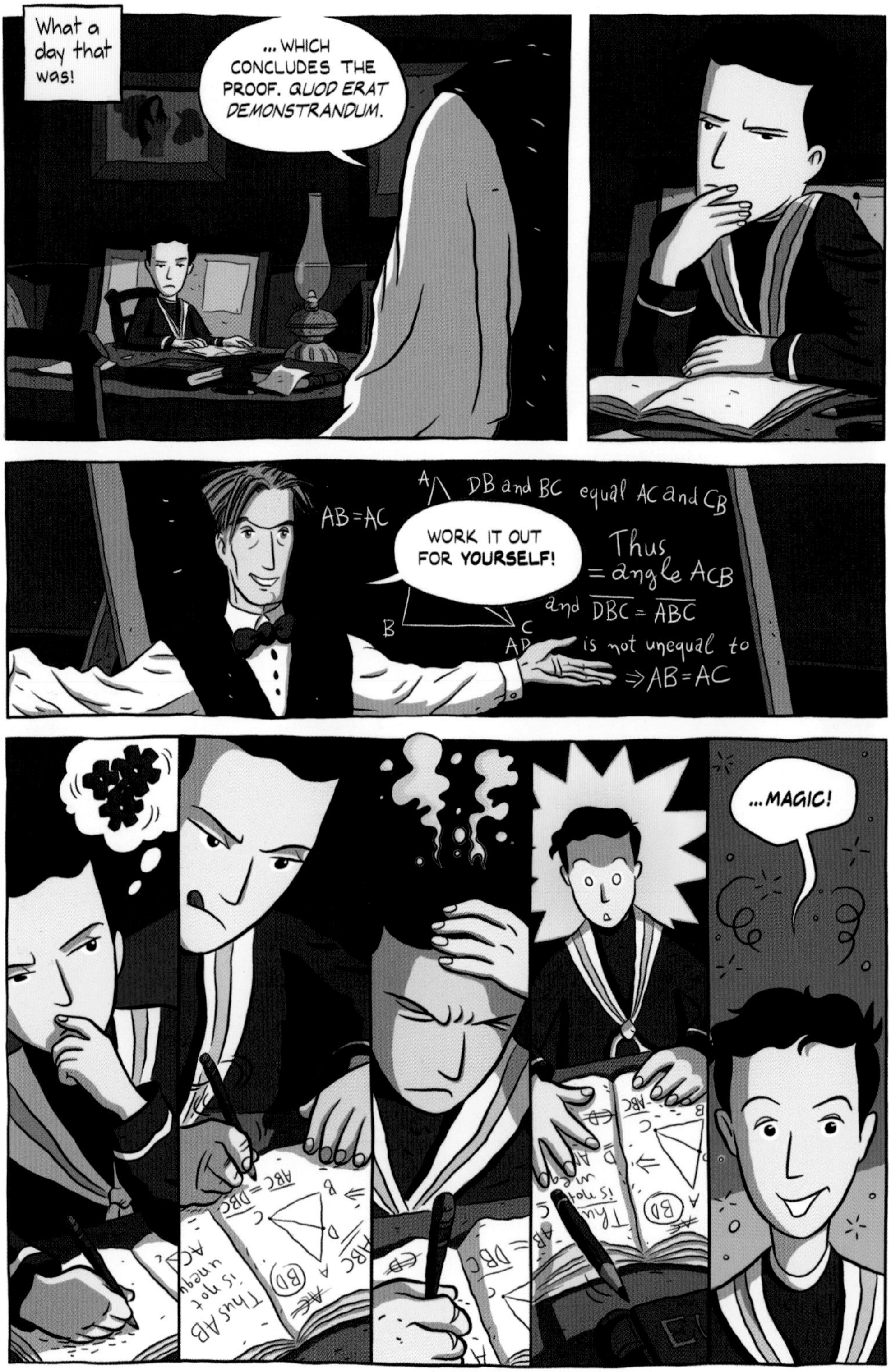
What a day that was!
...WHICH CONCLUDES THE PROOF. *QUOD ERAT DEMONSTRANDUM.*
AB=AC
DB and BC equal AC and CB
WORK IT OUT FOR **YOURSELF!**
Thus
= angle ACB
and DBC = ABC
is not unequal to
⇒AB=AC
...MAGIC!

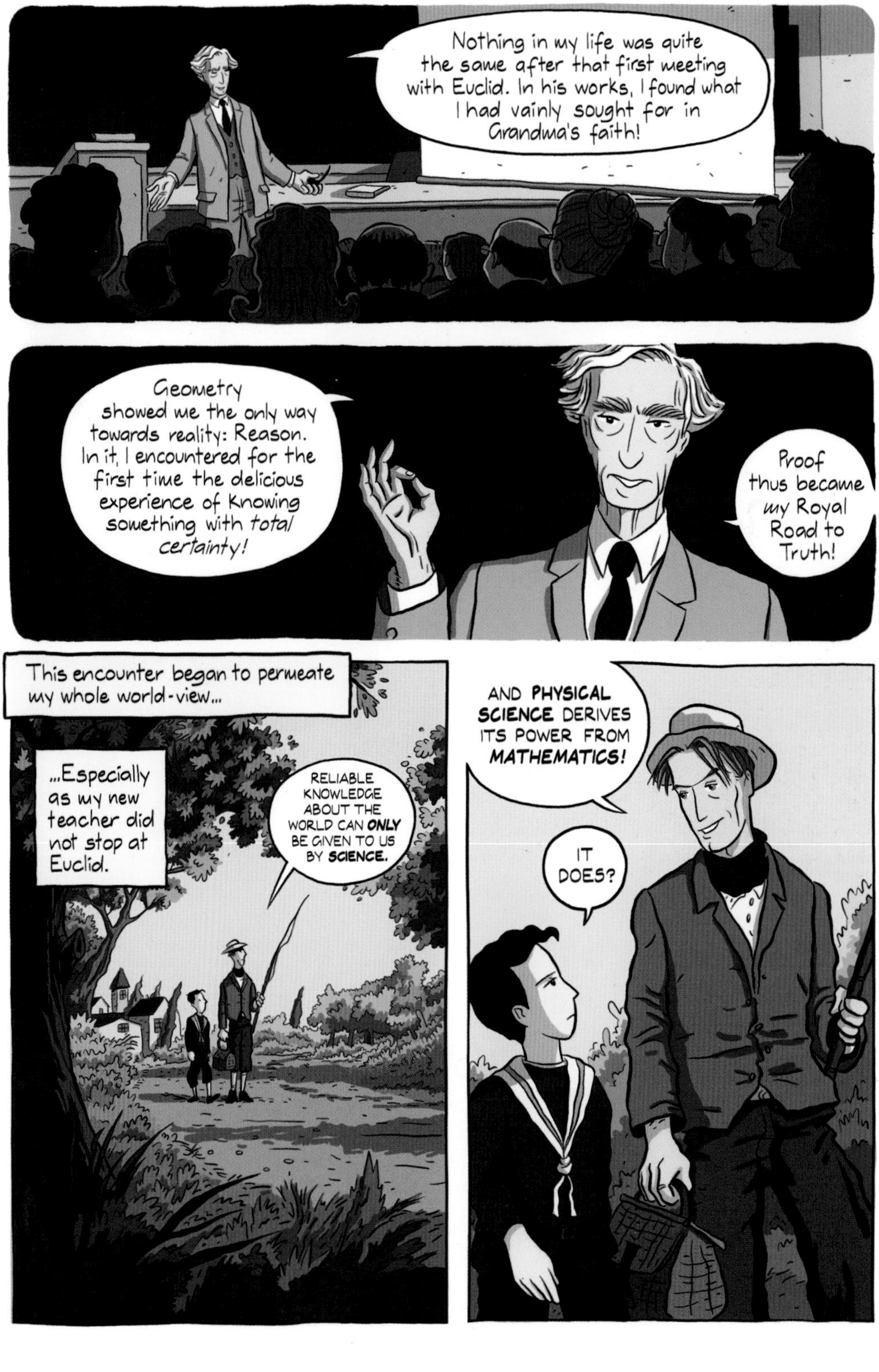

Nothing in my life was quite the same after that first meeting with Euclid. In his works, I found what I had vainly sought for in Grandma's faith!
Geometry showed me the only way towards reality: Reason. In it, I encountered for the first time the delicious experience of Knowing something with *total certainty*!
Proof thus became my Royal Road to Truth!
This encounter began to permeate my whole world-view...
...Especially as my new teacher did not stop at Euclid.
RELIABLE KNOWLEDGE ABOUT THE WORLD CAN **ONLY** BE GIVEN TO US BY **SCIENCE.**
AND **PHYSICAL SCIENCE** DERIVES ITS POWER FROM **MATHEMATICS!**
IT DOES?

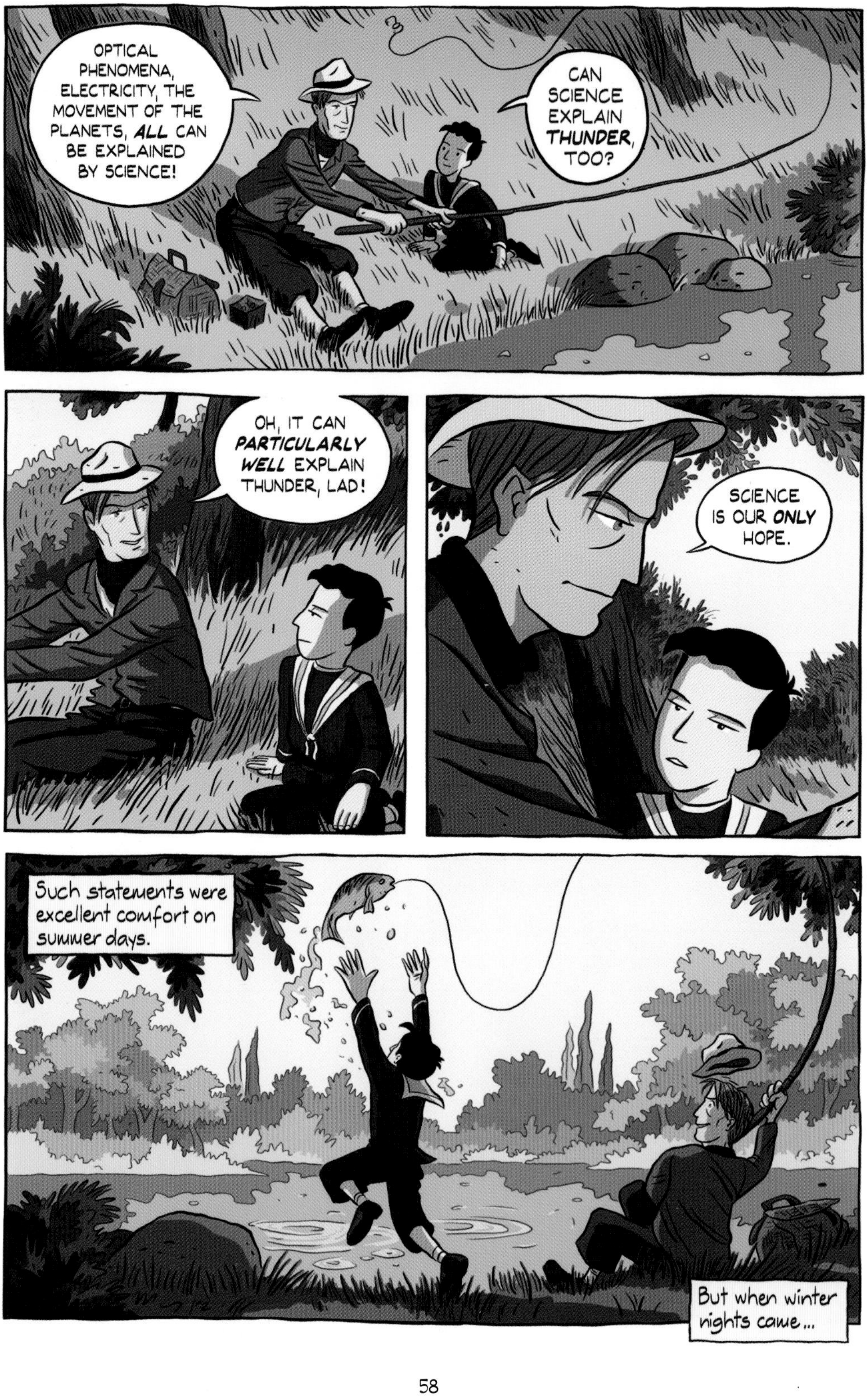
OPTICAL PHENOMENA, ELECTRICITY, THE MOVEMENT OF THE PLANETS, **ALL** CAN BE EXPLAINED BY SCIENCE!
CAN SCIENCE EXPLAIN **THUNDER**, TOO?
OH, IT CAN **PARTICULARLY WELL** EXPLAIN THUNDER, LAD!
SCIENCE IS OUR **ONLY** HOPE.
Such statements were excellent comfort on summer days.
But when winter nights came...

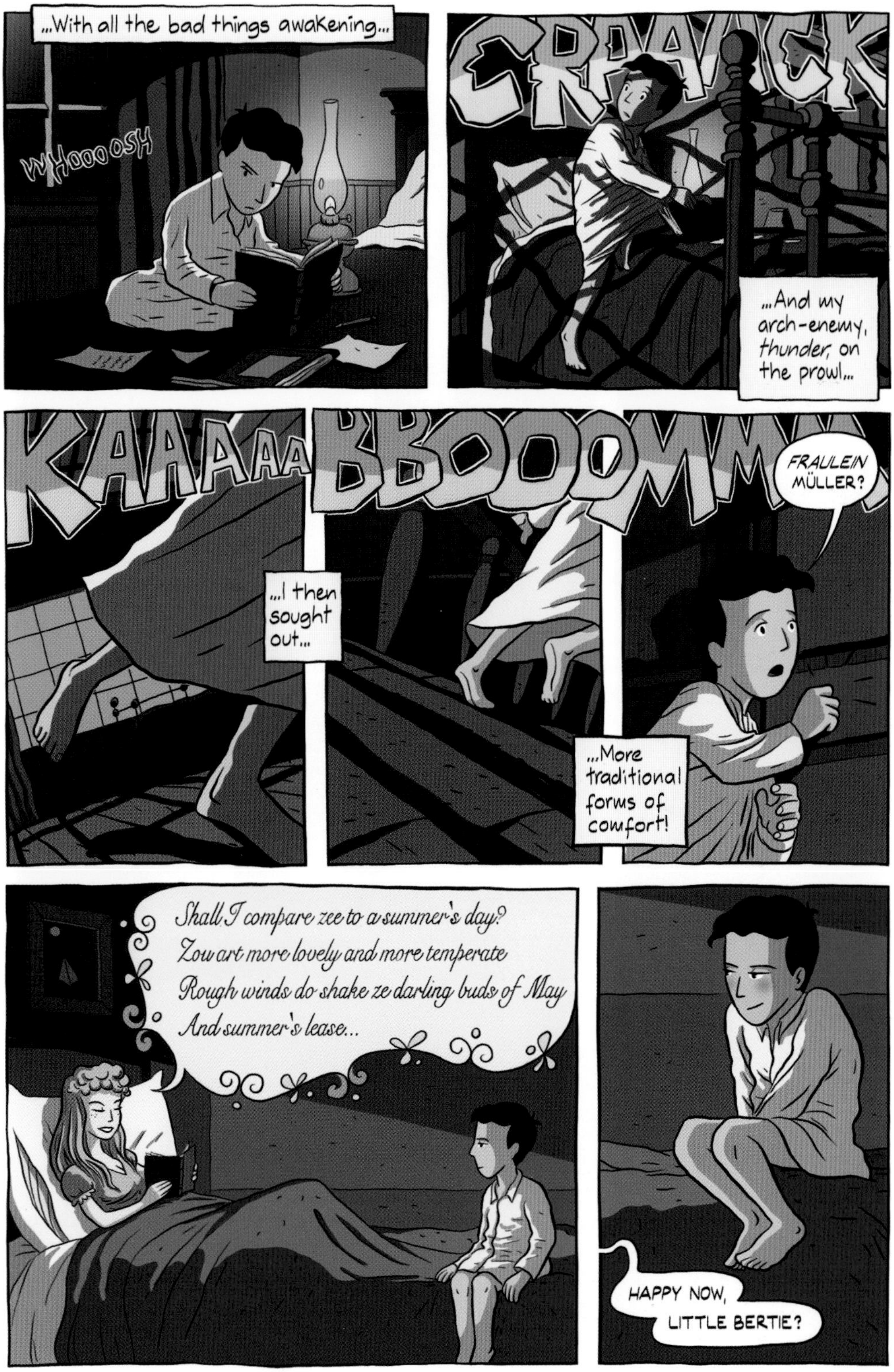
...With all the bad things awakening...
WHOOOOSH
CRAAACK
...And my arch-enemy, thunder, on the prowl...
KAAAAA
BBOOOMMM
...I then sought out...
FRAULEIN MÜLLER?
...More traditional forms of comfort!
Shall I compare zee to a summer's day?
Zou art more lovely and more temperate
Rough winds do shake ze darling buds of May
And summer's lease...
HAPPY NOW, LITTLE BERTIE?

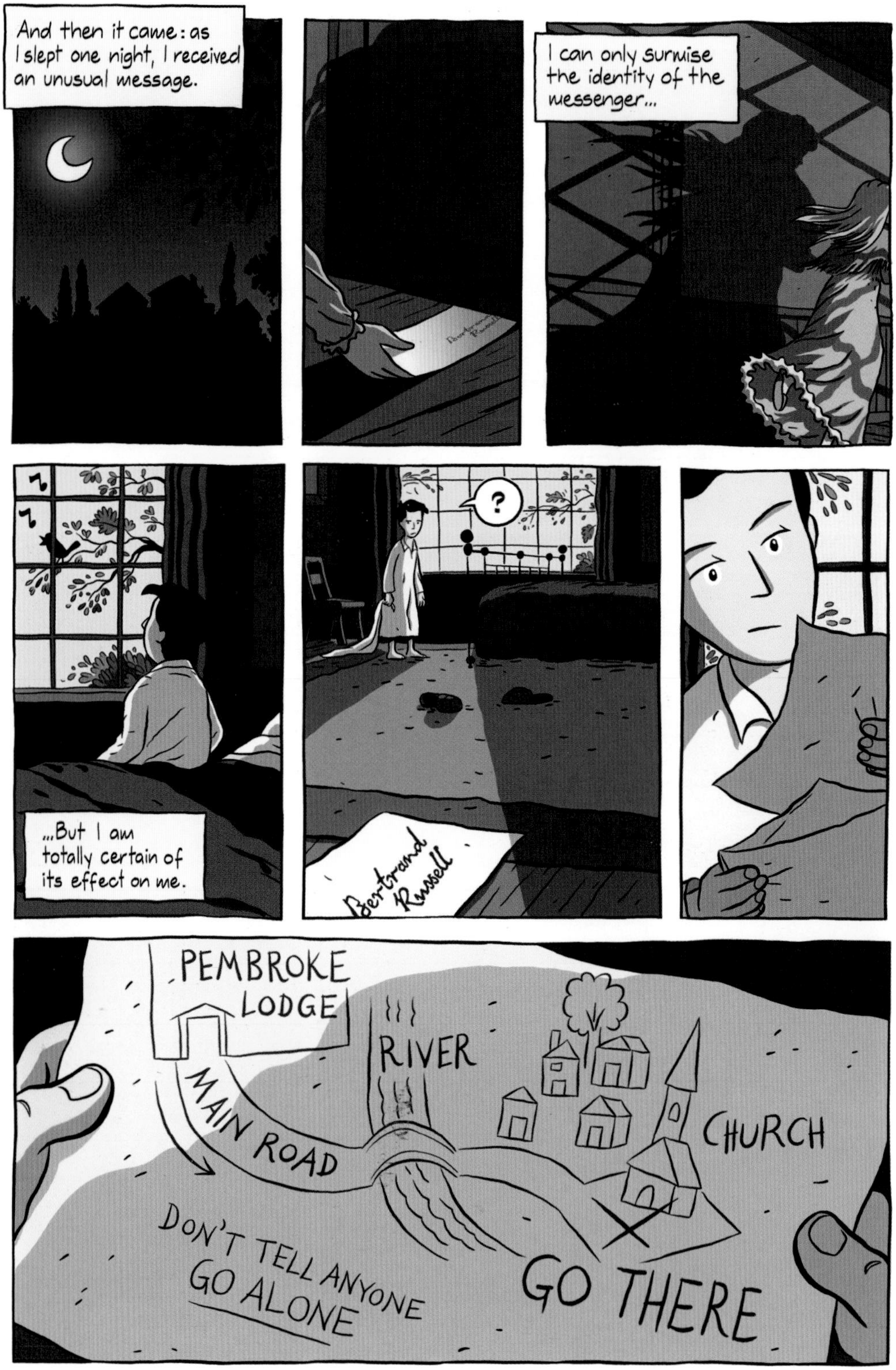
And then it came: as I slept one night, I received an unusual message.
I can only surmise the identity of the messenger...
...But I am totally certain of its effect on me.
?
Bertrand Russell
PEMBROKE LODGE
RIVER
MAIN ROAD
CHURCH
DON'T TELL ANYONE
GO ALONE
GO THERE

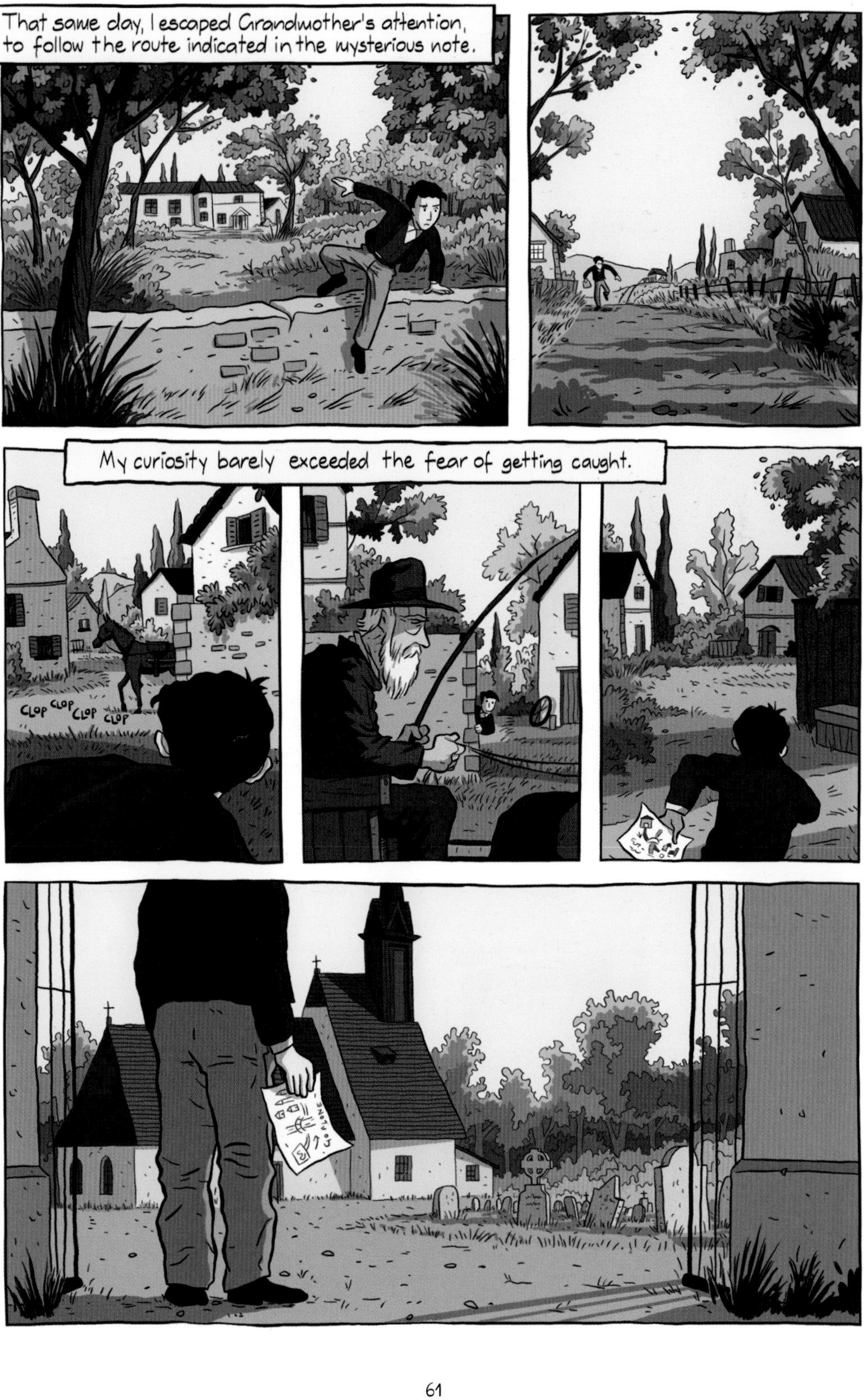
That same day, I escaped Grandmother's attention, to follow the route indicated in the mysterious note.
My curiosity barely exceeded the fear of getting caught.
CLOP CLOP CLOP CLOP

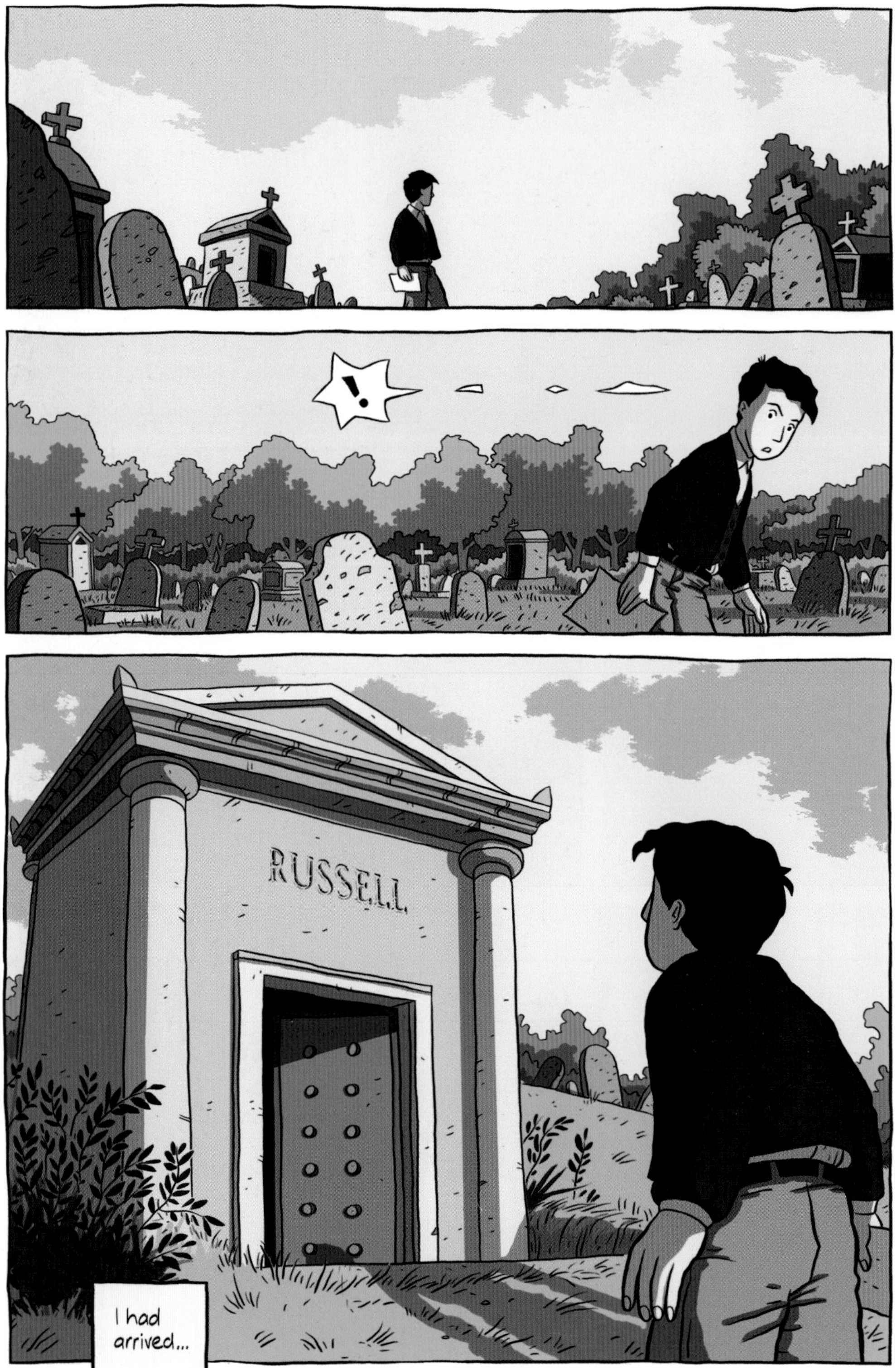
!
RUSSELL
I had arrived...

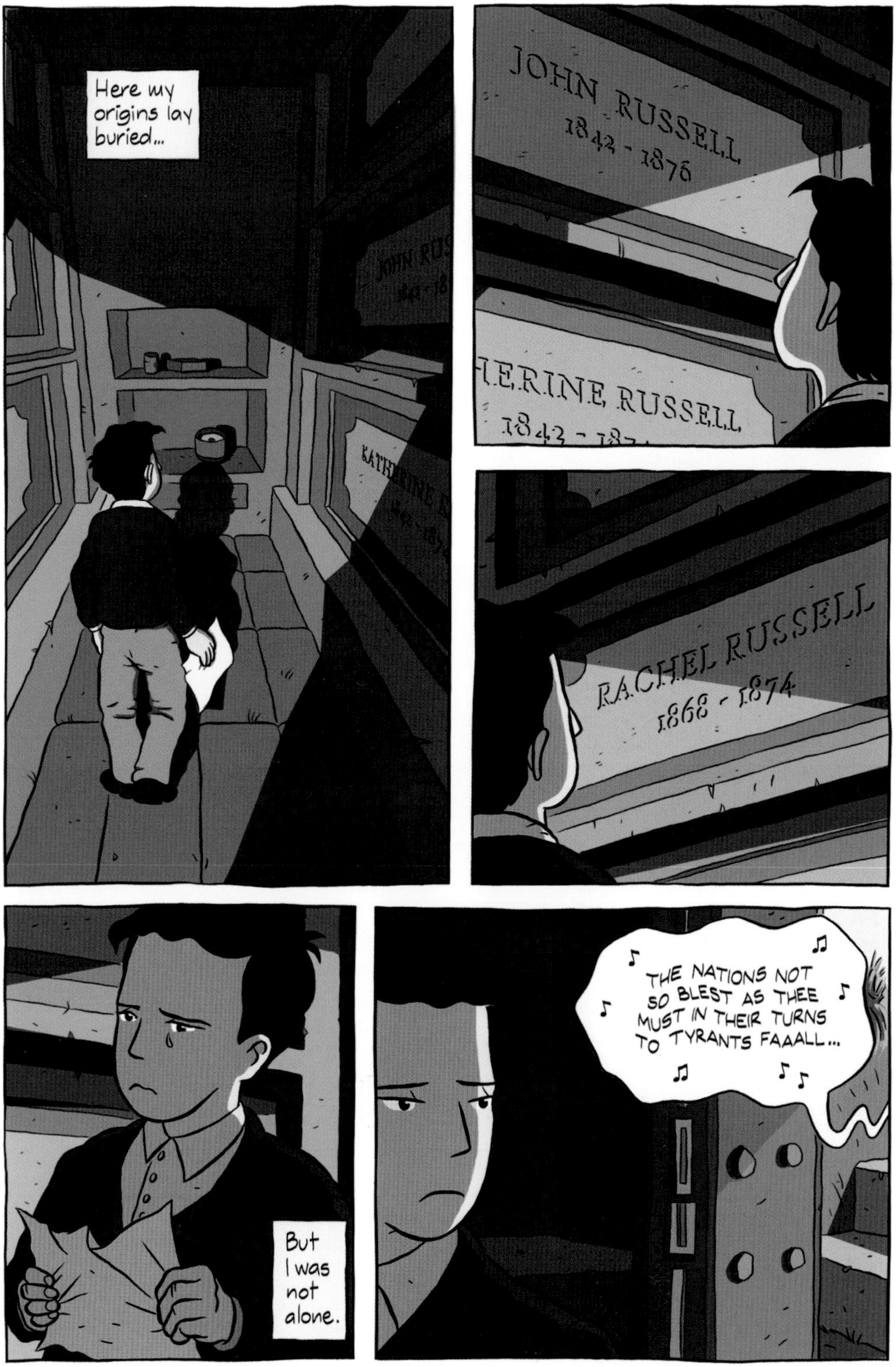
Here my origins lay buried...
JOHN RUSSELL
1842 - 1876
RACHEL RUSSELL
1868 - 1874
But I was not alone.
THE NATIONS NOT SO BLEST AS THEE MUST IN THEIR TURNS TO TYRANTS FAAALL...

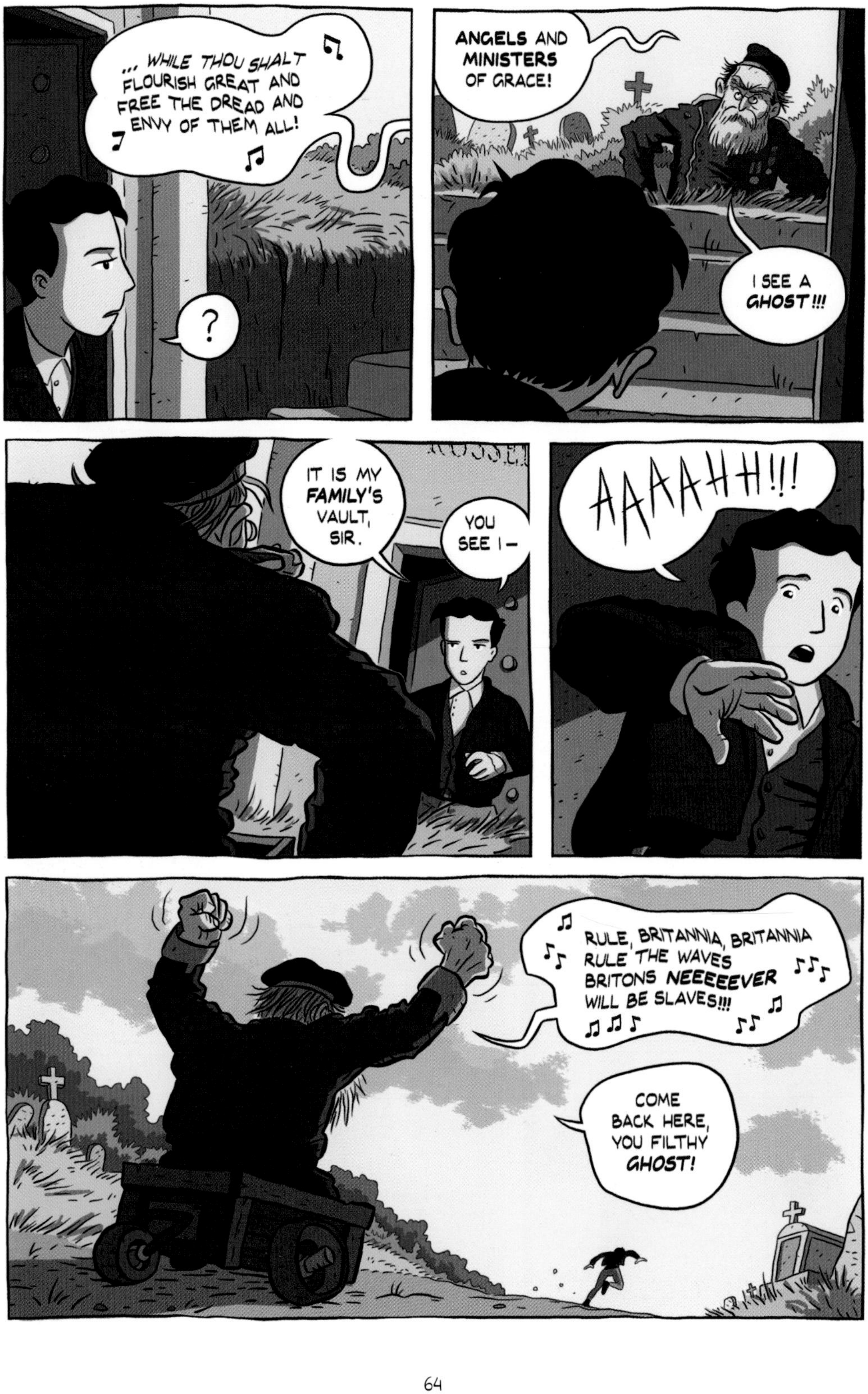

... WHILE THOU SHALT FLOURISH GREAT AND FREE THE DREAD AND ENVY OF THEM ALL!
?
ANGELS AND MINISTERS OF GRACE!
I SEE A GHOST!!!
IT IS MY FAMILY'S VAULT, SIR.
YOU SEE I—
AAAAHH!!!
RULE, BRITANNIA, BRITANNIA RULE THE WAVES BRITONS NEEEEEVER WILL BE SLAVES!!!
COME BACK HERE, YOU FILTHY GHOST!

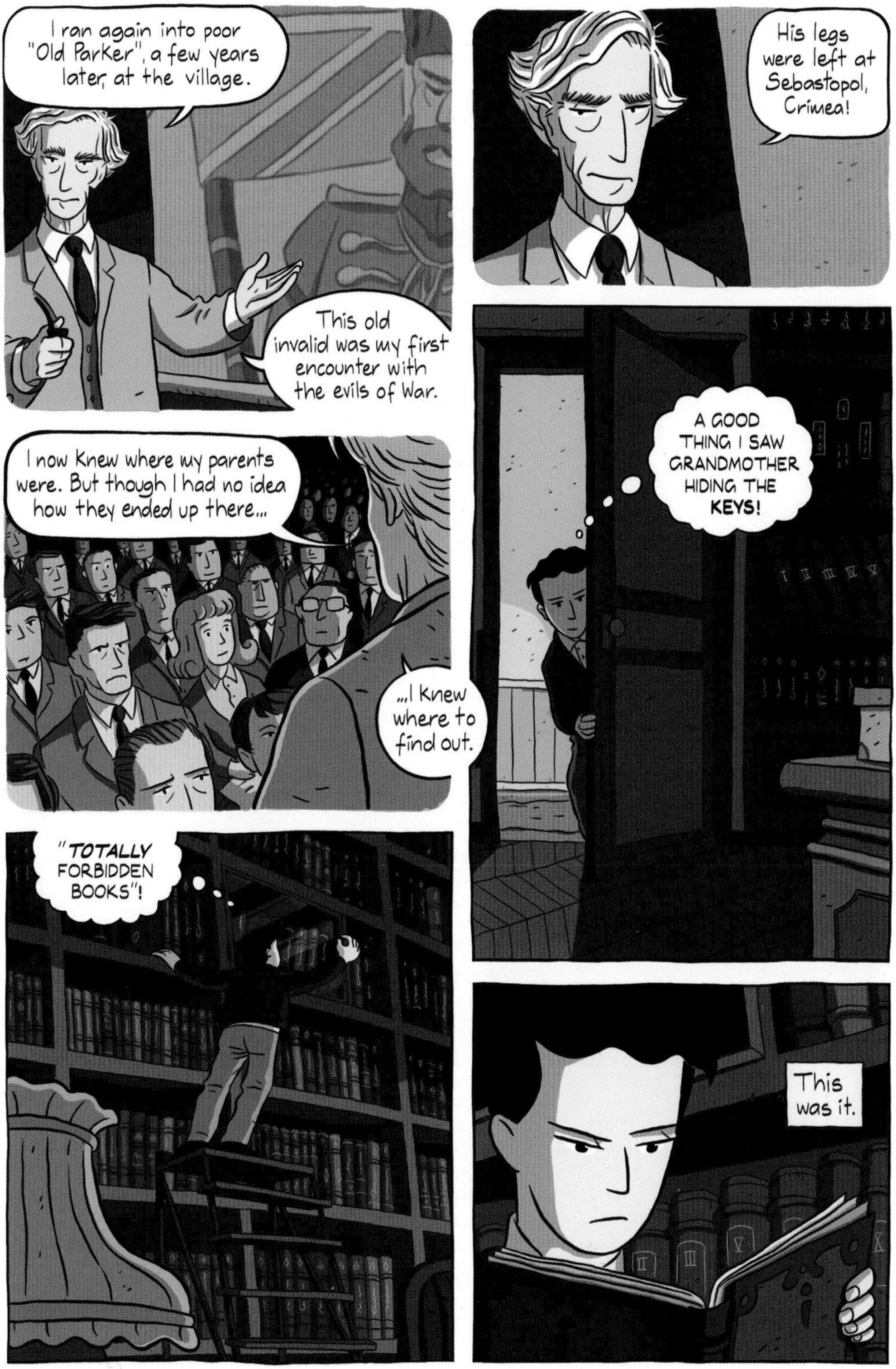
I ran again into poor "Old Parker", a few years later, at the village.
This old invalid was my first encounter with the evils of War.
His legs were left at Sebastopol, Crimea!
I now knew where my parents were. But though I had no idea how they ended up there...
...I knew where to find out.
A GOOD THING I SAW GRANDMOTHER HIDING THE **KEYS!**
"***TOTALLY*** FORBIDDEN BOOKS"!
This was it.

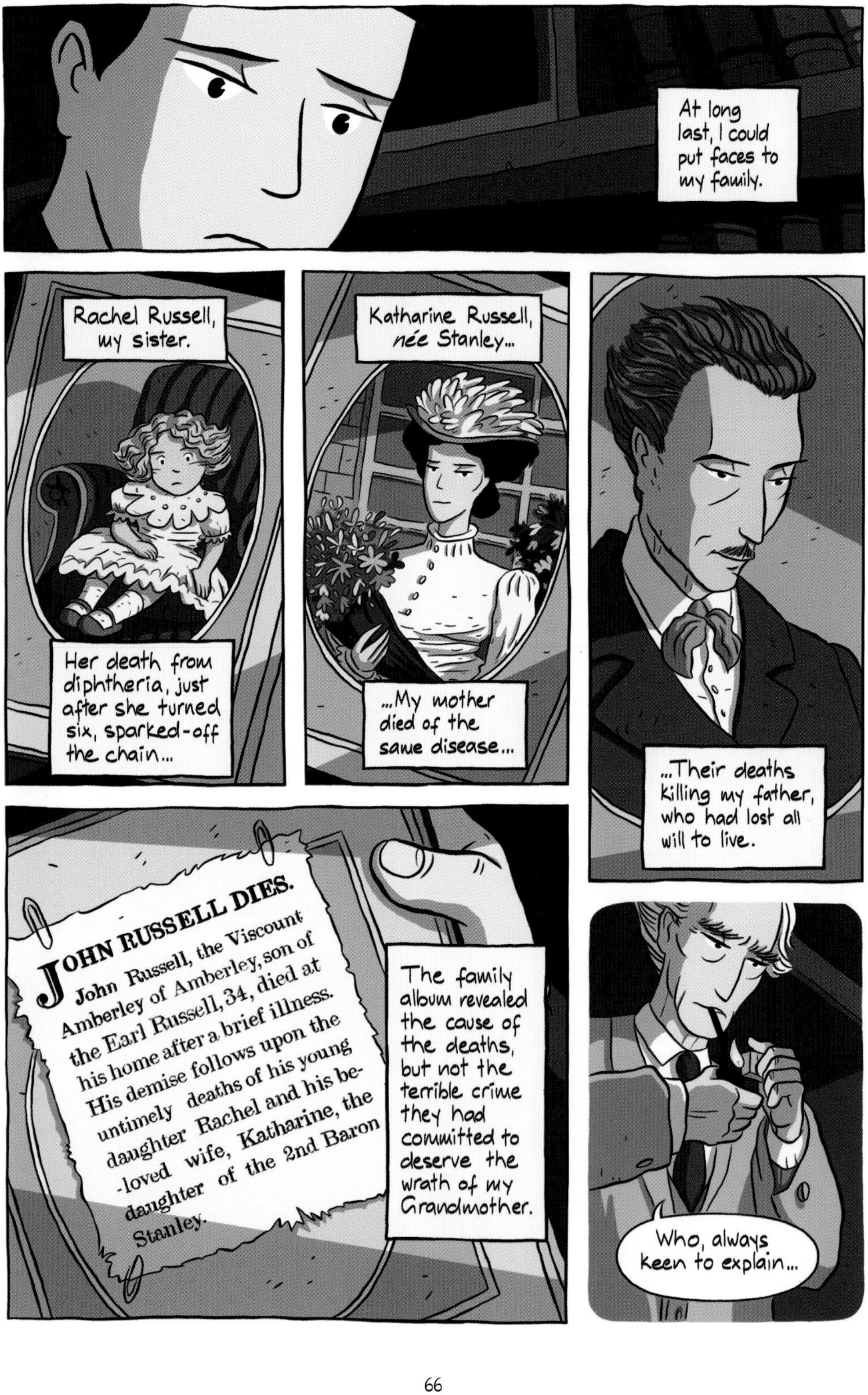
At long last, I could put faces to my family.
Rachel Russell, my sister.
Her death from diphtheria, just after she turned six, sparked-off the chain...
Katharine Russell, *née* Stanley...
...My mother died of the same disease...
...Their deaths killing my father, who had lost all will to live.
JOHN RUSSELL DIES.
JOHN RUSSELL, the Viscount John Russell, the Viscount Amberley of Amberley, son of the Earl Russell, 34, died at his home after a brief illness. His demise follows upon the untimely deaths of his young daughter Rachel and his be-loved wife, Katharine, the daughter of the 2nd Baron Stanley.
The family album revealed the cause of the deaths, but not the terrible crime they had committed to deserve the wrath of my Grandmother.
Who, always keen to explain...

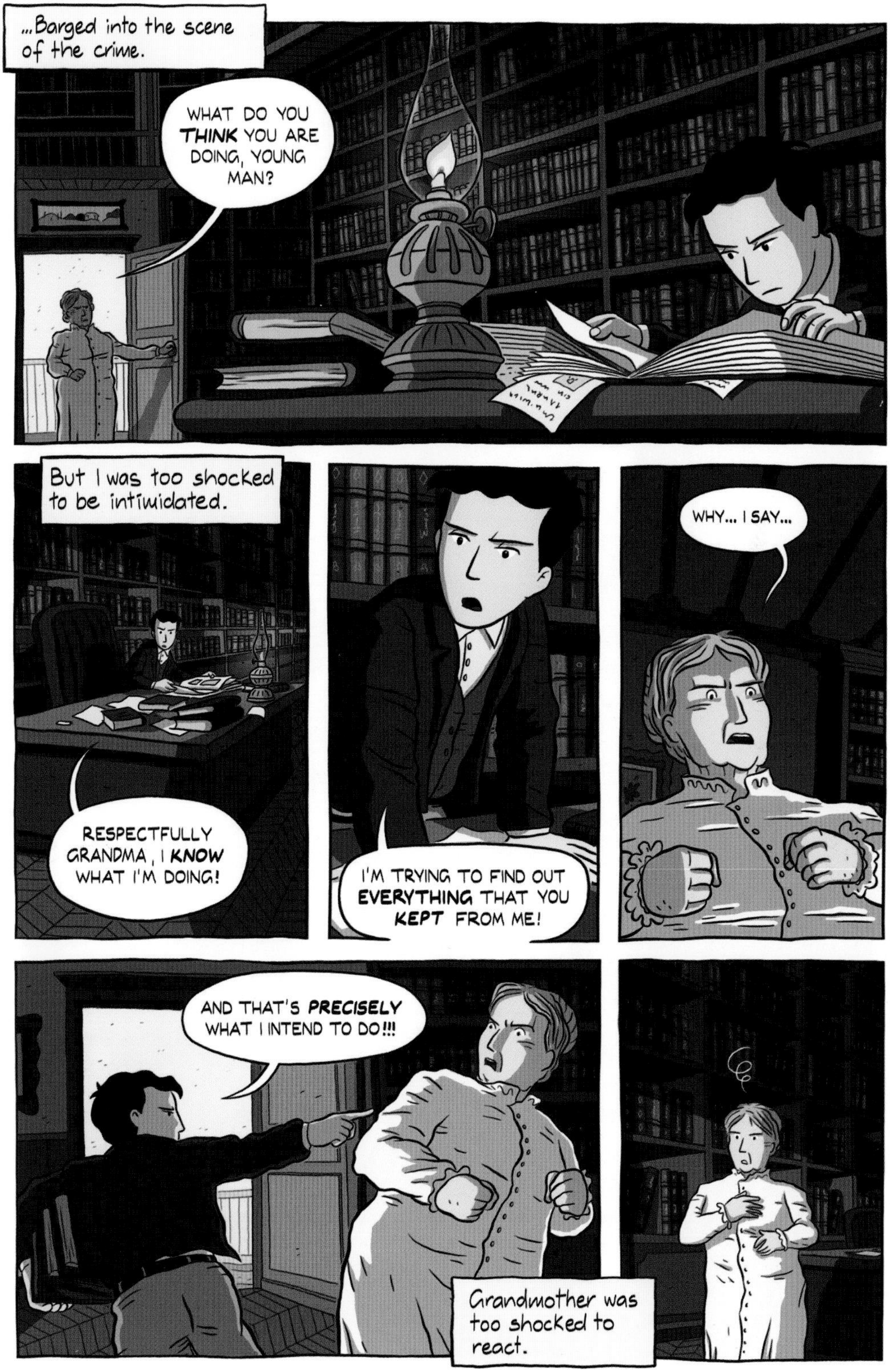
...Barged into the scene of the crime.
WHAT DO YOU **THINK** YOU ARE DOING, YOUNG MAN?
But I was too shocked to be intimidated.
RESPECTFULLY GRANDMA, I **KNOW** WHAT I'M DOING!
I'M TRYING TO FIND OUT **EVERYTHING** THAT YOU **KEPT** FROM ME!
WHY... I SAY...
AND THAT'S **PRECISELY** WHAT I INTEND TO DO!!!
Grandmother was too shocked to react.

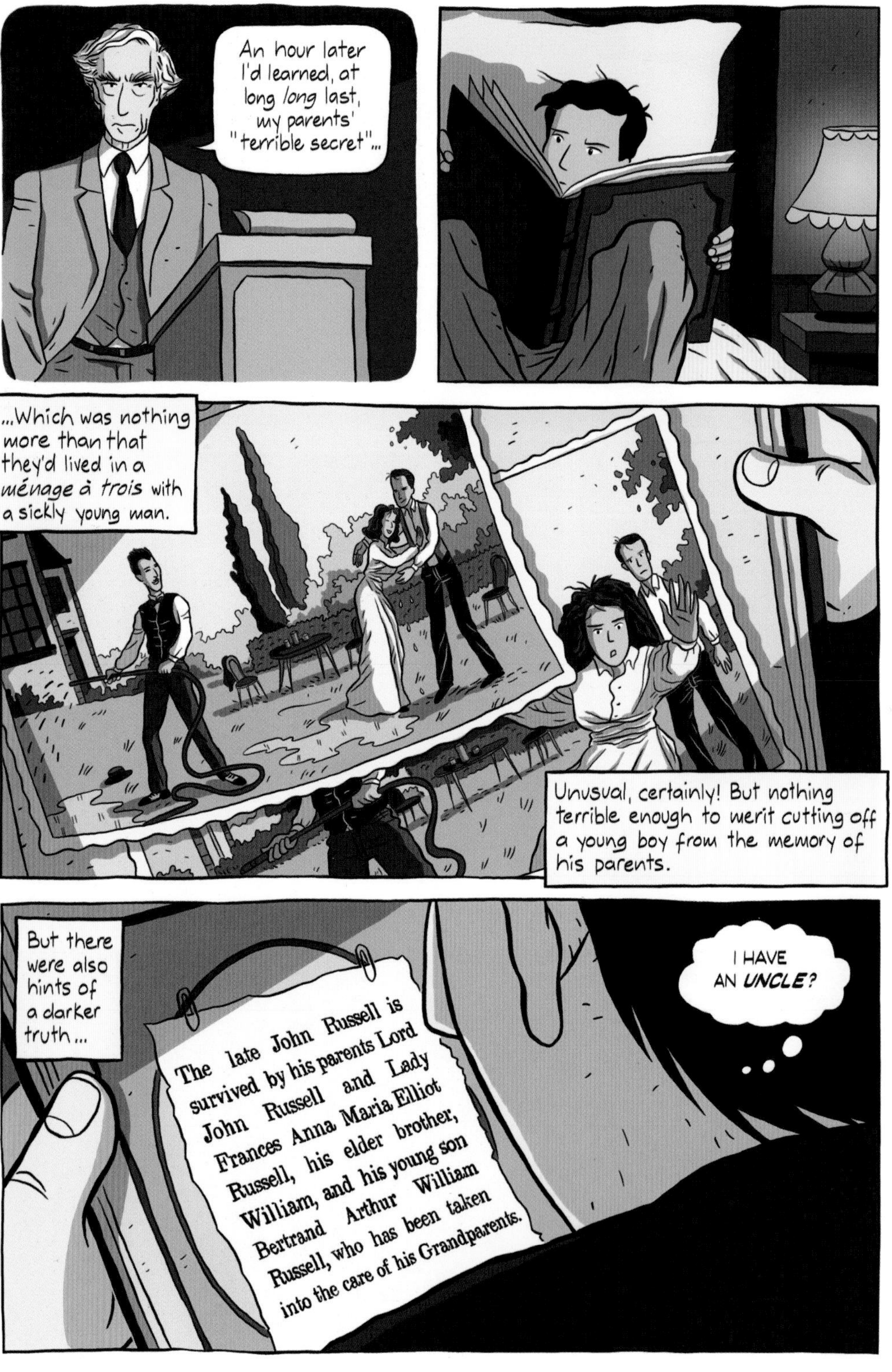

An hour later I'd learned, at long *long* last, my parents' "terrible secret"...
...Which was nothing more than that they'd lived in a *ménage à trois* with a sickly young man.
Unusual, certainly! But nothing terrible enough to merit cutting off a young boy from the memory of his parents.
But there were also hints of a darker truth...
The late John Russell is survived by his parents Lord John Russell and Lady Frances Anna Maria Elliot Russell, his elder brother, William, and his young son Bertrand Arthur William Russell, who has been taken into the care of his Grandparents.
I HAVE AN ***UNCLE?***

That year, I began my first philosophical work: an intimate diary!

But as I was now at war with Grandmother, it was in code.

Simply, a transliteration into Greek, a language she did not know.

This reads: "Most humans behave irrationally. All the more reason to pursue the study of Logic... Of course I am also human and thus am no stranger to fits of non-logical thinking. But also I can discern these tendencies in myself and thus am more able to resist them..."

"Greek Exercises", as my notebook was called for reasons of subterfuge, became a haven for all my secret, forbidden thoughts.

GREEK EX

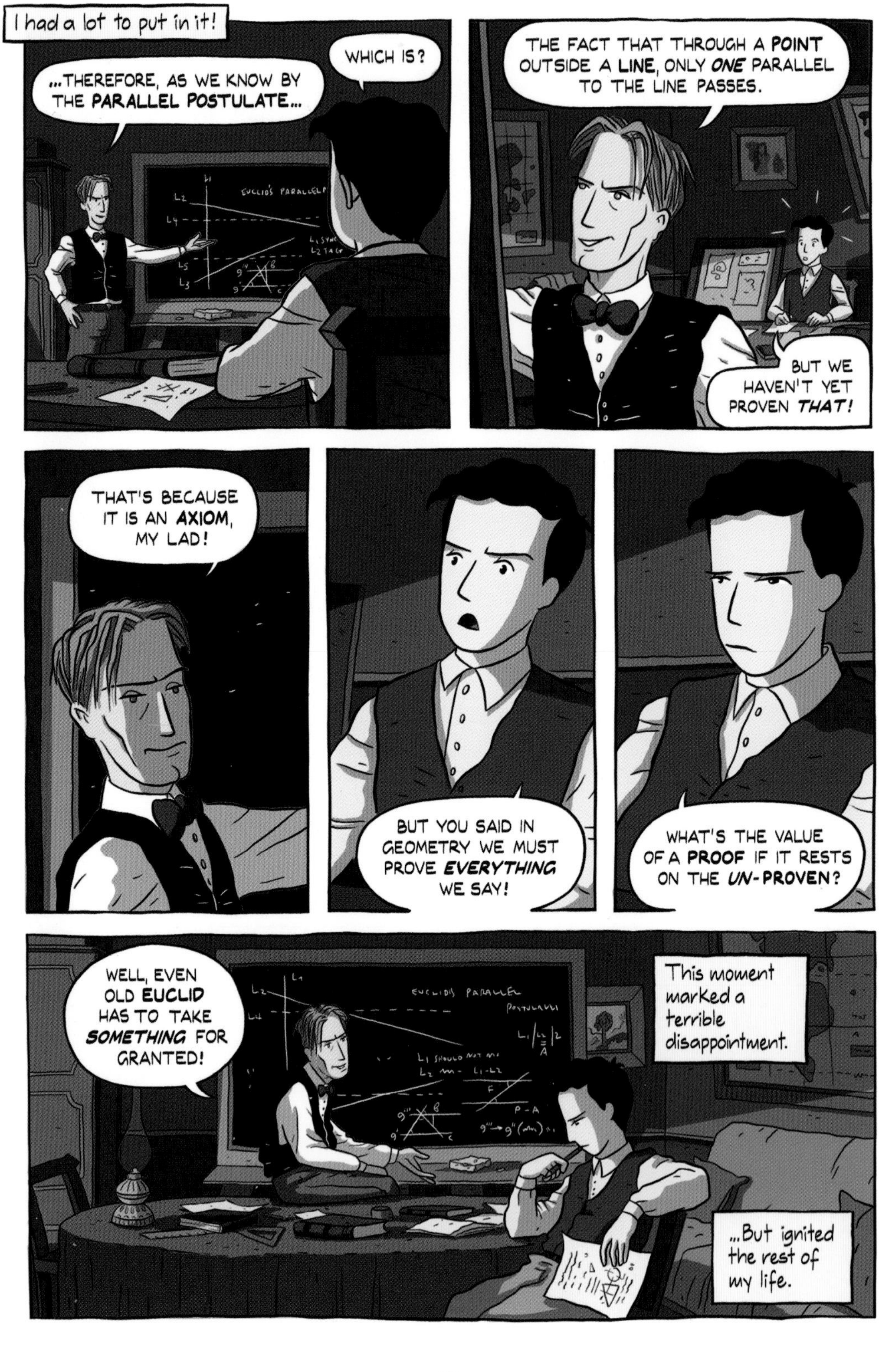
I had a lot to put in it!
...THEREFORE, AS WE KNOW BY THE PARALLEL POSTULATE...
WHICH IS?
THE FACT THAT THROUGH A POINT OUTSIDE A LINE, ONLY ONE PARALLEL TO THE LINE PASSES.
BUT WE HAVEN'T YET PROVEN THAT!
THAT'S BECAUSE IT IS AN AXIOM, MY LAD!
BUT YOU SAID IN GEOMETRY WE MUST PROVE EVERYTHING WE SAY!
WHAT'S THE VALUE OF A PROOF IF IT RESTS ON THE UN-PROVEN?
WELL, EVEN OLD EUCLID HAS TO TAKE SOMETHING FOR GRANTED!
This moment marked a terrible disappointment.
...But ignited the rest of my life.

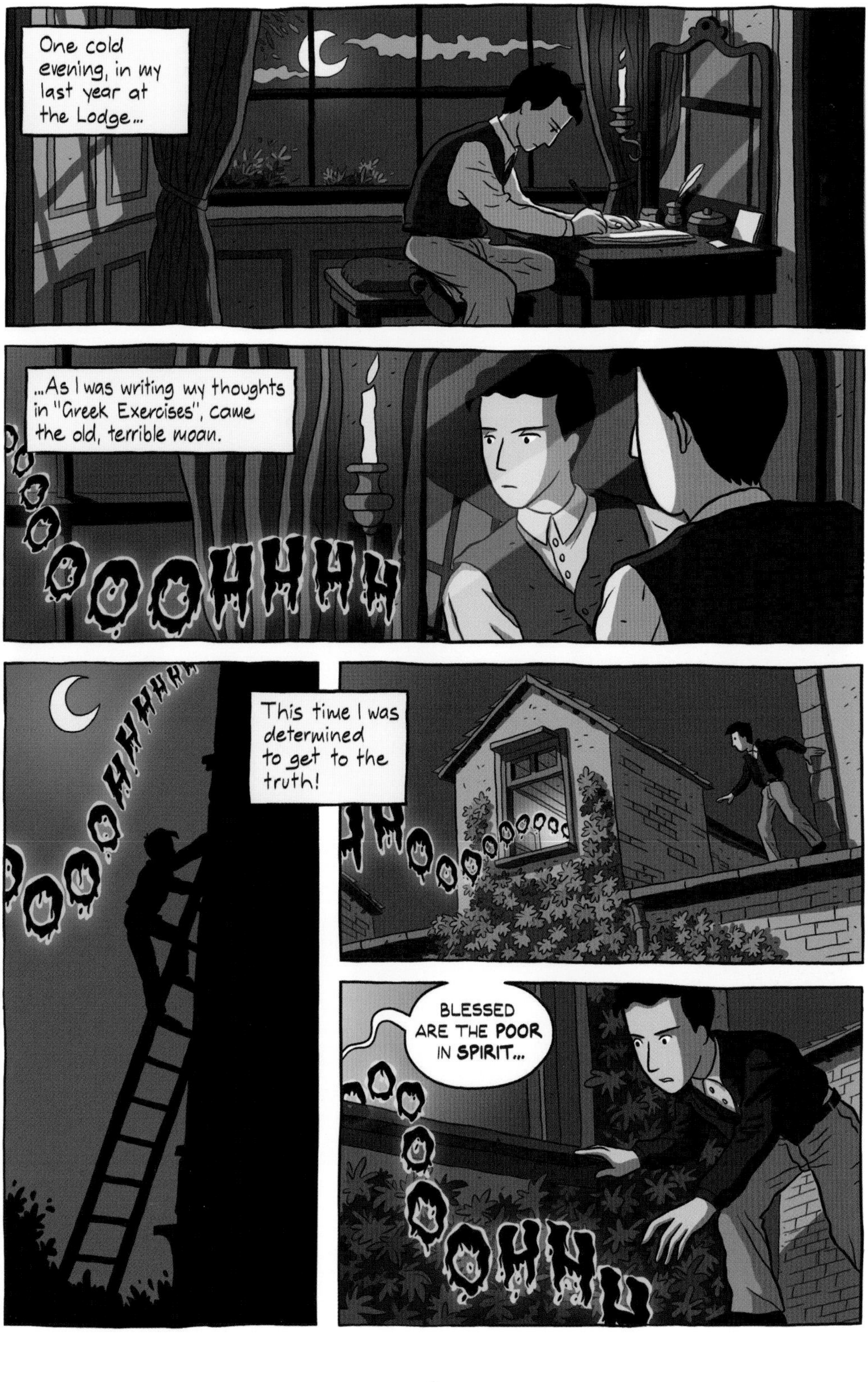

One cold evening, in my last year at the Lodge...
...As I was writing my thoughts in "Greek Exercises", came the old, terrible moan.
OOOOOOOHHHHH
OOOOOHHHHHH
This time I was determined to get to the truth!
HOOOOOOOOOO
BLESSED ARE THE POOR IN SPIRIT...
OOOOOHHH

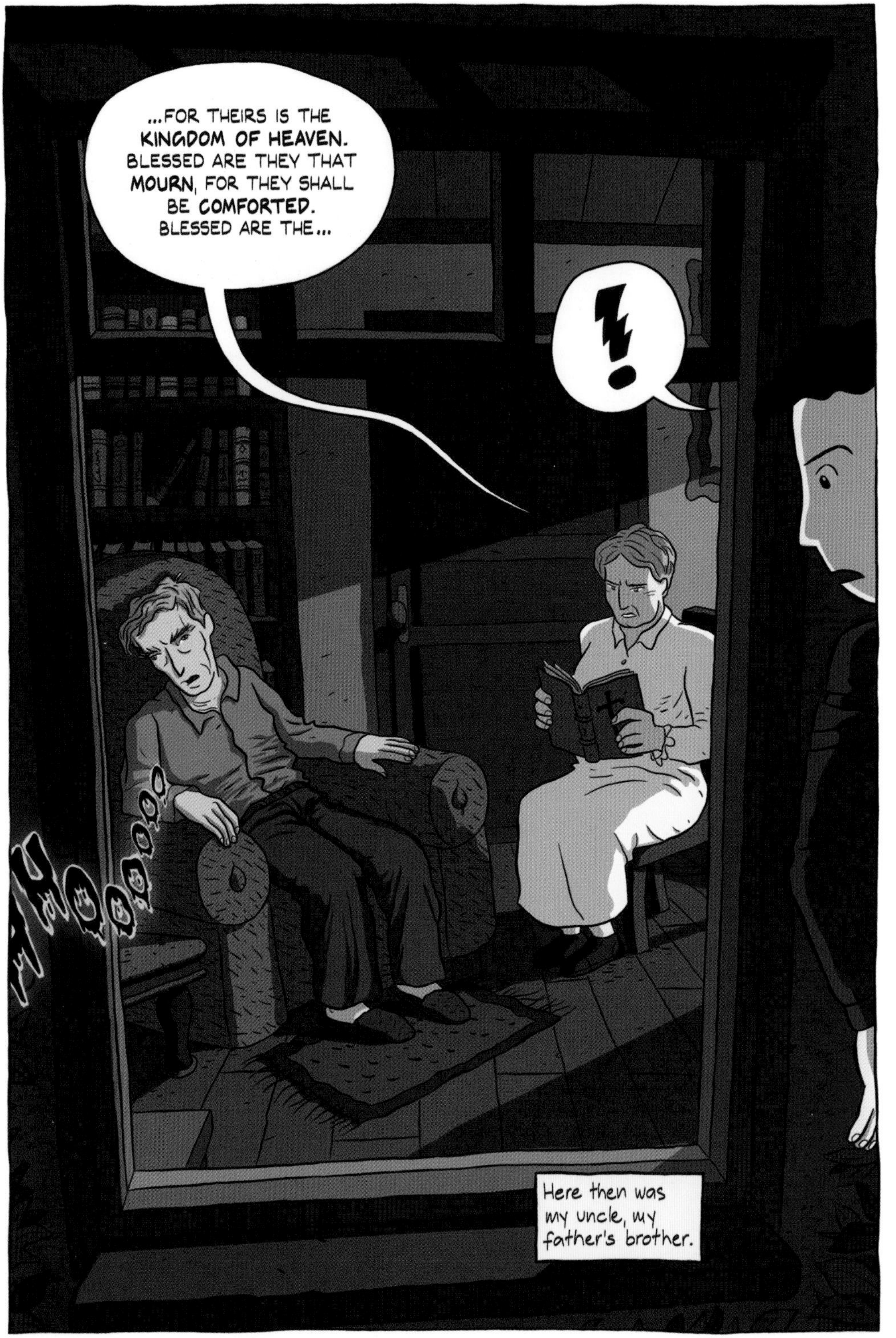
...FOR THEIRS IS THE KINGDOM OF HEAVEN. BLESSED ARE THEY THAT MOURN, FOR THEY SHALL BE COMFORTED. BLESSED ARE THE...
!
Here then was my uncle, my father's brother.

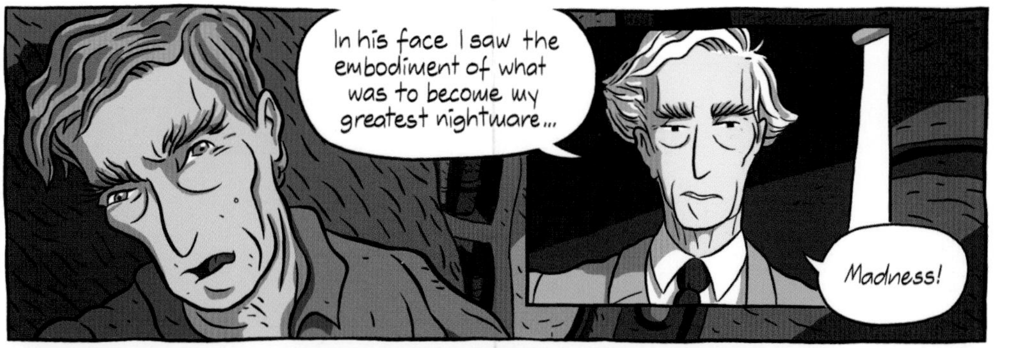
In his face I saw the embodiment of what was to become my greatest nightmare...
Madness!

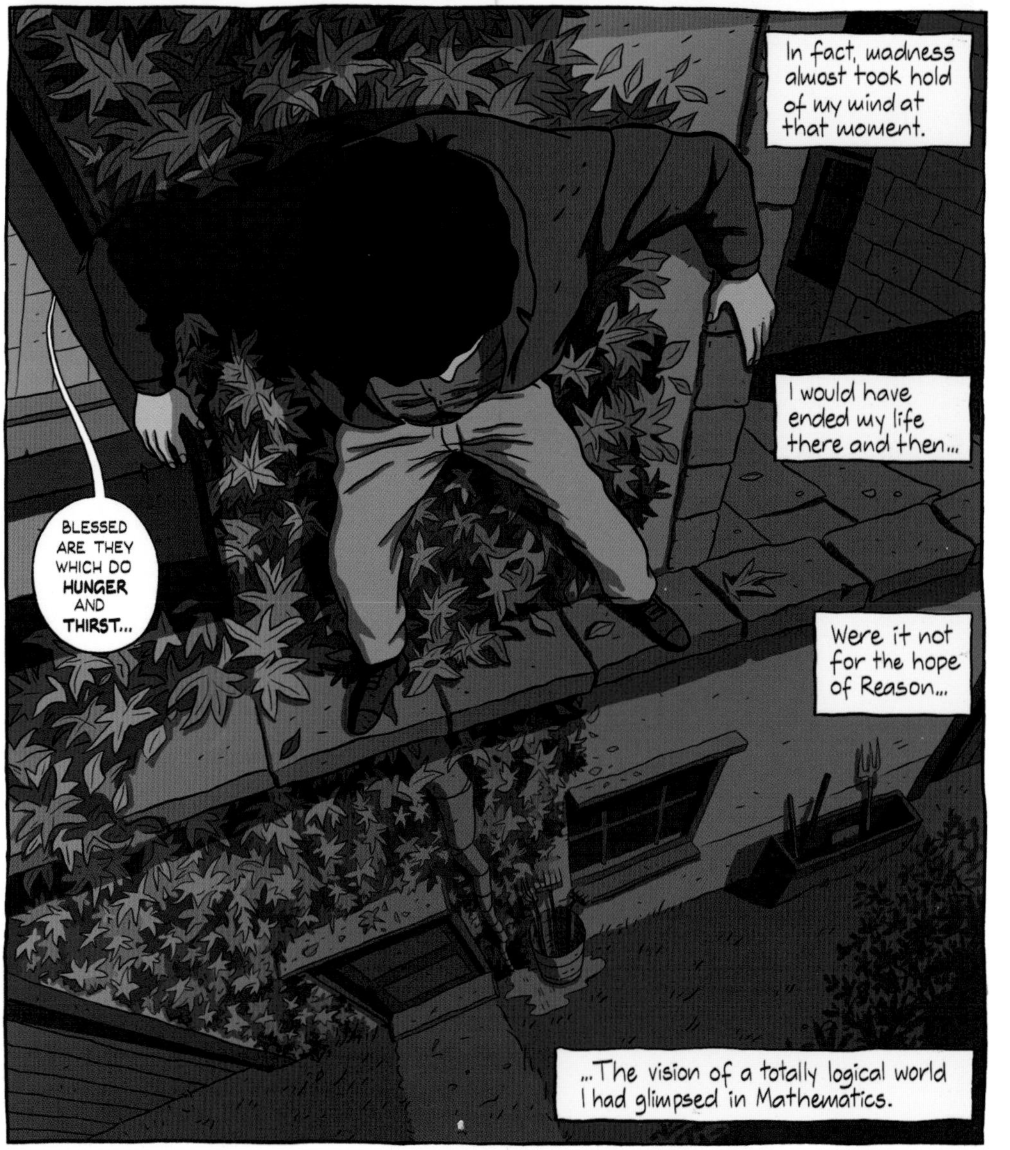
In fact, madness almost took hold of my mind at that moment.
BLESSED ARE THEY WHICH DO HUNGER AND THIRST...
I would have ended my life there and then...
Were it not for the hope of Reason...
...The vision of a totally logical world I had glimpsed in Mathematics.

2. THE SORCERER'S APPRENTICE

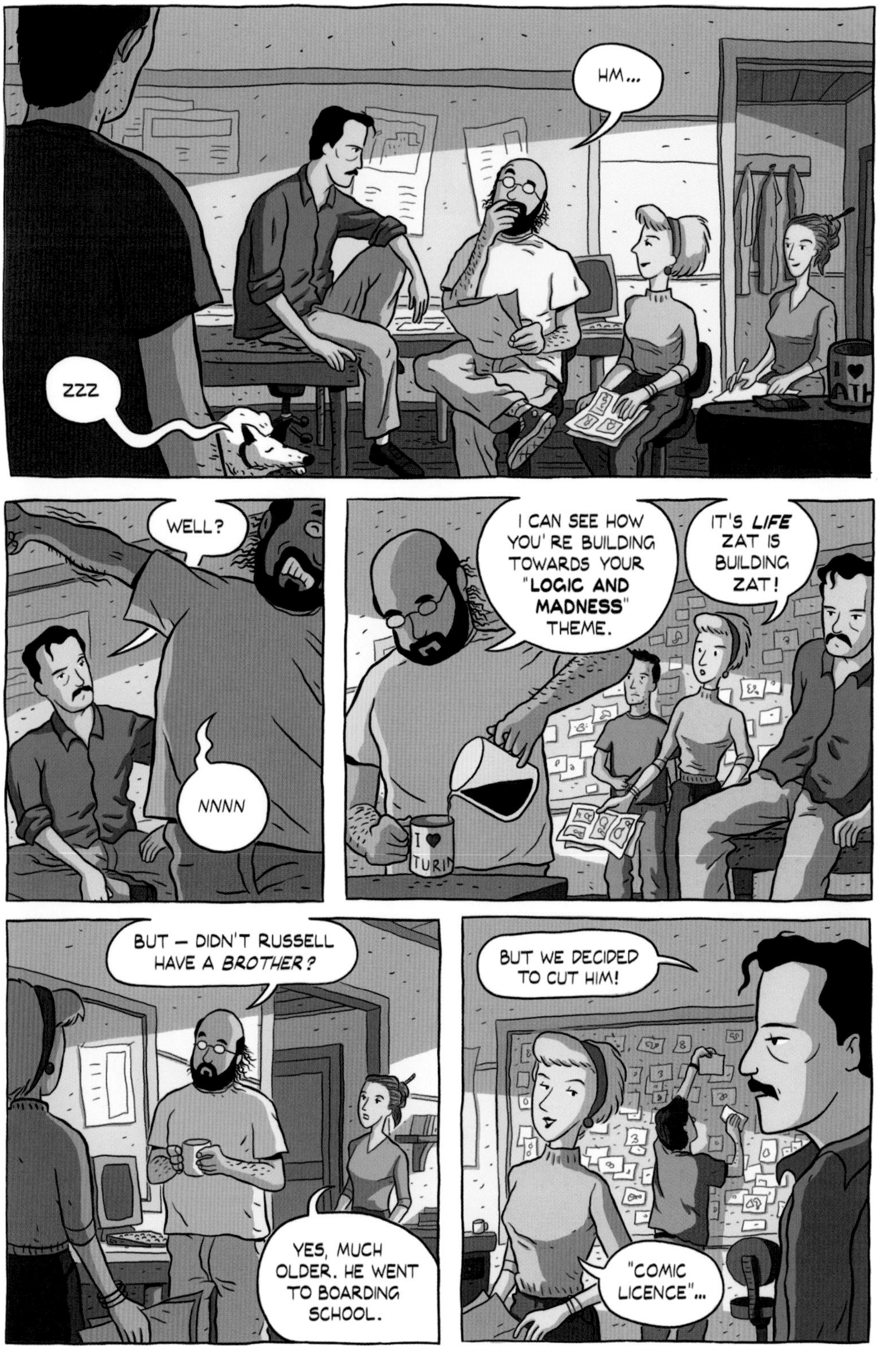
HM...
ZZZ
WELL?
NNNN
I CAN SEE HOW YOU'RE BUILDING TOWARDS YOUR "LOGIC AND MADNESS" THEME.
IT'S LIFE ZAT IS BUILDING ZAT!
I ♥ TURIN
BUT — DIDN'T RUSSELL HAVE A BROTHER?
YES, MUCH OLDER. HE WENT TO BOARDING SCHOOL.
BUT WE DECIDED TO CUT HIM!
"COMIC LICENCE"...

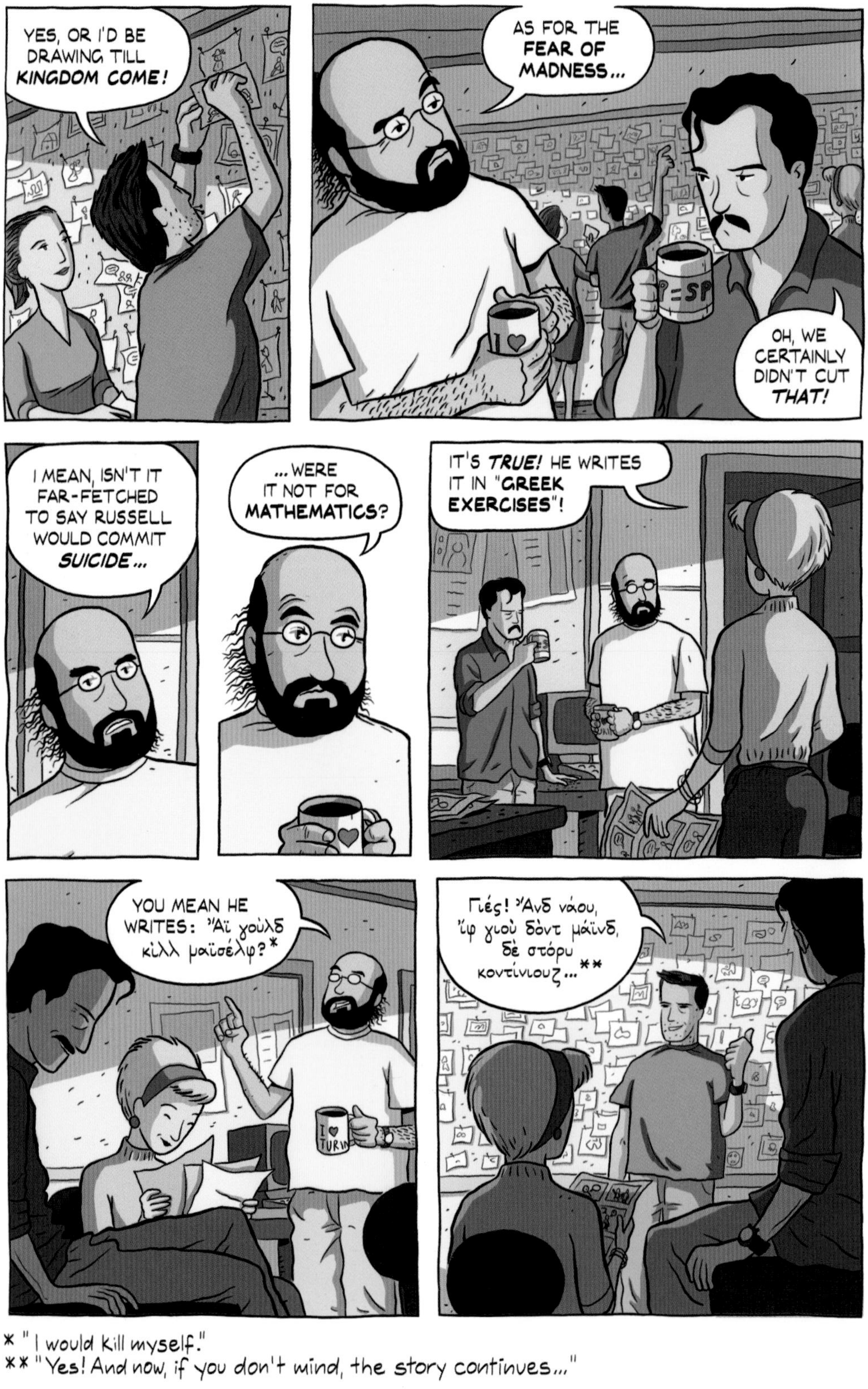
YES, OR I'D BE DRAWING TILL **KINGDOM COME!**
AS FOR THE **FEAR OF MADNESS...**
P=SP
OH, WE CERTAINLY DIDN'T CUT ***THAT!***
I MEAN, ISN'T IT FAR-FETCHED TO SAY RUSSELL WOULD COMMIT ***SUICIDE...***
...WERE IT NOT FOR **MATHEMATICS?**
IT'S ***TRUE!*** HE WRITES IT IN "**GREEK EXERCISES**"!
YOU MEAN HE WRITES: "Ἄϊ γοὺλδ κὶλλ μαϊσέλφ?*
I ♥ TURIN
Γιές! Ἄνδ νάου, ἴφ γιοὺ δὸντ μάϊνδ, δὲ στόρυ κοντίνιουζ...**

* "I would kill myself."

** "Yes! And now, if you don't mind, the story continues..."

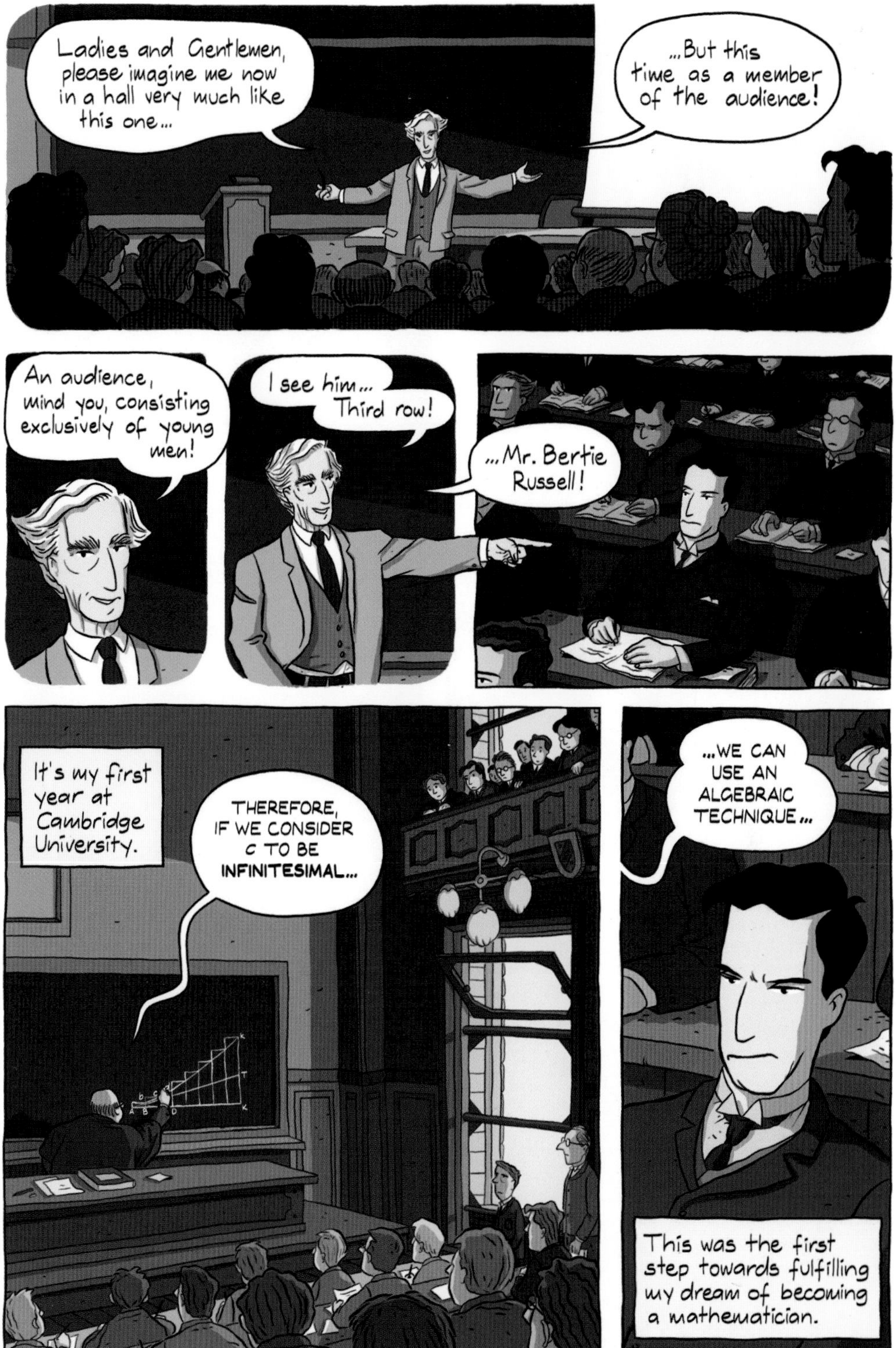
Ladies and Gentlemen, please imagine me now in a hall very much like this one...
...But this time as a member of the audience!
An audience, mind you, consisting exclusively of young men!
I see him... Third row!
...Mr. Bertie Russell!
It's my first year at Cambridge University.
THEREFORE, IF WE CONSIDER c TO BE INFINITESIMAL...
...WE CAN USE AN ALGEBRAIC TECHNIQUE...
This was the first step towards fulfilling my dream of becoming a mathematician.

EXCUSE ME, PROFESSOR!

I WANTED TO ASK YOU HOW YOU *DEFINE* "INFINITESIMAL"?

WHY, *OBVIOUSLY*, AS "THAT WHICH IS INFINITELY SMALL"!

But my introduction to the "Queen of the Sciences" was a sheer disappointment.

BUT, THAT IS *CIRCULAR!*

YOU *CAN'T* DEFINE A TERM BY *USING* IT IN THE *DEFINITION!*

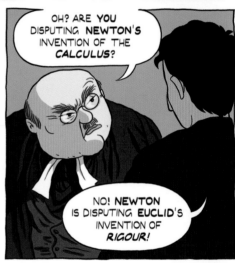

OH? ARE *YOU* DISPUTING *NEWTON'S* INVENTION OF THE *CALCULUS?*

NO! *NEWTON* IS DISPUTING *EUCLID'S* INVENTION OF *RIGOUR!*

MATHEMATICS IS THE *LAST RECOURSE* OF *REASON!* WE CANNOT UNDERMINE IT, WITH *SLOPPY THINKING!*

NOR CAN WE *CHANGE* IT, TO SUIT A PRESUMPTUOUS YOUNG MAN!

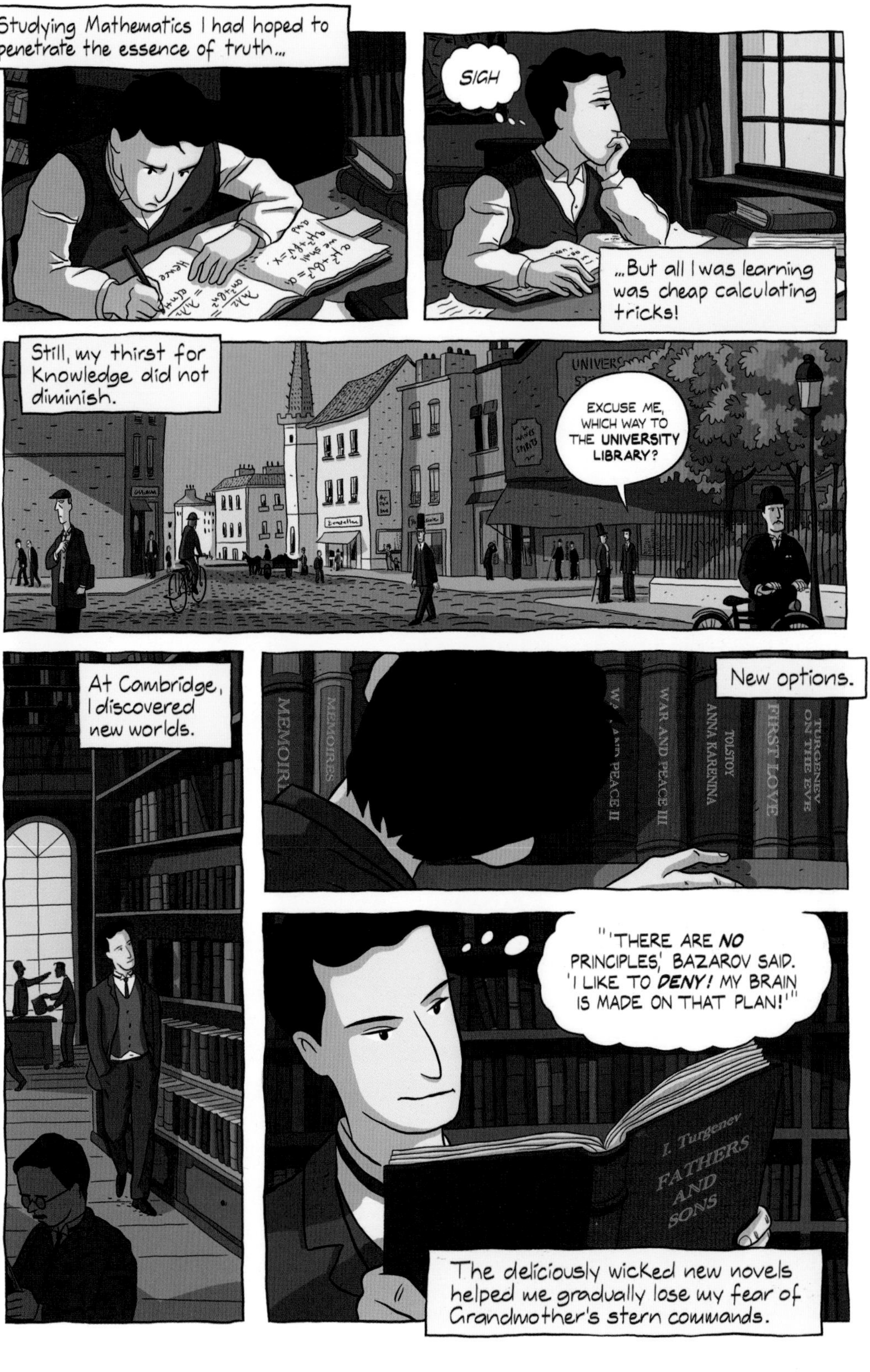
Studying Mathematics I had hoped to penetrate the essence of truth...
SIGH
...But all I was learning was cheap calculating tricks!
Still, my thirst for Knowledge did not diminish.
EXCUSE ME, WHICH WAY TO THE **UNIVERSITY LIBRARY**?
At Cambridge, I discovered new worlds.
New options.
WAR AND PEACE III
TOLSTOY
ANNA KARENINA
FIRST LOVE
TURGENEV
ON THE EVE
"'THERE ARE ***NO*** PRINCIPLES,' BAZAROV SAID. 'I LIKE TO ***DENY!*** MY BRAIN IS MADE ON THAT PLAN!'"
I. Turgenev
FATHERS AND SONS
The deliciously wicked new novels helped me gradually lose my fear of Grandmother's stern commands.

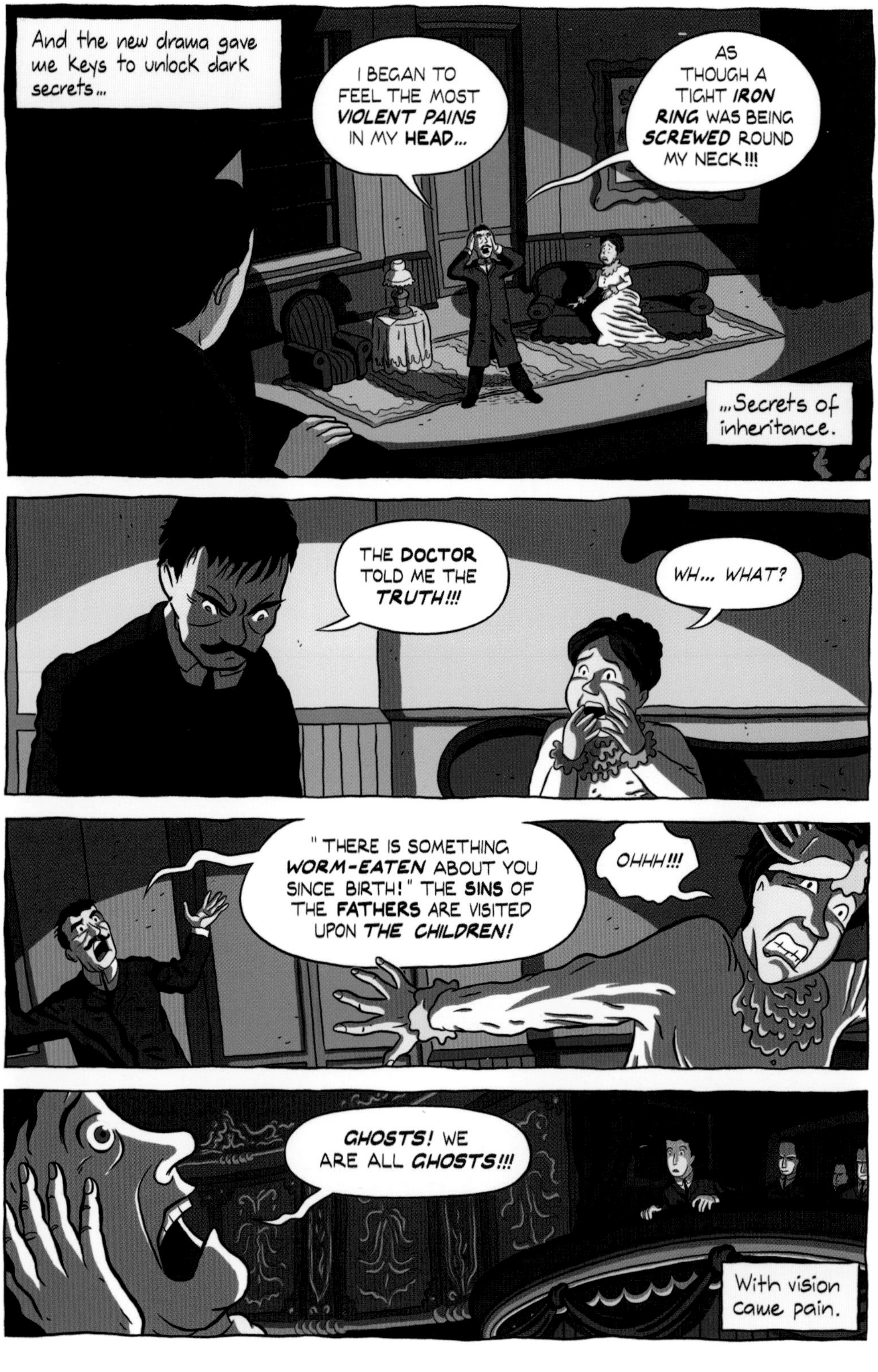
And the new drama gave me keys to unlock dark secrets...
I BEGAN TO FEEL THE MOST VIOLENT PAINS IN MY HEAD...
AS THOUGH A TIGHT IRON RING WAS BEING SCREWED ROUND MY NECK!!!
...Secrets of inheritance.
THE DOCTOR TOLD ME THE TRUTH!!!
WH... WHAT?
" THERE IS SOMETHING WORM-EATEN ABOUT YOU SINCE BIRTH! " THE SINS OF THE FATHERS ARE VISITED UPON THE CHILDREN!
OHHH!!!
GHOSTS! WE ARE ALL GHOSTS!!!
With vision came pain.

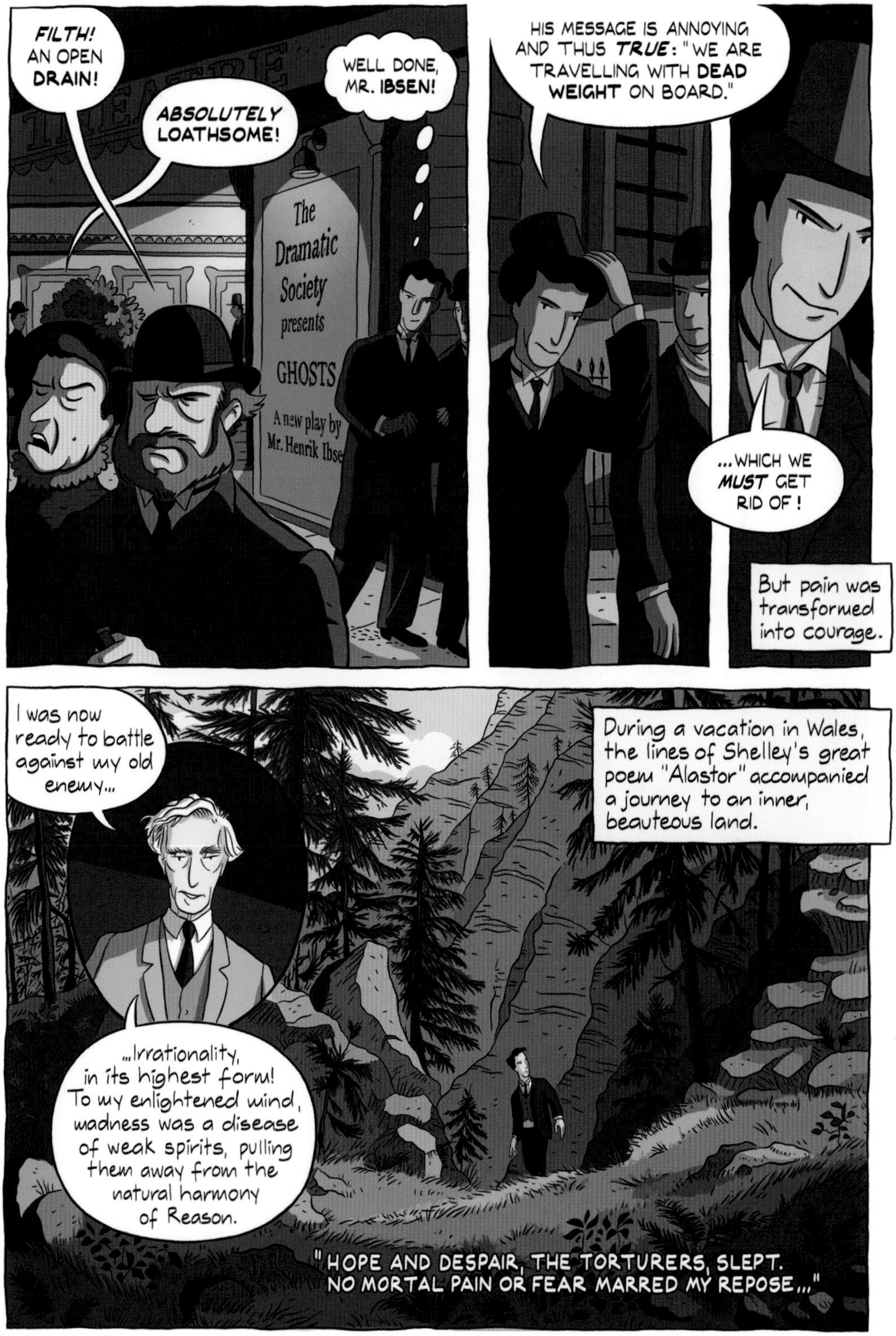
THEATRE
FILTH! AN OPEN DRAIN!
ABSOLUTELY LOATHSOME!
WELL DONE, MR. IBSEN!
The Dramatic Society presents GHOSTS A new play by Mr. Henrik Ibse
HIS MESSAGE IS ANNOYING AND THUS TRUE: "WE ARE TRAVELLING WITH DEAD WEIGHT ON BOARD."
...WHICH WE MUST GET RID OF!
But pain was transformed into courage.
I was now ready to battle against my old enemy...
During a vacation in Wales, the lines of Shelley's great poem "Alastor" accompanied a journey to an inner, beauteous land.
...Irrationality, in its highest form! To my enlightened mind, madness was a disease of weak spirits, pulling them away from the natural harmony of Reason.
"HOPE AND DESPAIR, THE TORTURERS, SLEPT. NO MORTAL PAIN OR FEAR MARRED MY REPOSE..."

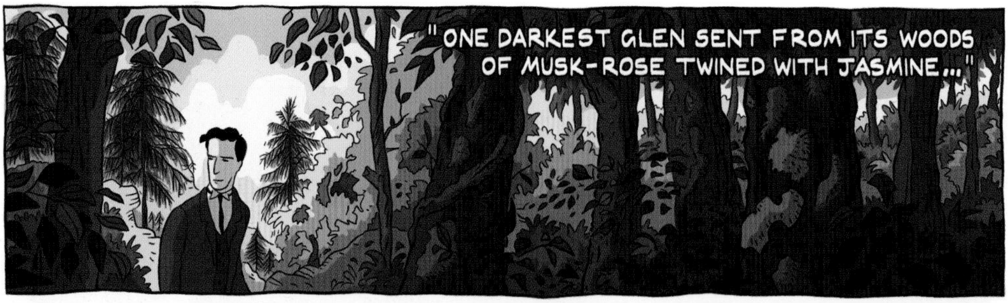
"ONE DARKEST GLEN SENT FROM ITS WOODS OF MUSK-ROSE TWINED WITH JASMINE..."
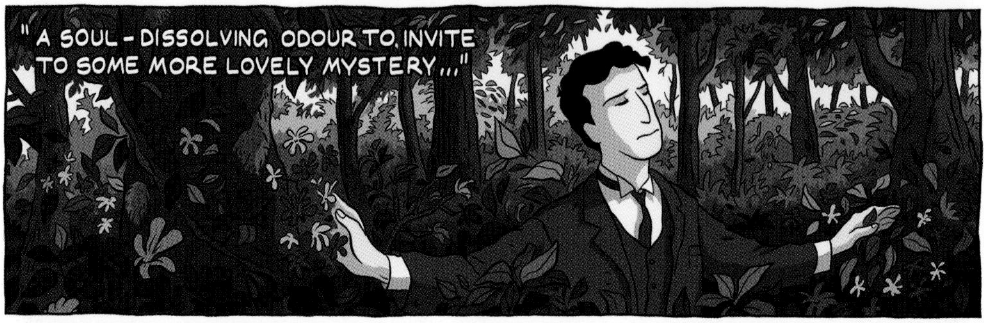
"A SOUL-DISSOLVING ODOUR TO INVITE TO SOME MORE LOVELY MYSTERY..."
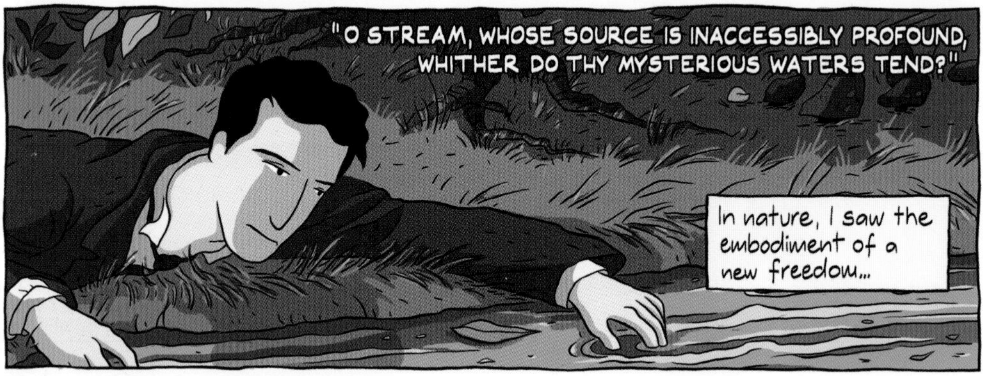
"O STREAM, WHOSE SOURCE IS INACCESSIBLY PROFOUND, WHITHER DO THY MYSTERIOUS WATERS TEND?"
In nature, I saw the embodiment of a new freedom...
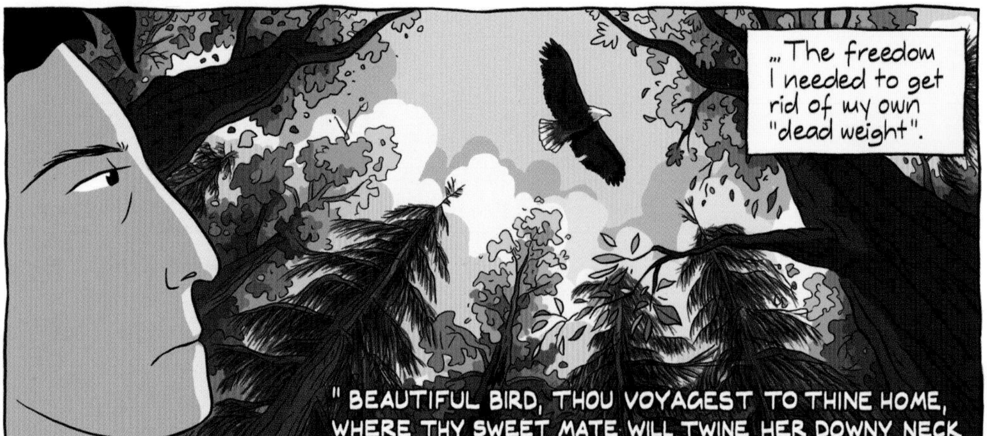
... The freedom I needed to get rid of my own "dead weight".
"BEAUTIFUL BIRD, THOU VOYAGEST TO THINE HOME, WHERE THY SWEET MATE WILL TWINE HER DOWNY NECK WITH THINE, AND WELCOME THY RETURN!"

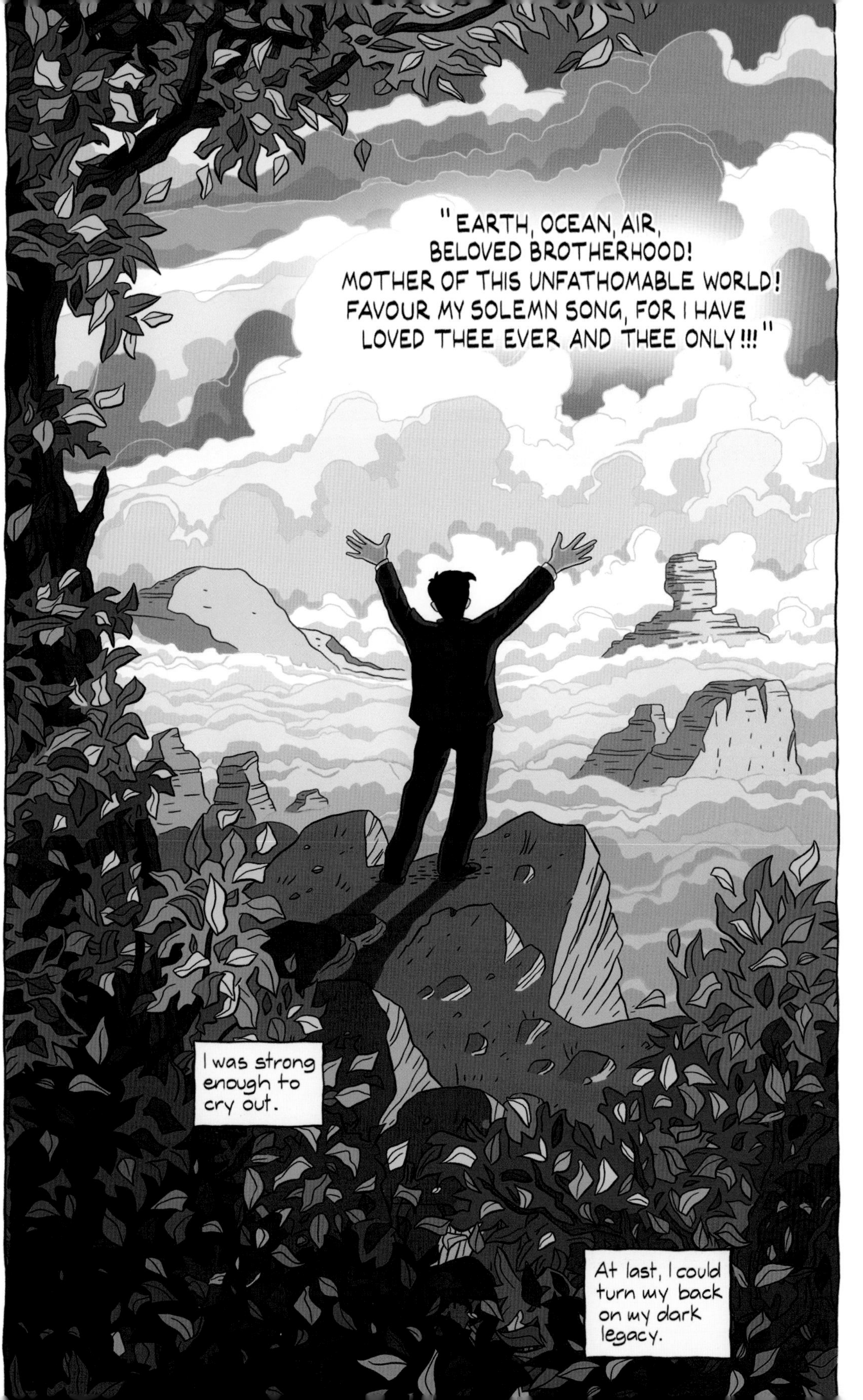
"EARTH, OCEAN, AIR, BELOVED BROTHERHOOD! MOTHER OF THIS UNFATHOMABLE WORLD! FAVOUR MY SOLEMN SONG, FOR I HAVE LOVED THEE EVER AND THEE ONLY!!!"
I was strong enough to cry out.
At last, I could turn my back on my dark legacy.

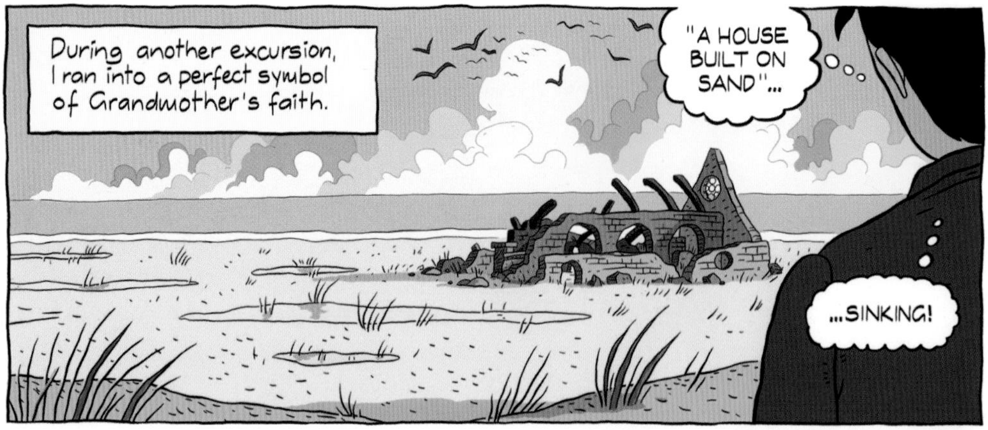
During another excursion, I ran into a perfect symbol of Grandmother's faith.
"A HOUSE BUILT ON SAND"...
...SINKING!
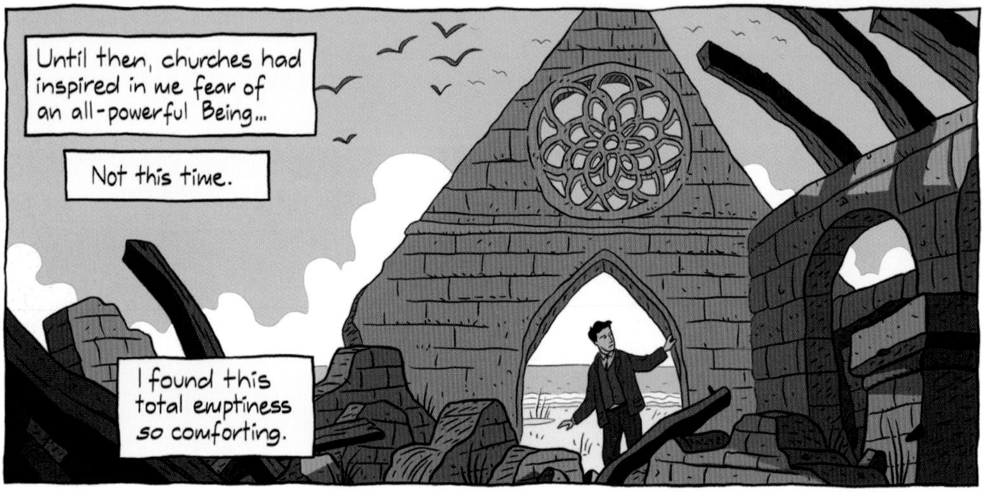
Until then, churches had inspired in me fear of an all-powerful Being...
Not this time.
I found this total emptiness so comforting.
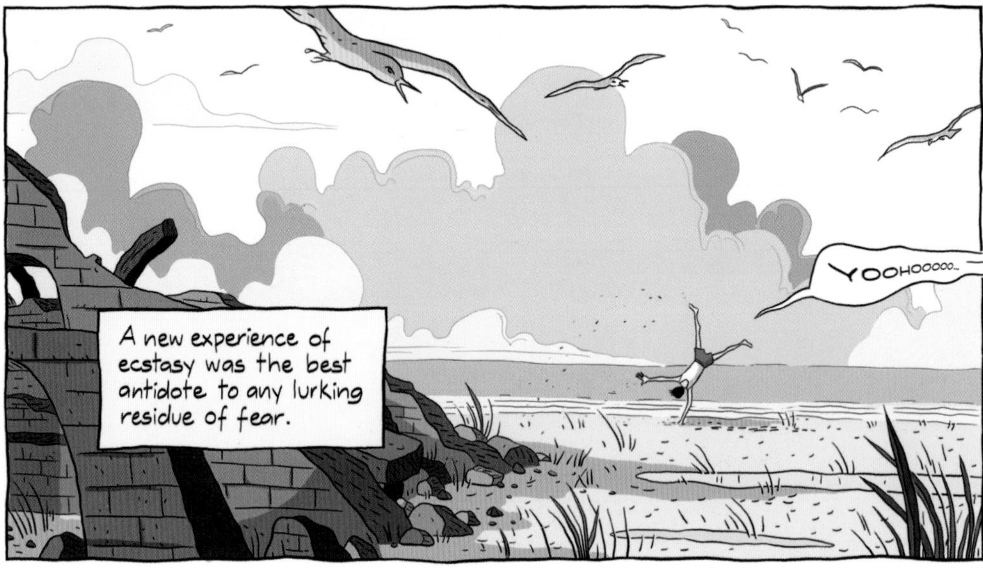
YOOHOOOOO...
A new experience of ecstasy was the best antidote to any lurking residue of fear.

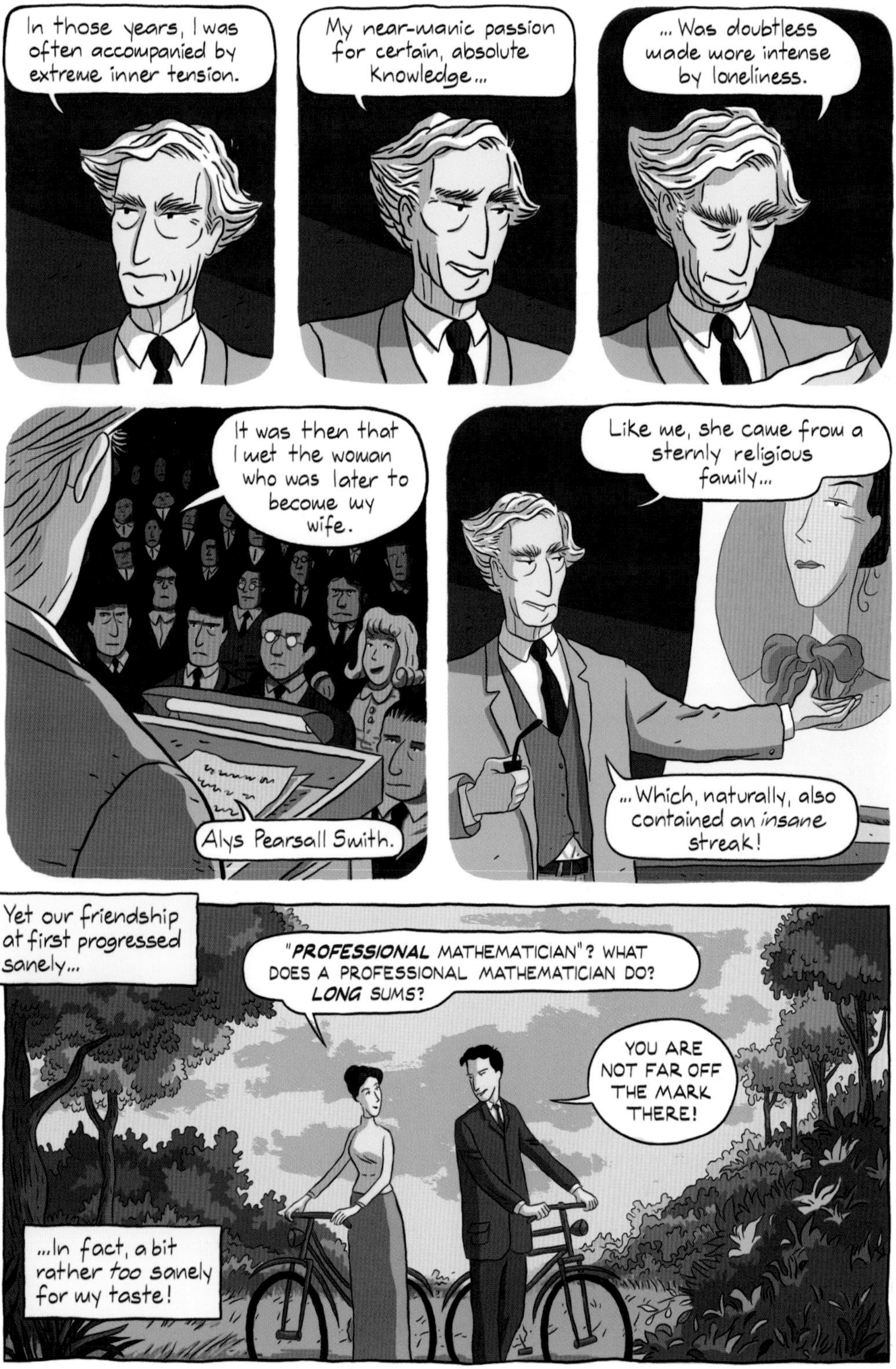
In those years, I was often accompanied by extreme inner tension.
My near-manic passion for certain, absolute knowledge...
...Was doubtless made more intense by loneliness.
It was then that I met the woman who was later to become my wife.
Alys Pearsall Smith.
Like me, she came from a sternly religious family...
...Which, naturally, also contained an *insane* streak!
Yet our friendship at first progressed sanely...
"***PROFESSIONAL*** MATHEMATICIAN"? WHAT DOES A PROFESSIONAL MATHEMATICIAN DO? ***LONG*** SUMS?
YOU ARE NOT FAR OFF THE MARK THERE!
...In fact, a bit rather *too* sanely for my taste!

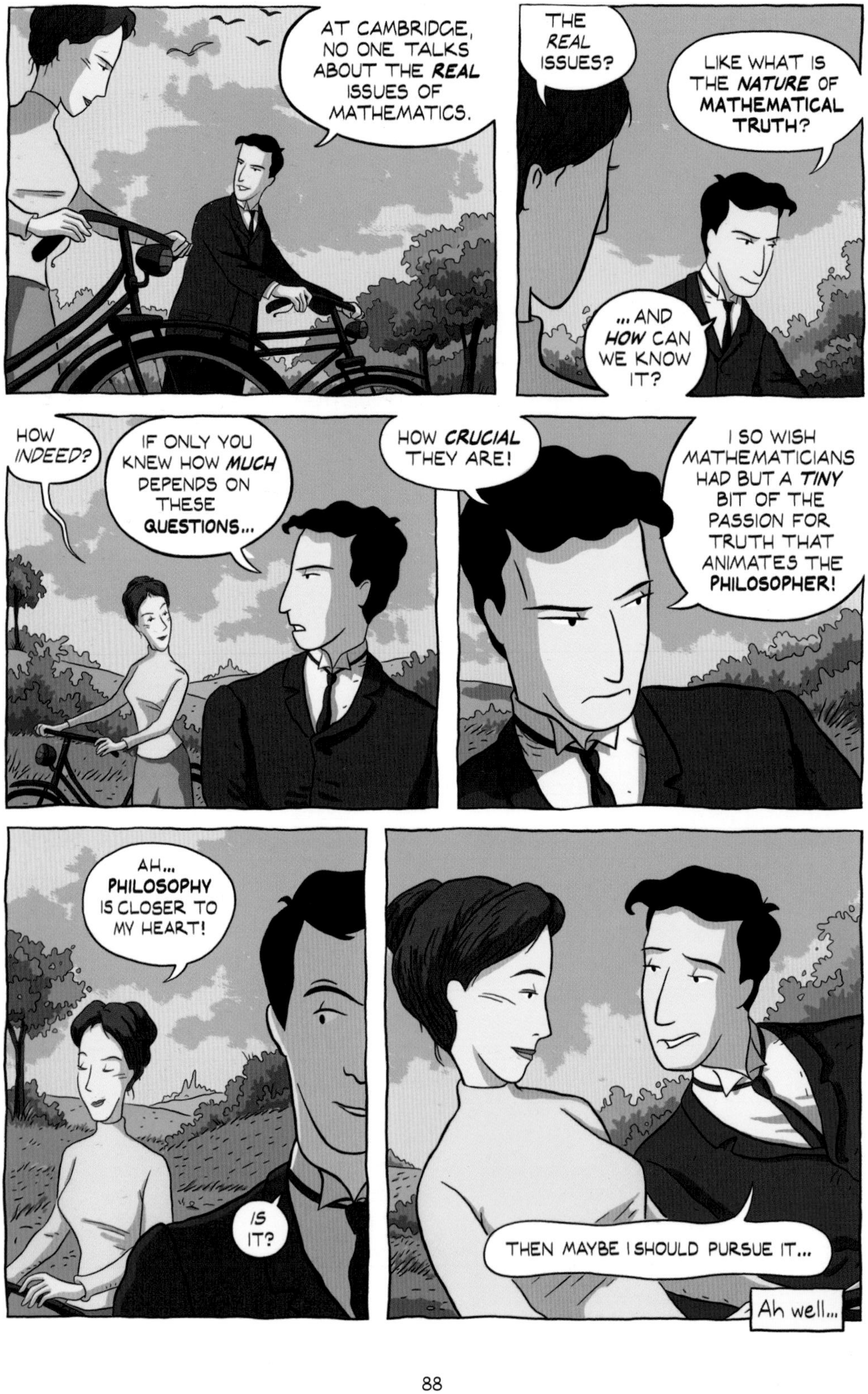
AT CAMBRIDGE, NO ONE TALKS ABOUT THE REAL ISSUES OF MATHEMATICS.
THE REAL ISSUES?
LIKE WHAT IS THE NATURE OF MATHEMATICAL TRUTH?
...AND HOW CAN WE KNOW IT?
HOW INDEED?
IF ONLY YOU KNEW HOW MUCH DEPENDS ON THESE QUESTIONS...
HOW CRUCIAL THEY ARE!
I SO WISH MATHEMATICIANS HAD BUT A TINY BIT OF THE PASSION FOR TRUTH THAT ANIMATES THE PHILOSOPHER!
AH... PHILOSOPHY IS CLOSER TO MY HEART!
IS IT?
THEN MAYBE I SHOULD PURSUE IT...
Ah well...

NOT ON MY ACCOUNT, YOU SHOULDN'T!
I was an absolute beginner in courtship!
With not the slightest bit of beginner's luck!
Yet there was no doubt...
GENTLEMEN, YOU MAY NOW BEGIN!
...I was infatuated!
Even during the "Tripos", the dreaded final exams...
Alys dominated my thoughts!

Though fortunately, with beneficial effects!

Mathematical Tripos

FIRST CLASS HONOURS... NOT *BAD!*

WELL DONE, MR. RUSSELL!

?

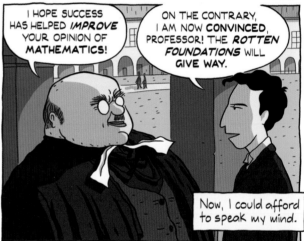

I HOPE SUCCESS HAS HELPED *IMPROVE* YOUR OPINION OF **MATHEMATICS!**

ON THE CONTRARY, I AM NOW **CONVINCED**, PROFESSOR! THE *ROTTEN FOUNDATIONS* WILL GIVE WAY.

Now, I could afford to speak my mind.

THE EDIFICE OF **MATHEMATICS** WILL **COLLAPSE!**

OH? AREN'T YOU CONCERNED THAT ITS FALL WILL *CRUSH* YOU TOO?

And to make my decisions.

NO! YOU SEE, I DON'T PLAN TO BE *INSIDE* IT.

!

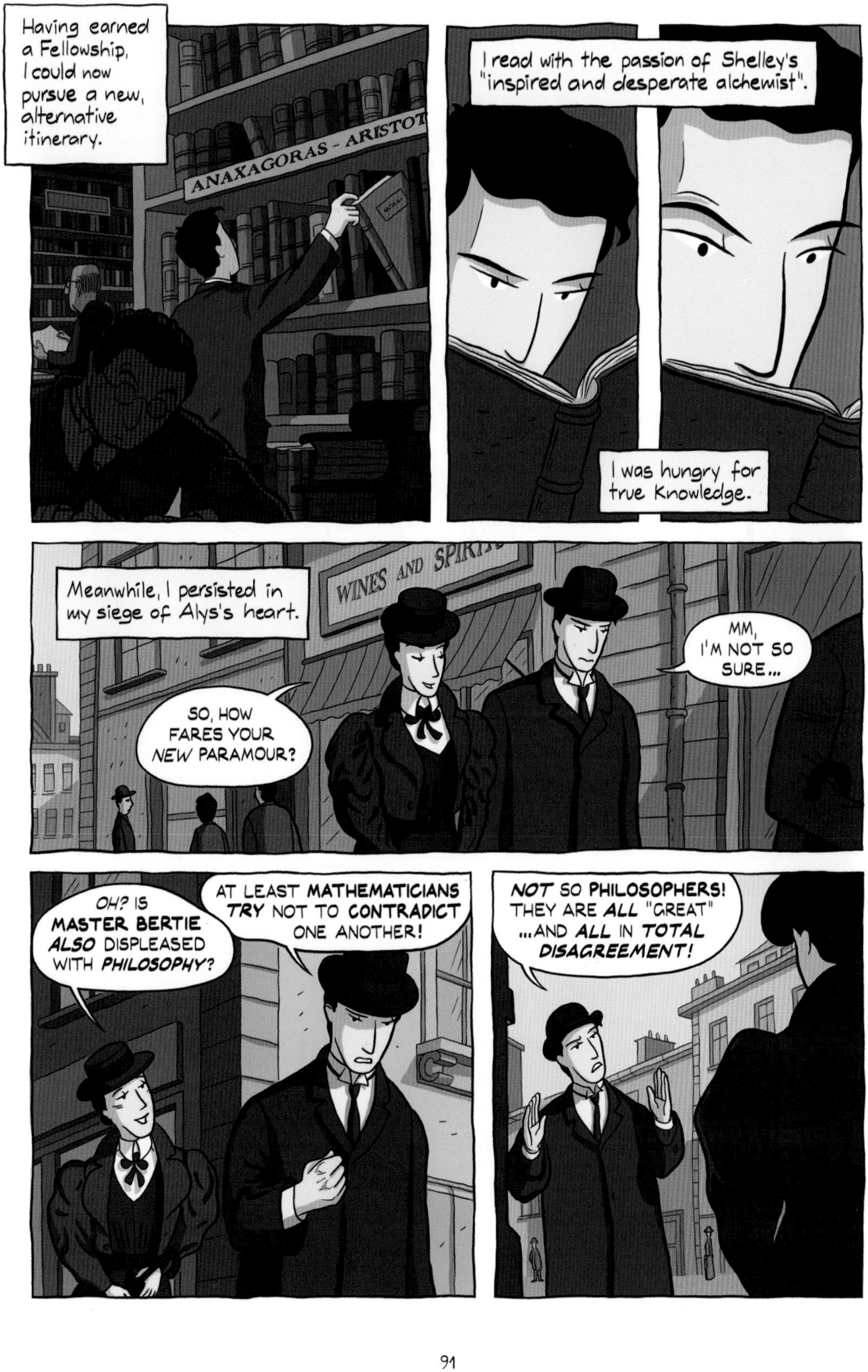
Having earned a Fellowship, I could now pursue a new, alternative itinerary.
ANAXAGORAS - ARISTOT
I read with the passion of Shelley's "inspired and desperate alchemist".
I was hungry for true knowledge.
Meanwhile, I persisted in my siege of Alys's heart.
WINES AND SPIRI
SO, HOW FARES YOUR *NEW* PARAMOUR?
MM, I'M NOT SO SURE...
OH? IS **MASTER BERTIE** ***ALSO*** DISPLEASED WITH ***PHILOSOPHY***?
AT LEAST **MATHEMATICIANS** ***TRY*** NOT TO **CONTRADICT** ONE ANOTHER!
NOT SO **PHILOSOPHERS!** THEY ARE ***ALL*** "GREAT" ...AND ***ALL*** IN ***TOTAL DISAGREEMENT!***

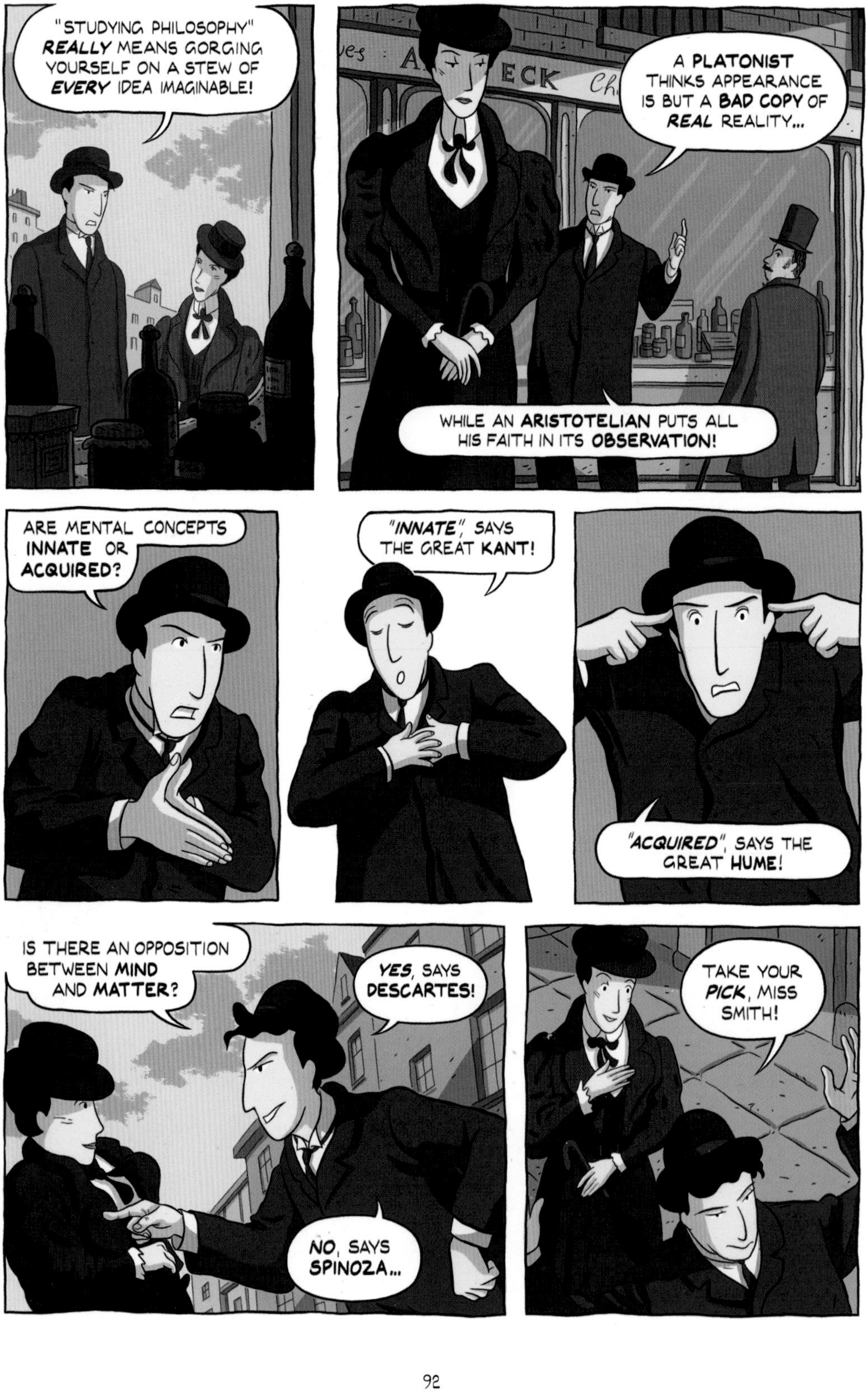
"STUDYING PHILOSOPHY" REALLY MEANS GORGING YOURSELF ON A STEW OF EVERY IDEA IMAGINABLE!
A PLATONIST THINKS APPEARANCE IS BUT A BAD COPY OF REAL REALITY...
WHILE AN ARISTOTELIAN PUTS ALL HIS FAITH IN ITS OBSERVATION!
ARE MENTAL CONCEPTS INNATE OR ACQUIRED?
"INNATE", SAYS THE GREAT KANT!
"ACQUIRED", SAYS THE GREAT HUME!
IS THERE AN OPPOSITION BETWEEN MIND AND MATTER?
YES, SAYS DESCARTES!
NO, SAYS SPINOZA...
TAKE YOUR PICK, MISS SMITH!

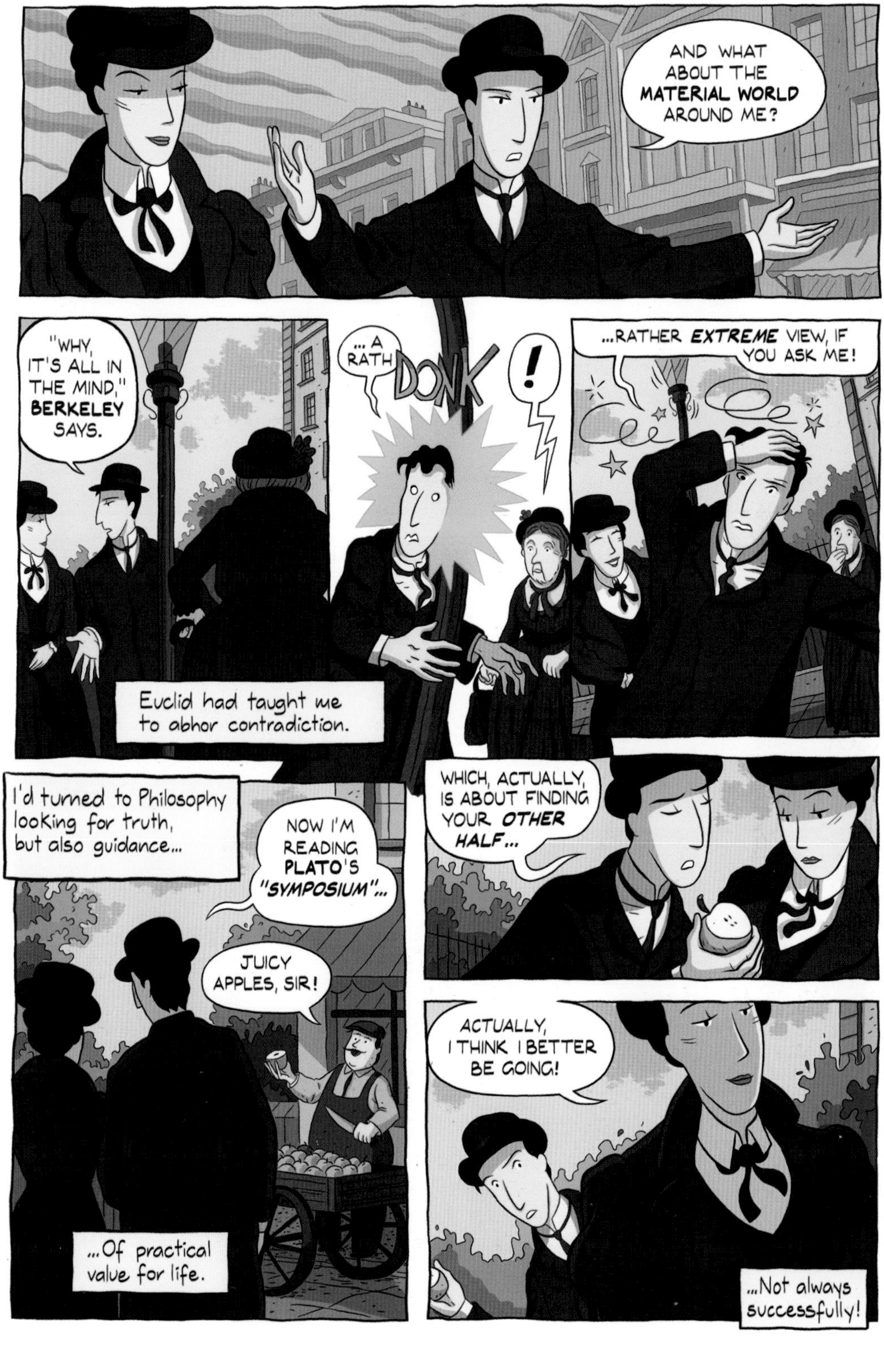
AND WHAT ABOUT THE **MATERIAL WORLD** AROUND ME?
"WHY, IT'S ALL IN THE MIND," **BERKELEY** SAYS.
Euclid had taught me to abhor contradiction.
... A RATH
DONK
!
...RATHER ***EXTREME*** VIEW, IF YOU ASK ME!
I'd turned to Philosophy looking for truth, but also guidance...
NOW I'M READING **PLATO**'S ***"SYMPOSIUM"...***
JUICY APPLES, SIR!
...Of practical value for life.
WHICH, ACTUALLY, IS ABOUT FINDING YOUR ***OTHER HALF...***
ACTUALLY, I THINK I BETTER BE GOING!
...Not always successfully!

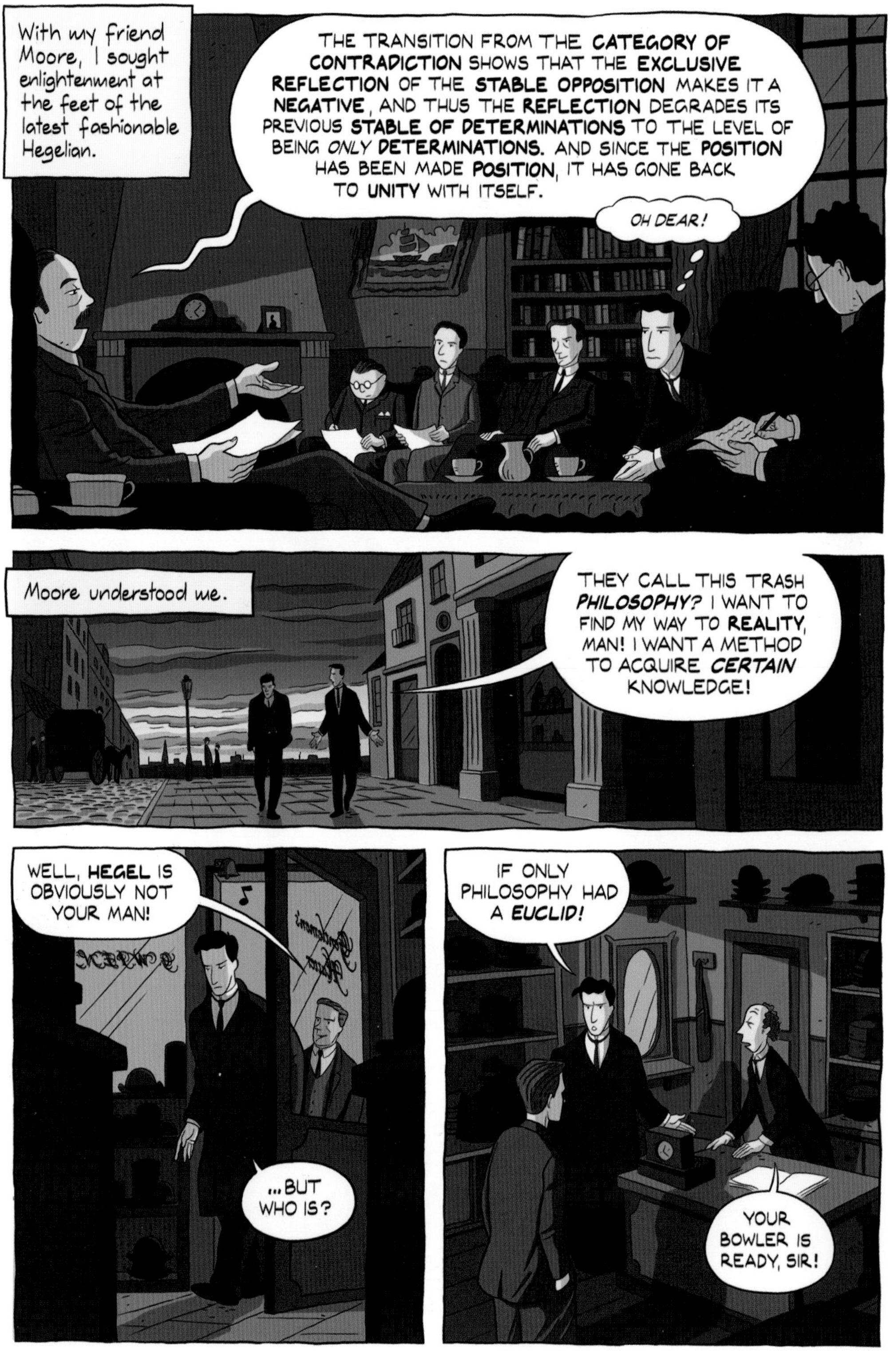
With my friend Moore, I sought enlightenment at the feet of the latest fashionable Hegelian.
THE TRANSITION FROM THE CATEGORY OF CONTRADICTION SHOWS THAT THE EXCLUSIVE REFLECTION OF THE STABLE OPPOSITION MAKES IT A NEGATIVE, AND THUS THE REFLECTION DEGRADES ITS PREVIOUS STABLE OF DETERMINATIONS TO THE LEVEL OF BEING ONLY DETERMINATIONS. AND SINCE THE POSITION HAS BEEN MADE POSITION, IT HAS GONE BACK TO UNITY WITH ITSELF.
OH DEAR!
Moore understood me.
THEY CALL THIS TRASH PHILOSOPHY? I WANT TO FIND MY WAY TO REALITY, MAN! I WANT A METHOD TO ACQUIRE CERTAIN KNOWLEDGE!
WELL, HEGEL IS OBVIOUSLY NOT YOUR MAN!
...BUT WHO IS?
IF ONLY PHILOSOPHY HAD A EUCLID!
YOUR BOWLER IS READY, SIR!

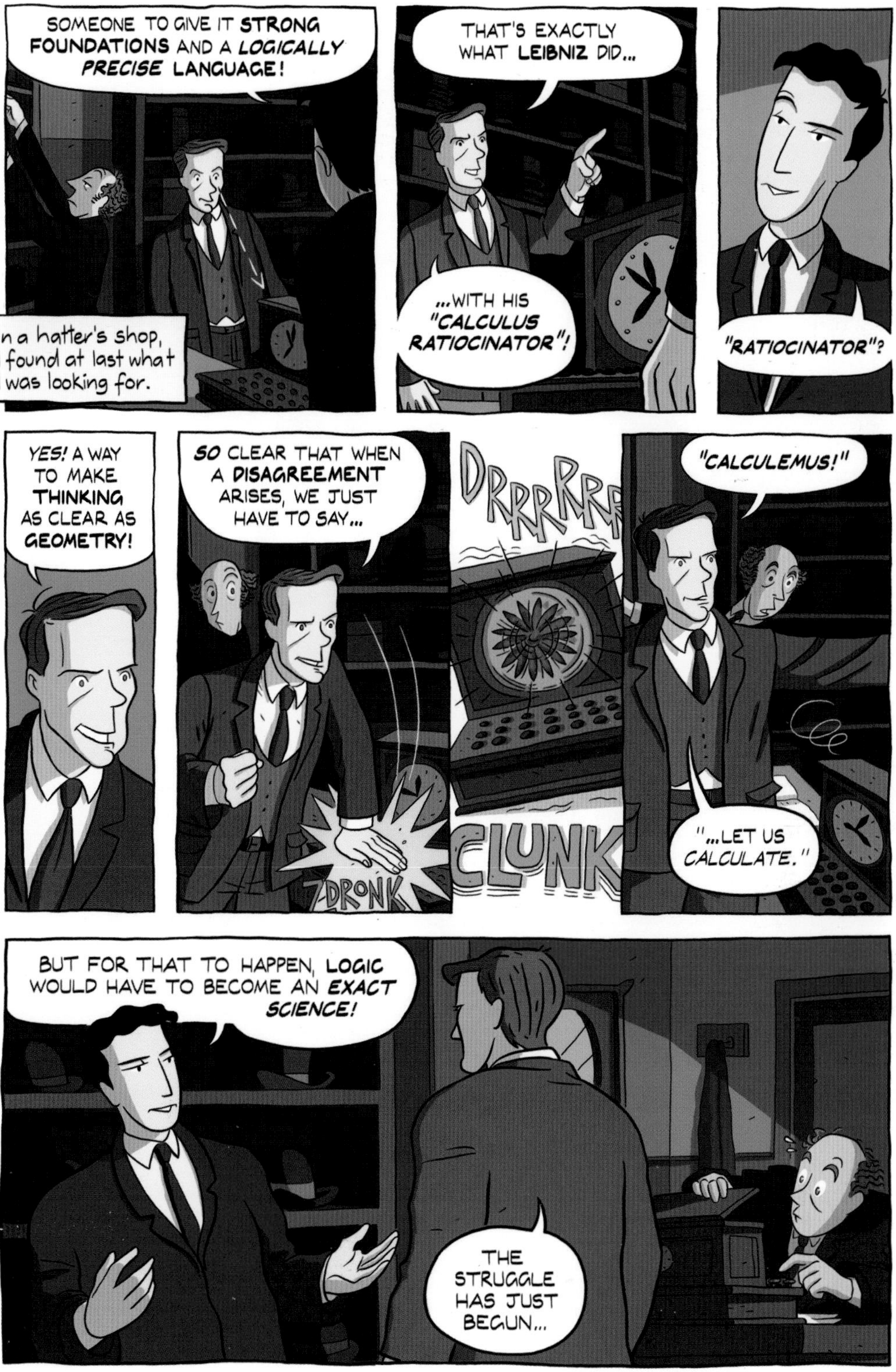
SOMEONE TO GIVE IT **STRONG FOUNDATIONS** AND A ***LOGICALLY PRECISE*** **LANGUAGE!**
In a hatter's shop, I found at last what I was looking for.
THAT'S EXACTLY WHAT **LEIBNIZ** DID...
...WITH HIS ***"CALCULUS RATIOCINATOR"!***
"RATIOCINATOR"?
YES! A WAY TO MAKE **THINKING** AS CLEAR AS **GEOMETRY!**
SO CLEAR THAT WHEN A **DISAGREEMENT** ARISES, WE JUST HAVE TO SAY...
ZDRONK
DRRRRRR
CLUNK
"CALCULEMUS!"
"...LET US CALCULATE."
BUT FOR THAT TO HAPPEN, **LOGIC** WOULD HAVE TO BECOME AN ***EXACT SCIENCE!***
THE STRUGGLE HAS JUST BEGUN...

WHAT **LEIBNIZ** DREAMT OF WITH THE "*CALCULUS RATIOCINATOR*", **GEORGE BOOLE** CONTINUED WITH HIS *"LAWS"*.

?

PERHAPS YOU'D LIKE TO TRY IT ON, SIR?

WHY DON'T I *KNOW* THIS?

BECAUSE **PHILOSOPHERS** THINK IT'S *MATHEMATICS* AND **MATHEMATICIANS** *PHILOSOPHY!*

LOOK...

!

My first meeting with Euclid had planted a seed...

THE **ATOMS** OF **LOGIC** ARE THE PROPOSITIONS.

...But hearing about Leibniz's dream was the actual calling.

WHICH WE *COMBINE* THROUGH CERTAIN LAWS.

After that day I knew: I was a...

...LOGICIAN!

AHA... AND NOW THE PLOT *THICKENS!*

Then and there, Moore introduced us to a new, extraordinary world.

ALL THAT **LOGIC** REALLY IS, IS USING **COMBINATIONS** OF THE KNOWN, TO REACH THE *UNKNOWN*.

THE **GREEKS** KNEW ALL THAT!

BUT **LEIBNIZ** USED A **FORMAL**, *SYMBOLIC* LANGUAGE TO SAY THINGS LIKE THIS...

$$A \oplus A = A$$

"A **TAUTOLOGY** ADDED TO ITSELF IS A **TAUTOLOGY**."

TIME OUT!

"A TAUTOLOGY ADDED TO ITSELF IS A TAUTOLOGY." ISN'T THIS A LITTLE *TOO* TECHNICAL?

I MEAN, *WHAT'S* A "TAUTOLOGY"?

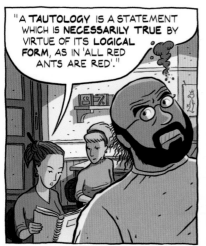

"A **TAUTOLOGY** IS A STATEMENT WHICH IS **NECESSARILY TRUE** BY VIRTUE OF ITS **LOGICAL FORM**, AS IN 'ALL RED ANTS ARE RED'."

I *KNOW* WHAT A TAUTOLOGY IS, THANK YOU!

BUT DOES THE **AVERAGE READER** KNOW THIS?

IS THERE SUCH A BEING?

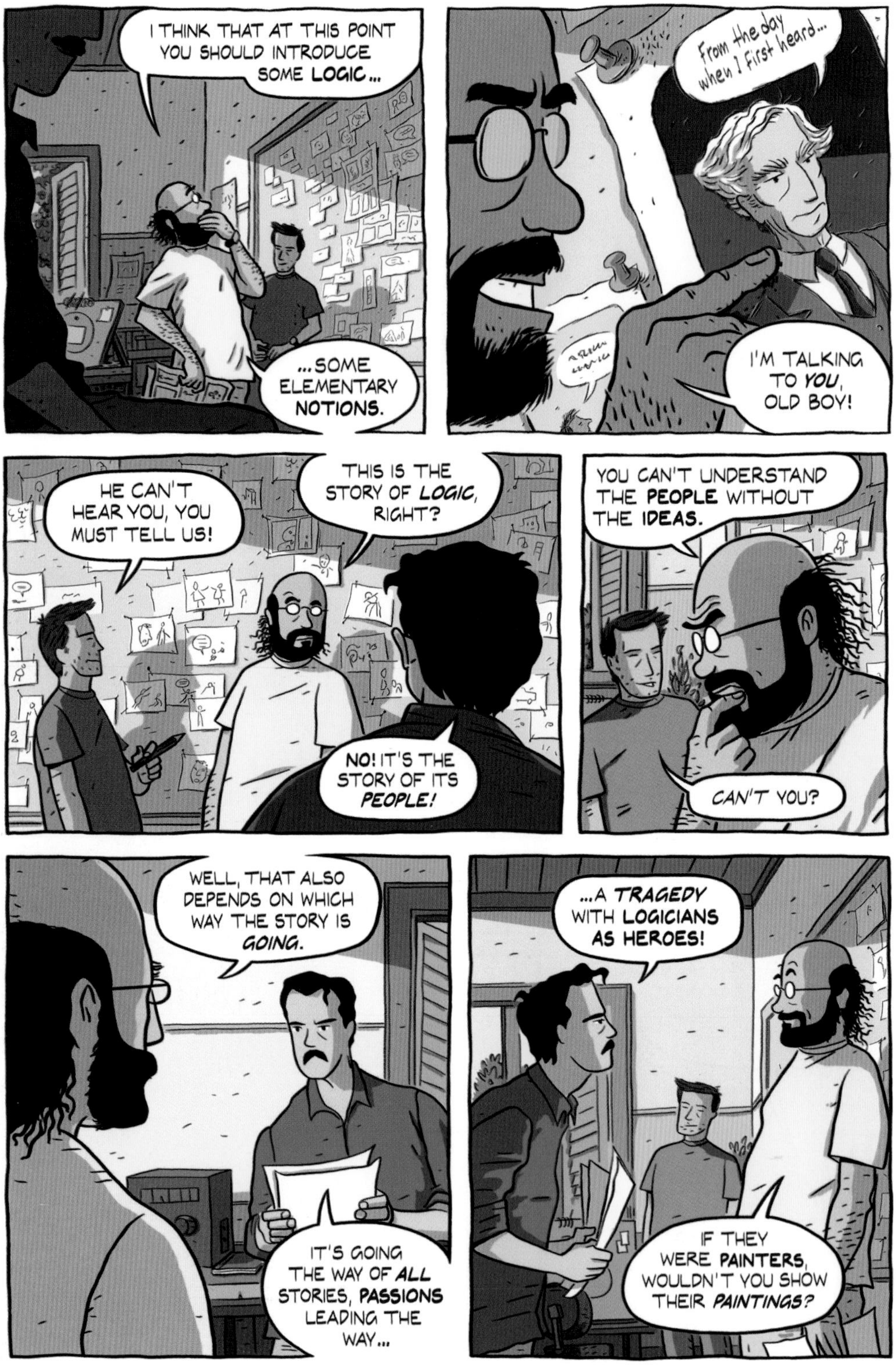
I THINK THAT AT THIS POINT YOU SHOULD INTRODUCE SOME **LOGIC**...
...SOME ELEMENTARY **NOTIONS**.
From the day when I first heard...
I'M TALKING TO ***YOU***, OLD BOY!
HE CAN'T HEAR YOU, YOU MUST TELL US!
THIS IS THE STORY OF ***LOGIC***, RIGHT?
NO! IT'S THE STORY OF ITS ***PEOPLE!***
YOU CAN'T UNDERSTAND THE **PEOPLE** WITHOUT THE **IDEAS**.
CAN'T YOU?
WELL, THAT ALSO DEPENDS ON WHICH WAY THE STORY IS ***GOING***.
IT'S GOING THE WAY OF ***ALL*** STORIES, **PASSIONS** LEADING THE WAY...
...A ***TRAGEDY*** WITH **LOGICIANS AS HEROES!**
IF THEY WERE **PAINTERS**, WOULDN'T YOU SHOW THEIR ***PAINTINGS?***

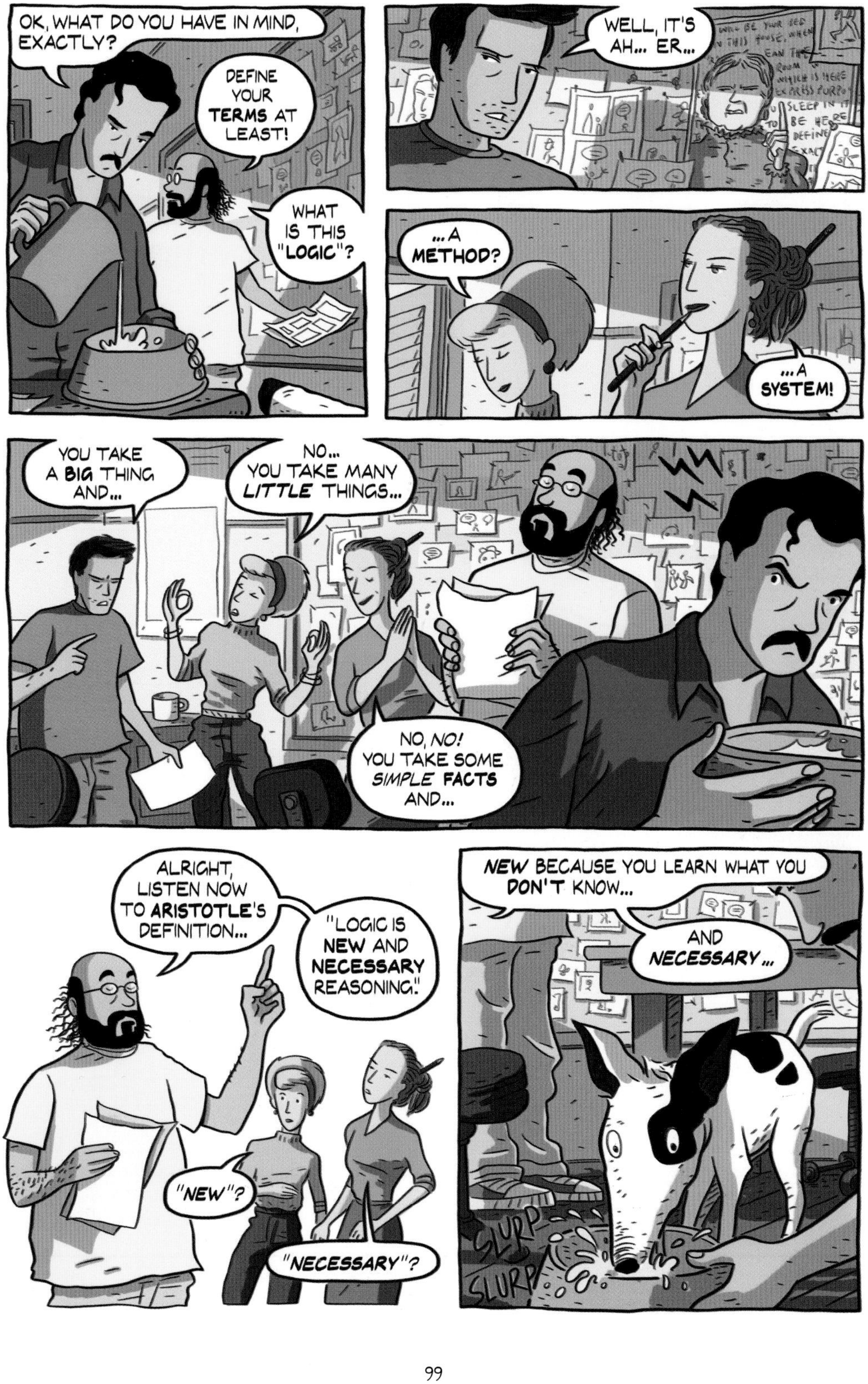
OK, WHAT DO YOU HAVE IN MIND, EXACTLY?
DEFINE YOUR TERMS AT LEAST!
WHAT IS THIS "LOGIC"?
WELL, IT'S AH... ER...
...A METHOD?
...A SYSTEM!
YOU TAKE A BIG THING AND...
NO... YOU TAKE MANY LITTLE THINGS...
NO, NO! YOU TAKE SOME SIMPLE FACTS AND...
ALRIGHT, LISTEN NOW TO ARISTOTLE'S DEFINITION...
"LOGIC IS NEW AND NECESSARY REASONING."
"NEW"?
"NECESSARY"?
NEW BECAUSE YOU LEARN WHAT YOU DON'T KNOW...
AND NECESSARY...
SLURP
SLURP

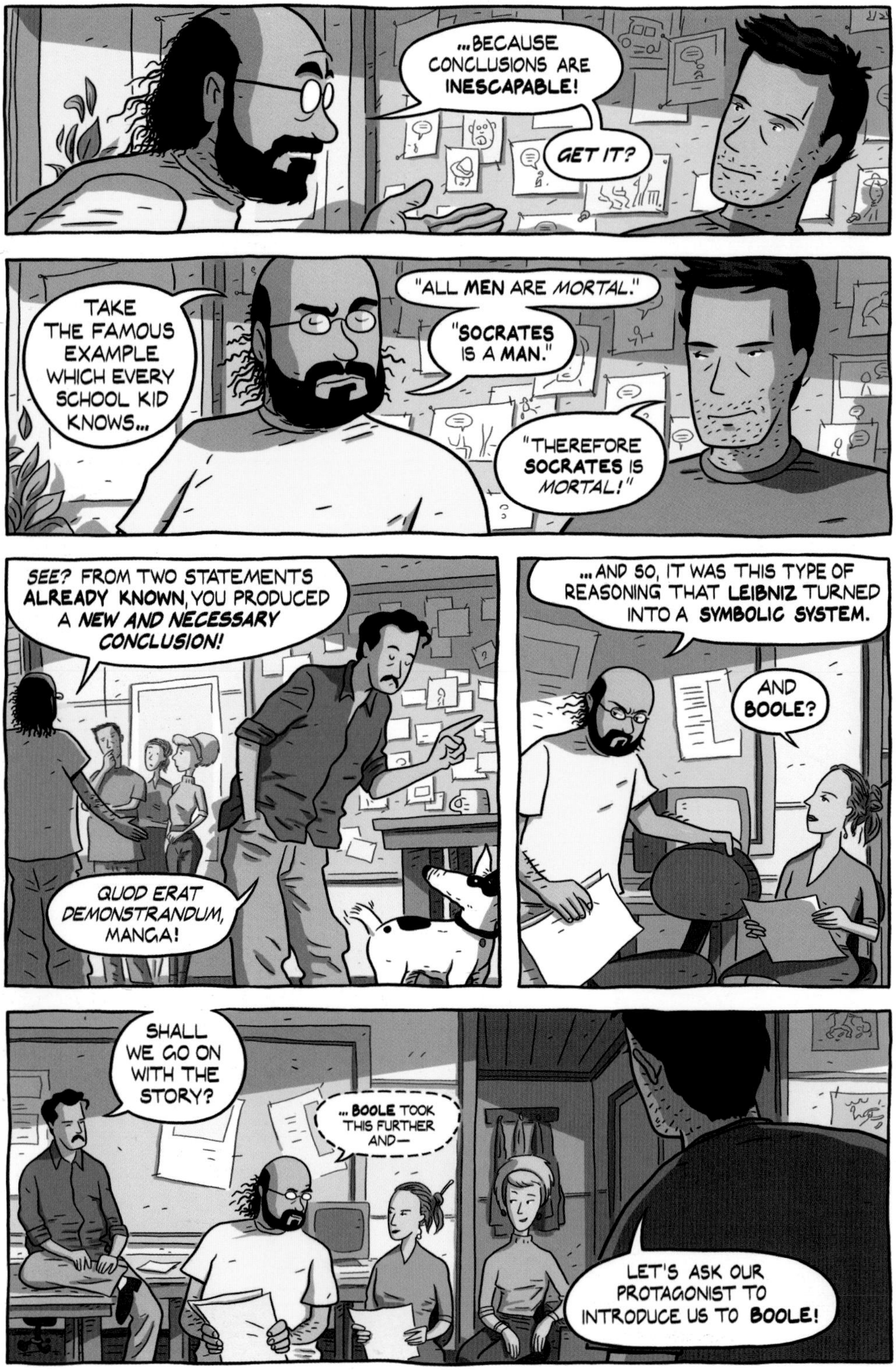
...BECAUSE CONCLUSIONS ARE INESCAPABLE!
GET IT?
TAKE THE FAMOUS EXAMPLE WHICH EVERY SCHOOL KID KNOWS...
"ALL MEN ARE MORTAL."
"SOCRATES IS A MAN."
"THEREFORE SOCRATES IS MORTAL!"
SEE? FROM TWO STATEMENTS ALREADY KNOWN, YOU PRODUCED A NEW AND NECESSARY CONCLUSION!
QUOD ERAT DEMONSTRANDUM, MANGA!
...AND SO, IT WAS THIS TYPE OF REASONING THAT LEIBNIZ TURNED INTO A SYMBOLIC SYSTEM.
AND BOOLE?
SHALL WE GO ON WITH THE STORY?
...BOOLE TOOK THIS FURTHER AND—
LET'S ASK OUR PROTAGONIST TO INTRODUCE US TO BOOLE!

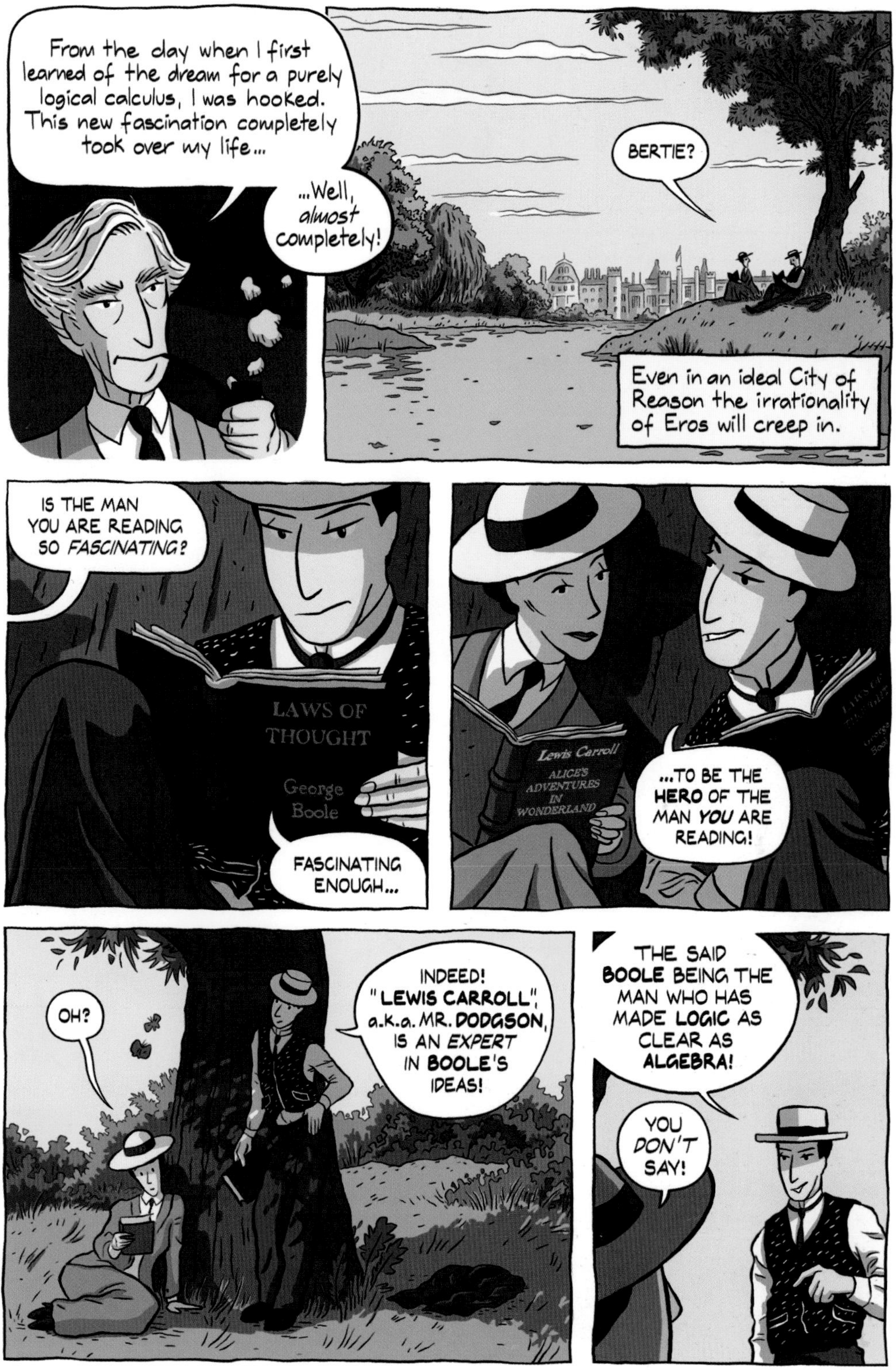
From the day when I first learned of the dream for a purely logical calculus, I was hooked. This new fascination completely took over my life...
...Well, *almost* completely!
BERTIE?
Even in an ideal City of Reason the irrationality of Eros will creep in.
IS THE MAN YOU ARE READING SO *FASCINATING*?
LAWS OF THOUGHT
George Boole
FASCINATING ENOUGH...
Lewis Carroll
ALICE'S ADVENTURES IN WONDERLAND
...TO BE THE **HERO** OF THE MAN ***YOU*** ARE READING!
OH?
INDEED! "**LEWIS CARROLL**", a.k.a. MR. **DODGSON**, IS AN *EXPERT* IN **BOOLE**'S IDEAS!
THE SAID **BOOLE** BEING THE MAN WHO HAS MADE **LOGIC** AS CLEAR AS **ALGEBRA!**
YOU *DON'T* SAY!

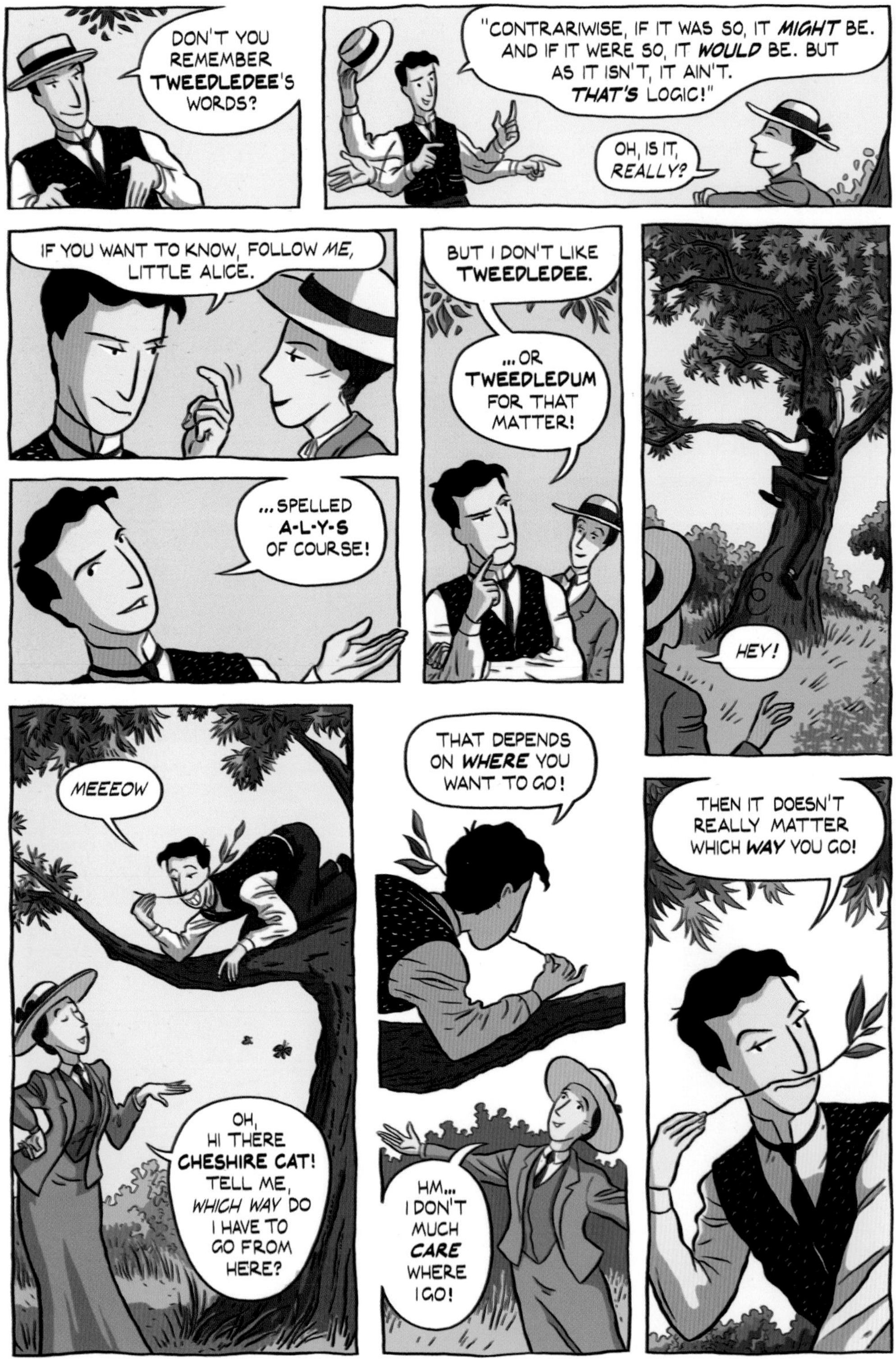
DON'T YOU REMEMBER **TWEEDLEDEE**'S WORDS?
"CONTRARIWISE, IF IT WAS SO, IT ***MIGHT*** BE. AND IF IT WERE SO, IT ***WOULD*** BE. BUT AS IT ISN'T, IT AIN'T. ***THAT'S*** LOGIC!"
OH, IS IT, *REALLY?*
IF YOU WANT TO KNOW, FOLLOW *ME,* LITTLE ALICE.
...SPELLED **A-L-Y-S** OF COURSE!
BUT I DON'T LIKE **TWEEDLEDEE.**
...OR **TWEEDLEDUM** FOR THAT MATTER!
HEY!
MEEEOW
OH, HI THERE **CHESHIRE CAT!** TELL ME, *WHICH WAY* DO I HAVE TO GO FROM HERE?
THAT DEPENDS ON ***WHERE*** YOU WANT TO GO!
HM... I DON'T MUCH ***CARE*** WHERE I GO!
THEN IT DOESN'T REALLY MATTER WHICH ***WAY*** YOU GO!

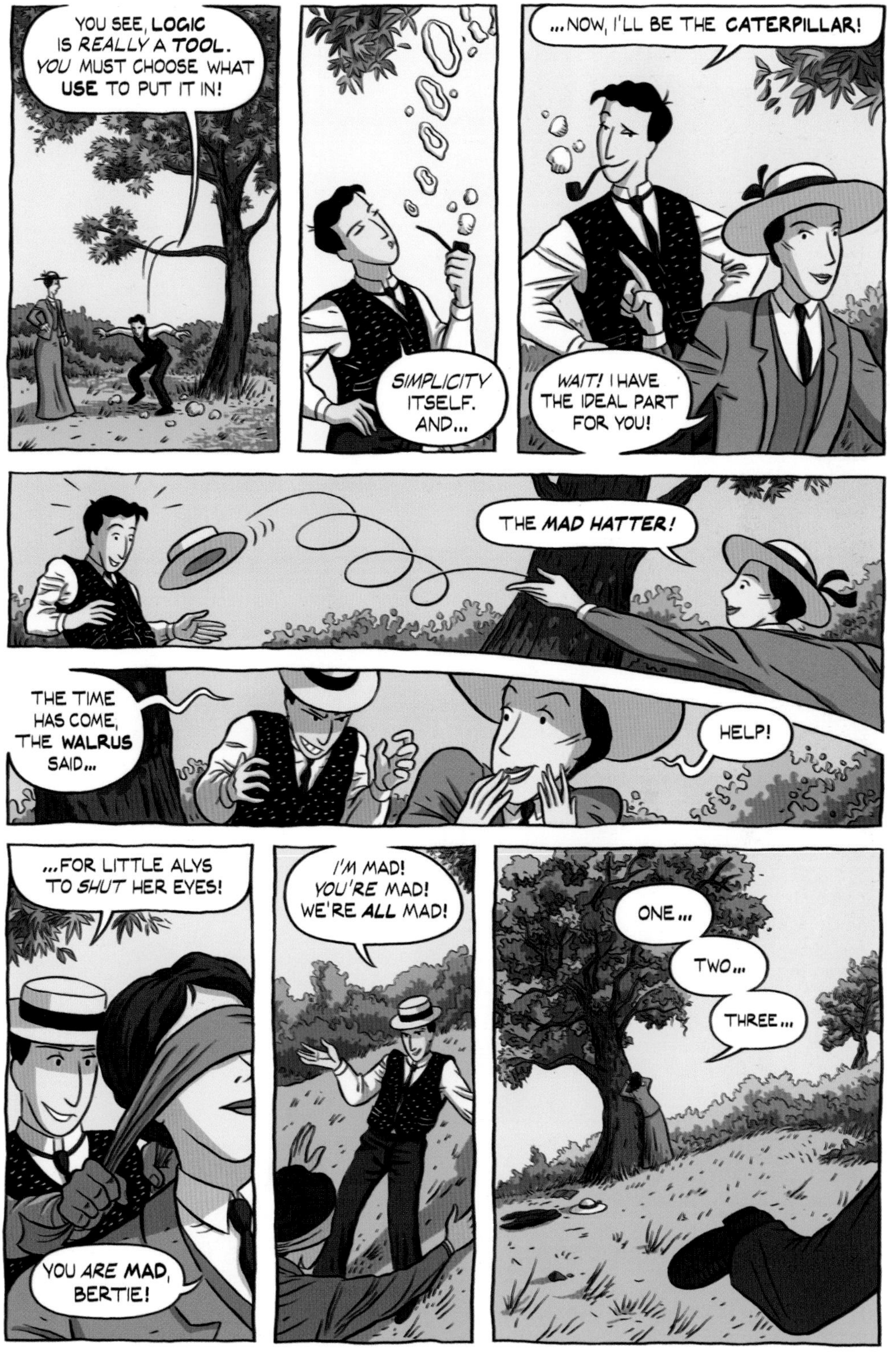
YOU SEE, LOGIC IS REALLY A TOOL. YOU MUST CHOOSE WHAT USE TO PUT IT IN!
SIMPLICITY ITSELF. AND...
...NOW, I'LL BE THE CATERPILLAR!
WAIT! I HAVE THE IDEAL PART FOR YOU!
THE MAD HATTER!
THE TIME HAS COME, THE WALRUS SAID...
HELP!
...FOR LITTLE ALYS TO SHUT HER EYES!
YOU ARE MAD, BERTIE!
I'M MAD! YOU'RE MAD! WE'RE ALL MAD!
ONE...
TWO...
THREE...

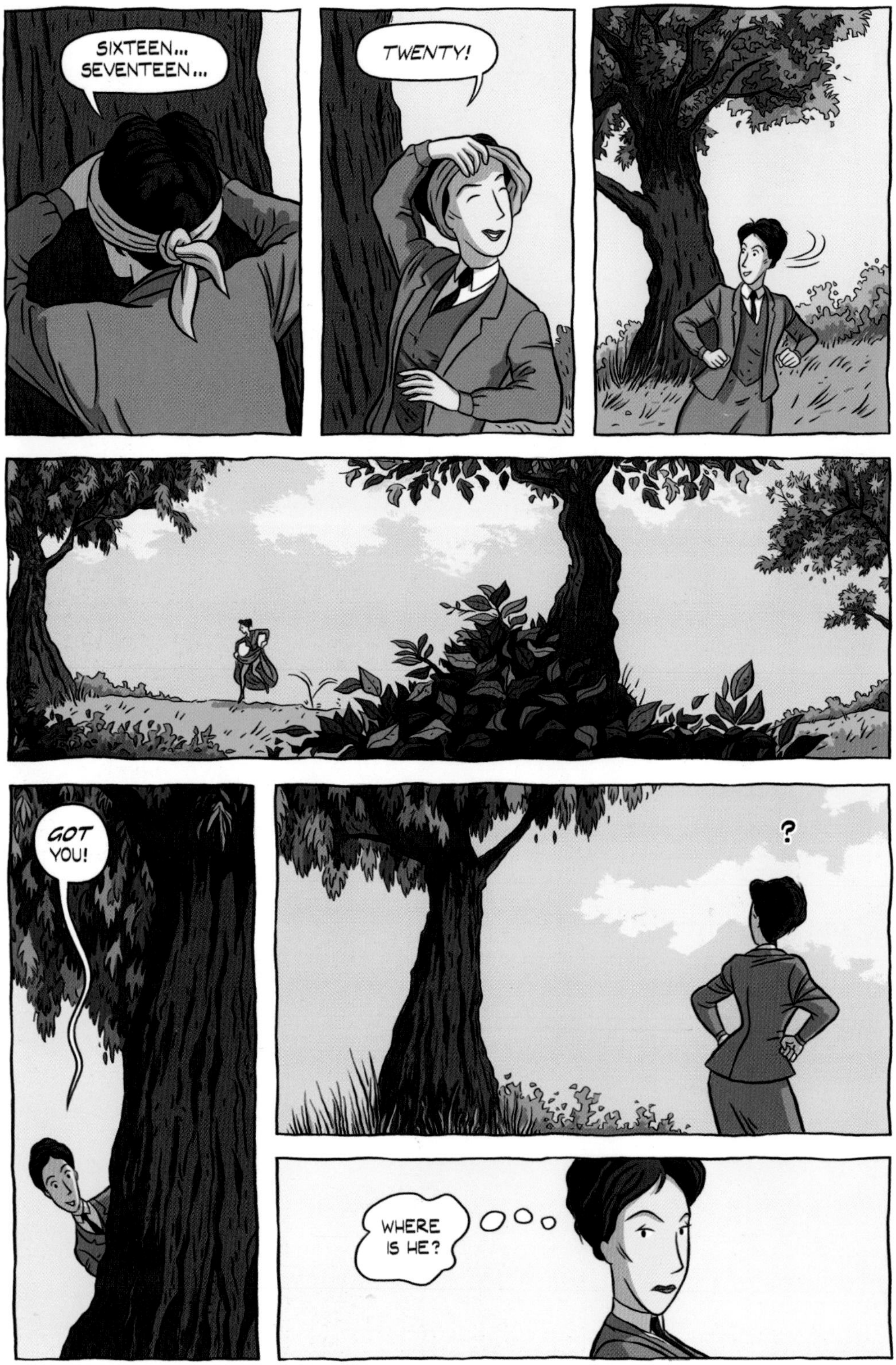
SIXTEEN... SEVENTEEN...
TWENTY!
GOT YOU!
?
WHERE IS HE?

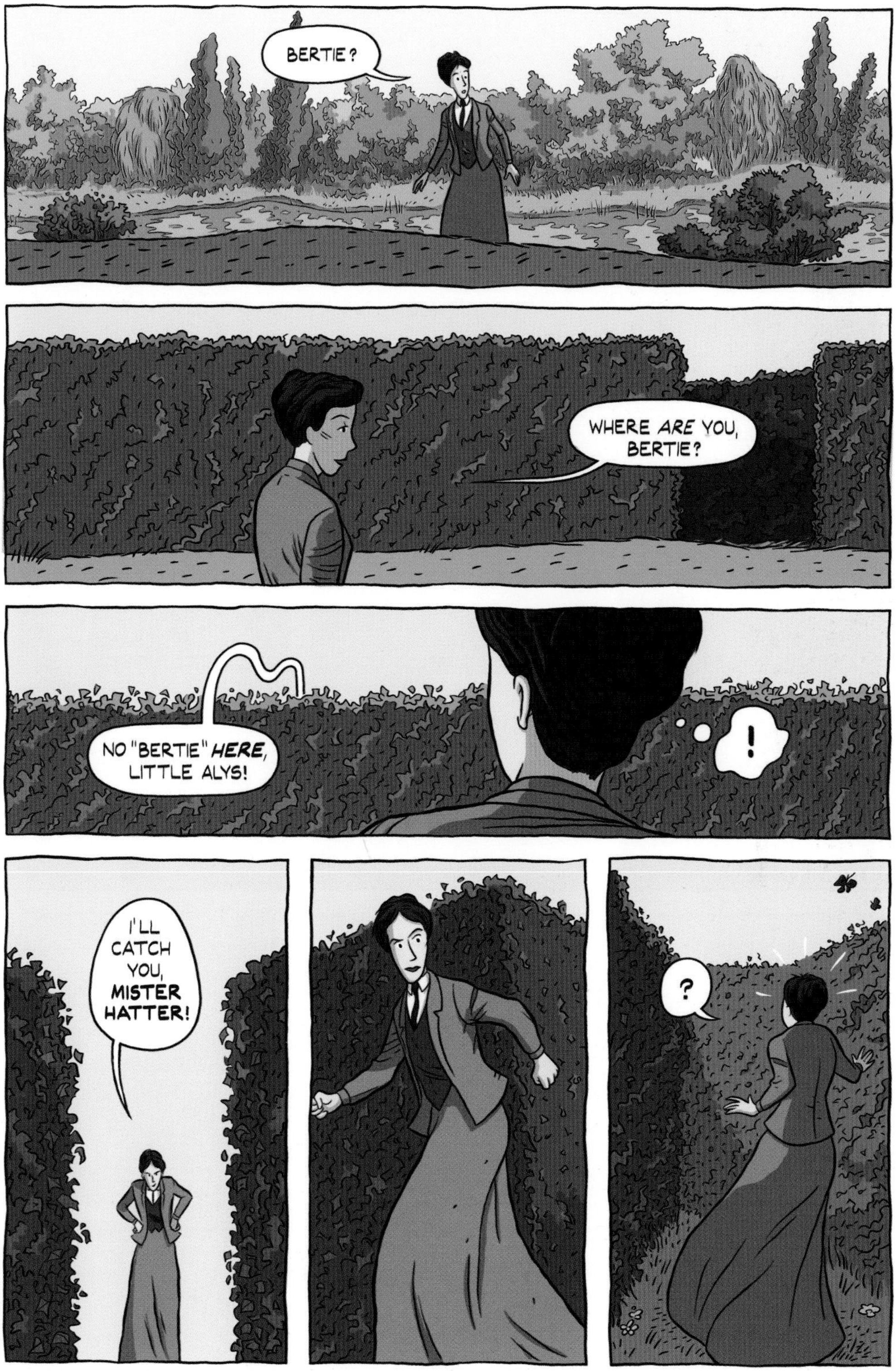
BERTIE?
WHERE *ARE* YOU, BERTIE?
NO "BERTIE" ***HERE***, LITTLE ALYS!
!
I'LL CATCH YOU, **MISTER HATTER!**
?

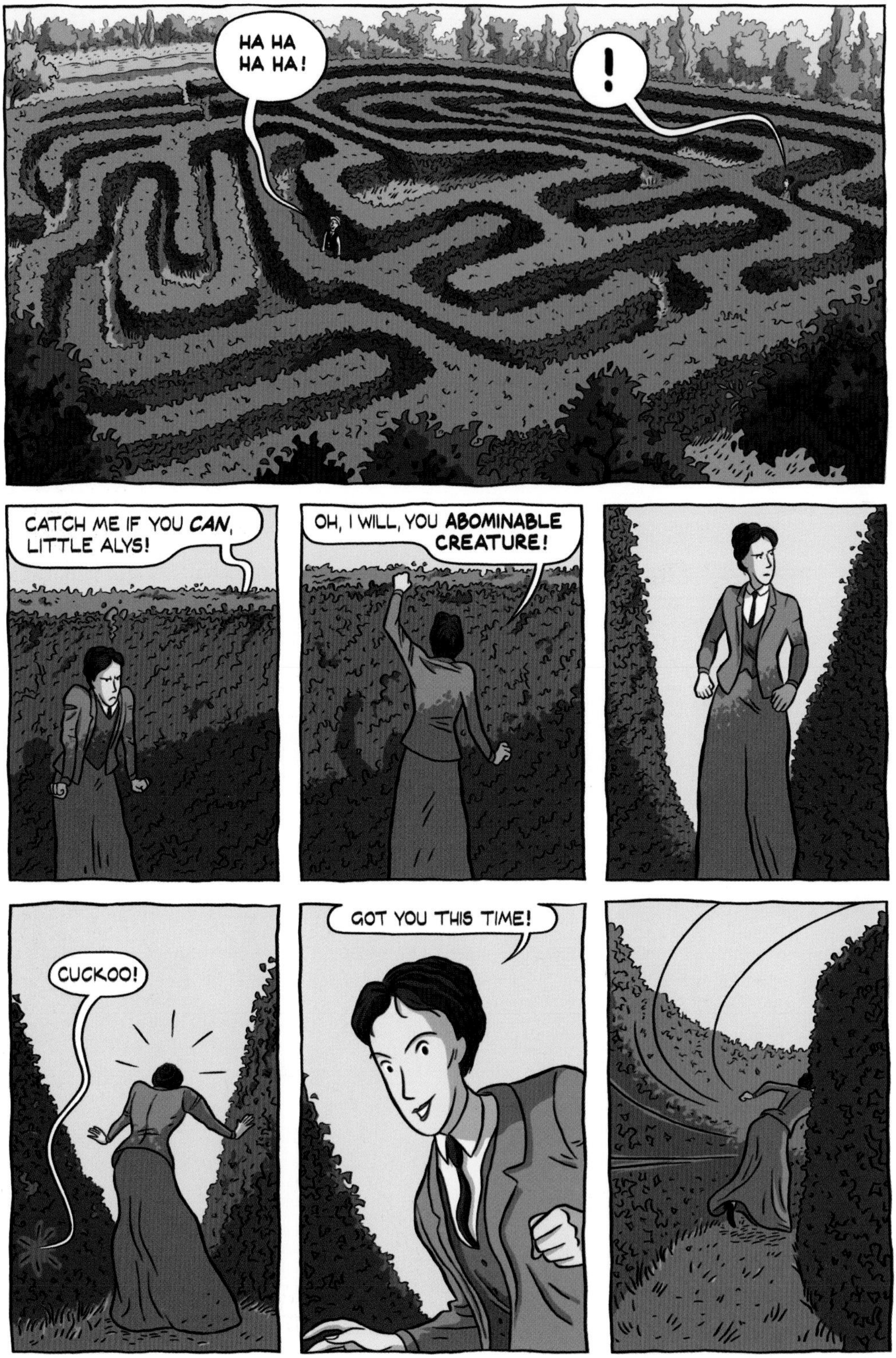
HA HA
HA HA!
!
CATCH ME IF YOU CAN, LITTLE ALYS!
OH, I WILL, YOU ABOMINABLE CREATURE!
CUCKOO!
GOT YOU THIS TIME!

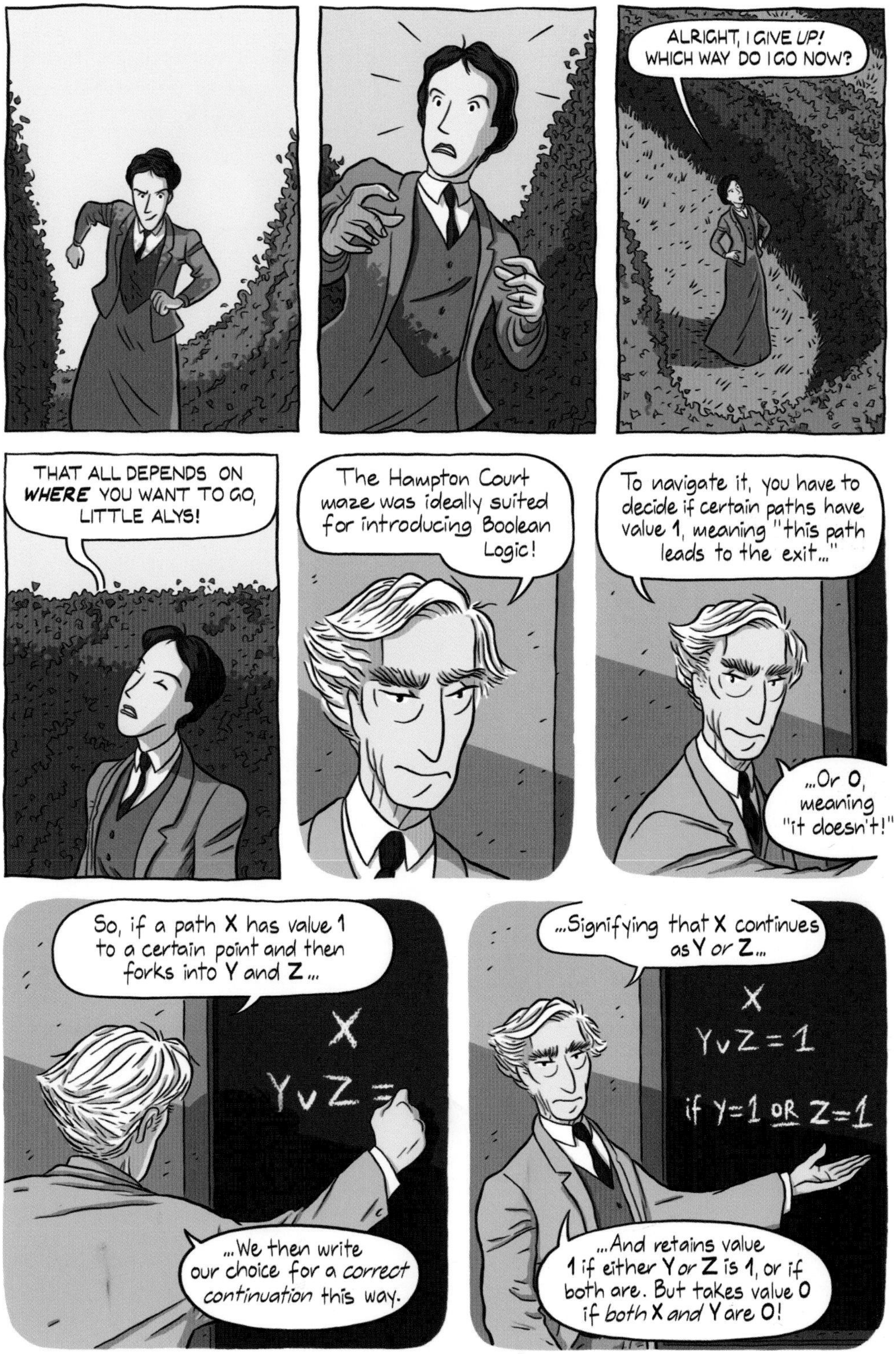
ALRIGHT, I GIVE UP! WHICH WAY DO I GO NOW?
THAT ALL DEPENDS ON WHERE YOU WANT TO GO, LITTLE ALYS!
The Hampton Court maze was ideally suited for introducing Boolean Logic!
To navigate it, you have to decide if certain paths have value 1, meaning "this path leads to the exit..."
...Or 0, meaning "it doesn't!"
So, if a path X has value 1 to a certain point and then forks into Y and Z...
X
Y v Z =
...We then write our choice for a correct continuation this way.
...Signifying that X continues as Y or Z...
X
Y v Z = 1
if Y=1 OR Z=1
...And retains value 1 if either Y or Z is 1, or if both are. But takes value 0 if both X and Y are 0!

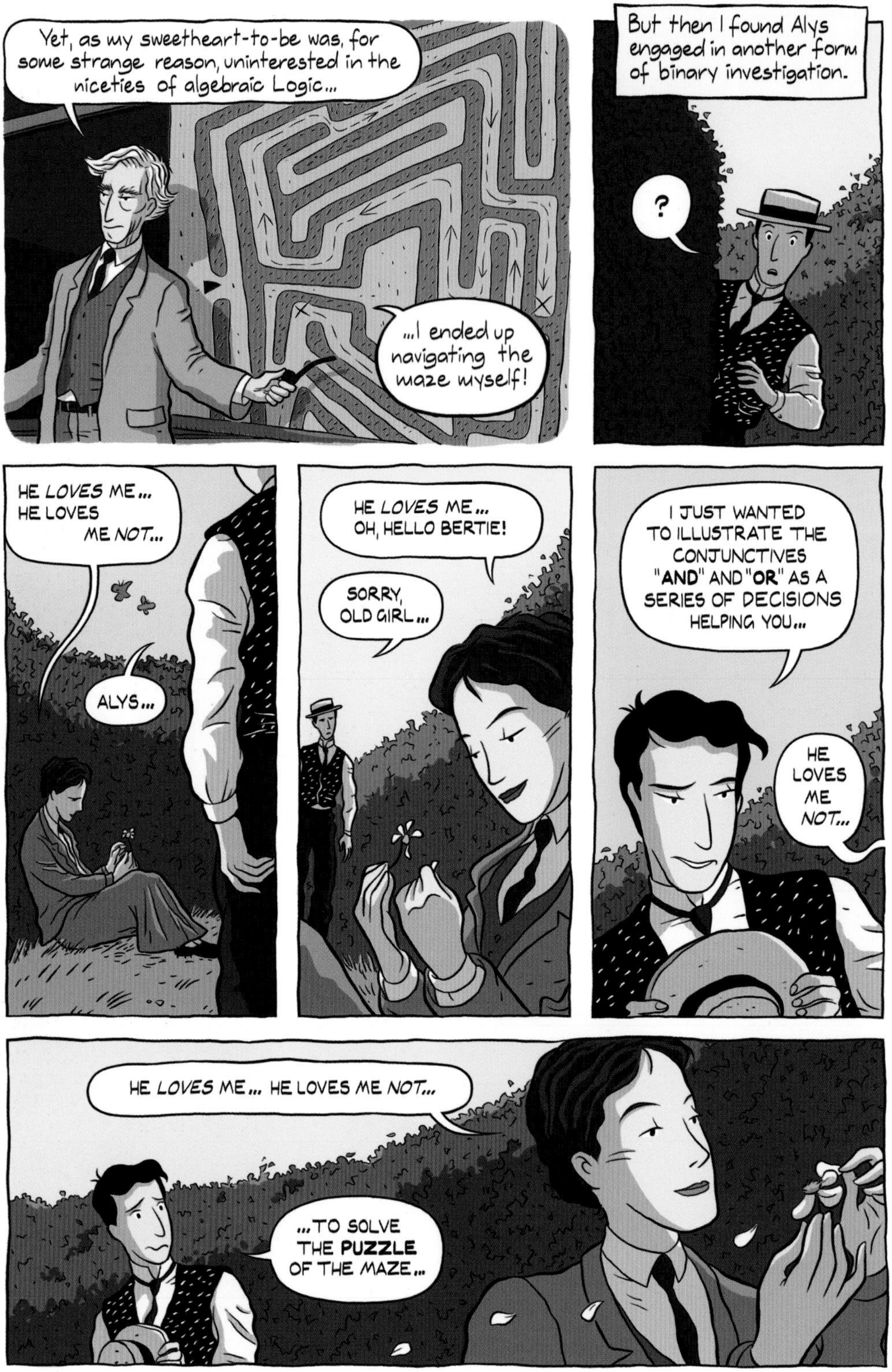
Yet, as my sweetheart-to-be was, for some strange reason, uninterested in the niceties of algebraic Logic...
...I ended up navigating the maze myself!
But then I found Alys engaged in another form of binary investigation.
?
HE *LOVES* ME... HE LOVES ME *NOT*...
ALYS...
HE *LOVES* ME... OH, HELLO BERTIE!
SORRY, OLD GIRL...
I JUST WANTED TO ILLUSTRATE THE CONJUNCTIVES "**AND**" AND "**OR**" AS A SERIES OF DECISIONS HELPING YOU...
HE LOVES ME *NOT*...
HE *LOVES* ME... HE LOVES ME *NOT*...
...TO SOLVE THE **PUZZLE** OF THE MAZE...

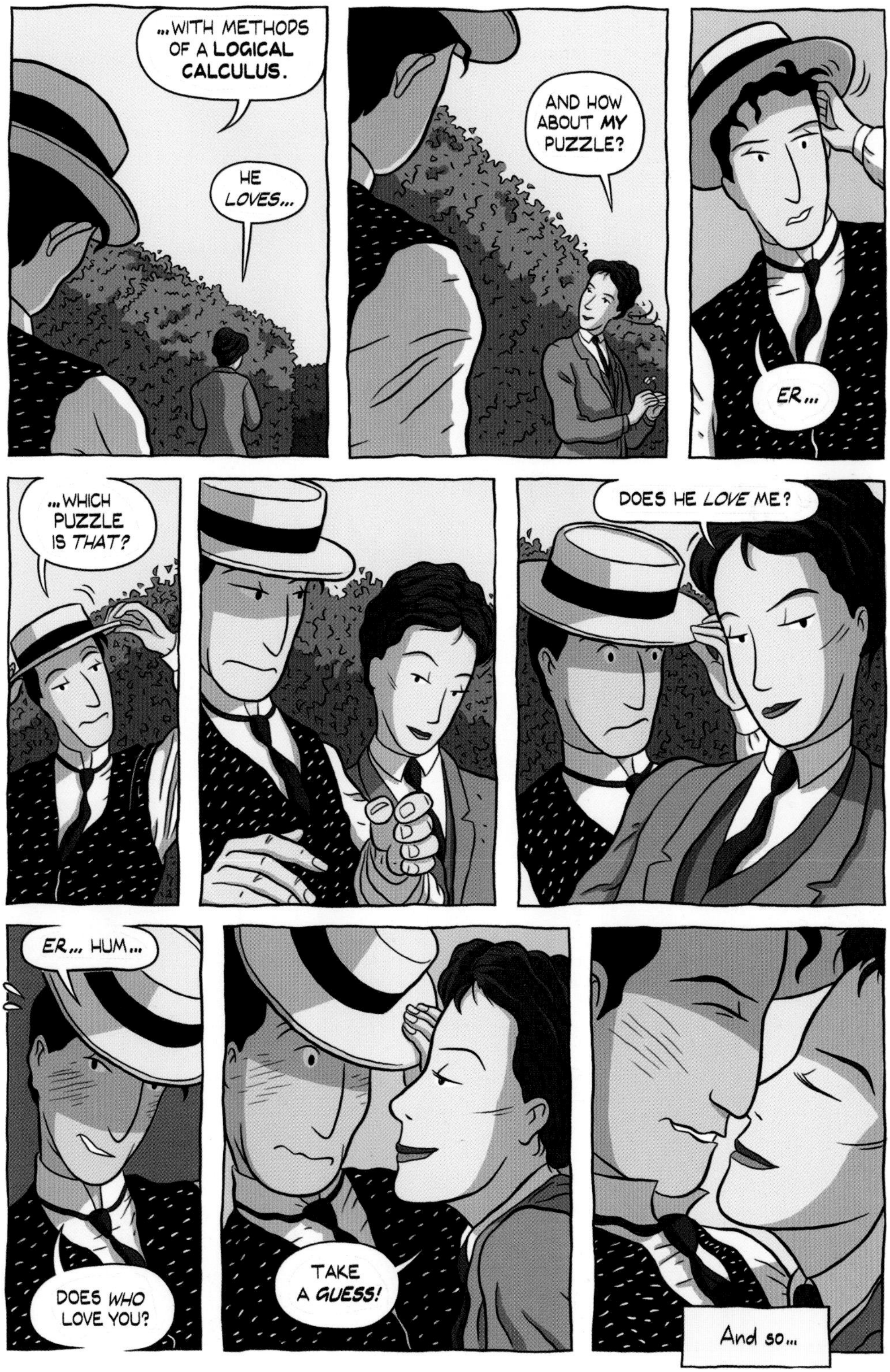
...WITH METHODS OF A **LOGICAL CALCULUS**.
HE *LOVES*...
AND HOW ABOUT **MY** PUZZLE?
ER...
...WHICH PUZZLE IS *THAT*?
DOES HE *LOVE* ME?
ER... HUM...
DOES *WHO* LOVE YOU?
TAKE A ***GUESS!***
And so...

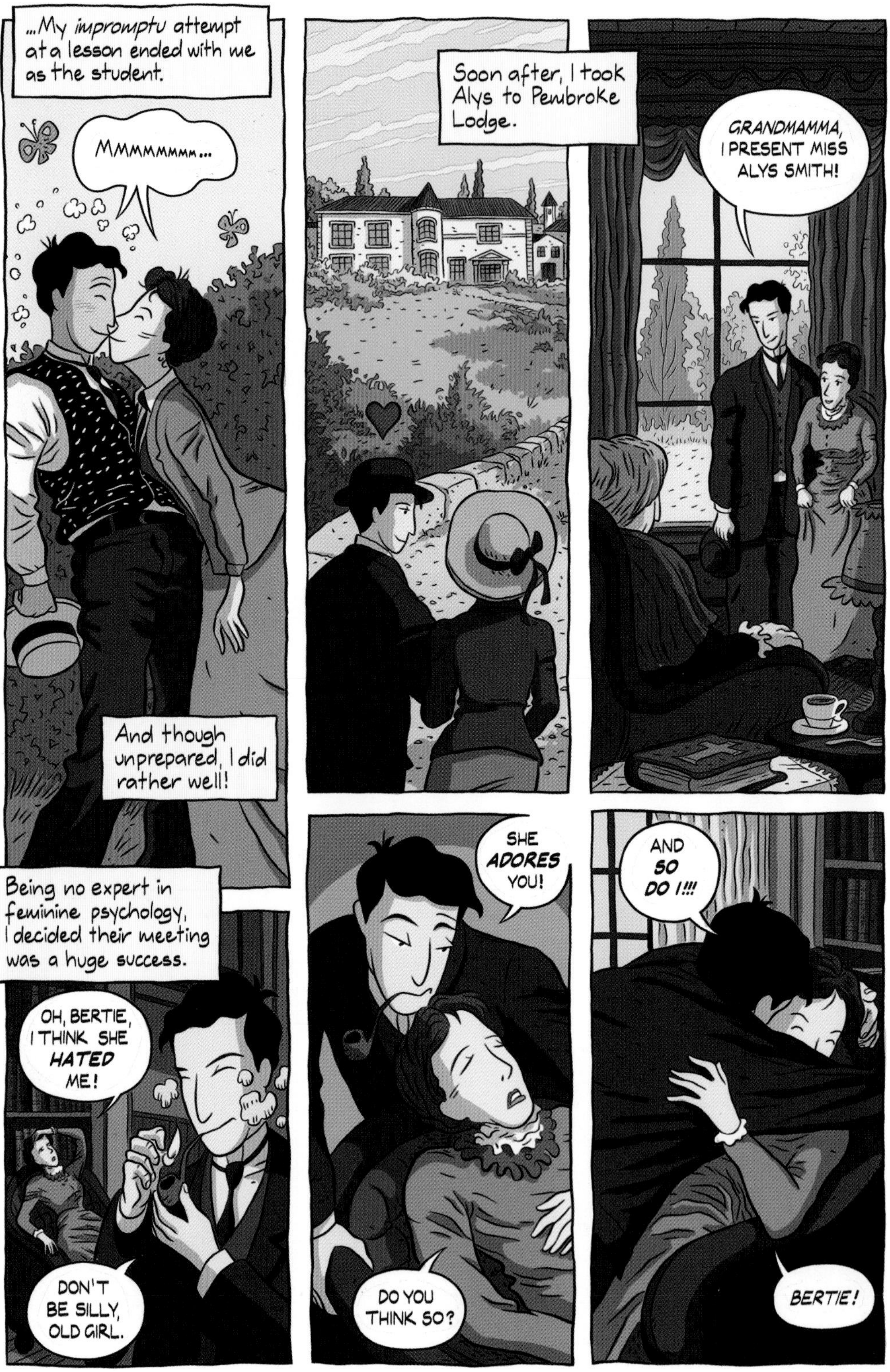
...My *impromptu* attempt at a lesson ended with me as the student.
MMMMMMMM...
And though unprepared, I did rather well!
Soon after, I took Alys to Pembroke Lodge.
GRANDMAMMA, I PRESENT MISS ALYS SMITH!
Being no expert in feminine psychology, I decided their meeting was a huge success.
OH, BERTIE, I THINK SHE **HATED** ME!
DON'T BE SILLY, OLD GIRL.
SHE **ADORES** YOU!
DO YOU THINK SO?
AND **SO DO I!!!**
BERTIE!

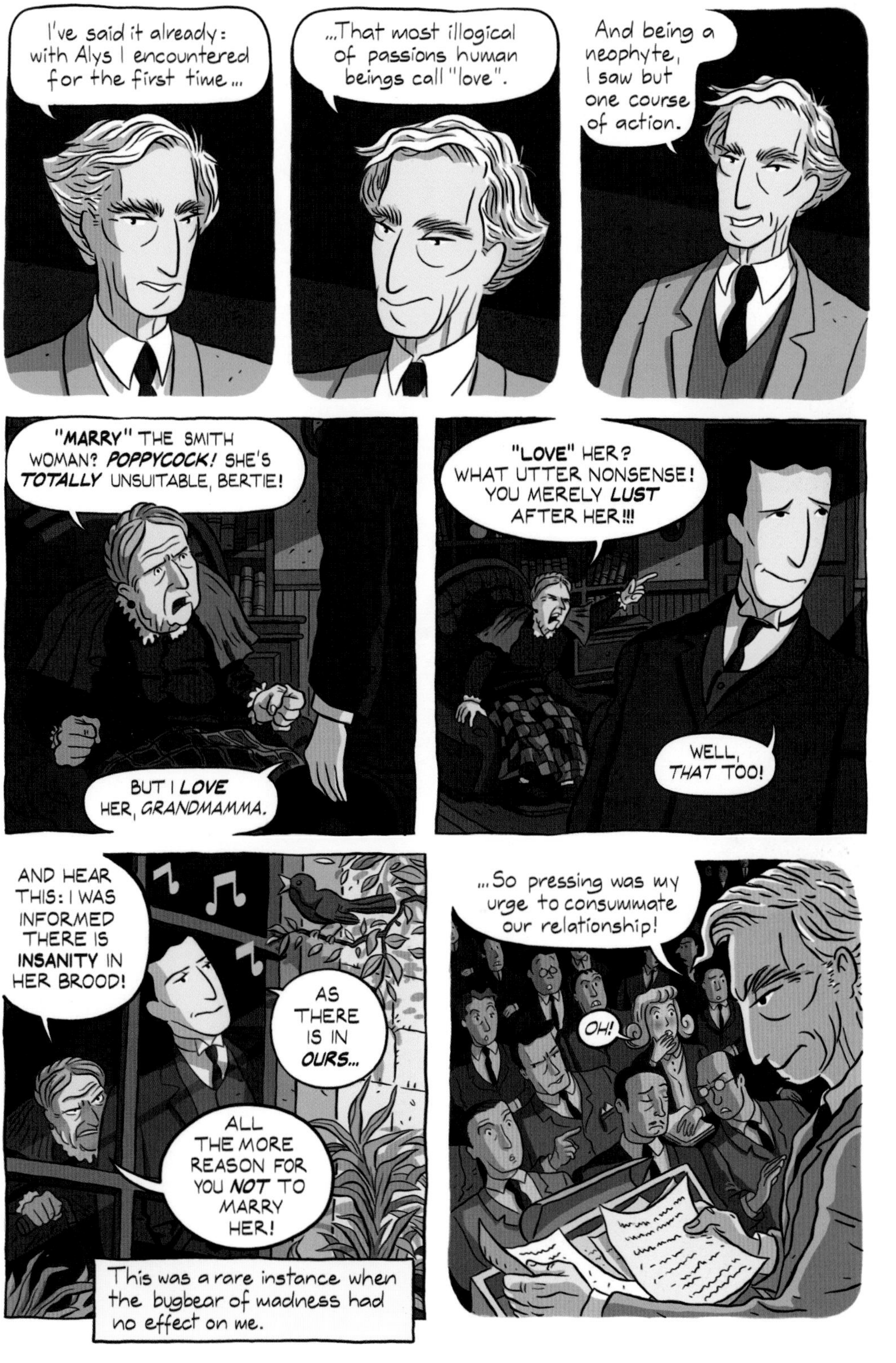
I've said it already: with Alys I encountered for the first time...
...That most illogical of passions human beings call "love".
And being a neophyte, I saw but one course of action.
"MARRY" THE SMITH WOMAN? POPPYCOCK! SHE'S TOTALLY UNSUITABLE, BERTIE!
BUT I LOVE HER, GRANDMAMMA.
"LOVE" HER? WHAT UTTER NONSENSE! YOU MERELY LUST AFTER HER!!!
WELL, THAT TOO!
AND HEAR THIS: I WAS INFORMED THERE IS INSANITY IN HER BROOD!
AS THERE IS IN OURS...
ALL THE MORE REASON FOR YOU NOT TO MARRY HER!
This was a rare instance when the bugbear of madness had no effect on me.
...So pressing was my urge to consummate our relationship!
OH!

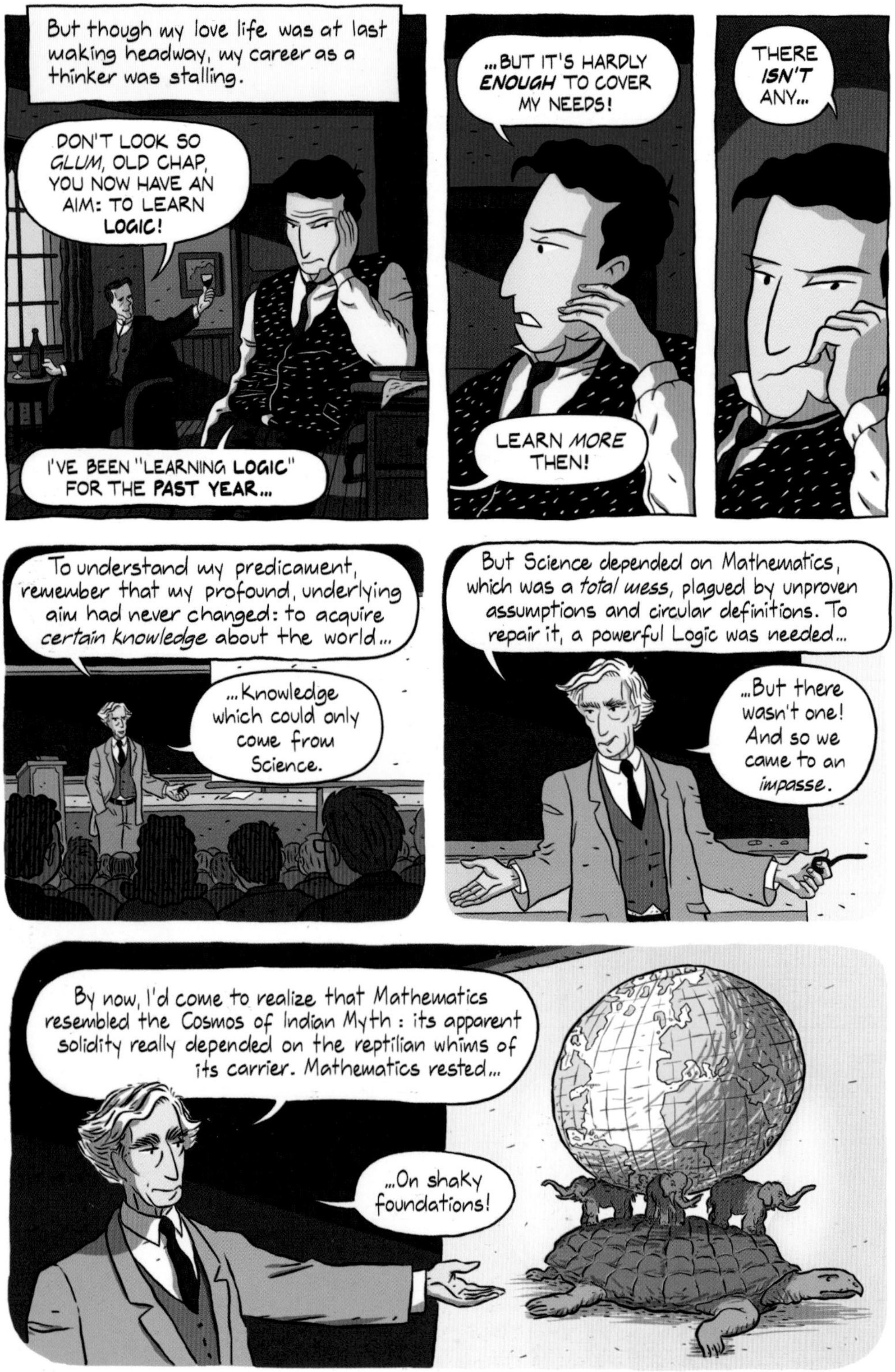
But though my love life was at last making headway, my career as a thinker was stalling.
DON'T LOOK SO GLUM, OLD CHAP, YOU NOW HAVE AN AIM: TO LEARN LOGIC!
I'VE BEEN "LEARNING LOGIC" FOR THE PAST YEAR...
...BUT IT'S HARDLY ENOUGH TO COVER MY NEEDS!
LEARN MORE THEN!
THERE ISN'T ANY...
To understand my predicament, remember that my profound, underlying aim had never changed: to acquire certain knowledge about the world...
...Knowledge which could only come from Science.
But Science depended on Mathematics, which was a total mess, plagued by unproven assumptions and circular definitions. To repair it, a powerful Logic was needed...
...But there wasn't one! And so we came to an impasse.
By now, I'd come to realize that Mathematics resembled the Cosmos of Indian Myth : its apparent solidity really depended on the reptilian whims of its carrier. Mathematics rested...
...On shaky foundations!

The sorry state of the "Queen of the Sciences" was made even worse by the successes of Physics.

THE WORK OF **THOMSON** AND **RUTHERFORD** IS *TRULY* REVOLUTIONARY!

WE ARE APPROACHING THE **DEMOCRITEAN VISION,** THE DISCOVERY OF THE *ATOMS* OF MATTER!

...BUT OUR POOR **MATHEMATICS** LAGS LAMELY BEHIND!

AND WHAT'S WORSE IS THAT MATHEMATICIANS *DON'T* FACE UP TO THE PROBLEM!

With Moore, I dreamed of great discoveries.

WE *MUST* MAKE THEM REALIZE THE **TERRIBLE MESS!**

ONLY *THEN* CAN WE BEGIN TO SET THE HOUSE OF **MATHEMATICS** IN **ORDER**!!!

AWFULLY SORRY TO BARGE IN, CHAPS, BUT YOU ARE TALKING *BALDERDASH!* MATHEMATICS *IS* IN PERFECT ORDER!

The situation shocked me: most mathematicians were painfully unaware of the flimsiness of the foundations.

THEY ARE *FOOLS*, MOORE!

DON'T LOSE HEART, OLD MAN!

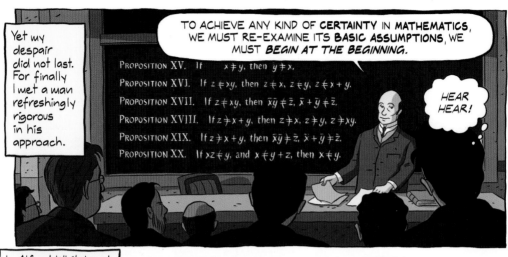

Yet my despair did not last. For finally I met a man refreshingly rigorous in his approach.

TO ACHIEVE ANY KIND OF **CERTAINTY** IN **MATHEMATICS**, WE MUST RE-EXAMINE ITS **BASIC ASSUMPTIONS**, WE MUST *BEGIN AT THE BEGINNING*.

PROPOSITION XV.   If $x \neq y$, then $\bar{y} \neq x$.

PROPOSITION XVI.   If $z \neq xy$, then $z \neq x$, $z \neq y$, $z \neq x + y$.

PROPOSITION XVII.   If $z \neq xy$, then $\bar{x}\bar{y} \neq \bar{z}$, $\bar{x} + \bar{y} \neq \bar{z}$.

PROPOSITION XVIII.   If $z \neq x + y$, then $z \neq x$, $z \neq y$, $z \neq xy$.

PROPOSITION XIX.   If $z \neq x + y$, then $\bar{x}\bar{y} \neq \bar{z}$, $\bar{x} + \bar{y} \neq \bar{z}$.

PROPOSITION XX.   If $xz \neq y$, and $x \neq y + z$, then $x \neq y$.

HEAR HEAR!

In Alfred Whitehead, I found a strong and kindred spirit.

...A mentor.

OH, THERE *ARE* PEOPLE WHO CAN SEE THE SITUATION CLEARLY, RUSSELL. BUT *ALL* OF THEM ARE, ALAS, ON THE **CONTINENT**.

IF WE UNITE THE HEALTHY PARTS OF **MATHEMATICS** AND THE CONCEPTUAL SOPHISTICATION OF THE **NEW LOGIC**, WE CAN LAUNCH A POWERFUL ATTACK.

"NEW"? **LOGIC** HAS NOT ADVANCED ONE *BIT* AFTER **BOOLE!**

SURE, IT HAS COME A CERTAIN WAY SINCE OLD **ARISTOTLE**...

BUT IS IT STRONG ENOUGH YET TO DEAL WITH **MATHEMATICS?**

THERE IS AN OLD GERMAN SAYING: "IF YOU WANT TO *LEARN* SOMETHING, GO ON A *JOURNEY!*"

It was Whitehead, more than anyone else, who helped me see beyond the provincialism of the English mathematical establishment.

114

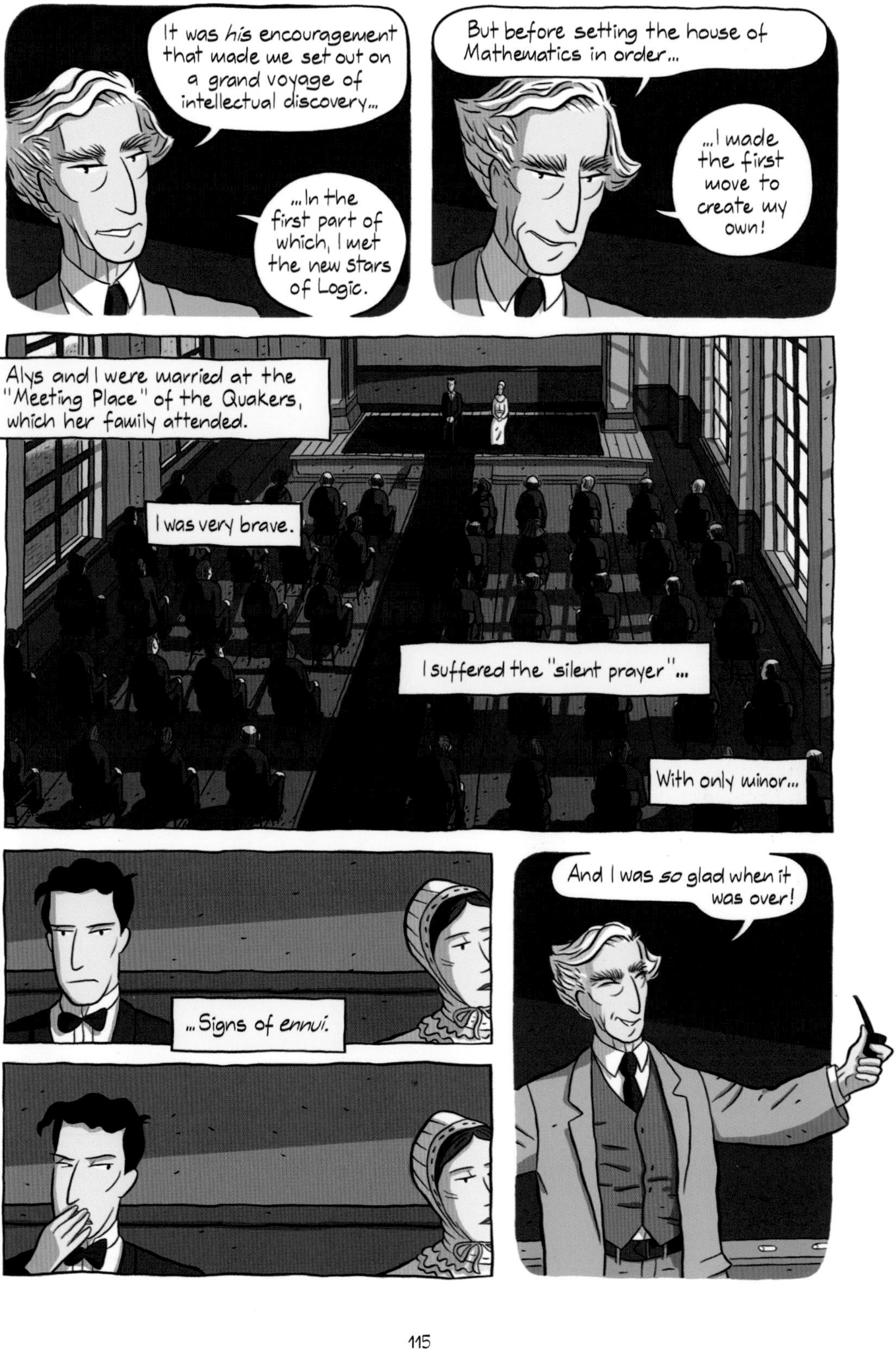
It was *his* encouragement that made me set out on a grand voyage of intellectual discovery...
...In the first part of which, I met the new stars of Logic.
But before setting the house of Mathematics in order...
...I made the first move to create my own!
Alys and I were married at the "Meeting Place" of the Quakers, which her family attended.
I was very brave.
I suffered the "silent prayer"...
With only minor...
...Signs of *ennui*.
And I was *so* glad when it was over!

3. WANDERJAHRE

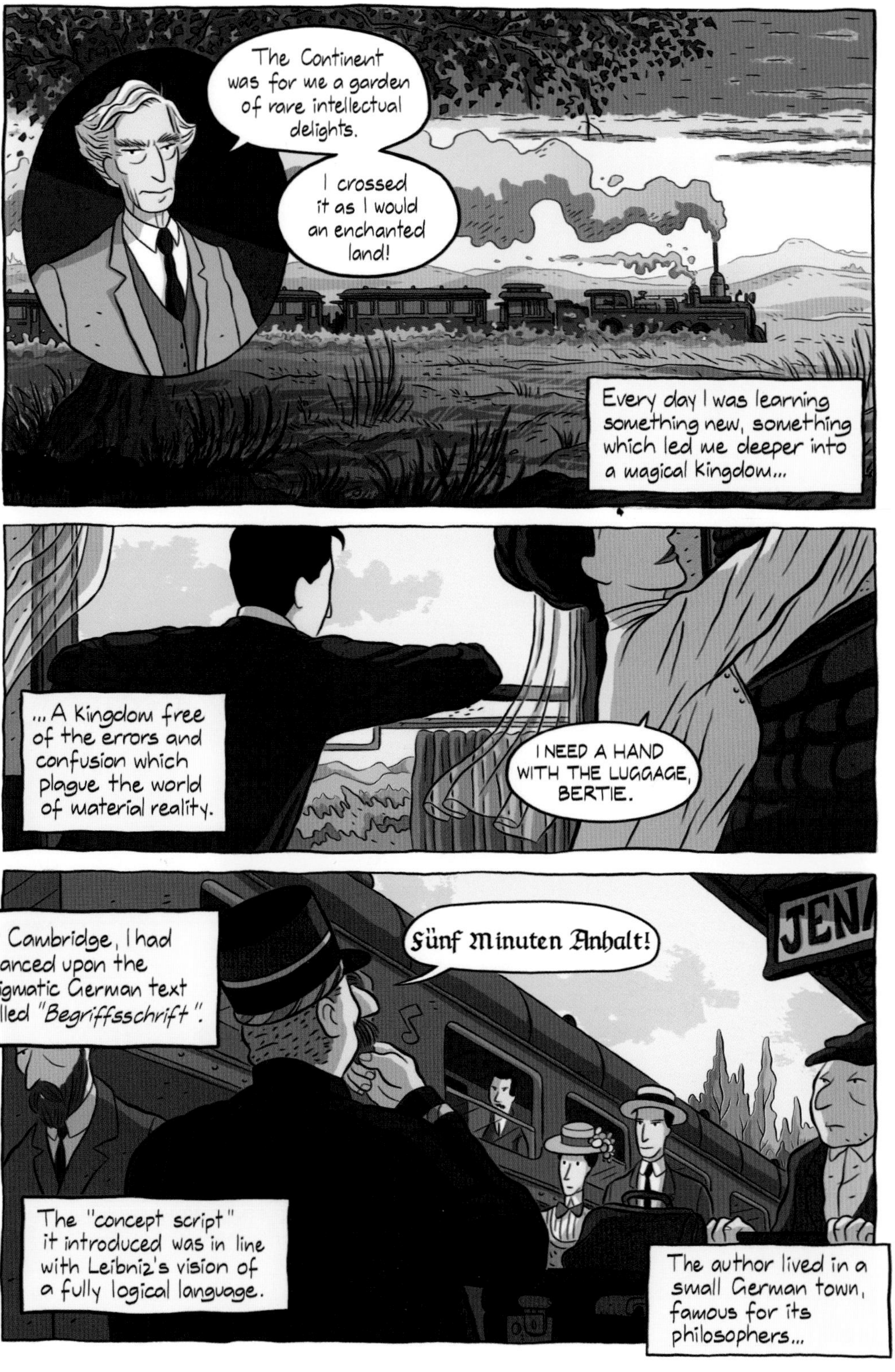
The Continent was for me a garden of rare intellectual delights.
I crossed it as I would an enchanted land!
Every day I was learning something new, something which led me deeper into a magical kingdom...
...A kingdom free of the errors and confusion which plague the world of material reality.
I NEED A HAND WITH THE LUGGAGE, BERTIE.
At Cambridge, I had chanced upon the enigmatic German text called "Begriffsschrift".
Fünf Minuten Anhalt!
JENA
The "concept script" it introduced was in line with Leibniz's vision of a fully logical language.
The author lived in a small German town, famous for its philosophers...

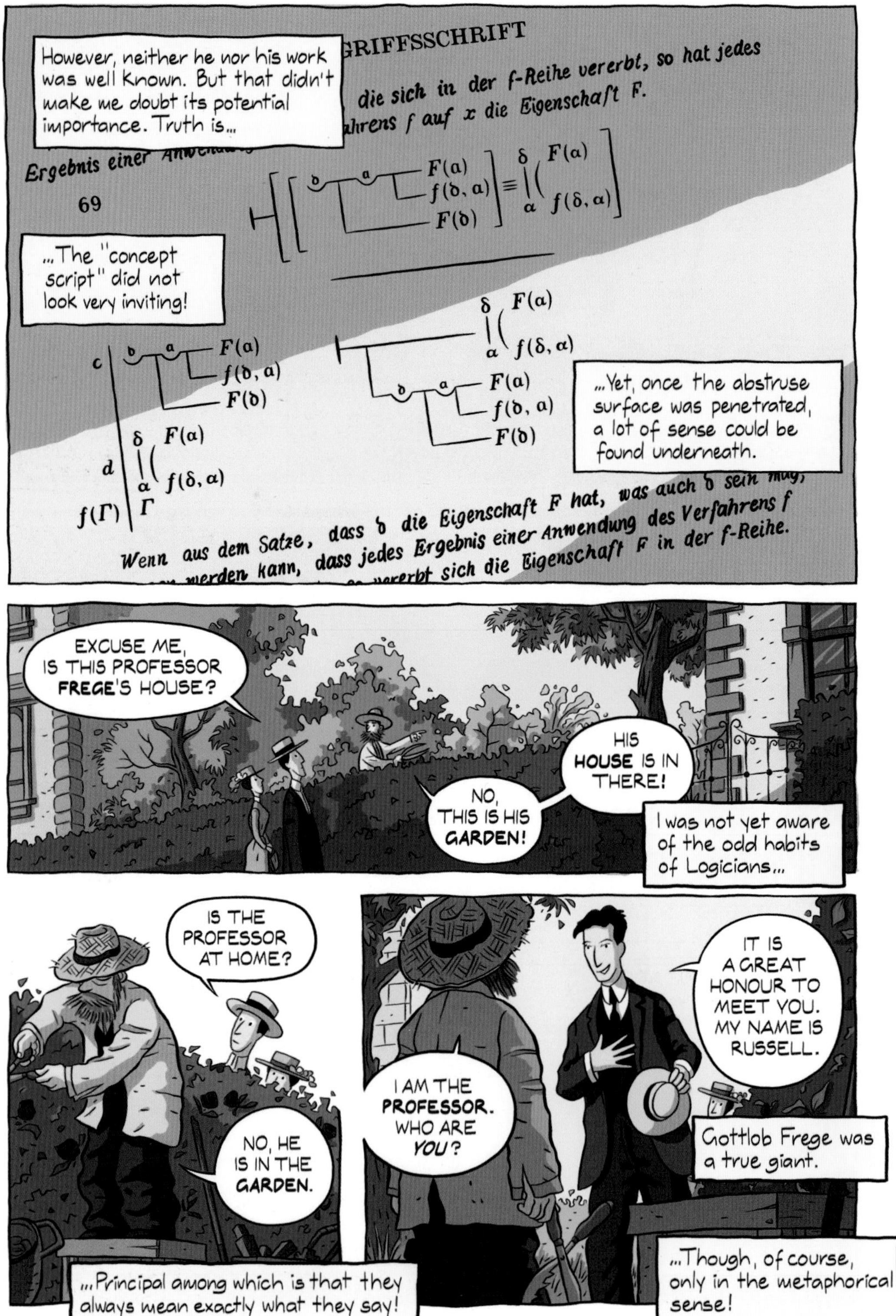
However, neither he nor his work was well known. But that didn't make me doubt its potential importance. Truth is...
GRIFFSSCHRIFT
die sich in der f-Reihe vererbt, so hat jedes
ahrens f auf x die Eigenschaft F.
Ergebnis einer
69
...The "concept script" did not look very inviting!
...Yet, once the abstruse surface was penetrated, a lot of sense could be found underneath.
Wenn aus dem Satze, dass δ die Eigenschaft F hat, was auch δ sein mag,
dass jedes Ergebnis einer Anwendung des Verfahrens f
werden kann, dass jedes
vererbt sich die Eigenschaft F in der f-Reihe.
EXCUSE ME, IS THIS PROFESSOR FREGE'S HOUSE?
NO, THIS IS HIS GARDEN!
HIS HOUSE IS IN THERE!
I was not yet aware of the odd habits of Logicians...
IS THE PROFESSOR AT HOME?
NO, HE IS IN THE GARDEN.
...Principal among which is that they always mean exactly what they say!
I AM THE PROFESSOR. WHO ARE YOU?
IT IS A GREAT HONOUR TO MEET YOU. MY NAME IS RUSSELL.
Gottlob Frege was a true giant.
...Though, of course, only in the metaphorical sense!

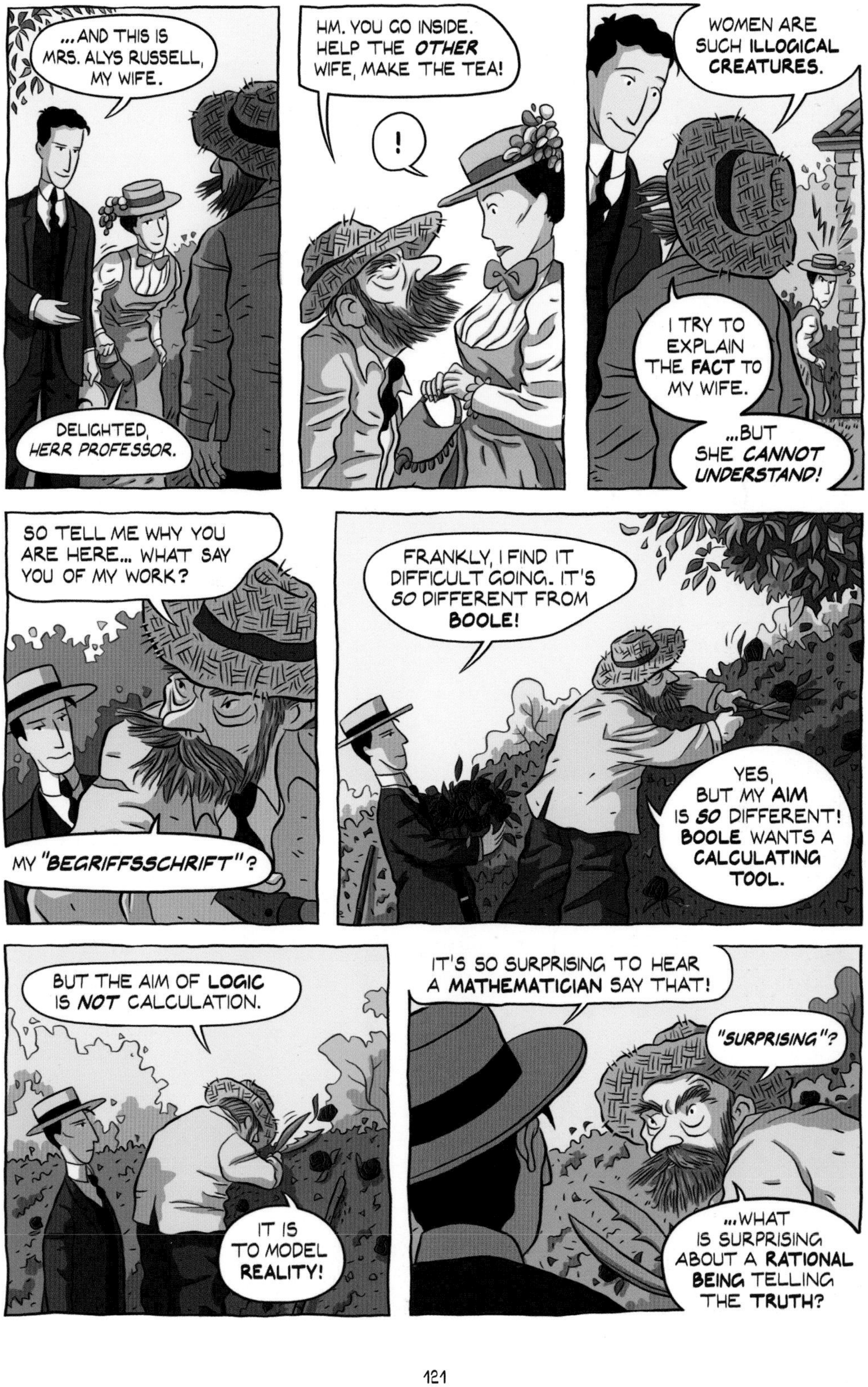
...AND THIS IS MRS. ALYS RUSSELL, MY WIFE.
DELIGHTED, HERR PROFESSOR.
HM. YOU GO INSIDE. HELP THE OTHER WIFE, MAKE THE TEA!
!
WOMEN ARE SUCH ILLOGICAL CREATURES.
I TRY TO EXPLAIN THE FACT TO MY WIFE.
...BUT SHE CANNOT UNDERSTAND!
SO TELL ME WHY YOU ARE HERE... WHAT SAY YOU OF MY WORK?
MY "BEGRIFFSSCHRIFT"?
FRANKLY, I FIND IT DIFFICULT GOING. IT'S SO DIFFERENT FROM BOOLE!
YES, BUT MY AIM IS SO DIFFERENT! BOOLE WANTS A CALCULATING TOOL.
BUT THE AIM OF LOGIC IS NOT CALCULATION.
IT IS TO MODEL REALITY!
IT'S SO SURPRISING TO HEAR A MATHEMATICIAN SAY THAT!
"SURPRISING"?
...WHAT IS SURPRISING ABOUT A RATIONAL BEING TELLING THE TRUTH?

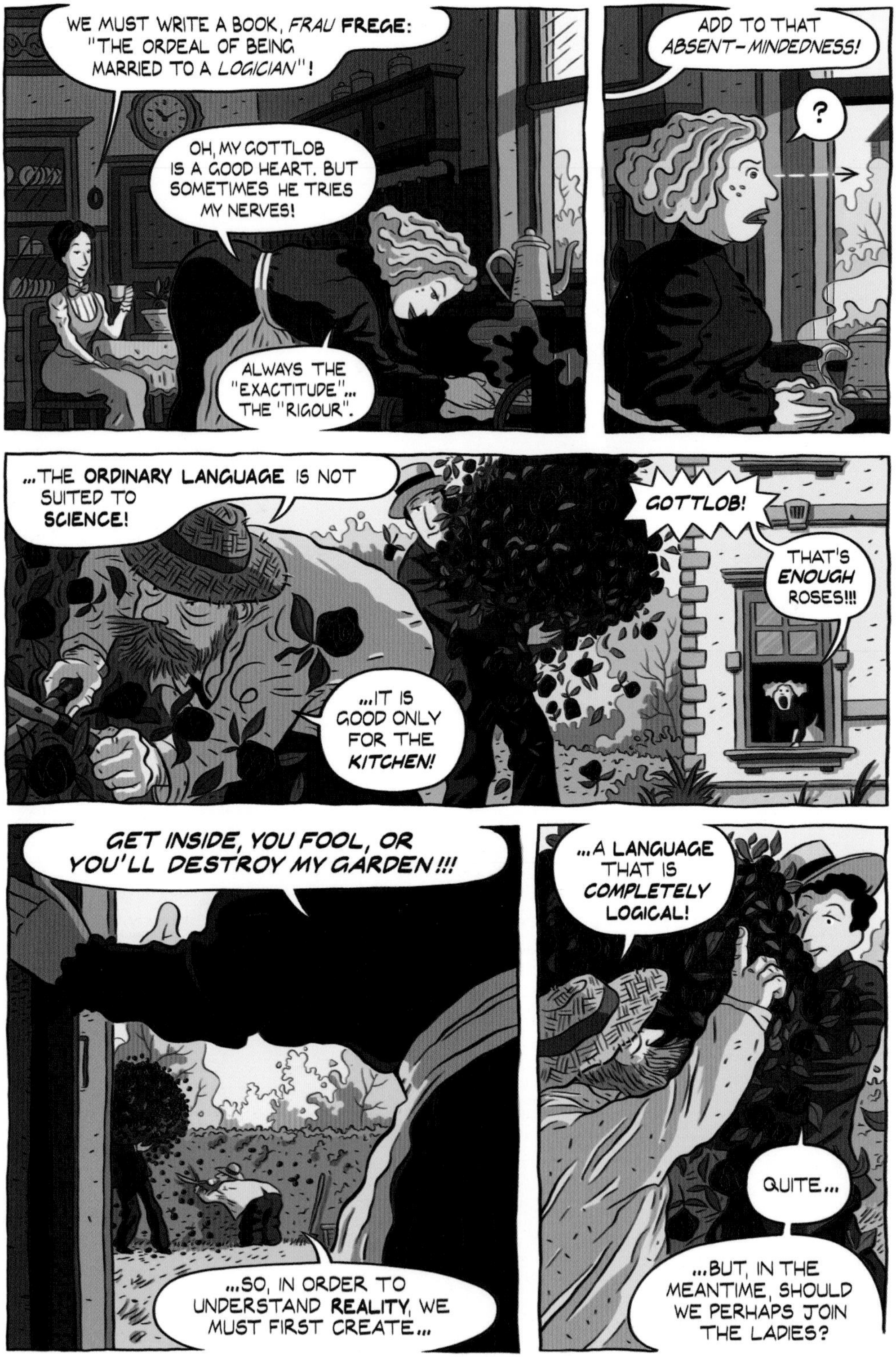
WE MUST WRITE A BOOK, *FRAU* **FREGE**: "THE ORDEAL OF BEING MARRIED TO A *LOGICIAN*"!
OH, MY GOTTLOB IS A GOOD HEART. BUT SOMETIMES HE TRIES MY NERVES!
ALWAYS THE "EXACTITUDE"... THE "RIGOUR".
ADD TO THAT *ABSENT-MINDEDNESS!*
?
...THE **ORDINARY LANGUAGE** IS NOT SUITED TO **SCIENCE!**
...IT IS GOOD ONLY FOR THE ***KITCHEN!***
GOTTLOB!
THAT'S ***ENOUGH*** ROSES!!!
GET INSIDE, YOU FOOL, OR YOU'LL DESTROY MY GARDEN!!!
...SO, IN ORDER TO UNDERSTAND **REALITY**, WE MUST FIRST CREATE...
...A **LANGUAGE** THAT IS ***COMPLETELY*** **LOGICAL!**
QUITE...
...BUT, IN THE MEANTIME, SHOULD WE PERHAPS JOIN THE LADIES?

ONLY SUCH A LANGUAGE CAN DEAL WITH THE **FOUNDATIONS OF MATHEMATICS!**

PRECISELY MY REASON FOR WANTING TO LEARN MORE ABOUT IT!

YOU ARE A **CENTAUR** *HERR* RUSSELL: HALF **MATHEMATICIAN,** AND HALF **PHILOSOPHER!**

YOU ARE LIKE ME IN THIS DICHOTOMY! INDEED, YOU AND I ARE...

...KINDRED SPIRITS!

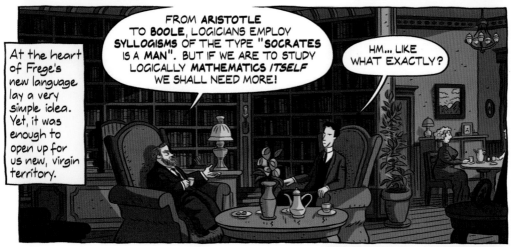

At the heart of Frege's new language lay a very simple idea. Yet, it was enough to open up for us new, virgin territory.

FROM **ARISTOTLE** TO **BOOLE,** LOGICIANS EMPLOY **SYLLOGISMS** OF THE TYPE "**SOCRATES IS A MAN**". BUT IF WE ARE TO STUDY LOGICALLY **MATHEMATICS** *ITSELF* WE SHALL NEED MORE!

HM... LIKE WHAT EXACTLY?

WE NEED TO INTRODUCE **VARIABLES!** WE HAVE TO BE ABLE TO SAY THINGS LIKE "X IS A MAN"...

...WHICH IS TRUE IF, FOR EXAMPLE, X IS EQUAL TO "**RUSSELL**" BUT FALSE IF IT IS ONE OF...

THESE THREE COOK—

ACH! WOMAN!

?

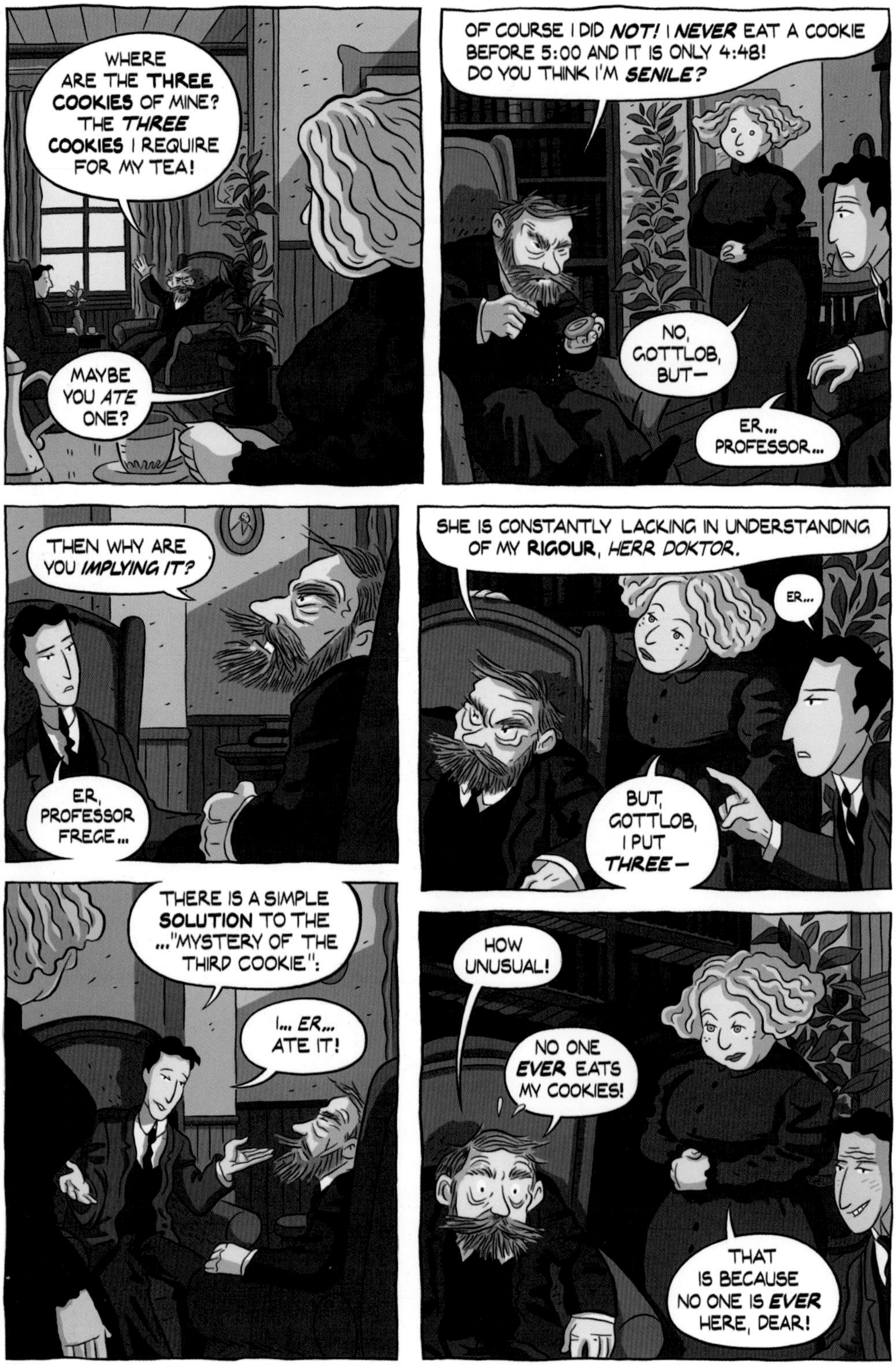
WHERE ARE THE THREE COOKIES OF MINE? THE THREE COOKIES I REQUIRE FOR MY TEA!
MAYBE YOU ATE ONE?
OF COURSE I DID NOT! I NEVER EAT A COOKIE BEFORE 5:00 AND IT IS ONLY 4:48! DO YOU THINK I'M SENILE?
NO, GOTTLOB, BUT—
ER... PROFESSOR...
THEN WHY ARE YOU IMPLYING IT?
ER, PROFESSOR FREGE...
SHE IS CONSTANTLY LACKING IN UNDERSTANDING OF MY RIGOUR, HERR DOKTOR.
ER...
BUT, GOTTLOB, I PUT THREE—
THERE IS A SIMPLE SOLUTION TO THE ..."MYSTERY OF THE THIRD COOKIE":
I... ER... ATE IT!
HOW UNUSUAL!
NO ONE EVER EATS MY COOKIES!
THAT IS BECAUSE NO ONE IS EVER HERE, DEAR!

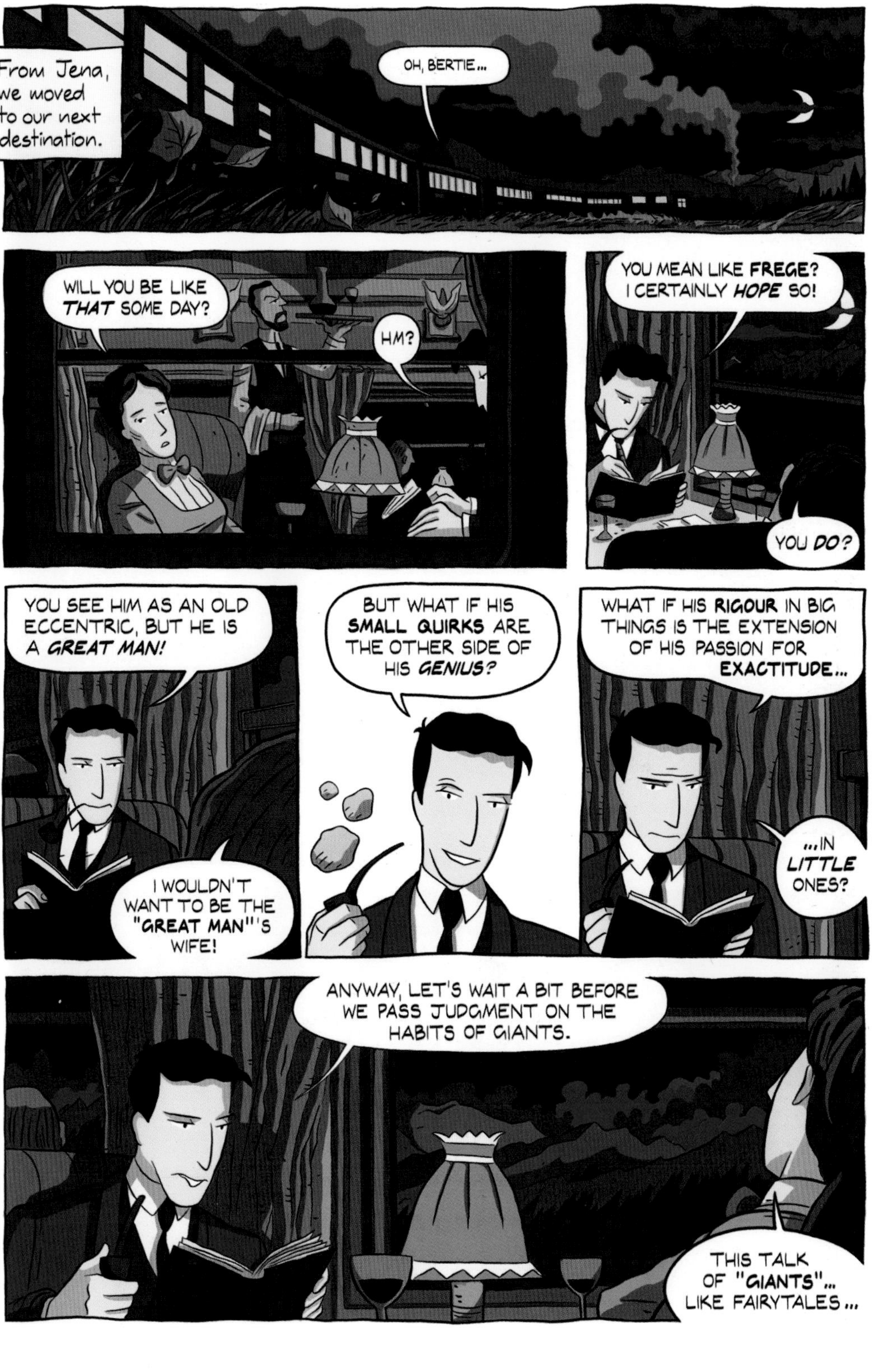
From Jena, we moved to our next destination.
OH, BERTIE...
WILL YOU BE LIKE THAT SOME DAY?
HM?
YOU MEAN LIKE FREGE? I CERTAINLY HOPE SO!
YOU DO?
YOU SEE HIM AS AN OLD ECCENTRIC, BUT HE IS A GREAT MAN!
I WOULDN'T WANT TO BE THE "GREAT MAN"'S WIFE!
BUT WHAT IF HIS SMALL QUIRKS ARE THE OTHER SIDE OF HIS GENIUS?
WHAT IF HIS RIGOUR IN BIG THINGS IS THE EXTENSION OF HIS PASSION FOR EXACTITUDE...
...IN LITTLE ONES?
ANYWAY, LET'S WAIT A BIT BEFORE WE PASS JUDGMENT ON THE HABITS OF GIANTS.
THIS TALK OF "GIANTS"... LIKE FAIRYTALES...

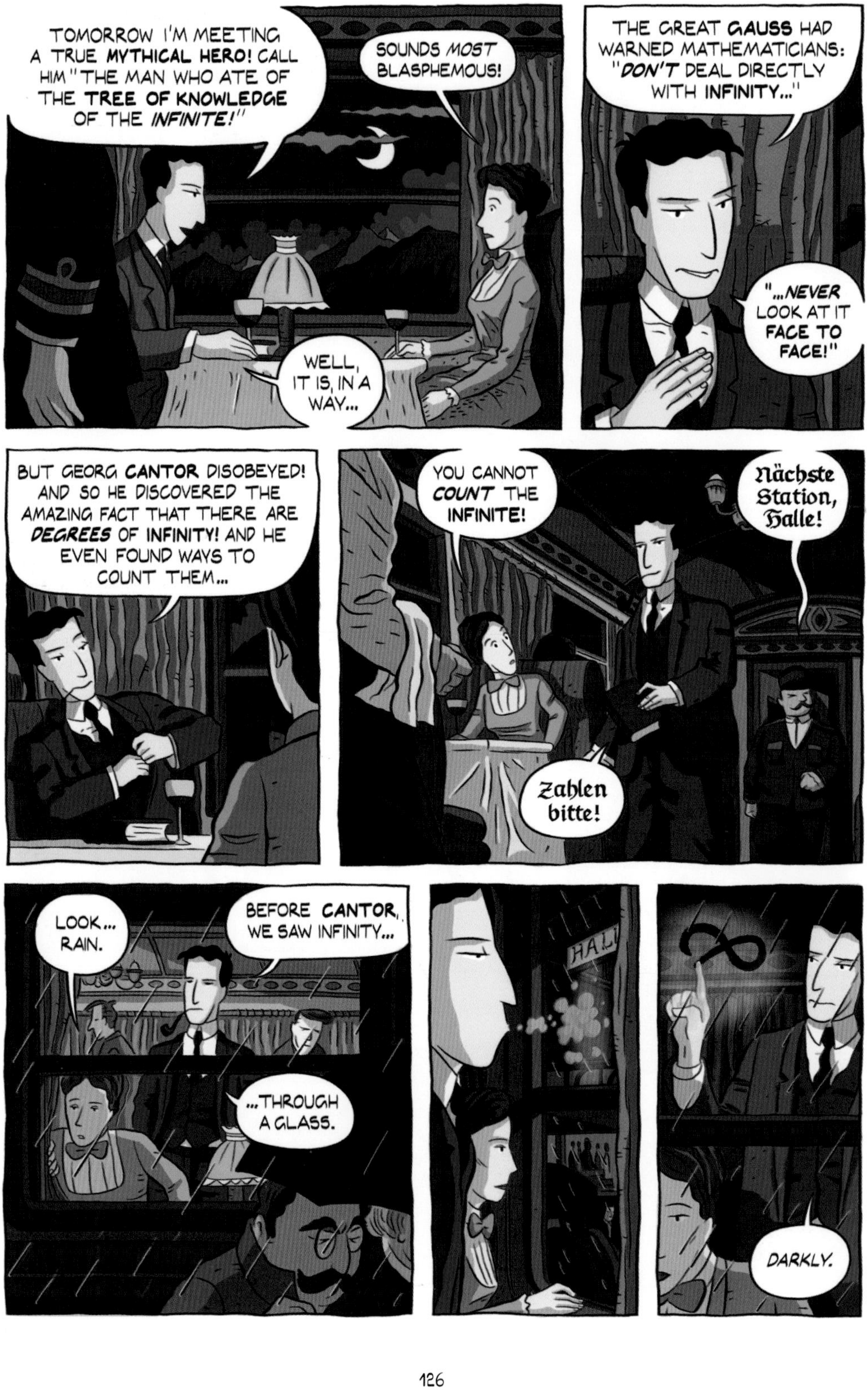
TOMORROW I'M MEETING A TRUE **MYTHICAL HERO!** CALL HIM "THE MAN WHO ATE OF THE **TREE OF KNOWLEDGE** OF THE ***INFINITE!***"
SOUNDS *MOST* BLASPHEMOUS!
WELL, IT IS, IN A WAY...
THE GREAT **GAUSS** HAD WARNED MATHEMATICIANS: "***DON'T*** DEAL DIRECTLY WITH **INFINITY...**"
"...***NEVER*** LOOK AT IT **FACE TO FACE!**"
BUT GEORG **CANTOR** DISOBEYED! AND SO HE DISCOVERED THE AMAZING FACT THAT THERE ARE ***DEGREES*** OF **INFINITY!** AND HE EVEN FOUND WAYS TO COUNT THEM...
YOU CANNOT ***COUNT*** THE **INFINITE!**
Nächste Station, Halle!
Zahlen bitte!
LOOK... RAIN.
BEFORE **CANTOR**, WE SAW INFINITY...
...THROUGH A GLASS.
HALL
DARKLY.

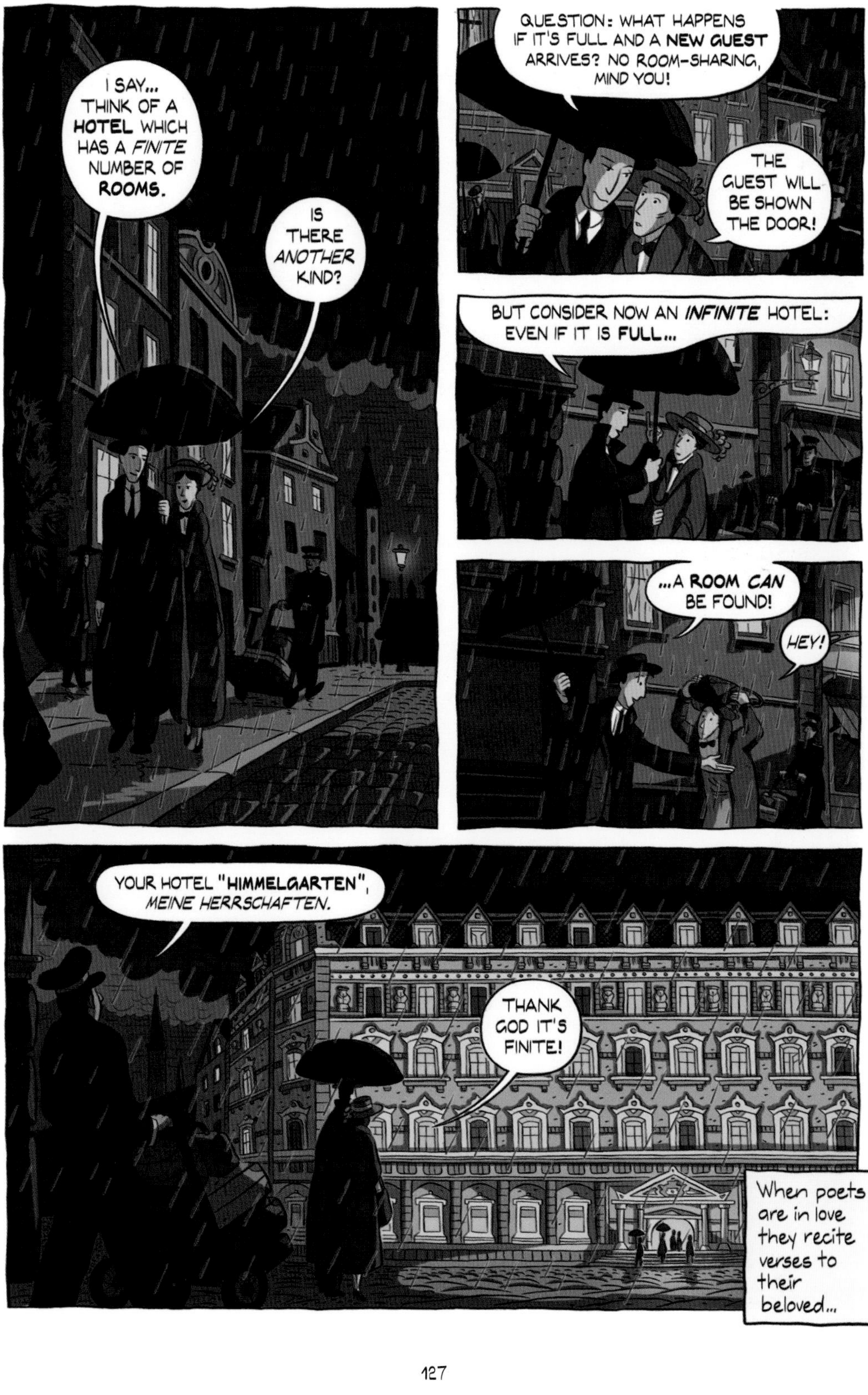
I SAY... THINK OF A **HOTEL** WHICH HAS A *FINITE* NUMBER OF **ROOMS**.
IS THERE *ANOTHER* KIND?
QUESTION: WHAT HAPPENS IF IT'S FULL AND A **NEW GUEST** ARRIVES? NO ROOM-SHARING, MIND YOU!
THE GUEST WILL BE SHOWN THE DOOR!
BUT CONSIDER NOW AN *INFINITE* HOTEL: EVEN IF IT IS **FULL**...
...A **ROOM** ***CAN*** BE FOUND!
HEY!
YOUR HOTEL "**HIMMELGARTEN**", *MEINE HERRSCHAFTEN.*
THANK GOD IT'S FINITE!
When poets are in love they recite verses to their beloved...

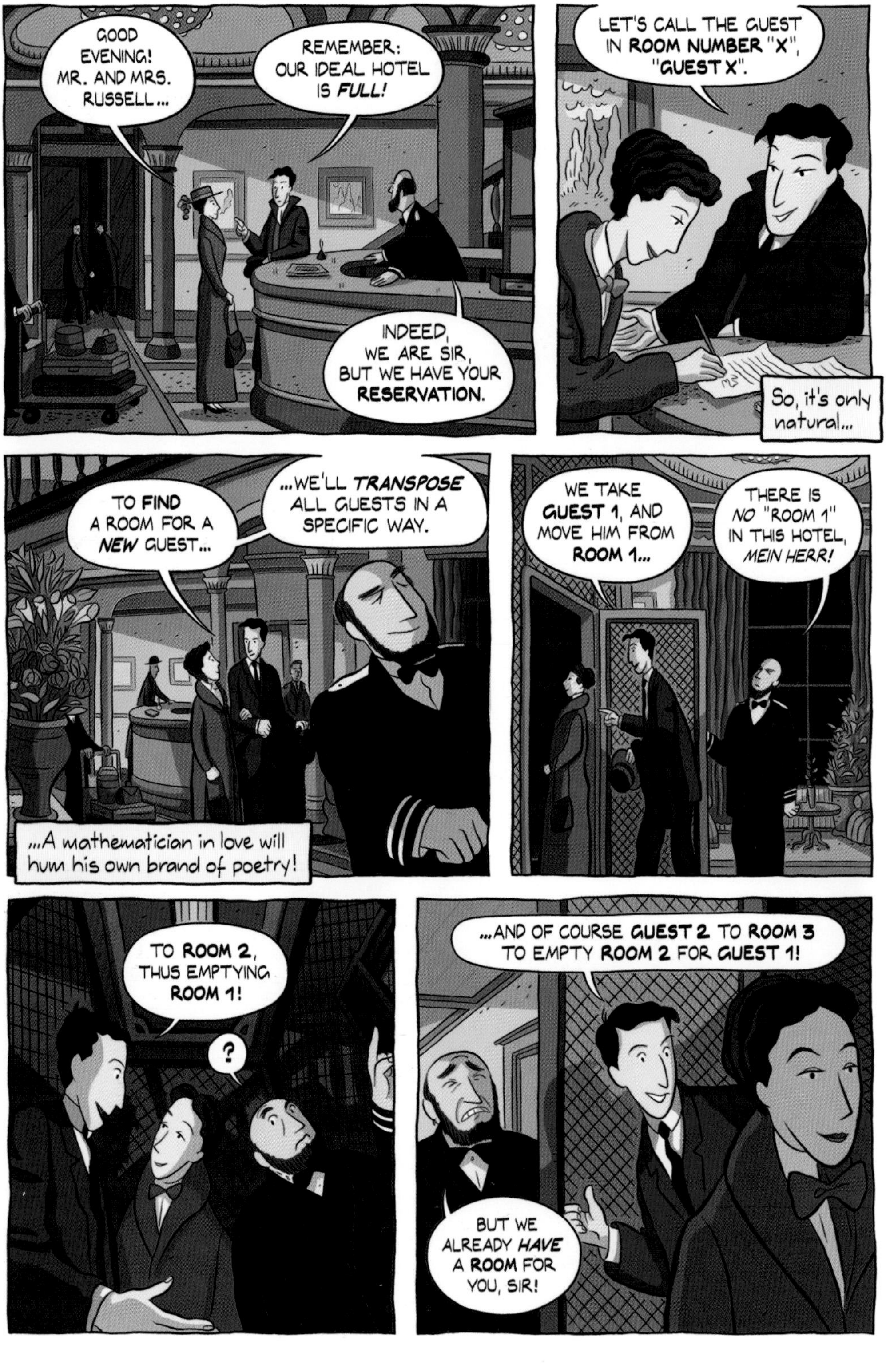
GOOD EVENING! MR. AND MRS. RUSSELL...
REMEMBER: OUR IDEAL HOTEL IS FULL!
INDEED, WE ARE SIR, BUT WE HAVE YOUR RESERVATION.
LET'S CALL THE GUEST IN ROOM NUMBER "X", "GUEST X".
So, it's only natural...
TO FIND A ROOM FOR A NEW GUEST...
...WE'LL TRANSPOSE ALL GUESTS IN A SPECIFIC WAY.
...A mathematician in love will hum his own brand of poetry!
WE TAKE GUEST 1, AND MOVE HIM FROM ROOM 1...
THERE IS NO "ROOM 1" IN THIS HOTEL, MEIN HERR!
TO ROOM 2, THUS EMPTYING ROOM 1!
?
...AND OF COURSE GUEST 2 TO ROOM 3 TO EMPTY ROOM 2 FOR GUEST 1!
BUT WE ALREADY HAVE A ROOM FOR YOU, SIR!

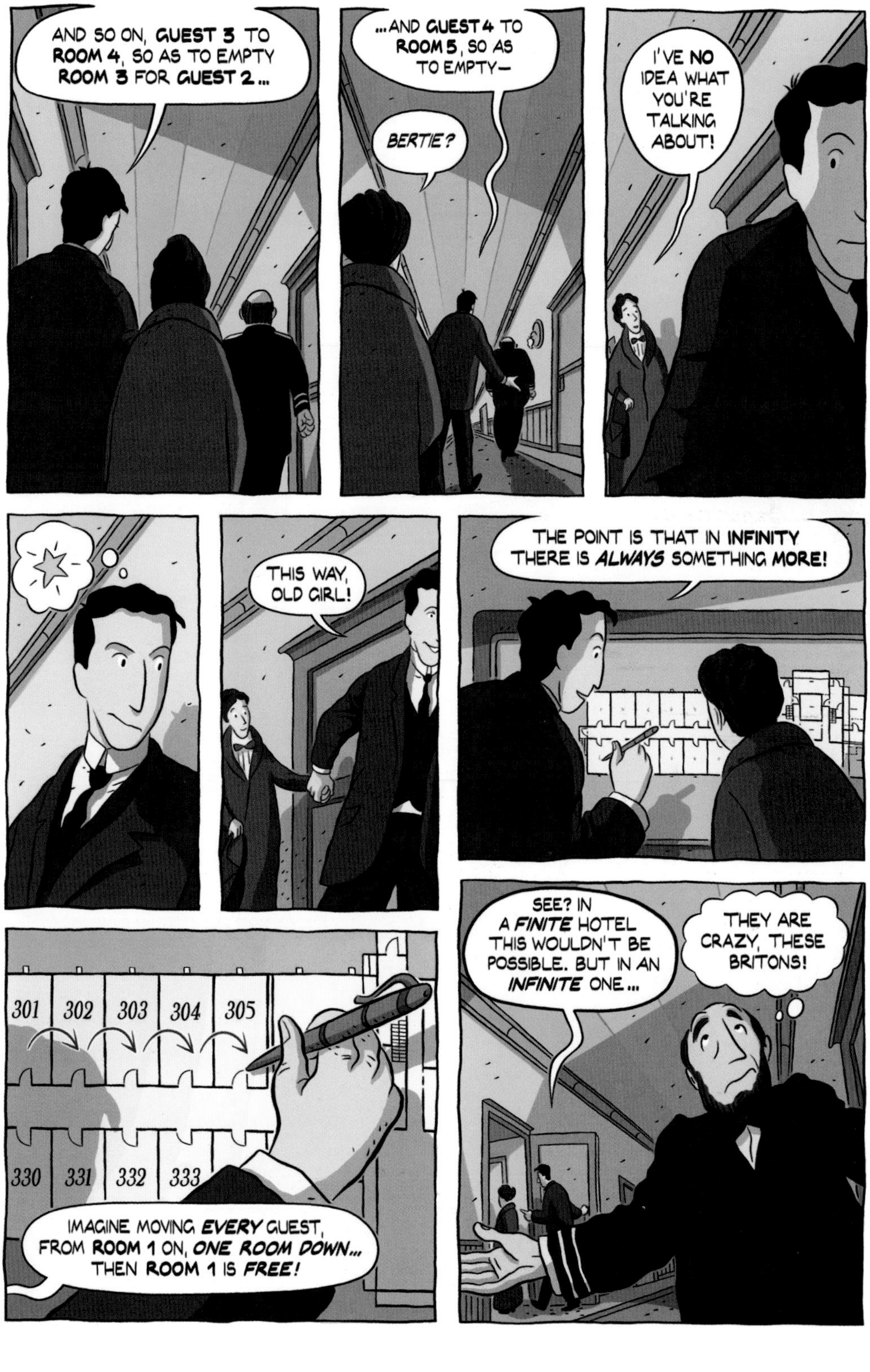
AND SO ON, GUEST 3 TO ROOM 4, SO AS TO EMPTY ROOM 3 FOR GUEST 2...
...AND GUEST 4 TO ROOM 5, SO AS TO EMPTY—
BERTIE?
I'VE NO IDEA WHAT YOU'RE TALKING ABOUT!
THIS WAY, OLD GIRL!
THE POINT IS THAT IN INFINITY THERE IS ALWAYS SOMETHING MORE!
301
302
303
304
305
330
331
332
333
IMAGINE MOVING EVERY GUEST, FROM ROOM 1 ON, ONE ROOM DOWN... THEN ROOM 1 IS FREE!
SEE? IN A FINITE HOTEL THIS WOULDN'T BE POSSIBLE. BUT IN AN INFINITE ONE...
THEY ARE CRAZY, THESE BRITONS!

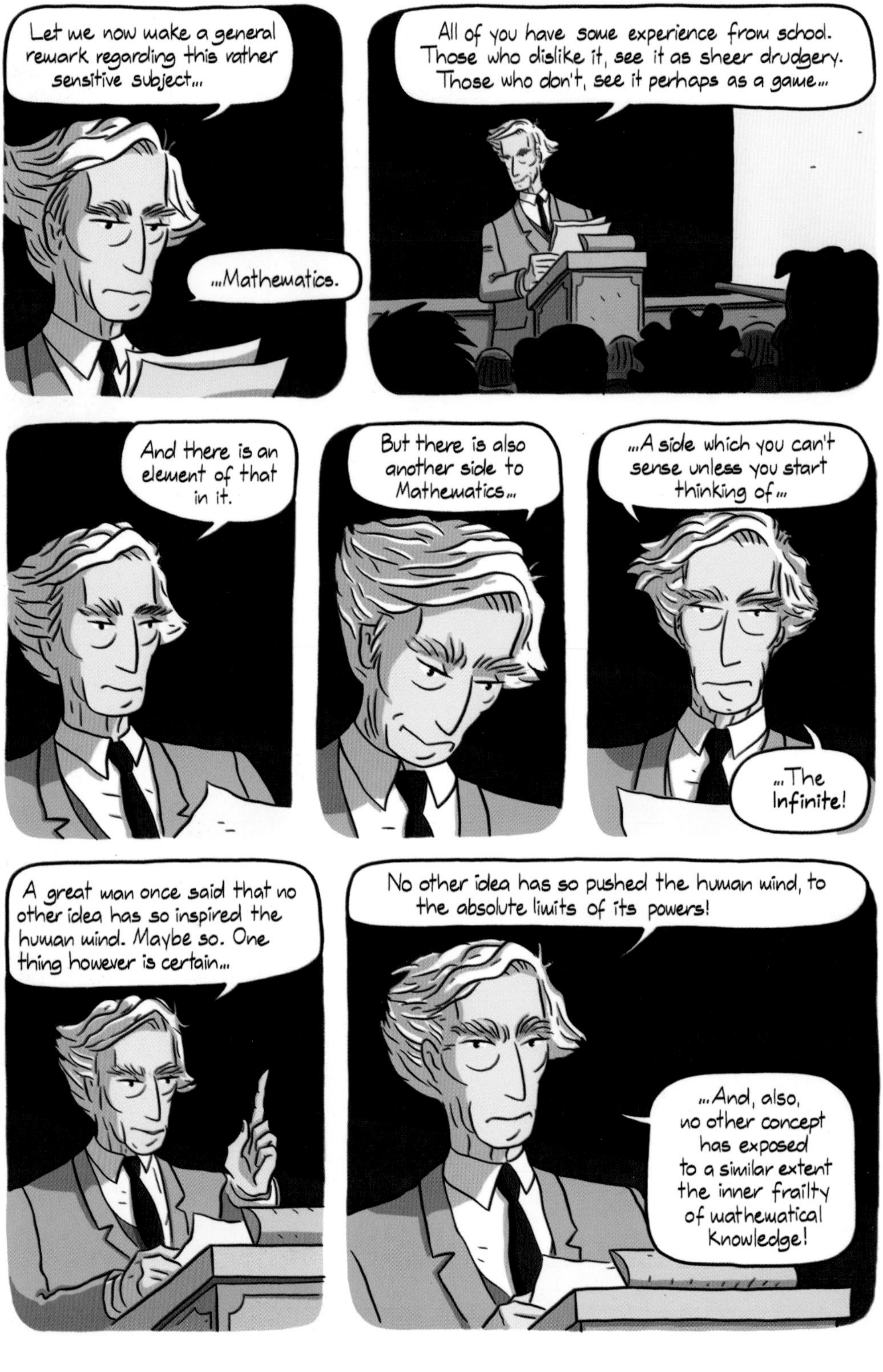

Let me now make a general remark regarding this rather sensitive subject...
...Mathematics.
All of you have some experience from school. Those who dislike it, see it as sheer drudgery. Those who don't, see it perhaps as a game...
And there is an element of that in it.
But there is also another side to Mathematics...
...A side which you can't sense unless you start thinking of...
...The Infinite!
A great man once said that no other idea has so inspired the human mind. Maybe so. One thing however is certain...
No other idea has so pushed the human mind, to the absolute limits of its powers!
...And, also, no other concept has exposed to a similar extent the inner frailty of mathematical knowledge!

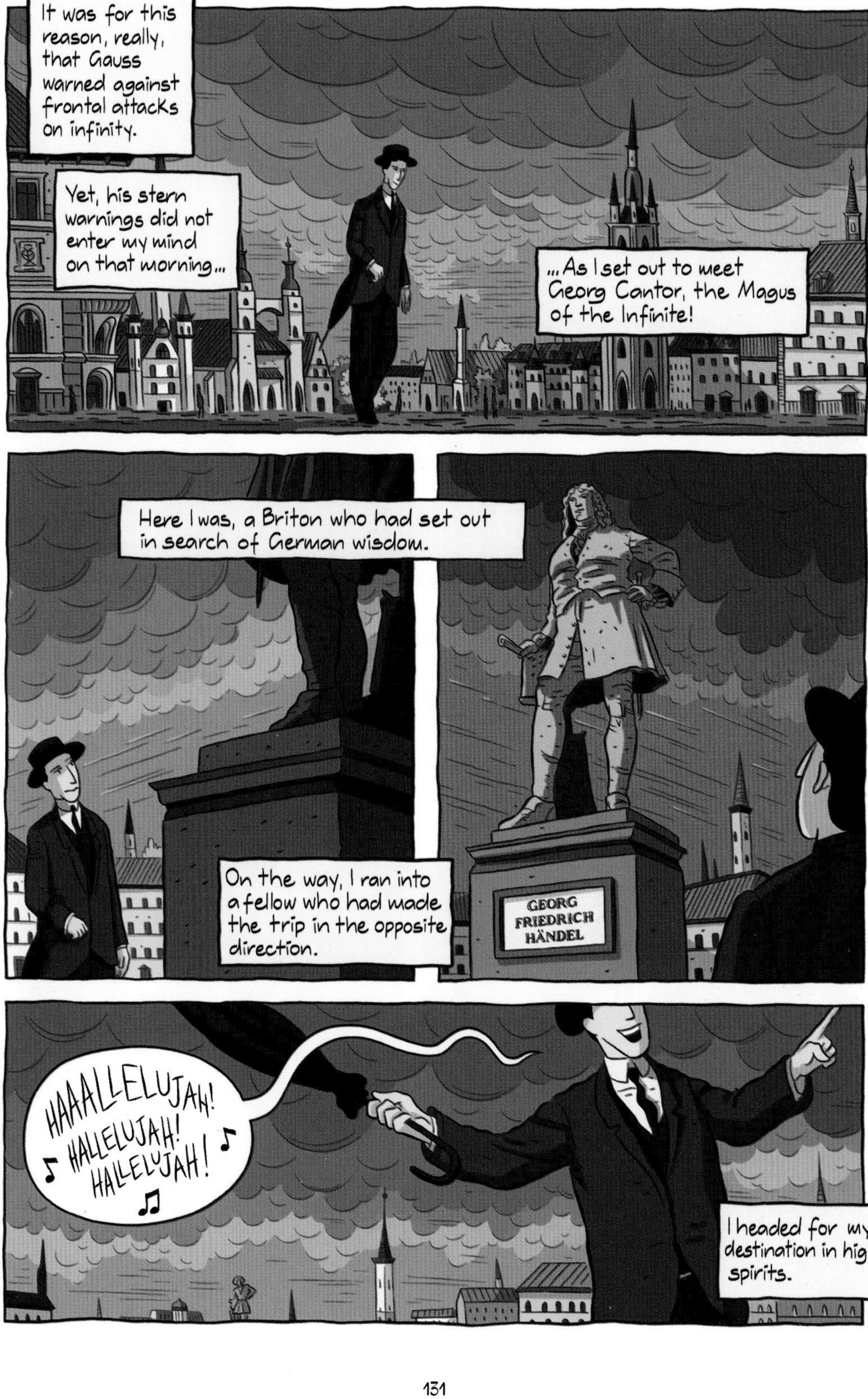
It was for this reason, really, that Gauss warned against frontal attacks on infinity.
Yet, his stern warnings did not enter my mind on that morning...
...As I set out to meet Georg Cantor, the Magus of the Infinite!
Here I was, a Briton who had set out in search of German wisdom.
On the way, I ran into a fellow who had made the trip in the opposite direction.
GEORG FRIEDRICH HÄNDEL
HAAALLELUJAH! HALLELUJAH! HALLELUJAH!
I headed for my destination in high spirits.

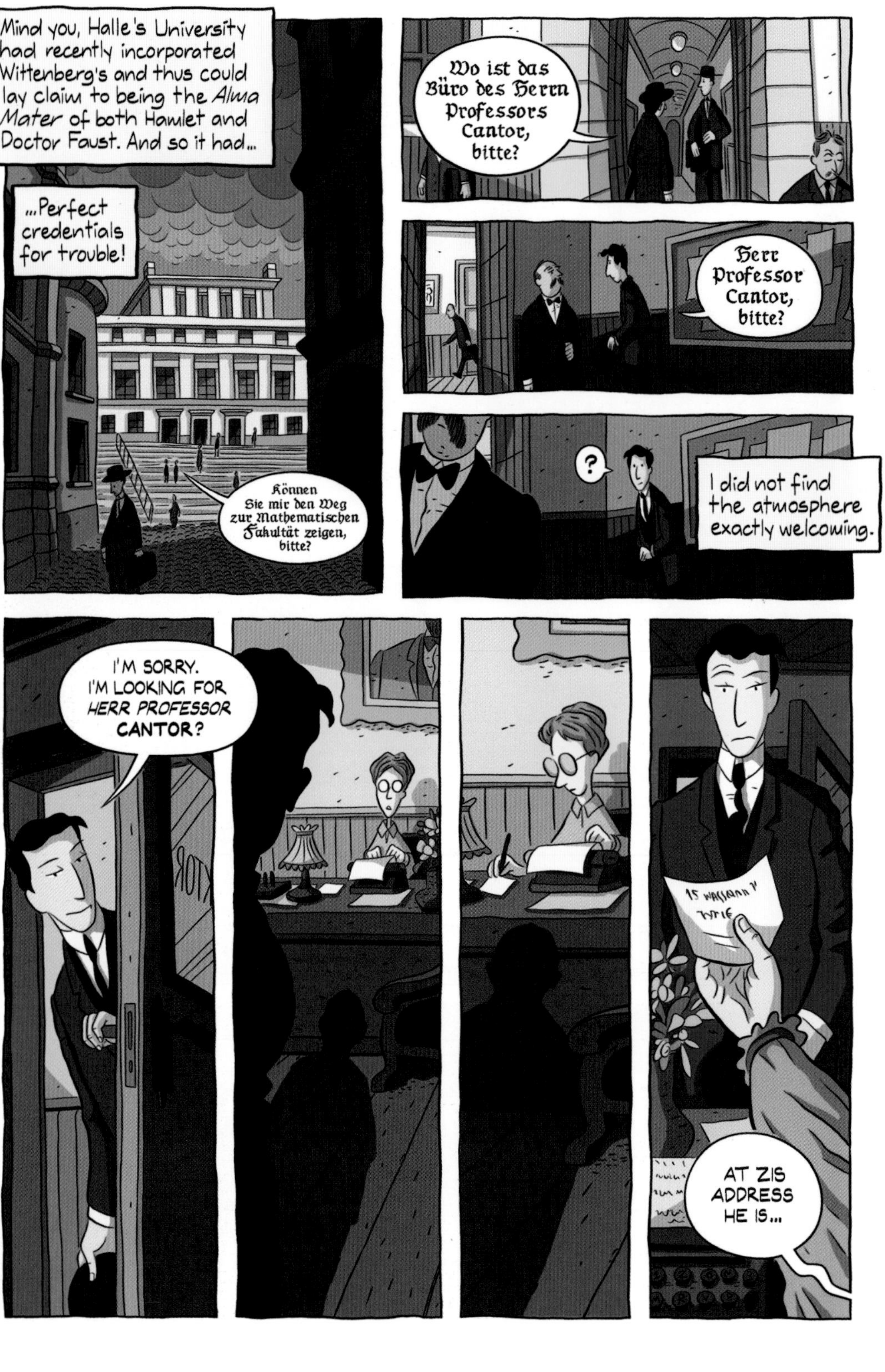
Mind you, Halle's University had recently incorporated Wittenberg's and thus could lay claim to being the *Alma Mater* of both Hamlet and Doctor Faust. And so it had...
...Perfect credentials for trouble!
Können Sie mir den Weg zur Mathematischen Fakultät zeigen, bitte?
Wo ist das Büro des Herrn Professors Cantor, bitte?
Herr Professor Cantor, bitte?
?
I did not find the atmosphere exactly welcoming.
I'M SORRY. I'M LOOKING FOR *HERR PROFESSOR* **CANTOR**?
AT ZIS ADDRESS HE IS...

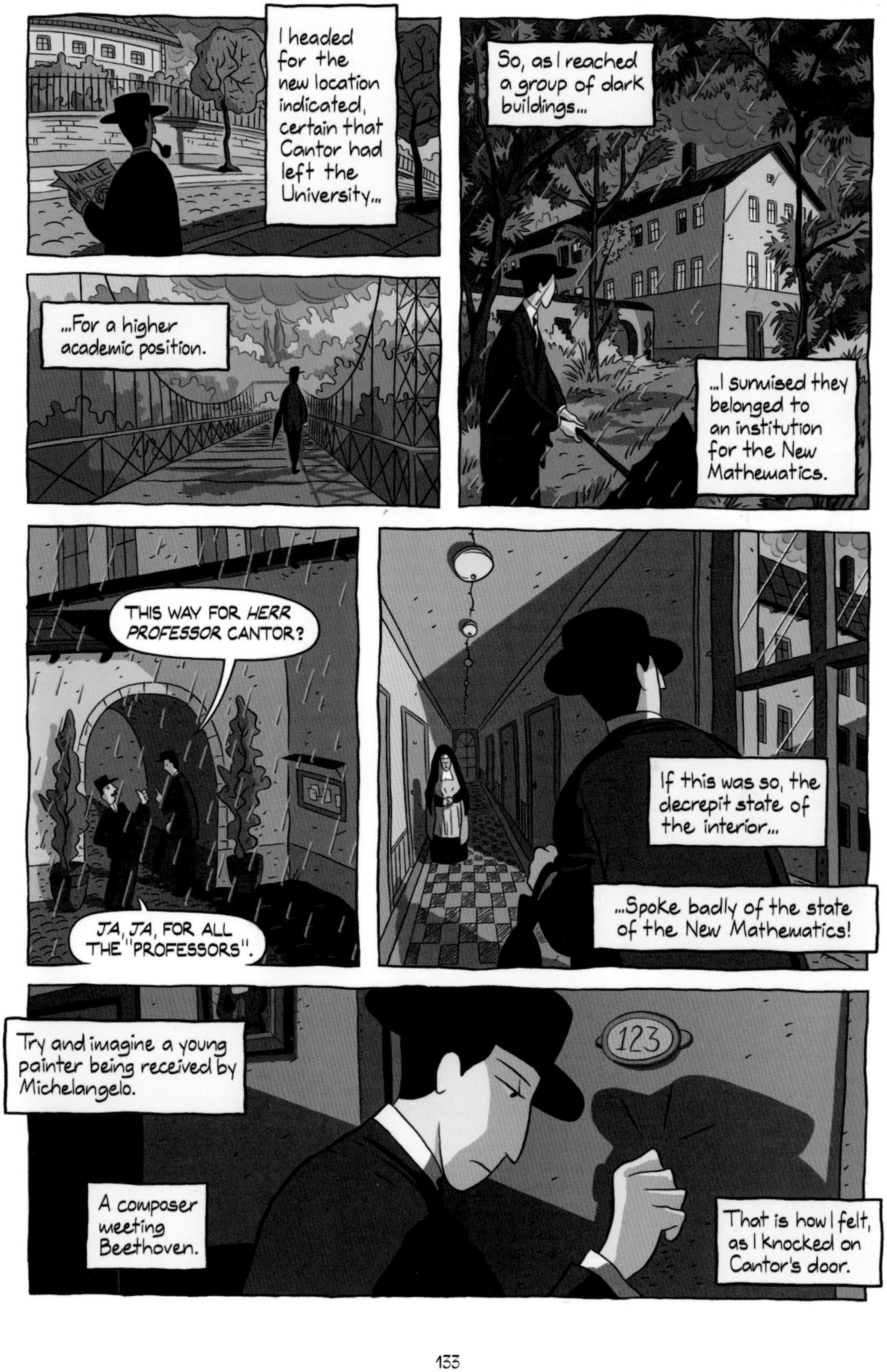

I headed for the new location indicated, certain that Cantor had left the University...
HALLE
...For a higher academic position.
So, as I reached a group of dark buildings...
...I surmised they belonged to an institution for the New Mathematics.
THIS WAY FOR *HERR PROFESSOR* CANTOR?
JA, JA, FOR ALL THE "PROFESSORS".
If this was so, the decrepit state of the interior...
...Spoke badly of the state of the New Mathematics!
Try and imagine a young painter being received by Michelangelo.
123
A composer meeting Beethoven.
That is how I felt, as I knocked on Cantor's door.

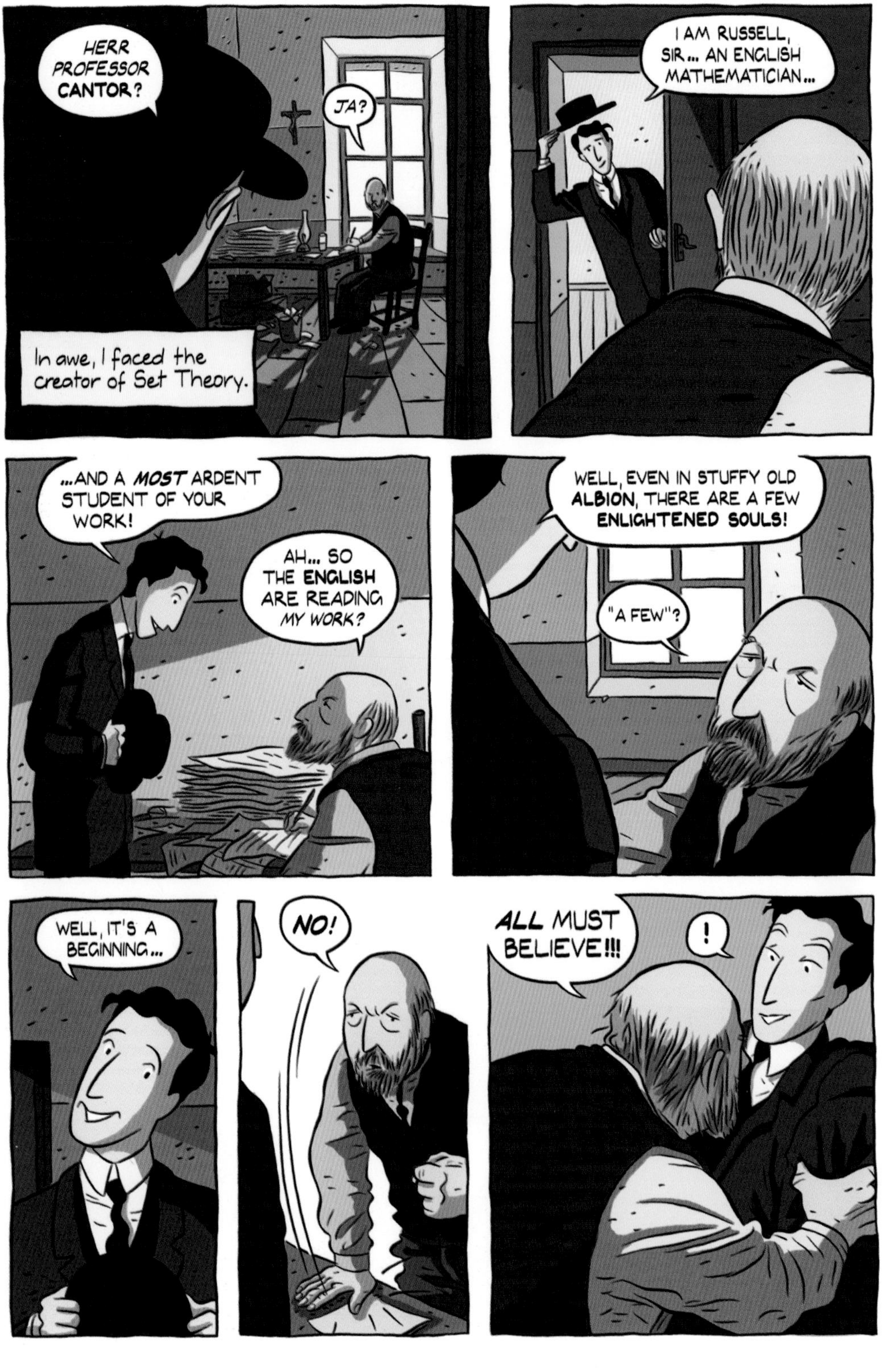
HERR PROFESSOR CANTOR?
JA?
In awe, I faced the creator of Set Theory.
I AM RUSSELL, SIR... AN ENGLISH MATHEMATICIAN...
...AND A MOST ARDENT STUDENT OF YOUR WORK!
AH... SO THE ENGLISH ARE READING MY WORK?
WELL, EVEN IN STUFFY OLD ALBION, THERE ARE A FEW ENLIGHTENED SOULS!
"A FEW"?
WELL, IT'S A BEGINNING...
NO!
ALL MUST BELIEVE!!!
!

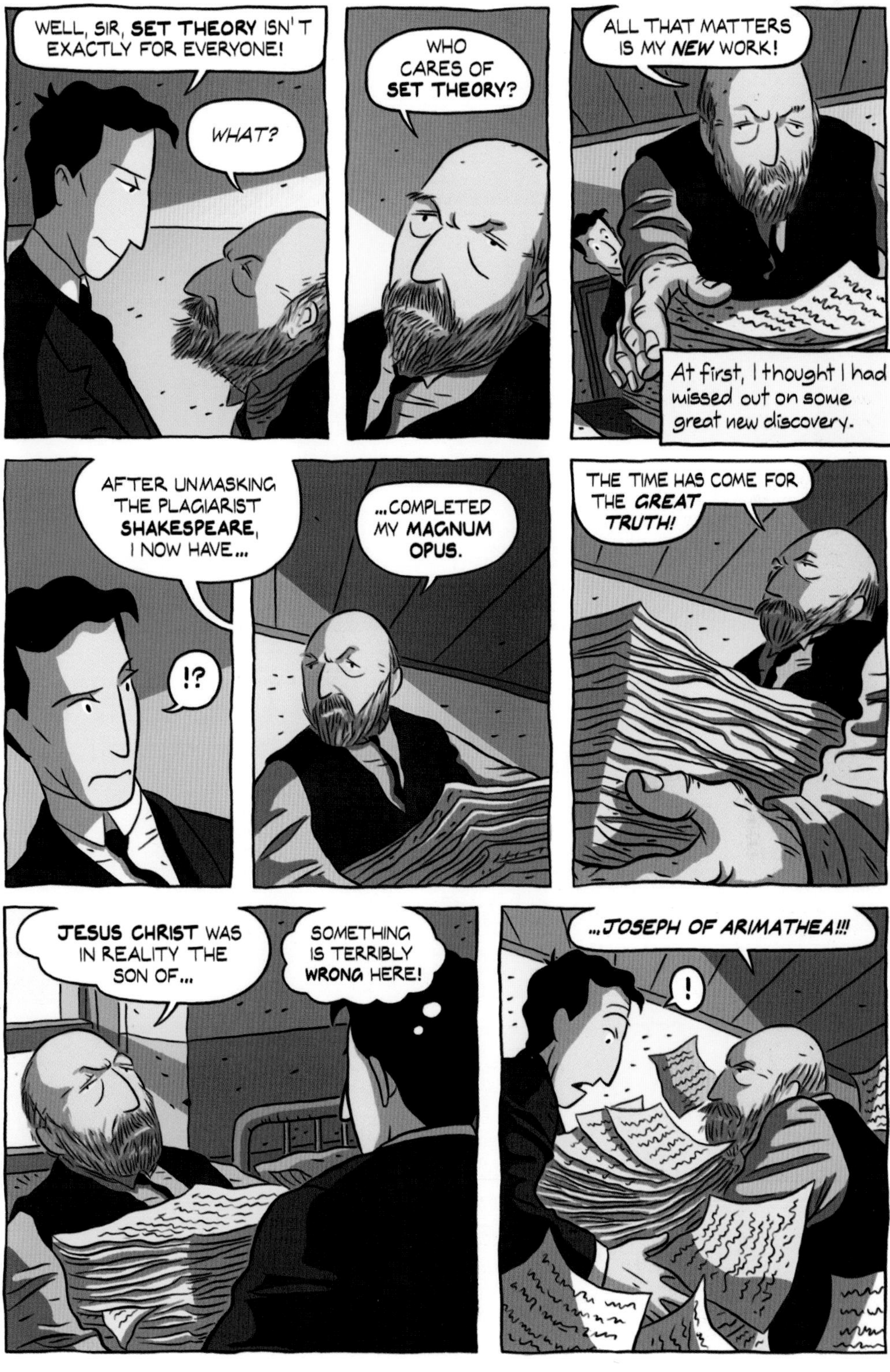
WELL, SIR, SET THEORY ISN'T EXACTLY FOR EVERYONE!
WHAT?
WHO CARES OF SET THEORY?
ALL THAT MATTERS IS MY NEW WORK!
At first, I thought I had missed out on some great new discovery.
AFTER UNMASKING THE PLAGIARIST SHAKESPEARE, I NOW HAVE...
!?
...COMPLETED MY MAGNUM OPUS.
THE TIME HAS COME FOR THE GREAT TRUTH!
JESUS CHRIST WAS IN REALITY THE SON OF...
SOMETHING IS TERRIBLY WRONG HERE!
...JOSEPH OF ARIMATHEA!!!
!

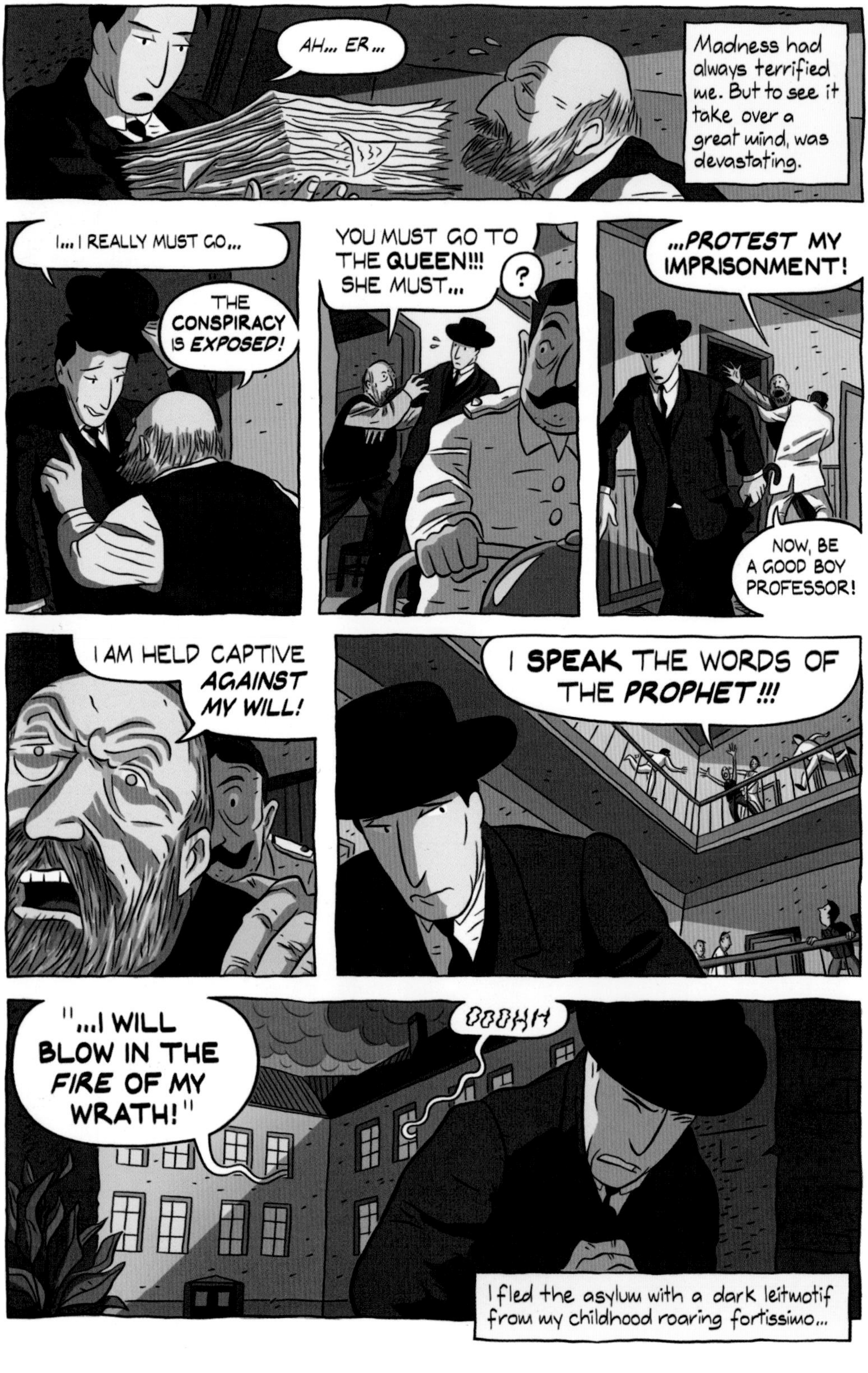
AH... ER...
Madness had always terrified me. But to see it take over a great mind, was devastating.
I... I REALLY MUST GO...
THE CONSPIRACY IS EXPOSED!
YOU MUST GO TO THE QUEEN!!! SHE MUST...
?
...PROTEST MY IMPRISONMENT!
NOW, BE A GOOD BOY PROFESSOR!
I AM HELD CAPTIVE AGAINST MY WILL!
I SPEAK THE WORDS OF THE PROPHET!!!
"...I WILL BLOW IN THE FIRE OF MY WRATH!"
OOOHH
I fled the asylum with a dark leitmotif from my childhood roaring fortissimo...

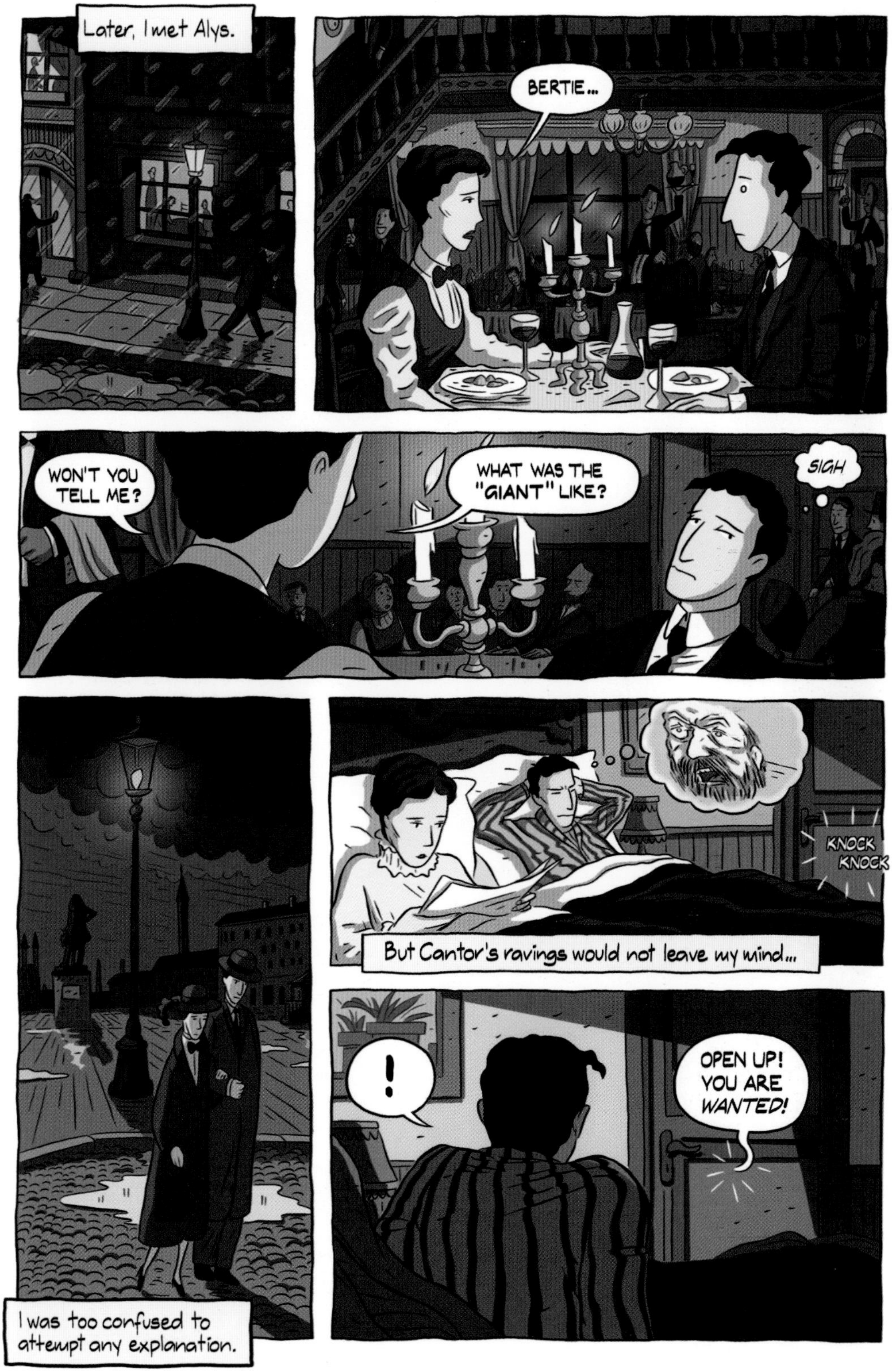

Later, I met Alys.
BERTIE...
WON'T YOU TELL ME?
WHAT WAS THE "GIANT" LIKE?
SIGH
I was too confused to attempt any explanation.
KNOCK KNOCK
But Cantor's ravings would not leave my mind...
!
OPEN UP! YOU ARE WANTED!

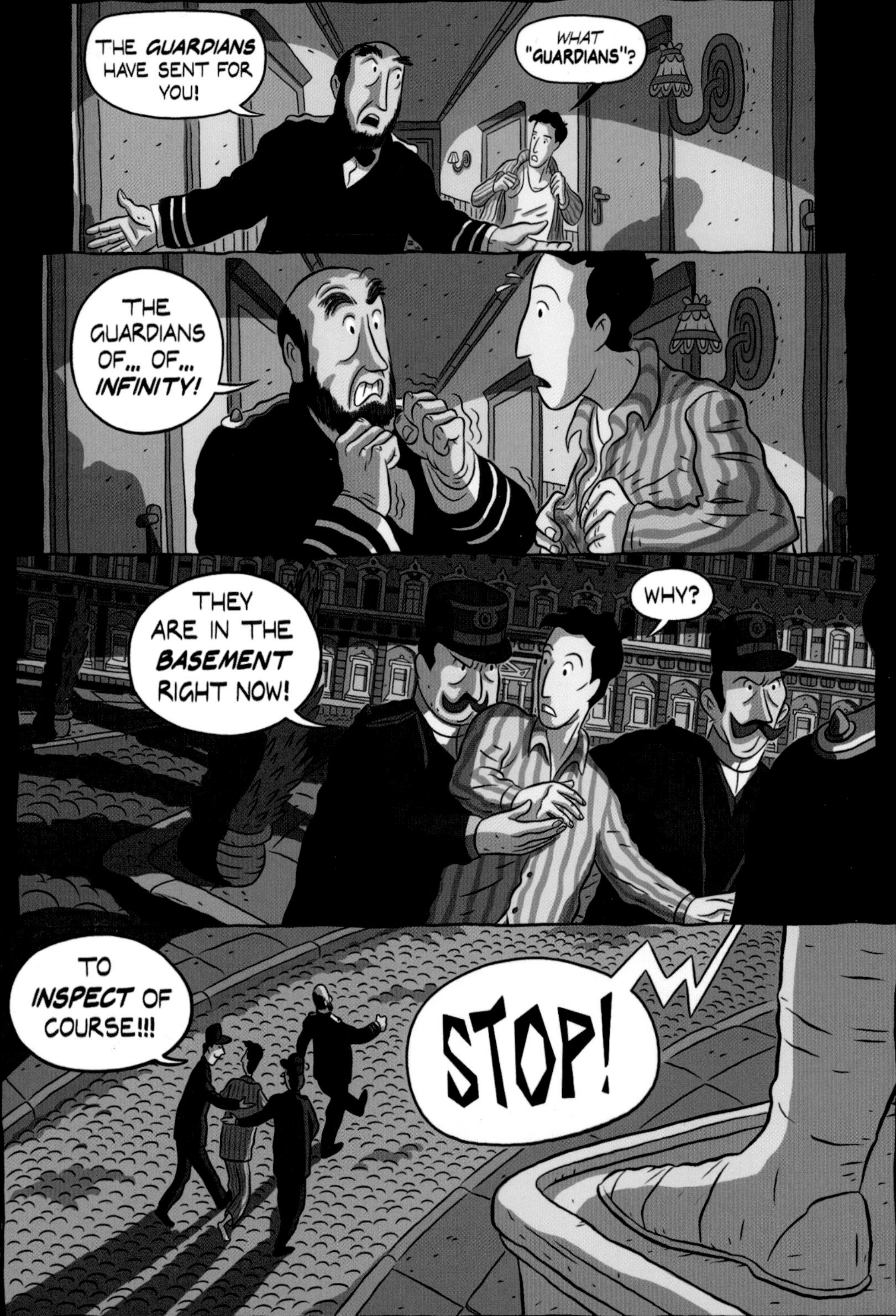
THE *GUARDIANS* HAVE SENT FOR YOU!
WHAT "GUARDIANS"?
THE GUARDIANS OF... OF... *INFINITY!*
THEY ARE IN THE *BASEMENT* RIGHT NOW!
WHY?
TO *INSPECT* OF COURSE!!!
STOP!

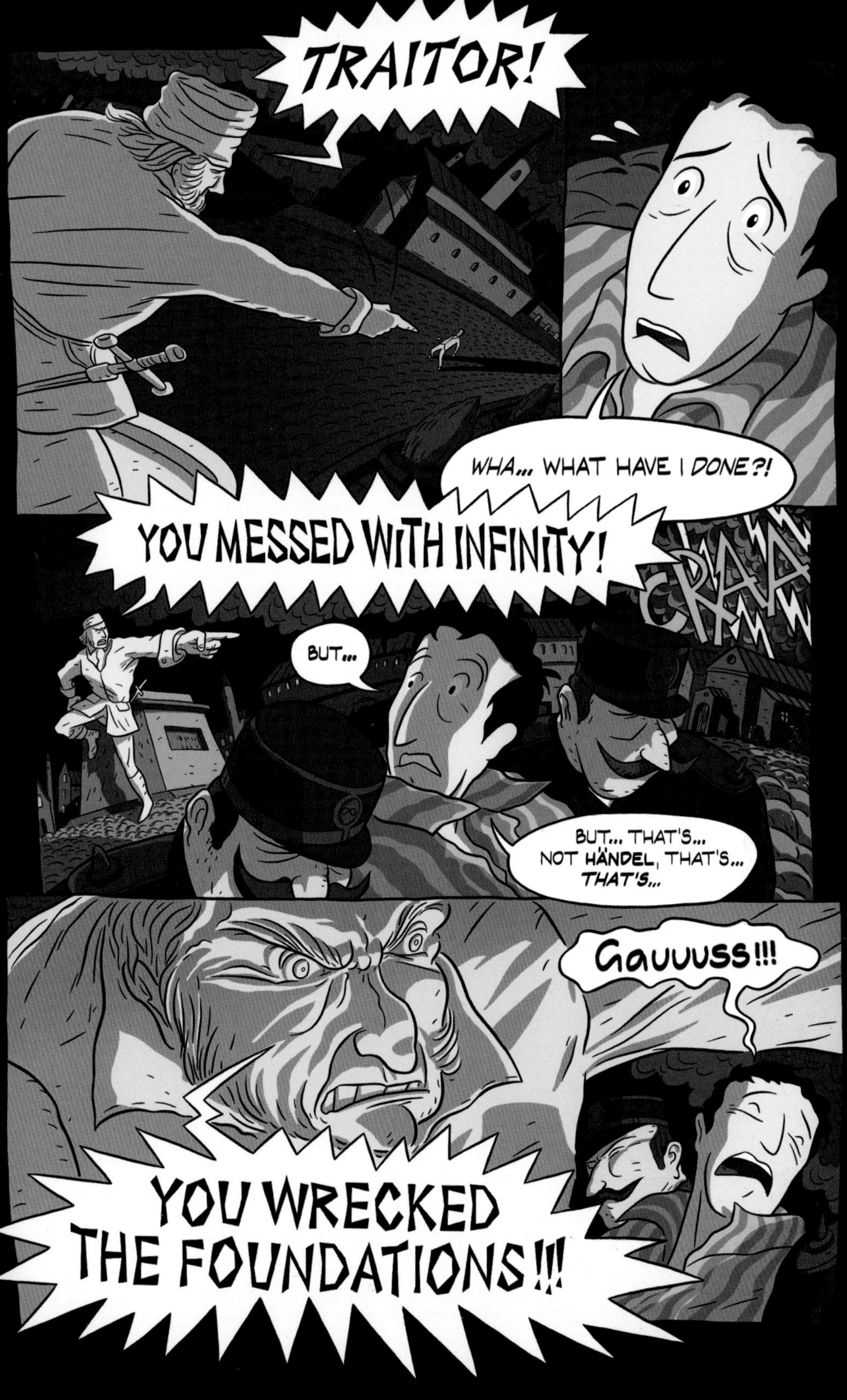
TRAITOR!
WHA... WHAT HAVE I DONE?!
YOU MESSED WITH INFINITY!
CRAA
BUT...
BUT... THAT'S... NOT HÄNDEL, THAT'S... THAT'S...
Gauuuss!!!
YOU WRECKED THE FOUNDATIONS!!!

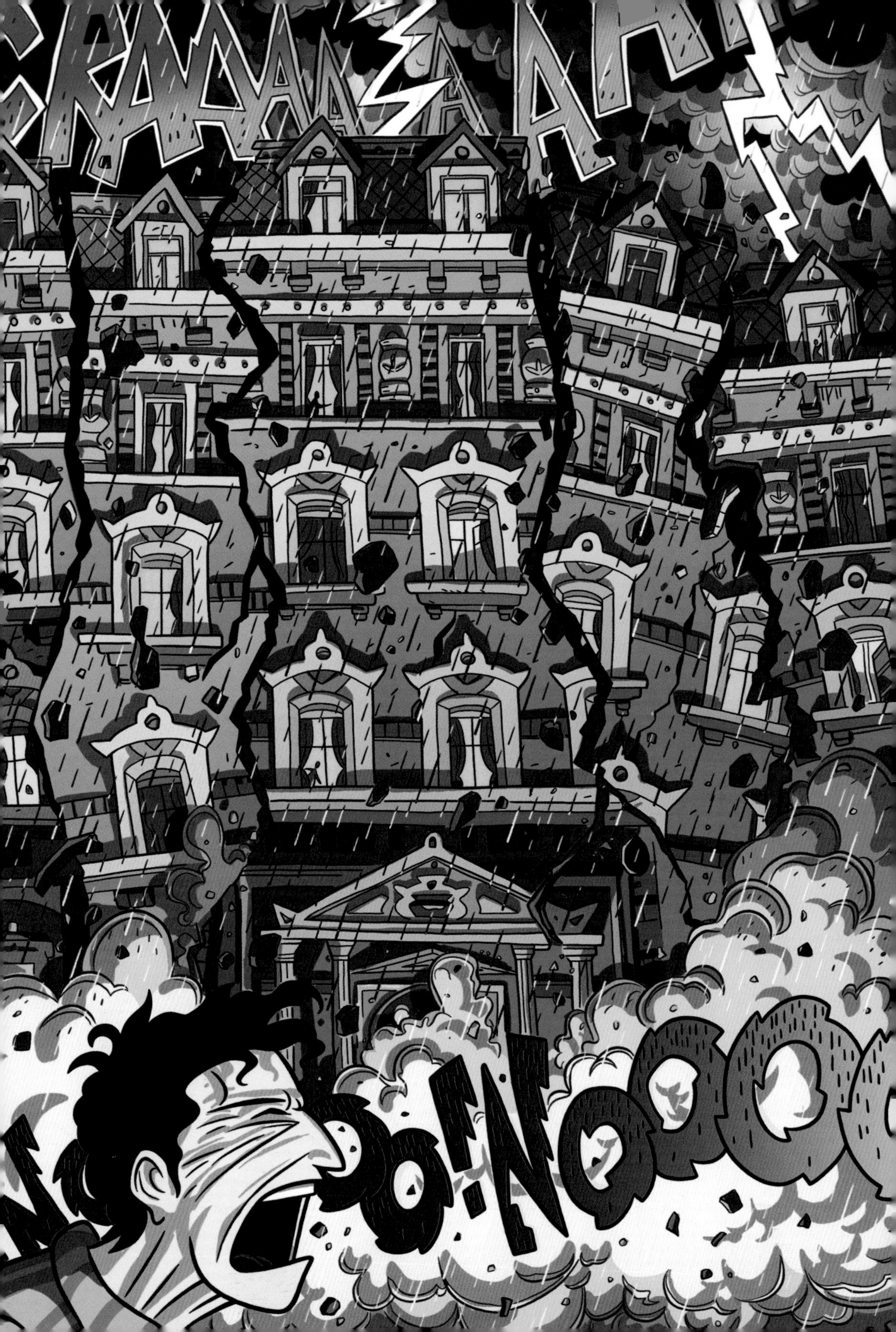

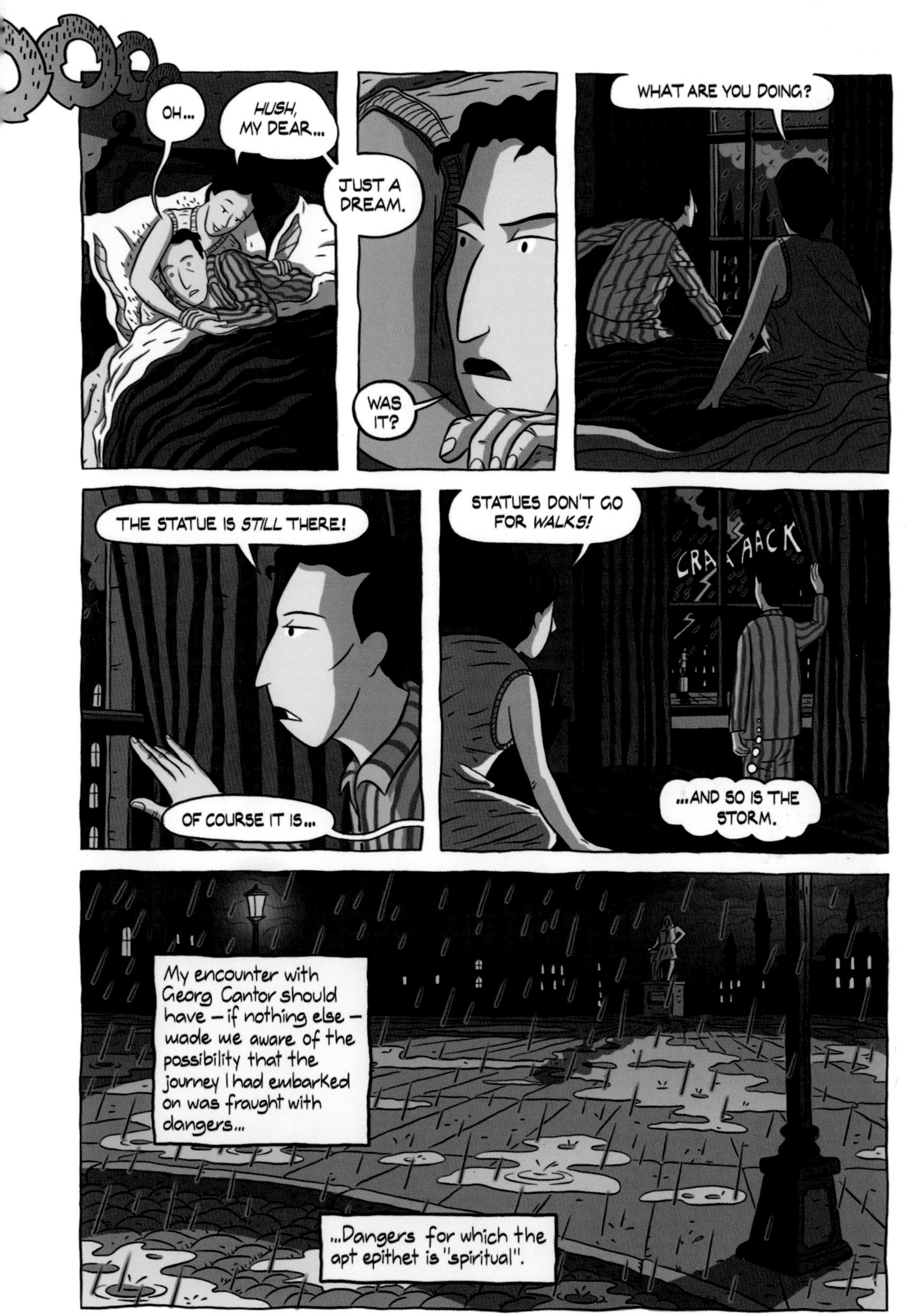
OH...
HUSH, MY DEAR...
JUST A DREAM.
WAS IT?
WHAT ARE YOU DOING?
THE STATUE IS STILL THERE!
OF COURSE IT IS...
STATUES DON'T GO FOR WALKS!
CRAAACK
...AND SO IS THE STORM.
My encounter with Georg Cantor should have – if nothing else – made me aware of the possibility that the journey I had embarked on was fraught with dangers...
...Dangers for which the apt epithet is "spiritual".

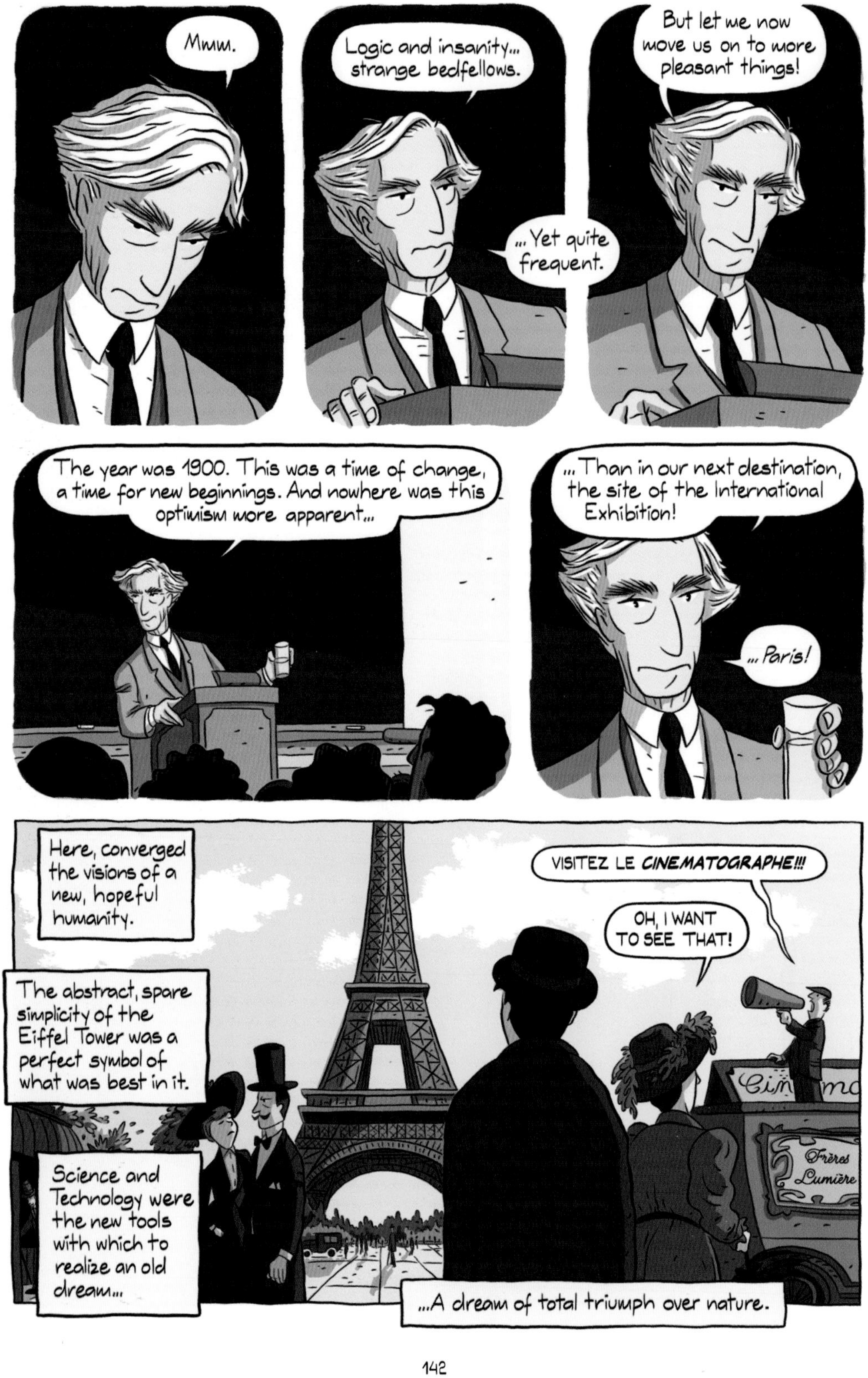
Mmm.
Logic and insanity... strange bedfellows.
...Yet quite frequent.
But let me now move us on to more pleasant things!
The year was 1900. This was a time of change, a time for new beginnings. And nowhere was this optimism more apparent...
...Than in our next destination, the site of the International Exhibition!
...Paris!
Here, converged the visions of a new, hopeful humanity.
The abstract, spare simplicity of the Eiffel Tower was a perfect symbol of what was best in it.
VISITEZ LE *CINEMATOGRAPHE*!!!
OH, I WANT TO SEE THAT!
Frères Lumière
Science and Technology were the new tools with which to realize an old dream...
...A dream of total triumph over nature.

The visitors to the Exhibition were like children, excited with the toys of a brave new world...

BERTIE, IT LOOKS SO *REAL*!

MIRACOLO!!!

...Toys which played in unexpected ways!

MAMMA MIA!!!

MON DIEU!!!

HEEEELP BERTIEEEE!!!

AAAAAAH!!!!

But we had not come to Paris for the Exhibition.

Or, rather, I personally hadn't!

LET'S GO UP THE "MECHANICAL STAIRCASE" ONCE AGAIN!

GOLLY!

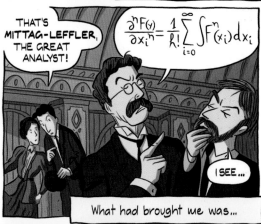

THAT'S MITTAG-LEFFLER, THE GREAT ANALYST!

$$\frac{\partial^n F_{(y)}}{\partial x_i^n} = \frac{1}{k!} \sum_{i=0}^{\infty} \int F_{(x_i)}^n dx_i$$

I SEE...

What had brought me was...

...The International Congress, the world's greatest rendez-vous of Mathematicians.

RUSSELL!

I SAY... WHITEHEAD!

Everybody who was anybody in Mathematics was there!

MONSIEUR LE PROFESSEUR HERMITE... HERR PROFESSOR MINKOWSKI...

HERR PROFESSOR FELIX KLEIN...

ENCHANTED!

AND LAST BUT NOT LEAST, HERR PROFESSOR DEDEKIND.

!

...DEEPLY HONOURED!

...All my new heroes under one roof!

THIS IS SO *AMAZING!* EVERY SINGLE *ONE* OF THESE PEOPLE IS A **MATHEMATICAL LEGEND!**

COME, SHE WILL BE WAITING AT THE COLONIAL PAVILION.

KLEIN CREATED A NEW GEOMETRY, DEDEKIND A —

I DON'T THINK YOU'VE MET...

MRS. EVELYN WHITEHEAD, MY WIFE!

HOW *ARE* YOU, MR. RUSSELL?

ER... I'M...

Ah well...

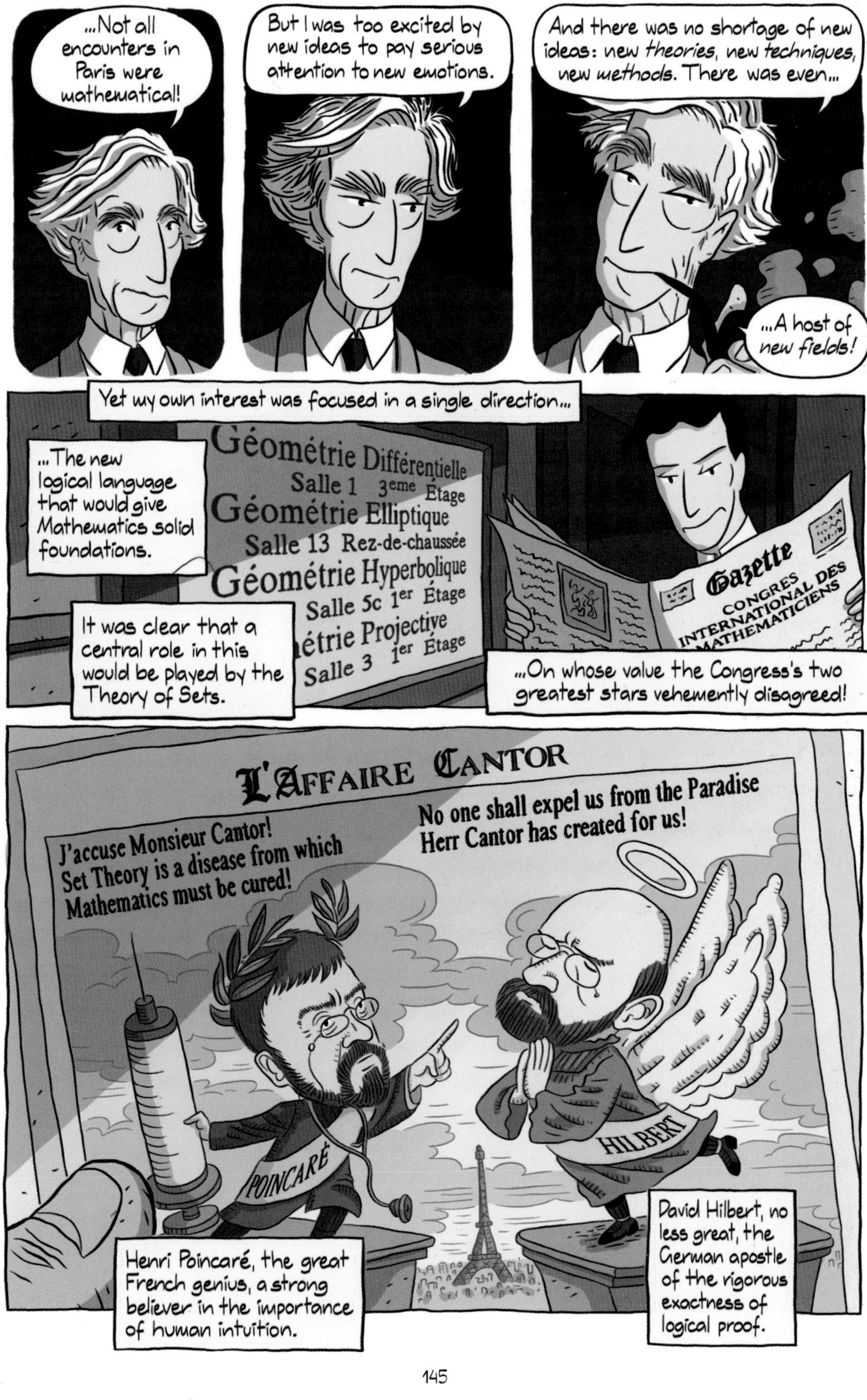
...Not all encounters in Paris were mathematical!
But I was too excited by new ideas to pay serious attention to new emotions.
And there was no shortage of new ideas: new *theories*, new *techniques*, new *methods*. There was even...
...A host of new *fields*!
Yet my own interest was focused in a single direction...
...The new logical language that would give Mathematics solid foundations.
Géométrie Différentielle Salle 1 3eme Étage
Géométrie Elliptique Salle 13 Rez-de-chaussée
Géométrie Hyperbolique Salle 5c 1er Étage
...étrie Projective Salle 3 1er Étage
It was clear that a central role in this would be played by the Theory of Sets.
Gazette
CONGRES INTERNATIONAL DES MATHEMATICIENS
...On whose value the Congress's two greatest stars vehemently disagreed!
L'AFFAIRE CANTOR
J'accuse Monsieur Cantor! Set Theory is a disease from which Mathematics must be cured!
No one shall expel us from the Paradise Herr Cantor has created for us!
POINCARÉ
HILBERT
Henri Poincaré, the great French genius, a strong believer in the importance of human intuition.
David Hilbert, no less great, the German apostle of the rigorous exactness of logical proof.

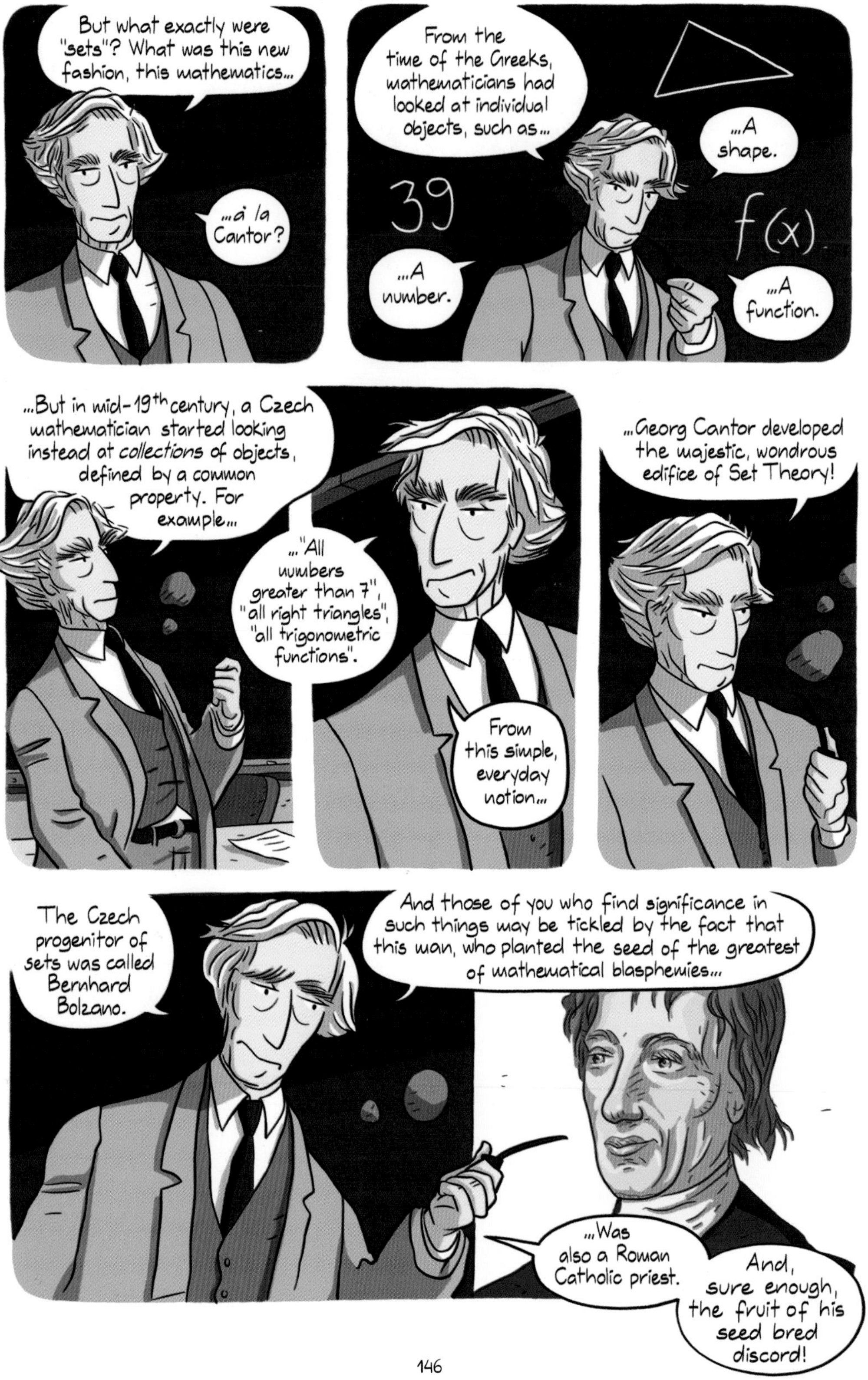
But what exactly were "sets"? What was this new fashion, this mathematics...
...à la Cantor?
From the time of the Greeks, mathematicians had looked at individual objects, such as...
...A shape.
39
f(x)
...A number.
...A function.
...But in mid-19th century, a Czech mathematician started looking instead at *collections* of objects, defined by a common property. For example...
..."All numbers greater than 7", "all right triangles", "all trigonometric functions".
From this simple, everyday notion...
...Georg Cantor developed the majestic, wondrous edifice of Set Theory!
The Czech progenitor of sets was called Bernhard Bolzano.
And those of you who find significance in such things may be tickled by the fact that this man, who planted the seed of the greatest of mathematical blasphemies...
...Was also a Roman Catholic priest.
And, sure enough, the fruit of his seed bred discord!

I LOVE THIS! MATHEMATICIANS ARE, AT LAST, IN SERIOUS CONFLICT ABOUT A THEORY!
SCONI
URES
...MAKING SPACE FOR US LOGICIANS TO AIR OUR VIEWS!
During the Congress, every restaurant and café was host to the new ideas.
BRASSERIE LIPP
LIPP
I WONDER IF DINERS SIT AS "PRO-" OR "CONTRA-SETS"!
S'IL VOUS PLAIT, MESSIEURS!
Of course, what made Set Theory so controversial was its centrality in the quest for secure foundations.
...ACCORDING TO POINCARÉ...
SHH, OR HE'LL HEAR YOU... HE'S OVER THERE!
M'SIEUR LE PROFESSEUR, HILBERT'S PROBLEM IS TOO MUCH GERMAN EFFICIENCY!
HAA HA HA HA! HA!
HA HA
HAA HA HA!
PINOT
NO! IT'S JUST THAT THE HERR PROFESSOR LIKES TOO MUCH THE SAUCISSES!
HOW EXACTLY?
?

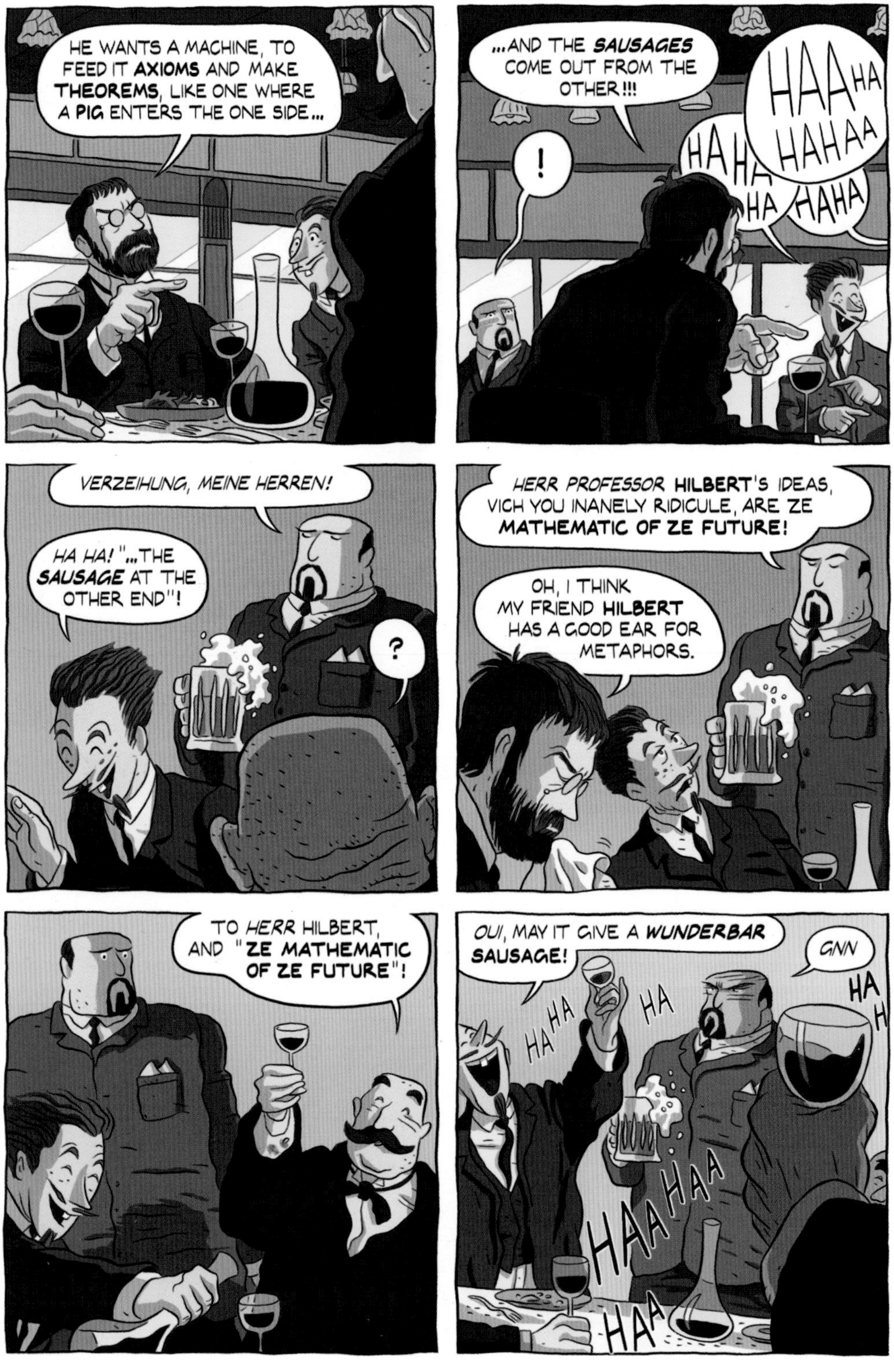

HE WANTS A MACHINE, TO FEED IT **AXIOMS** AND MAKE **THEOREMS**, LIKE ONE WHERE A **PIG** ENTERS THE ONE SIDE...
...AND THE ***SAUSAGES*** COME OUT FROM THE OTHER!!!
!
HAA HA HAHAA
HAHA
HA
HAHA
VERZEIHUNG, MEINE HERREN!
HA HA! "...THE ***SAUSAGE*** AT THE OTHER END"!
?
HERR PROFESSOR **HILBERT**'S IDEAS, VICH YOU INANELY RIDICULE, ARE ZE **MATHEMATIC OF ZE FUTURE!**
OH, I THINK *MY* FRIEND **HILBERT** HAS A GOOD EAR FOR METAPHORS.
TO *HERR* HILBERT, AND "**ZE MATHEMATIC OF ZE FUTURE**"!
OUI, MAY IT GIVE A ***WUNDERBAR*** **SAUSAGE!**
GNN
HA HA
HA
HA H
HAA HAA
HAA

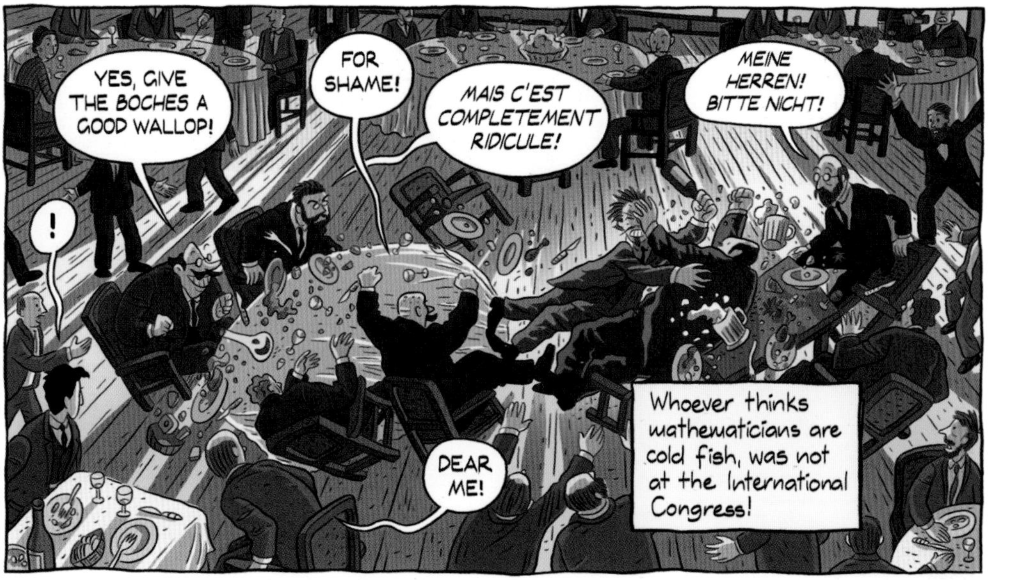
YES, GIVE THE BOCHES A GOOD WALLOP!
!
FOR SHAME!
MAIS C'EST COMPLETEMENT RIDICULE!
MEINE HERREN! BITTE NICHT!
DEAR ME!
Whoever thinks mathematicians are cold fish, was not at the International Congress!

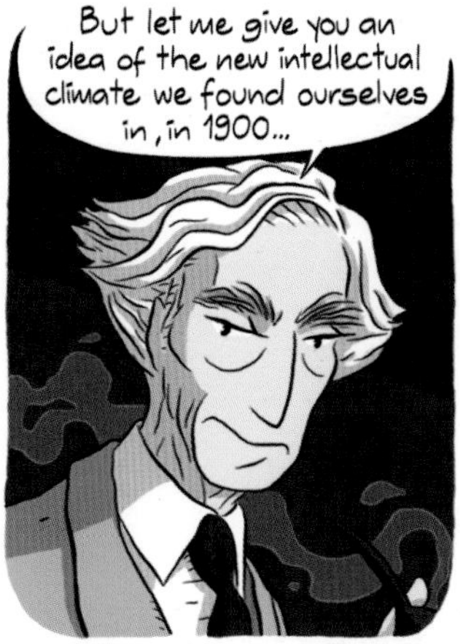
But let me give you an idea of the new intellectual climate we found ourselves in, in 1900...

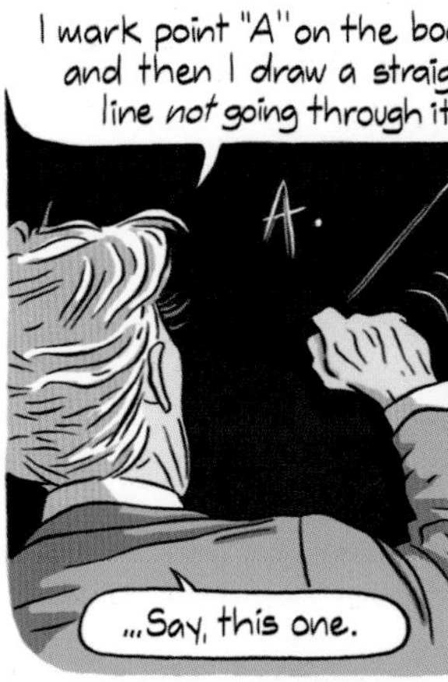
I mark point "A" on the board, and then I draw a straight line *not* going through it...
A.
...Say, this one.

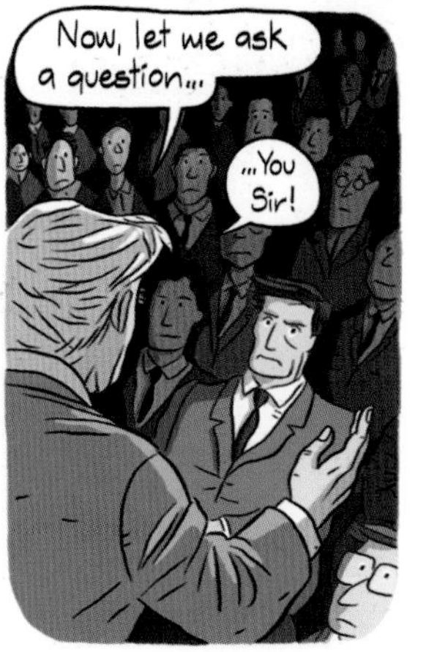
Now, let me ask a question...
...You Sir!

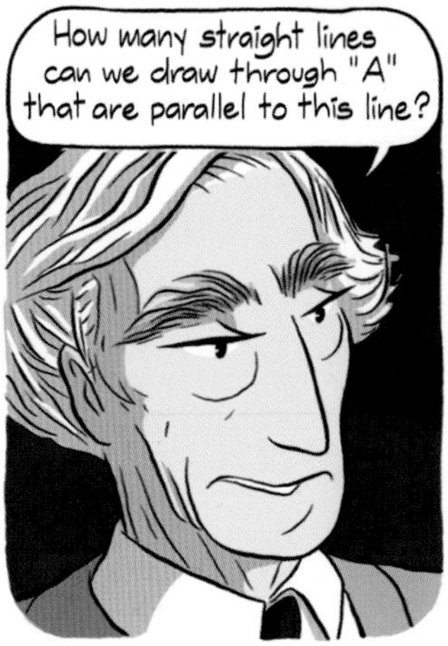
How many straight lines can we draw through "A" that are parallel to this line?

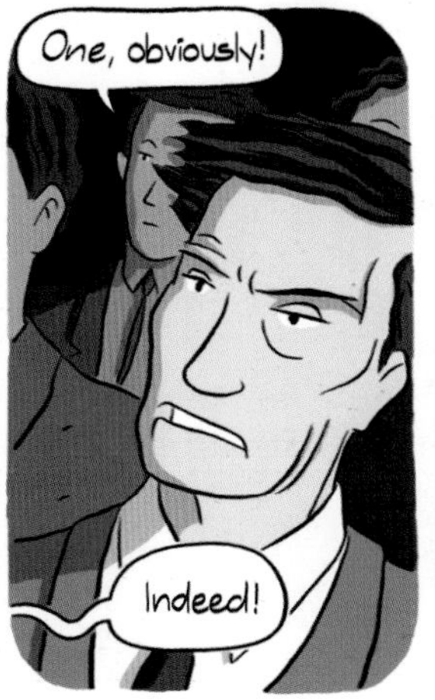
One, obviously!
Indeed!

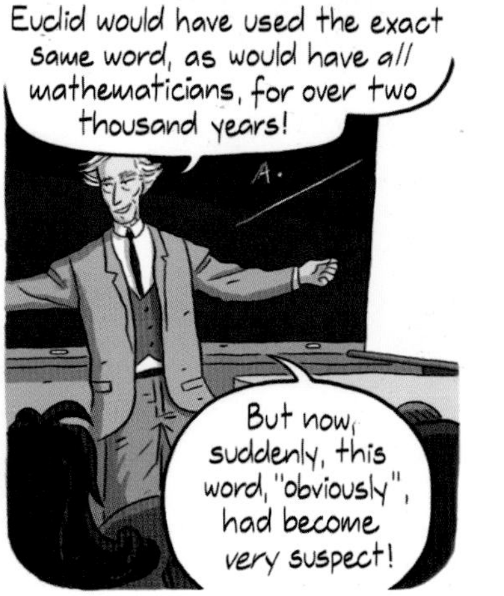
Euclid would have used the exact same word, as would have *all* mathematicians, for over two thousand years!
A.
But now, suddenly, this word, "obviously", had become *very* suspect!

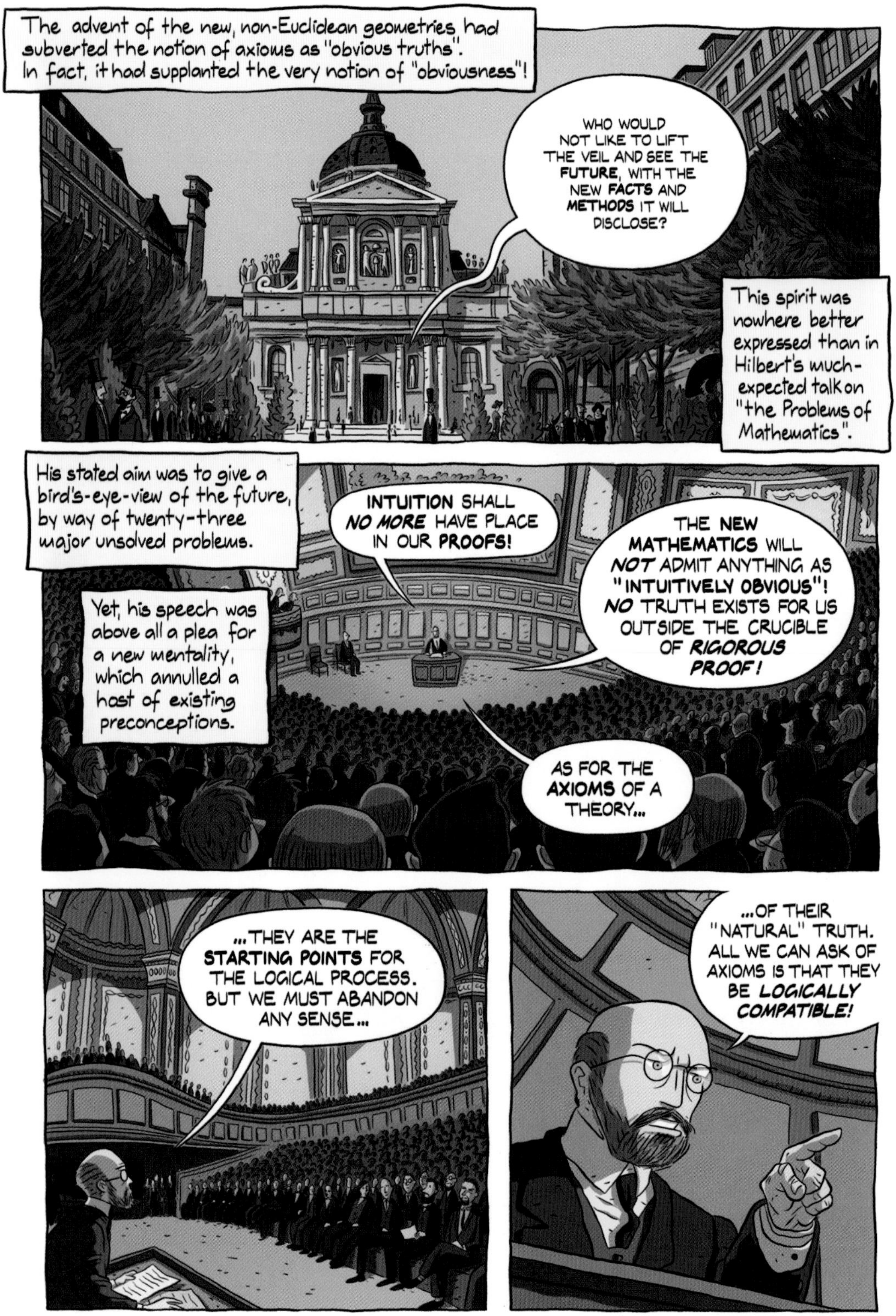
The advent of the new, non-Euclidean geometries had subverted the notion of axioms as "obvious truths". In fact, it had supplanted the very notion of "obviousness"!
WHO WOULD NOT LIKE TO LIFT THE VEIL AND SEE THE **FUTURE**, WITH THE NEW **FACTS** AND **METHODS** IT WILL DISCLOSE?
This spirit was nowhere better expressed than in Hilbert's much-expected talk on "the Problems of Mathematics".
His stated aim was to give a bird's-eye-view of the future, by way of twenty-three major unsolved problems.
INTUITION SHALL ***NO MORE*** HAVE PLACE IN OUR **PROOFS!**
THE **NEW MATHEMATICS** WILL ***NOT*** ADMIT ANYTHING AS **"INTUITIVELY OBVIOUS"!** ***NO*** TRUTH EXISTS FOR US OUTSIDE THE CRUCIBLE OF ***RIGOROUS PROOF!***
Yet, his speech was above all a plea for a new mentality, which annulled a host of existing preconceptions.
AS FOR THE **AXIOMS** OF A THEORY...
...THEY ARE THE **STARTING POINTS** FOR THE LOGICAL PROCESS. BUT WE MUST ABANDON ANY SENSE...
...OF THEIR "NATURAL" TRUTH. ALL WE CAN ASK OF AXIOMS IS THAT THEY BE ***LOGICALLY COMPATIBLE!***

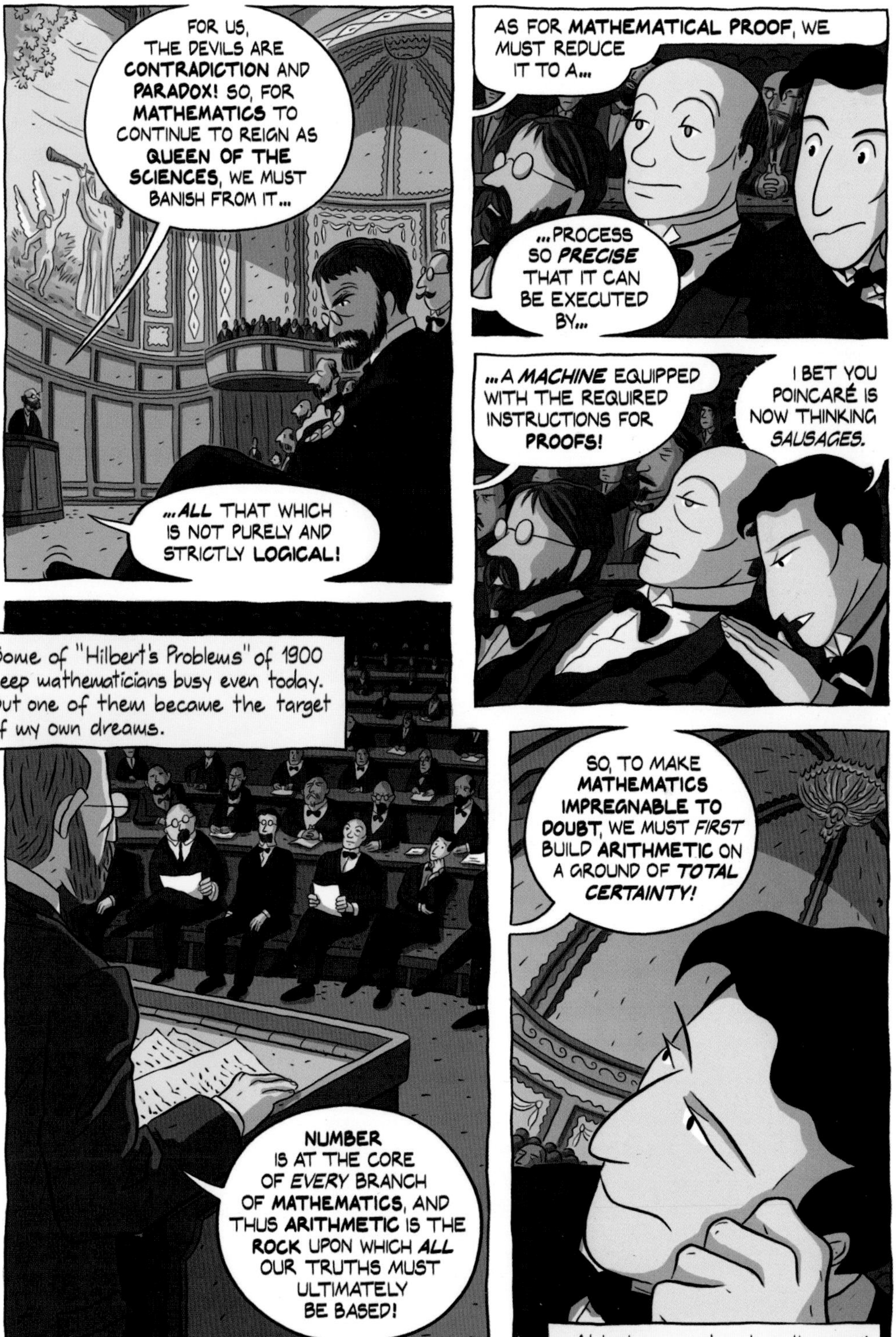
FOR US, THE DEVILS ARE CONTRADICTION AND PARADOX! SO, FOR MATHEMATICS TO CONTINUE TO REIGN AS QUEEN OF THE SCIENCES, WE MUST BANISH FROM IT...
...ALL THAT WHICH IS NOT PURELY AND STRICTLY LOGICAL!
AS FOR MATHEMATICAL PROOF, WE MUST REDUCE IT TO A...
...PROCESS SO PRECISE THAT IT CAN BE EXECUTED BY...
...A MACHINE EQUIPPED WITH THE REQUIRED INSTRUCTIONS FOR PROOFS!
I BET YOU POINCARÉ IS NOW THINKING SAUSAGES.
Some of "Hilbert's Problems" of 1900 keep mathematicians busy even today. But one of them became the target of my own dreams.
NUMBER IS AT THE CORE OF EVERY BRANCH OF MATHEMATICS, AND THUS ARITHMETIC IS THE ROCK UPON WHICH ALL OUR TRUTHS MUST ULTIMATELY BE BASED!
SO, TO MAKE MATHEMATICS IMPREGNABLE TO DOUBT, WE MUST FIRST BUILD ARITHMETIC ON A GROUND OF TOTAL CERTAINTY!
...At last, a grand and worthy goal!

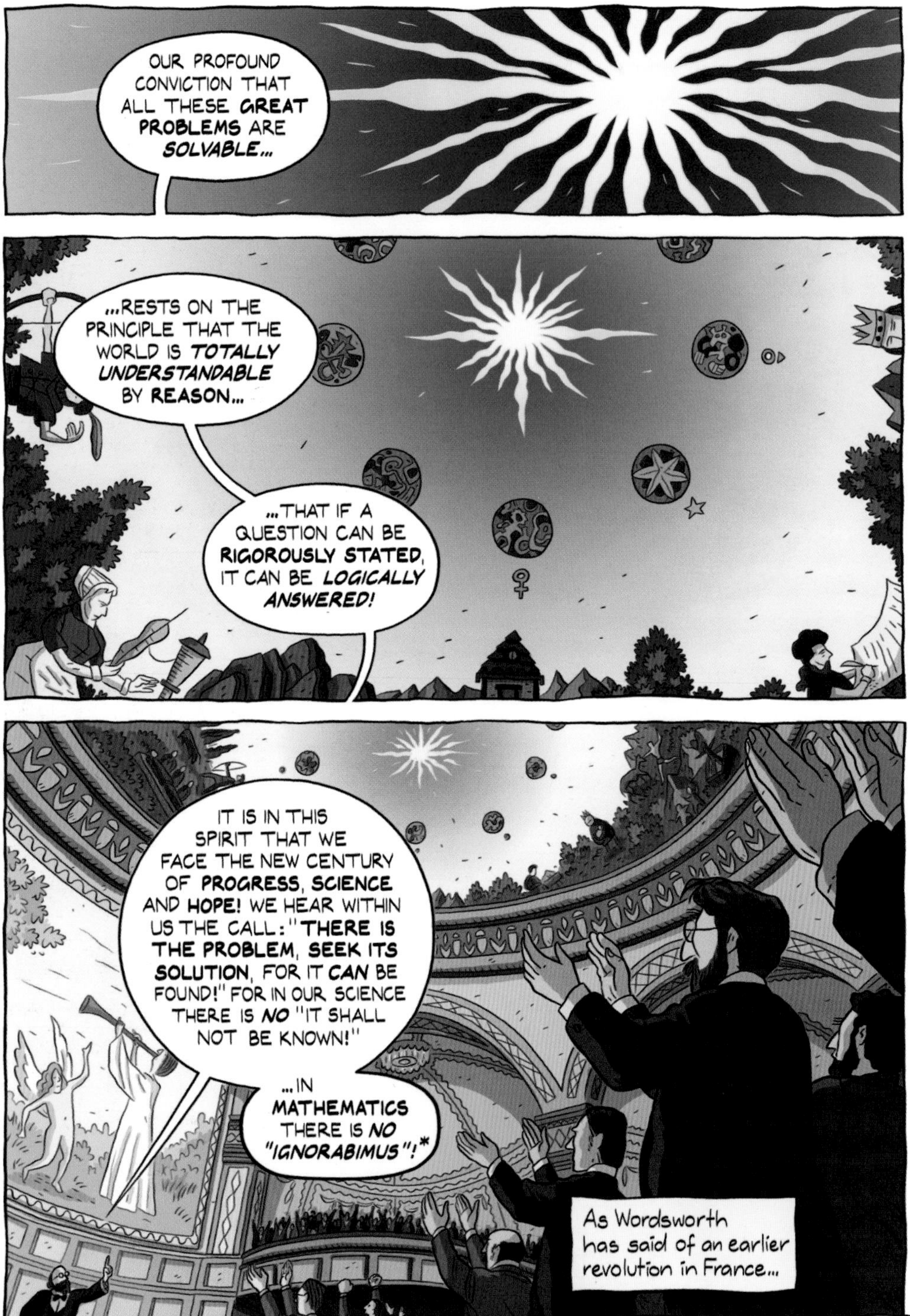
OUR PROFOUND CONVICTION THAT ALL THESE GREAT PROBLEMS ARE SOLVABLE...
...RESTS ON THE PRINCIPLE THAT THE WORLD IS TOTALLY UNDERSTANDABLE BY REASON...
...THAT IF A QUESTION CAN BE RIGOROUSLY STATED, IT CAN BE LOGICALLY ANSWERED!
IT IS IN THIS SPIRIT THAT WE FACE THE NEW CENTURY OF PROGRESS, SCIENCE AND HOPE! WE HEAR WITHIN US THE CALL: "THERE IS THE PROBLEM, SEEK ITS SOLUTION, FOR IT CAN BE FOUND!" FOR IN OUR SCIENCE THERE IS NO "IT SHALL NOT BE KNOWN!"
...IN MATHEMATICS THERE IS NO "IGNORABIMUS"!*
As Wordsworth has said of an earlier revolution in France...
"Bliss was it in that dawn to be alive. But to be young was very Heaven!"

*Latin for "we shall not know."

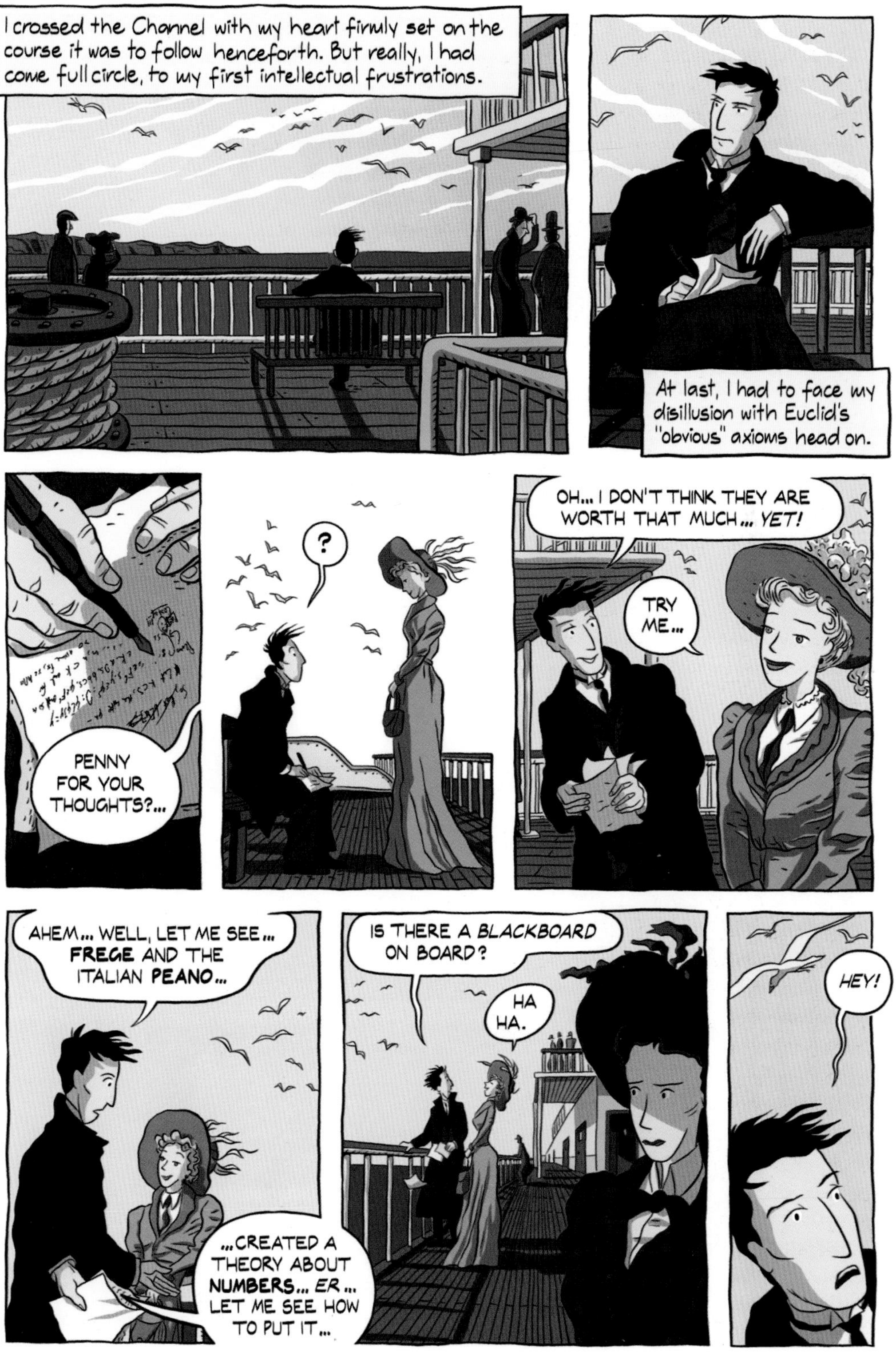
I crossed the Channel with my heart firmly set on the course it was to follow henceforth. But really, I had come full circle, to my first intellectual frustrations.
At last, I had to face my disillusion with Euclid's "obvious" axioms head on.
PENNY FOR YOUR THOUGHTS?...
?
OH... I DON'T THINK THEY ARE WORTH THAT MUCH... *YET!*
TRY ME...
AHEM... WELL, LET ME SEE... **FREGE** AND THE ITALIAN **PEANO**...
...CREATED A THEORY ABOUT **NUMBERS**... *ER*... LET ME SEE HOW TO PUT IT...
IS THERE A *BLACKBOARD* ON BOARD?
HA HA.
HEY!

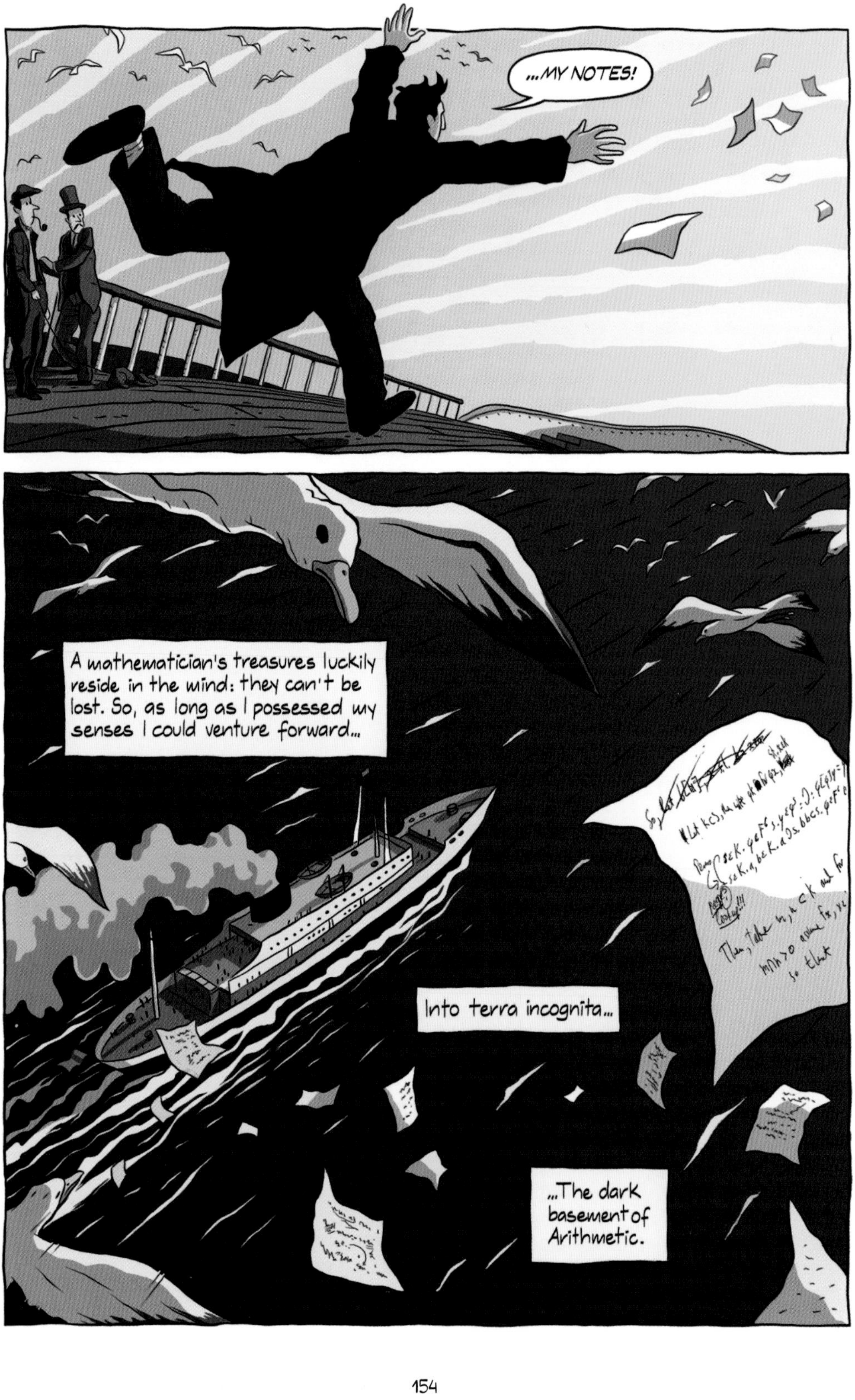
...MY NOTES!
A mathematician's treasures luckily reside in the mind: they can't be lost. So, as long as I possessed my senses I could venture forward...
Into terra incognita...
...The dark basement of Arithmetic.

4. PARADOXES

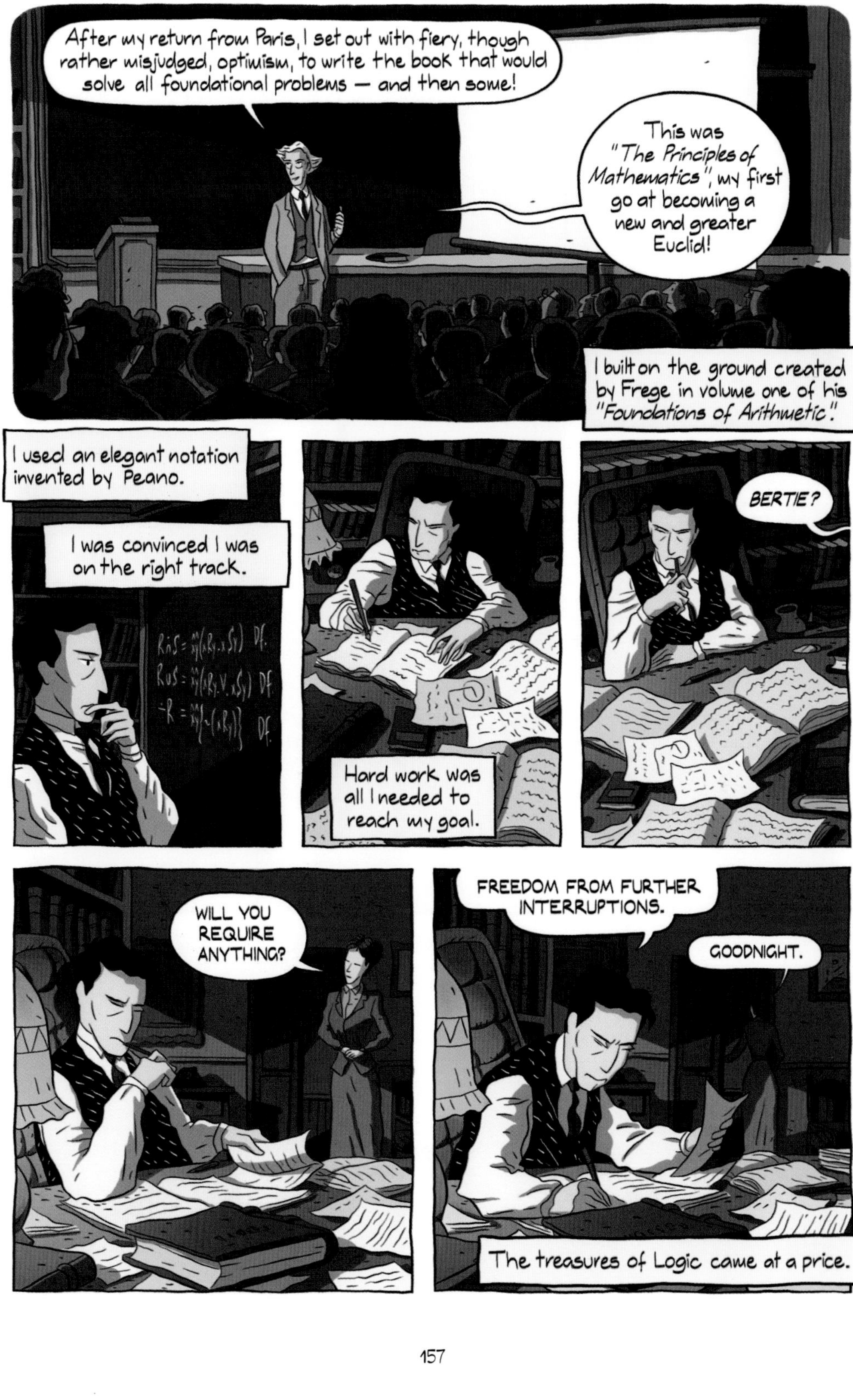
After my return from Paris, I set out with fiery, though rather misjudged, optimism, to write the book that would solve all foundational problems — and then some!
This was "The Principles of Mathematics", my first go at becoming a new and greater Euclid!
I built on the ground created by Frege in volume one of his "Foundations of Arithmetic."
I used an elegant notation invented by Peano.
I was convinced I was on the right track.
Hard work was all I needed to reach my goal.
BERTIE?
WILL YOU REQUIRE ANYTHING?
FREEDOM FROM FURTHER INTERRUPTIONS.
GOODNIGHT.
The treasures of Logic came at a price.

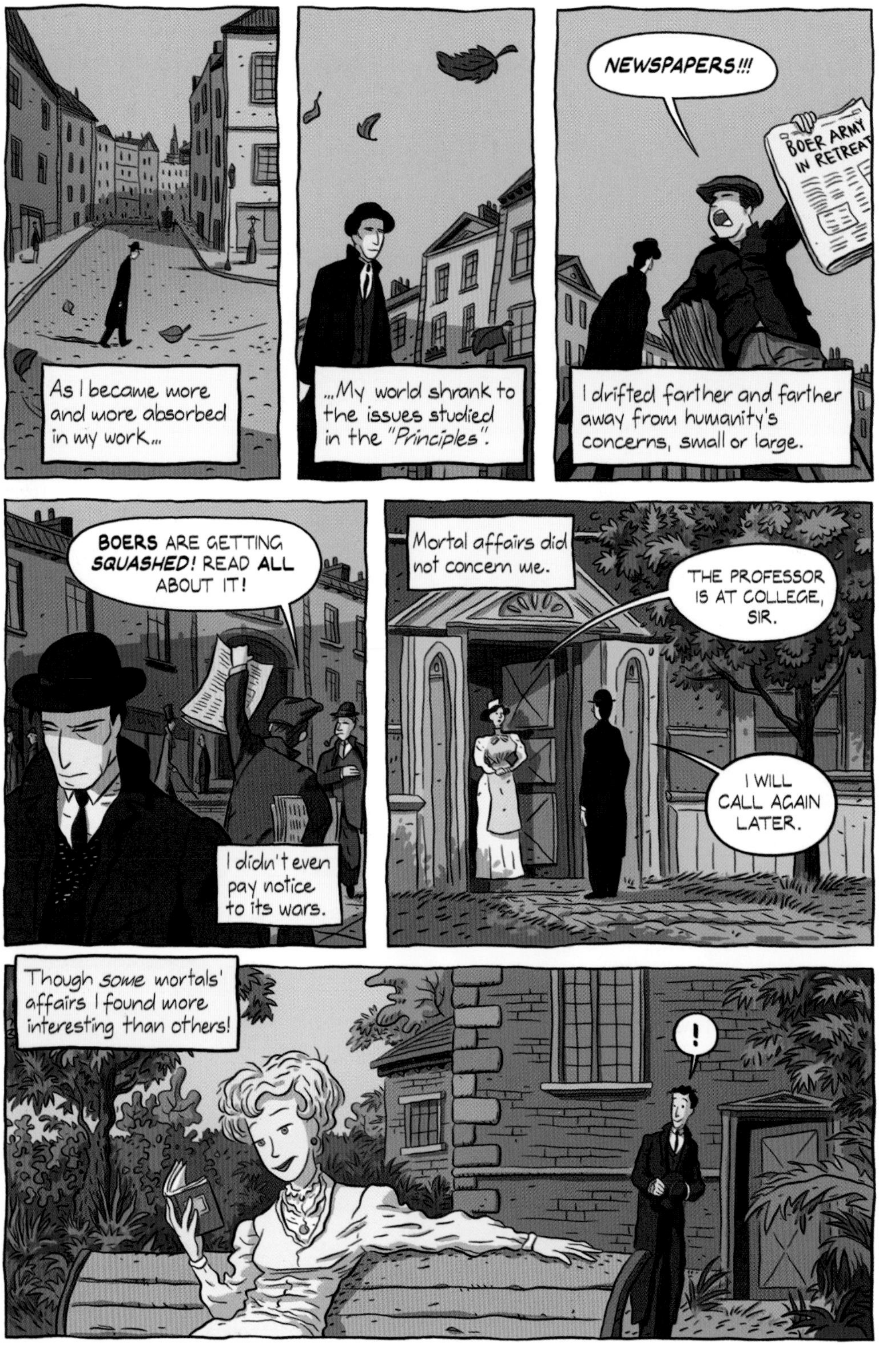
As I became more and more absorbed in my work...
...My world shrank to the issues studied in the "Principles".
NEWSPAPERS!!!
BOER ARMY IN RETREAT
I drifted farther and farther away from humanity's concerns, small or large.
BOERS ARE GETTING SQUASHED! READ ALL ABOUT IT!
I didn't even pay notice to its wars.
Mortal affairs did not concern me.
THE PROFESSOR IS AT COLLEGE, SIR.
I WILL CALL AGAIN LATER.
Though some mortals' affairs I found more interesting than others!
!

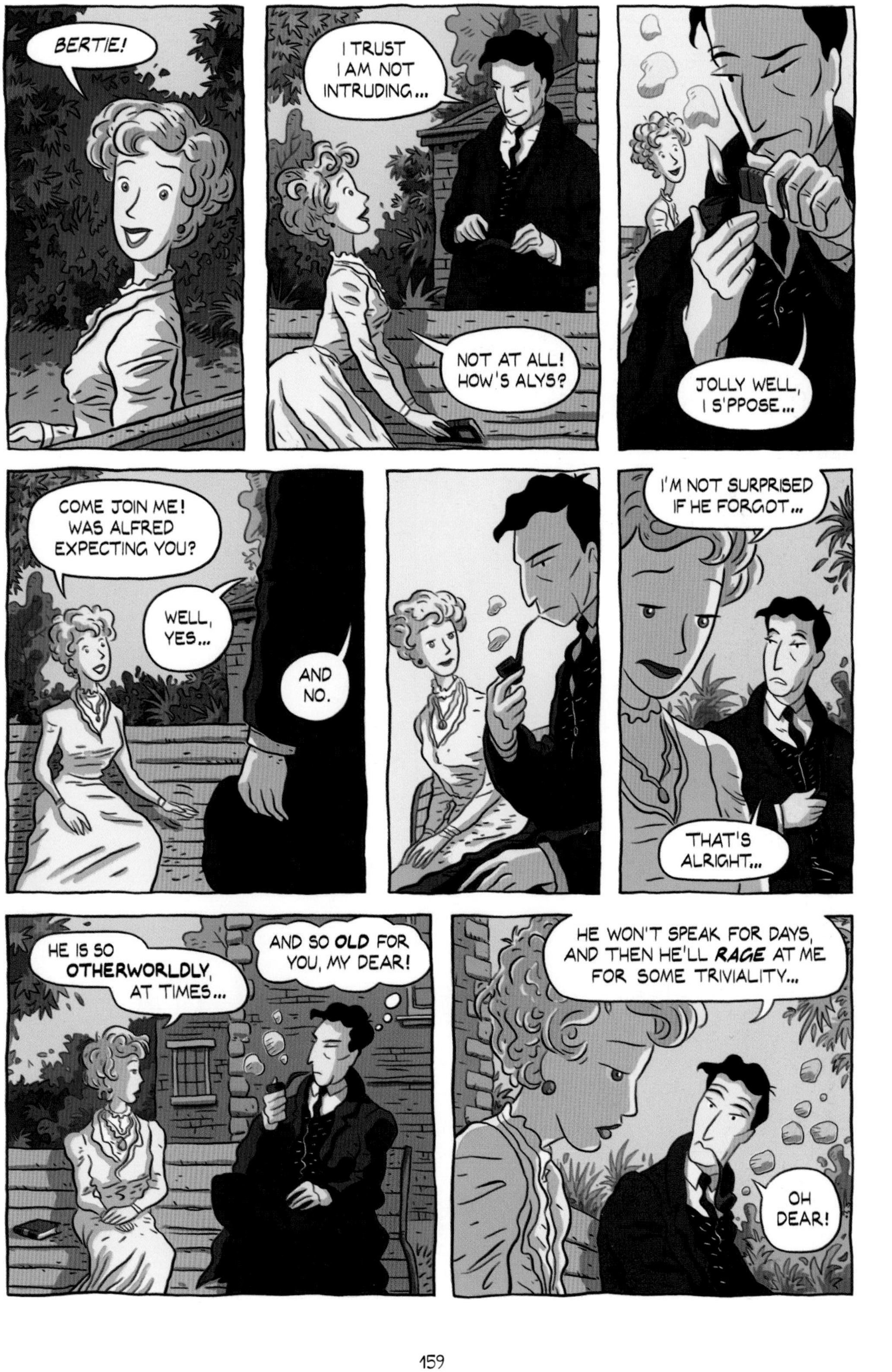
BERTIE!
I TRUST I AM NOT INTRUDING...
NOT AT ALL! HOW'S ALYS?
JOLLY WELL, I S'PPOSE...
COME JOIN ME! WAS ALFRED EXPECTING YOU?
WELL, YES...
AND NO.
I'M NOT SURPRISED IF HE FORGOT...
THAT'S ALRIGHT...
HE IS SO OTHERWORLDLY, AT TIMES...
AND SO OLD FOR YOU, MY DEAR!
HE WON'T SPEAK FOR DAYS, AND THEN HE'LL RAGE AT ME FOR SOME TRIVIALITY...
OH DEAR!

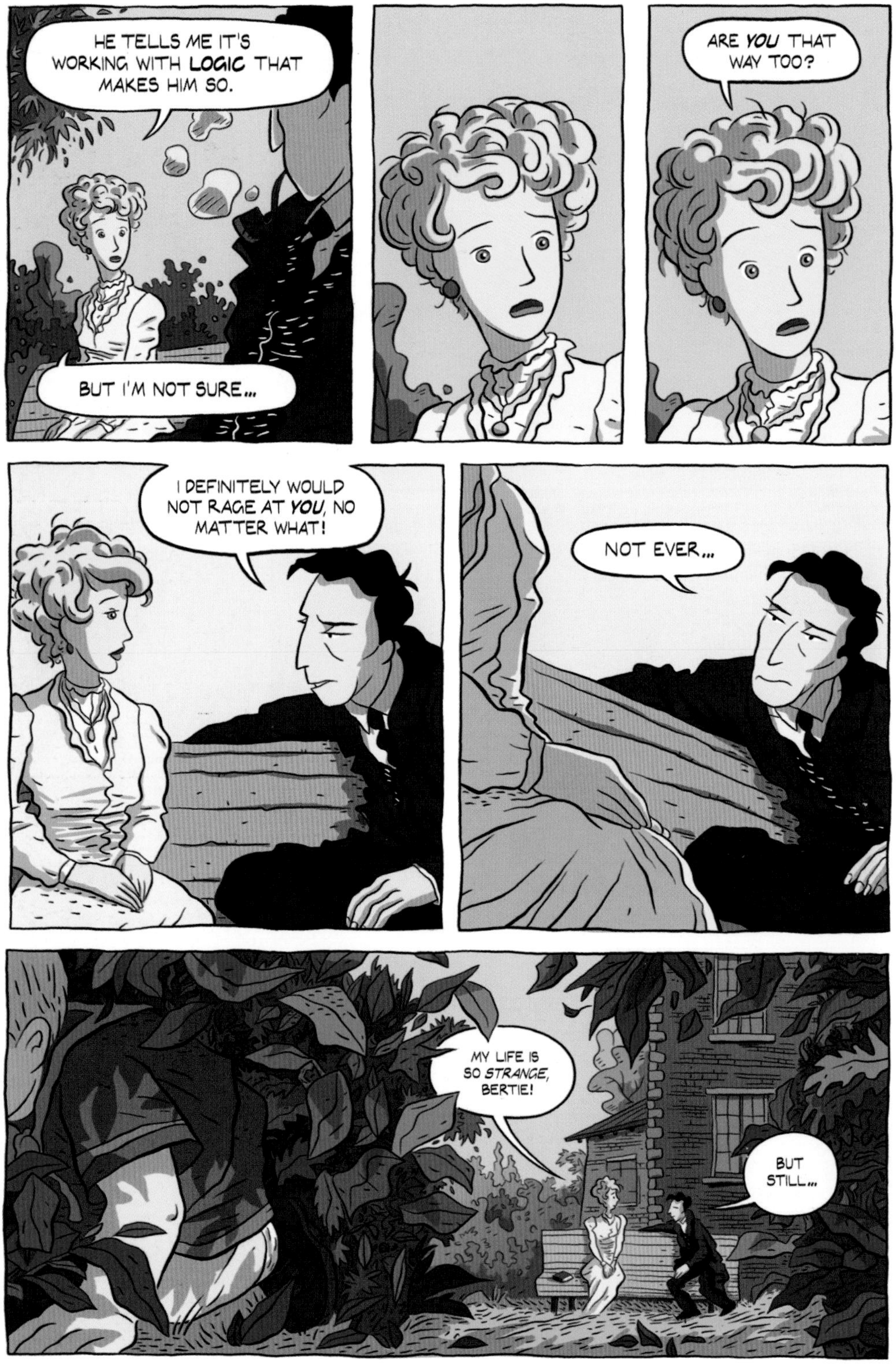
HE TELLS ME IT'S WORKING WITH **LOGIC** THAT MAKES HIM SO.
BUT I'M NOT SURE...
ARE ***YOU*** THAT WAY TOO?
I DEFINITELY WOULD NOT RAGE AT ***YOU***, NO MATTER WHAT!
NOT EVER...
MY LIFE IS SO *STRANGE*, BERTIE!
BUT STILL...

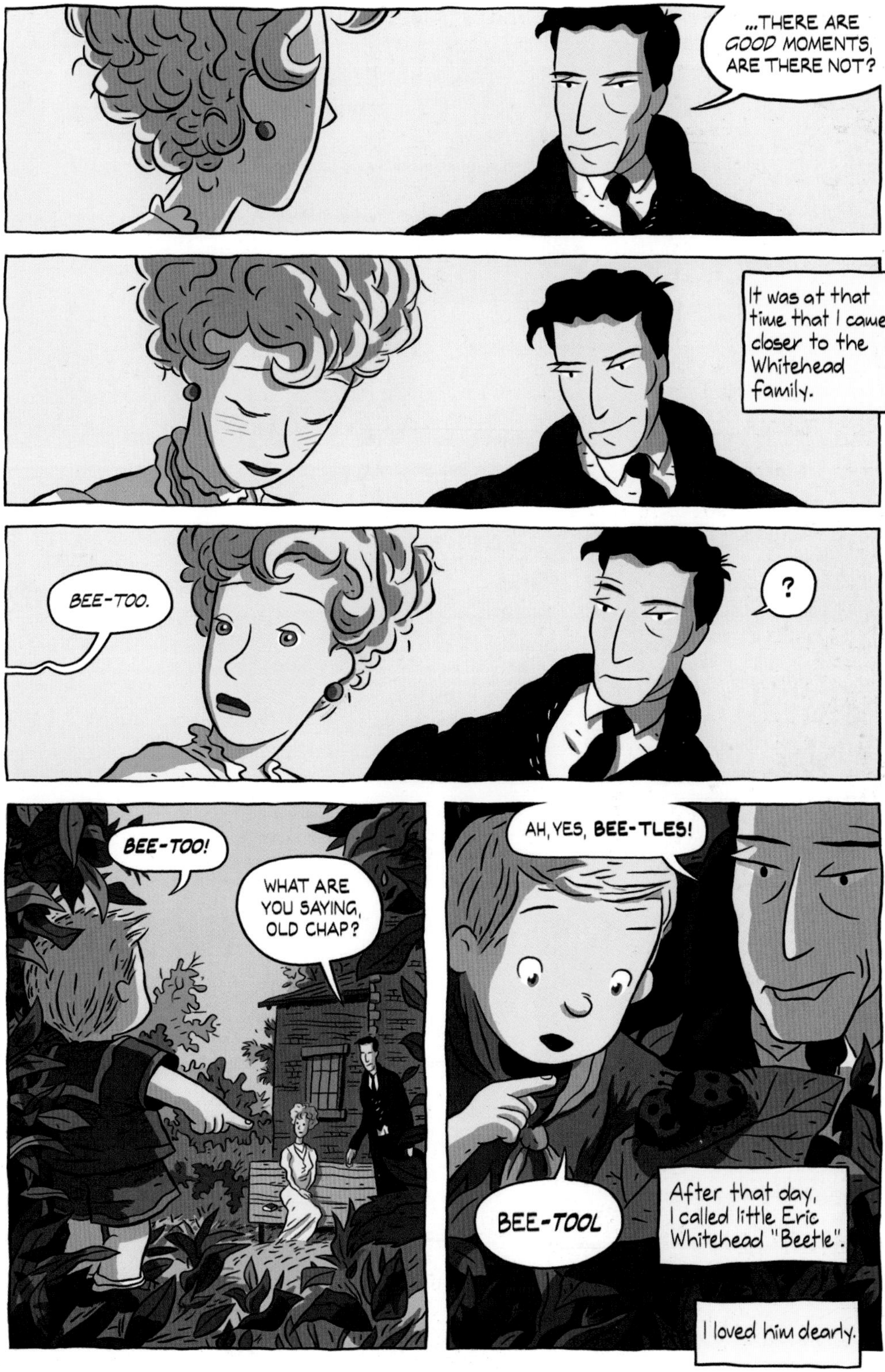
...THERE ARE *GOOD* MOMENTS, ARE THERE NOT?
It was at that time that I came closer to the Whitehead family.
BEE-TOO.
?
BEE-TOO!
WHAT ARE YOU SAYING, OLD CHAP?
AH, YES, **BEE-TLES!**
BEE-*TOOL*
After that day, I called little Eric Whitehead "Beetle".
I loved him dearly.

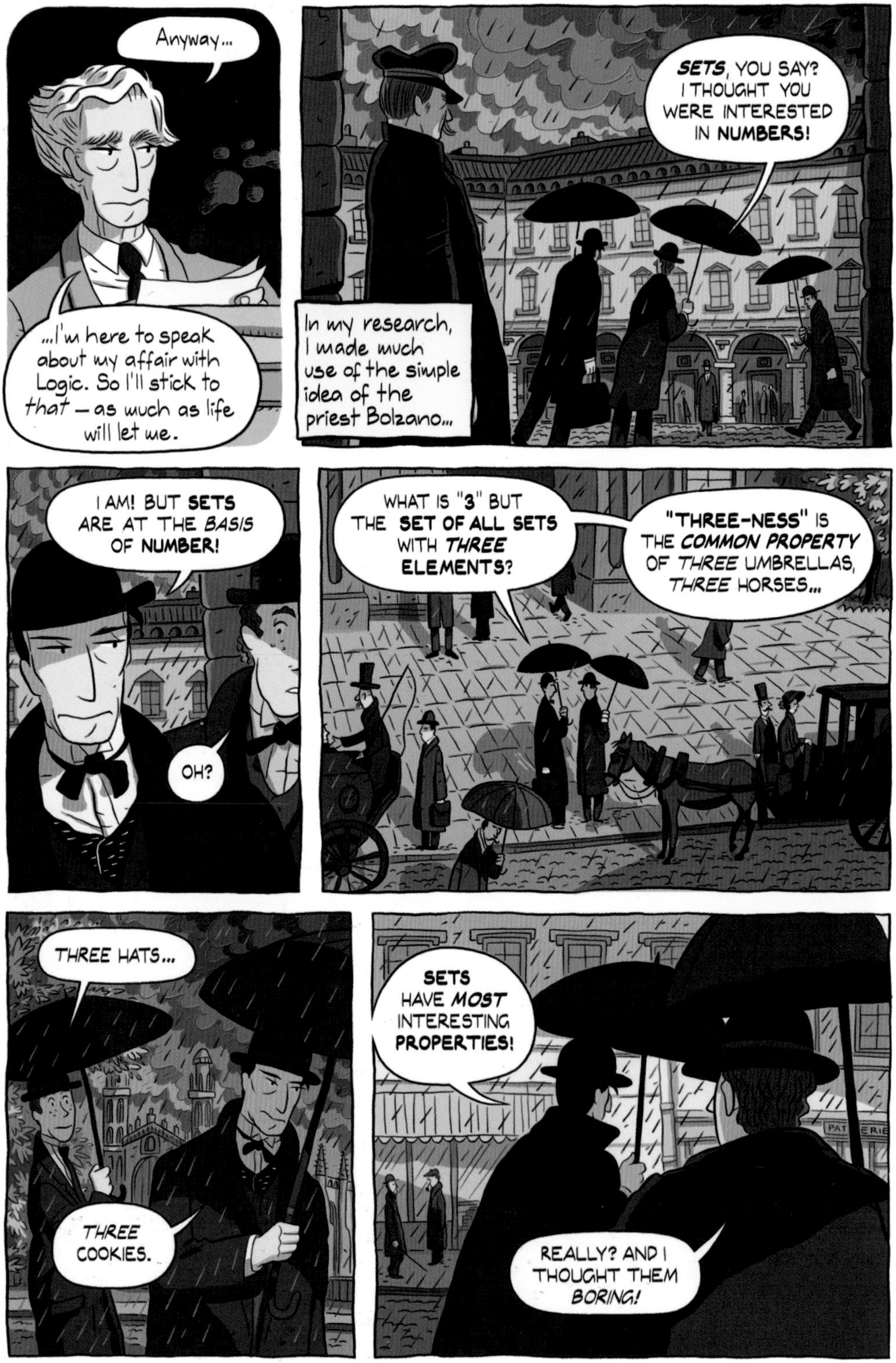
Anyway...
...I'm here to speak about my affair with Logic. So I'll stick to *that* — as much as life will let me.
In my research, I made much use of the simple idea of the priest Bolzano...
SETS, YOU SAY? I THOUGHT YOU WERE INTERESTED IN **NUMBERS!**
I AM! BUT **SETS** ARE AT THE *BASIS* OF **NUMBER!**
OH?
WHAT IS "**3**" BUT THE **SET OF ALL SETS** WITH ***THREE*** **ELEMENTS?**
"THREE-NESS" IS THE ***COMMON PROPERTY*** OF *THREE* UMBRELLAS, *THREE* HORSES...
THREE HATS...
THREE COOKIES.
SETS HAVE ***MOST*** INTERESTING **PROPERTIES!**
PATISSERIE
REALLY? AND I THOUGHT THEM *BORING!*

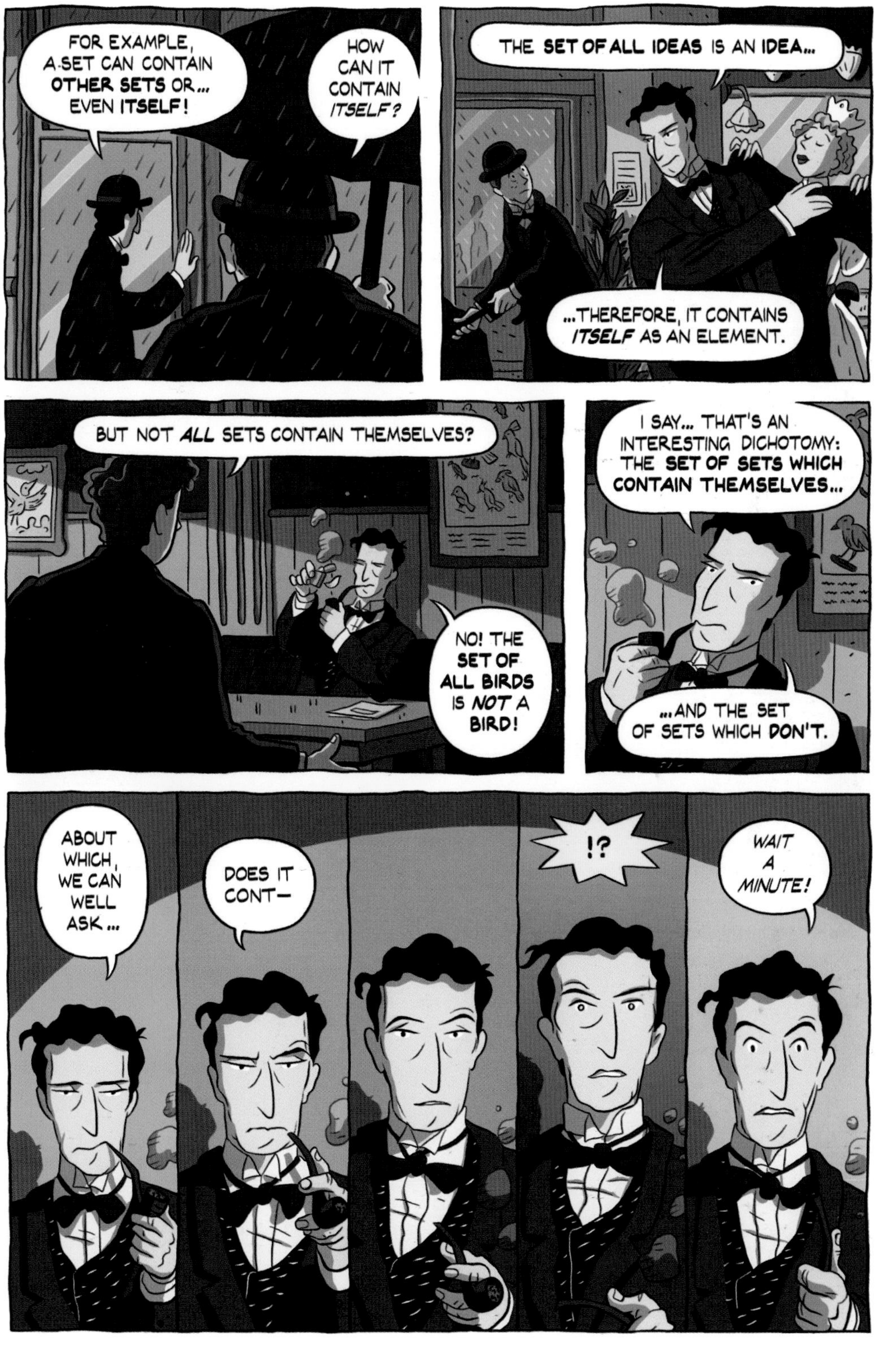
FOR EXAMPLE, A SET CAN CONTAIN **OTHER SETS** OR... EVEN **ITSELF**!
HOW CAN IT CONTAIN *ITSELF*?
THE **SET OF ALL IDEAS** IS AN **IDEA**...
...THEREFORE, IT CONTAINS *ITSELF* AS AN ELEMENT.
BUT NOT *ALL* SETS CONTAIN THEMSELVES?
NO! THE **SET OF ALL BIRDS** IS *NOT* A **BIRD**!
I SAY... THAT'S AN INTERESTING DICHOTOMY: THE **SET OF SETS WHICH CONTAIN THEMSELVES**...
...AND THE SET OF SETS WHICH **DON'T**.
ABOUT WHICH, WE CAN WELL ASK...
DOES IT CONT—
!?
WAIT A MINUTE!

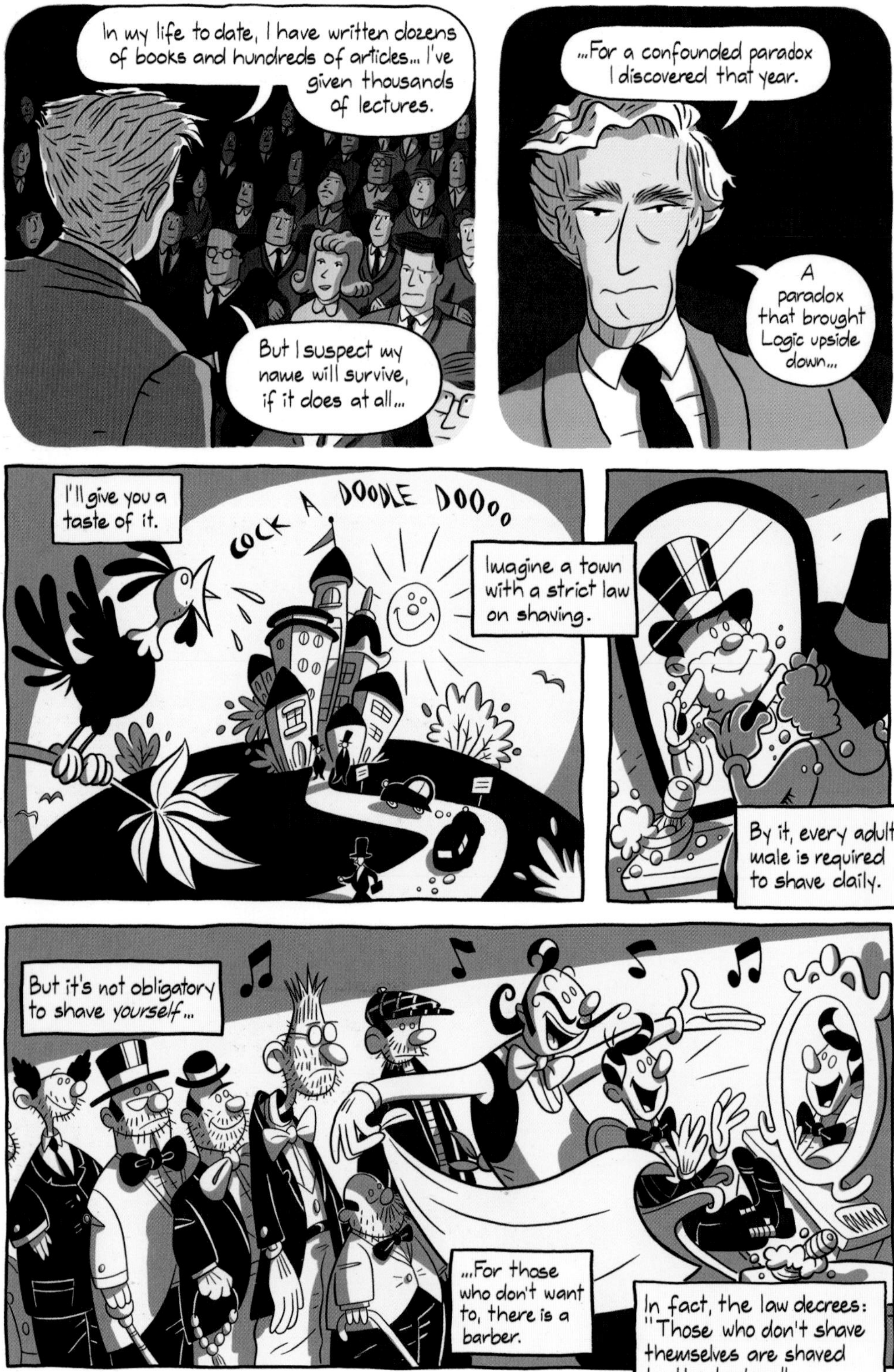
In my life to date, I have written dozens of books and hundreds of articles... I've given thousands of lectures.
But I suspect my name will survive, if it does at all...
...For a confounded paradox I discovered that year.
A paradox that brought Logic upside down...
I'll give you a taste of it.
COCK A DOODLE DOOoo
Imagine a town with a strict law on shaving.
By it, every adult male is required to shave daily.
But it's not obligatory to shave yourself...
...For those who don't want to, there is a barber.
In fact, the law decrees: "Those who don't shave themselves are shaved by the barber."

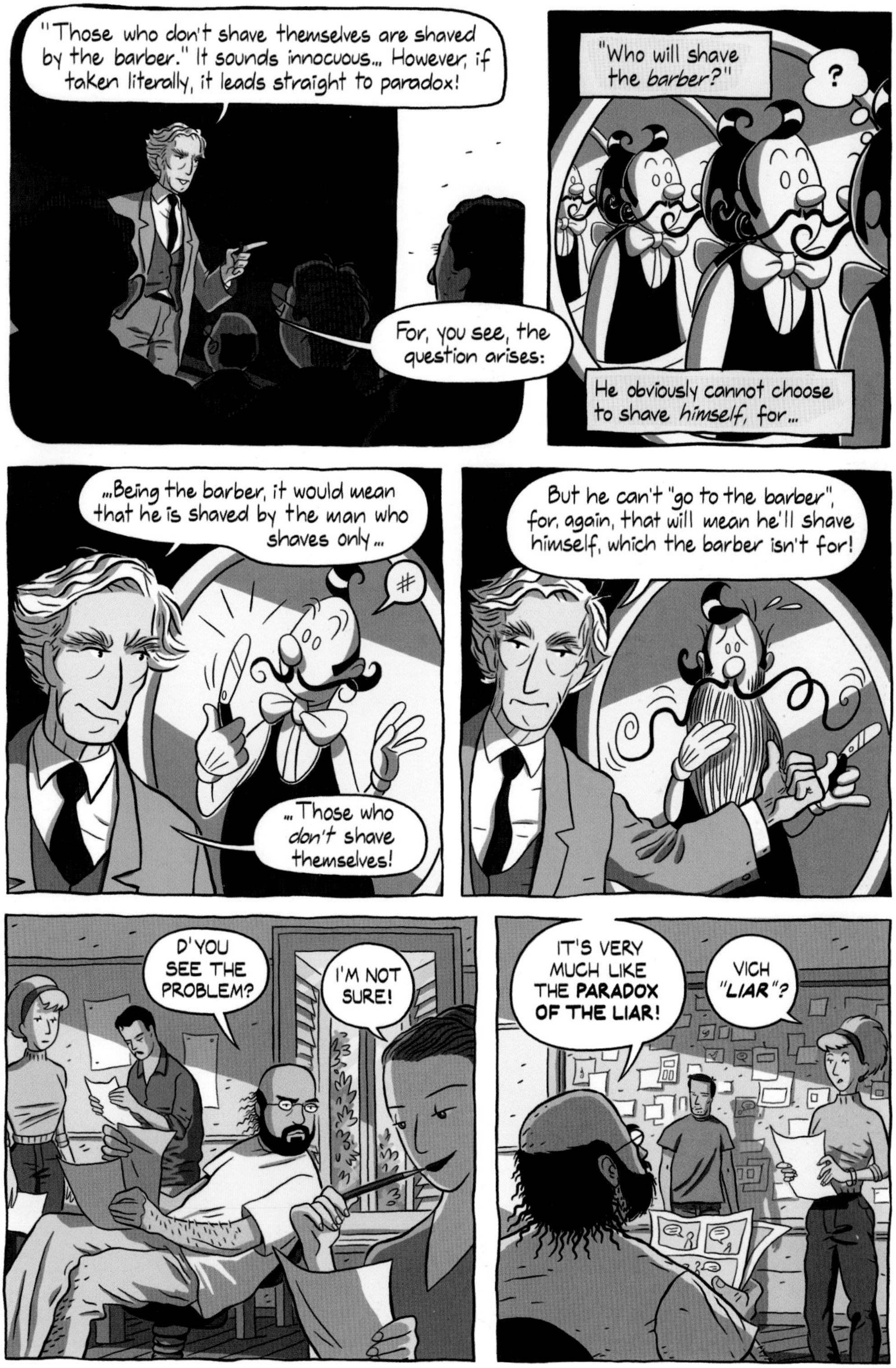
"Those who don't shave themselves are shaved by the barber." It sounds innocuous... However, if taken literally, it leads straight to paradox!
For, you see, the question arises:
"Who will shave the *barber?*"
He obviously cannot choose to shave *himself*, for...
...Being the barber, it would mean that he is shaved by the man who shaves only...
...Those who *don't* shave themselves!
But he can't "go to the barber", for, again, that will mean he'll shave himself, which the barber isn't for!
D'YOU SEE THE PROBLEM?
I'M NOT SURE!
IT'S VERY MUCH LIKE THE **PARADOX OF THE LIAR!**
VICH "***LIAR***"?

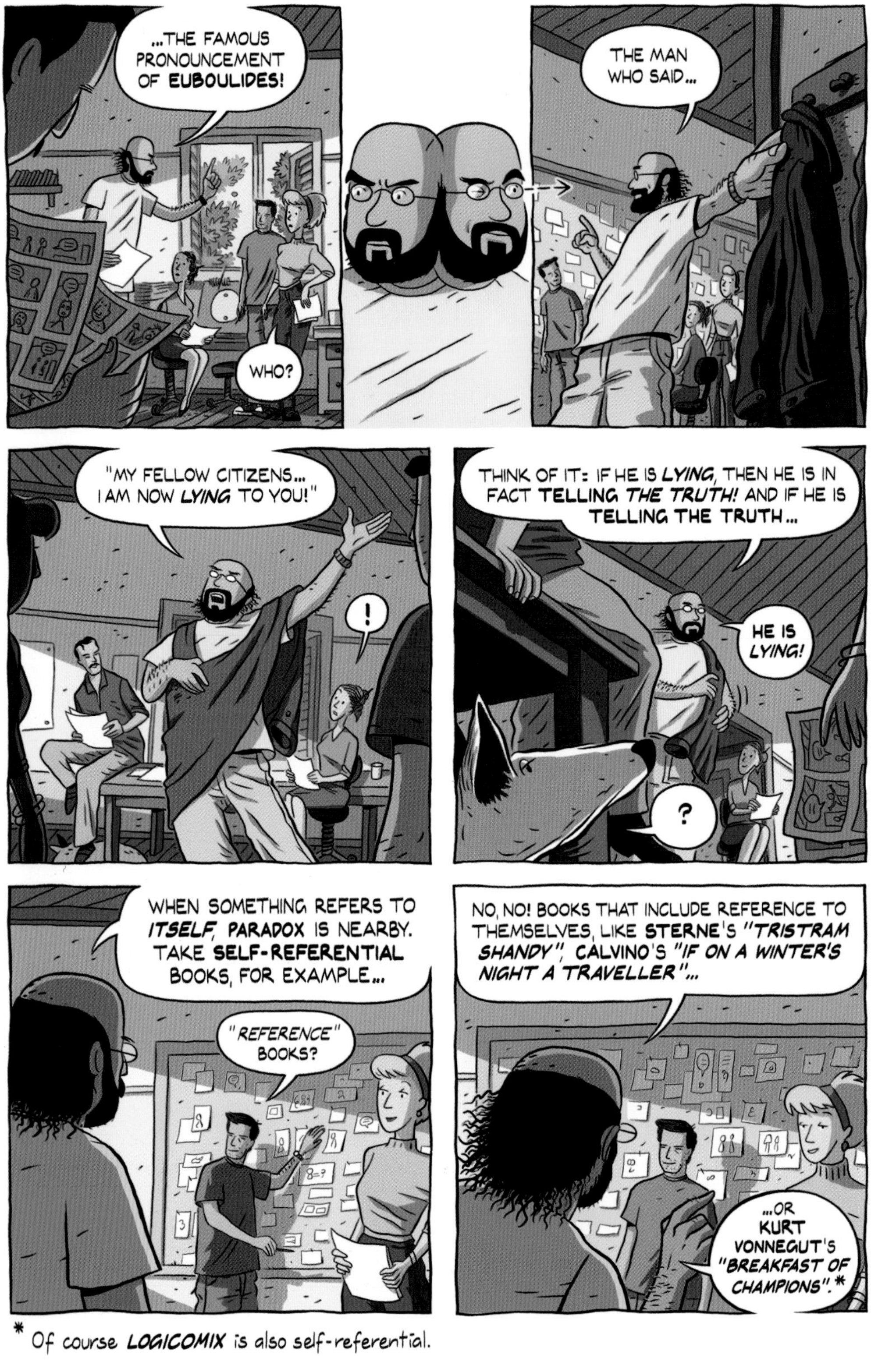
...THE FAMOUS PRONOUNCEMENT OF **EUBOULIDES!**
WHO?
THE MAN WHO SAID...
"MY FELLOW CITIZENS... I AM NOW ***LYING*** TO YOU!"
!
THINK OF IT: IF HE IS ***LYING***, THEN HE IS IN FACT **TELLING *THE TRUTH!*** AND IF HE IS **TELLING THE TRUTH...**
HE IS *LYING!*
?
WHEN SOMETHING REFERS TO ***ITSELF***, **PARADOX** IS NEARBY. TAKE **SELF-REFERENTIAL** BOOKS, FOR EXAMPLE...
"*REFERENCE*" BOOKS?
NO, NO! BOOKS THAT INCLUDE REFERENCE TO THEMSELVES, LIKE **STERNE**'S ***"TRISTRAM SHANDY"***, **CALVINO**'S ***"IF ON A WINTER'S NIGHT A TRAVELLER"***...
...OR **KURT VONNEGUT**'S ***"BREAKFAST OF CHAMPIONS"***.*

* Of course ***LOGICOMIX*** is also self-referential.

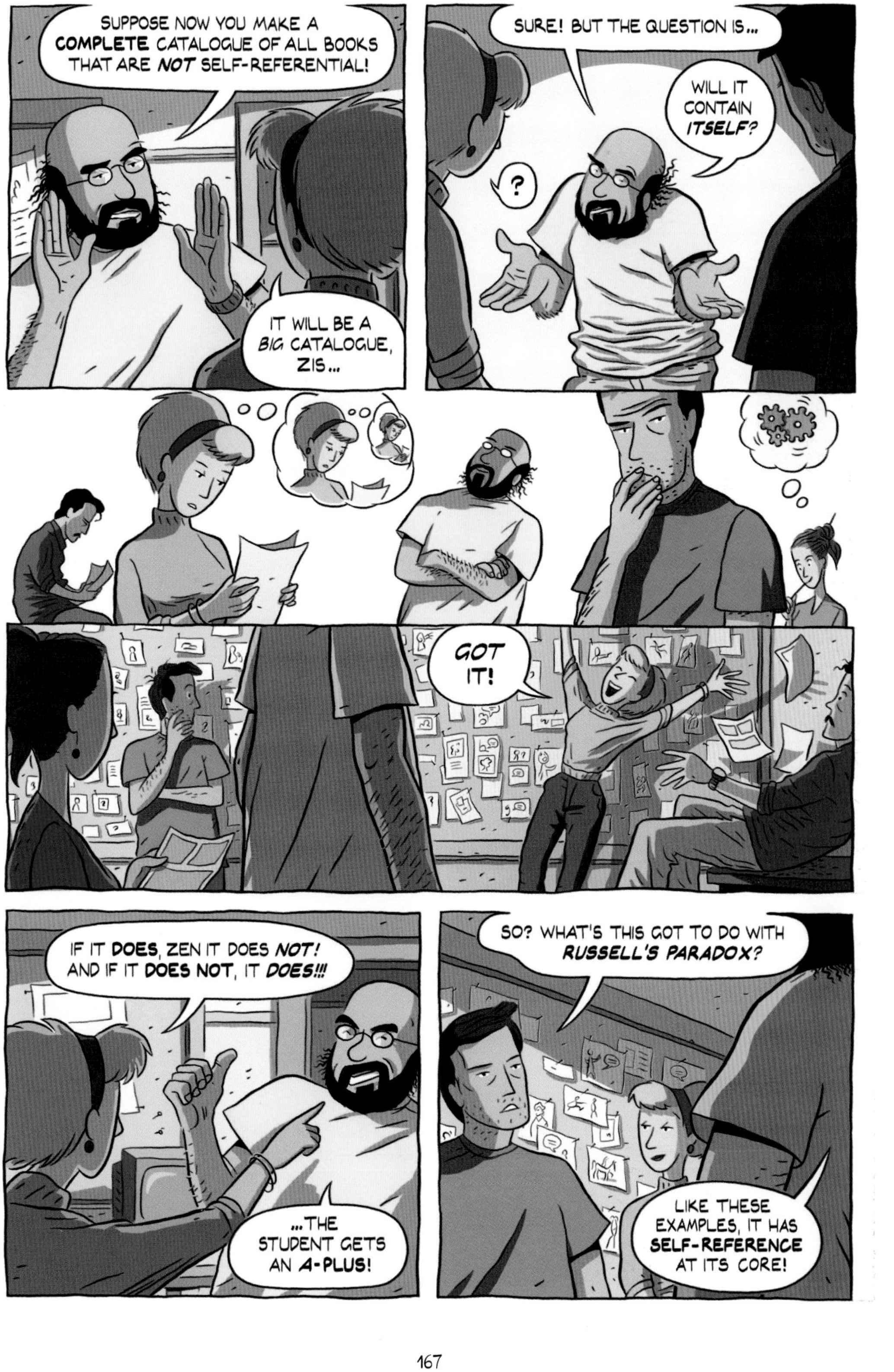
SUPPOSE NOW YOU MAKE A **COMPLETE** CATALOGUE OF ALL BOOKS THAT ARE ***NOT*** SELF-REFERENTIAL!
IT WILL BE A *BIG* CATALOGUE, ZIS...
SURE! BUT THE QUESTION IS...
?
WILL IT CONTAIN ***ITSELF***?
***GOT* IT!**
IF IT **DOES**, ZEN IT DOES ***NOT!*** AND IF IT **DOES NOT**, IT ***DOES!!!***
...THE STUDENT GETS AN ***A-PLUS!***
SO? WHAT'S THIS GOT TO DO WITH ***RUSSELL'S PARADOX?***
LIKE THESE EXAMPLES, IT HAS **SELF-REFERENCE** AT ITS CORE!

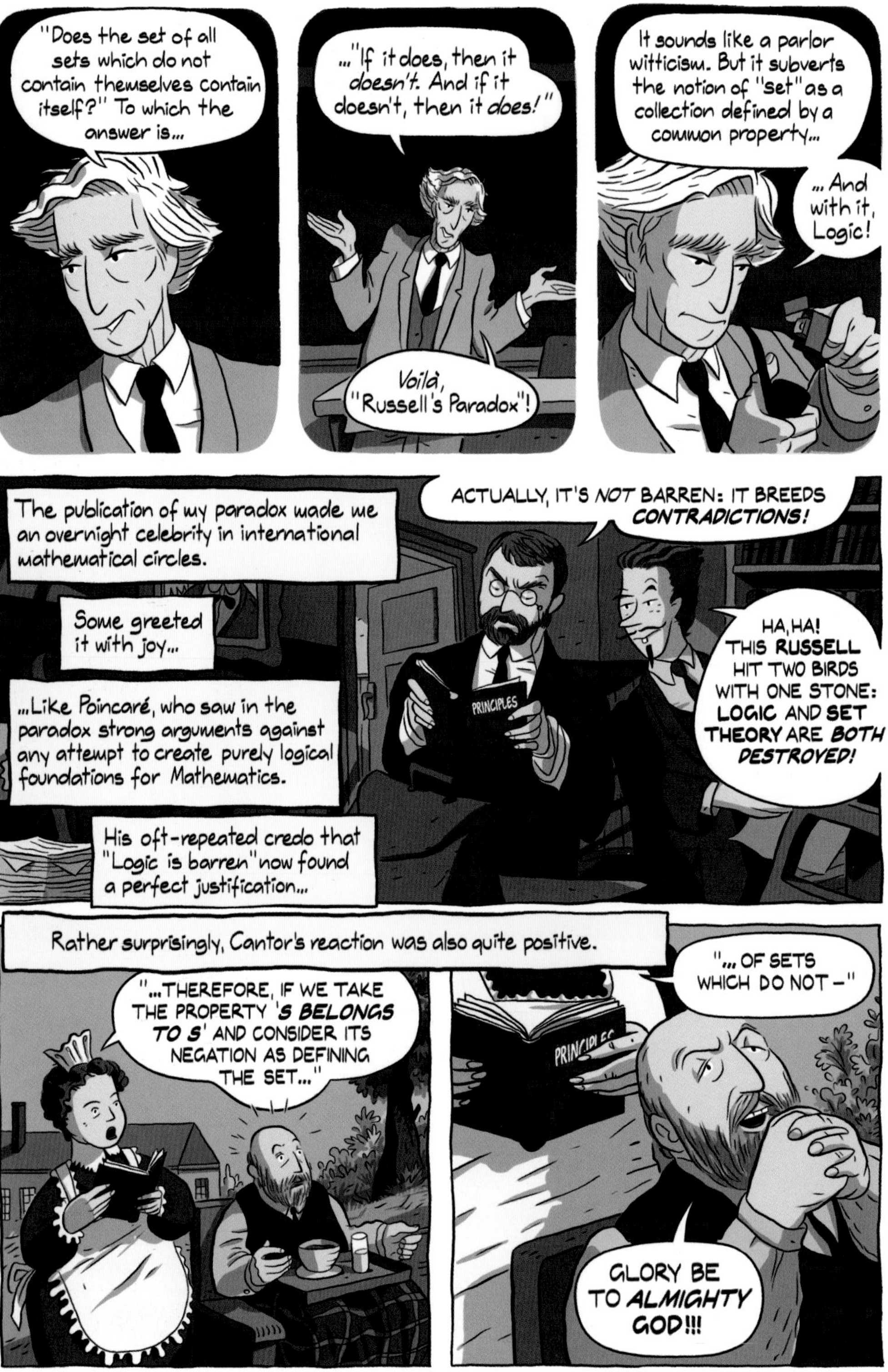
"Does the set of all sets which do not contain themselves contain itself?" To which the answer is...
..."If it does, then it doesn't. And if it doesn't, then it does!"
Voilà, "Russell's Paradox"!
It sounds like a parlor witticism. But it subverts the notion of "set" as a collection defined by a common property...
... And with it, Logic!
The publication of my paradox made me an overnight celebrity in international mathematical circles.
ACTUALLY, IT'S NOT BARREN: IT BREEDS CONTRADICTIONS!
Some greeted it with joy...
...Like Poincaré, who saw in the paradox strong arguments against any attempt to create purely logical foundations for Mathematics.
PRINCIPLES
HA, HA! THIS RUSSELL HIT TWO BIRDS WITH ONE STONE: LOGIC AND SET THEORY ARE BOTH DESTROYED!
His oft-repeated credo that "Logic is barren" now found a perfect justification...
Rather surprisingly, Cantor's reaction was also quite positive.
"...THEREFORE, IF WE TAKE THE PROPERTY 'S BELONGS TO S' AND CONSIDER ITS NEGATION AS DEFINING THE SET..."
"... OF SETS WHICH DO NOT –"
GLORY BE TO ALMIGHTY GOD!!!

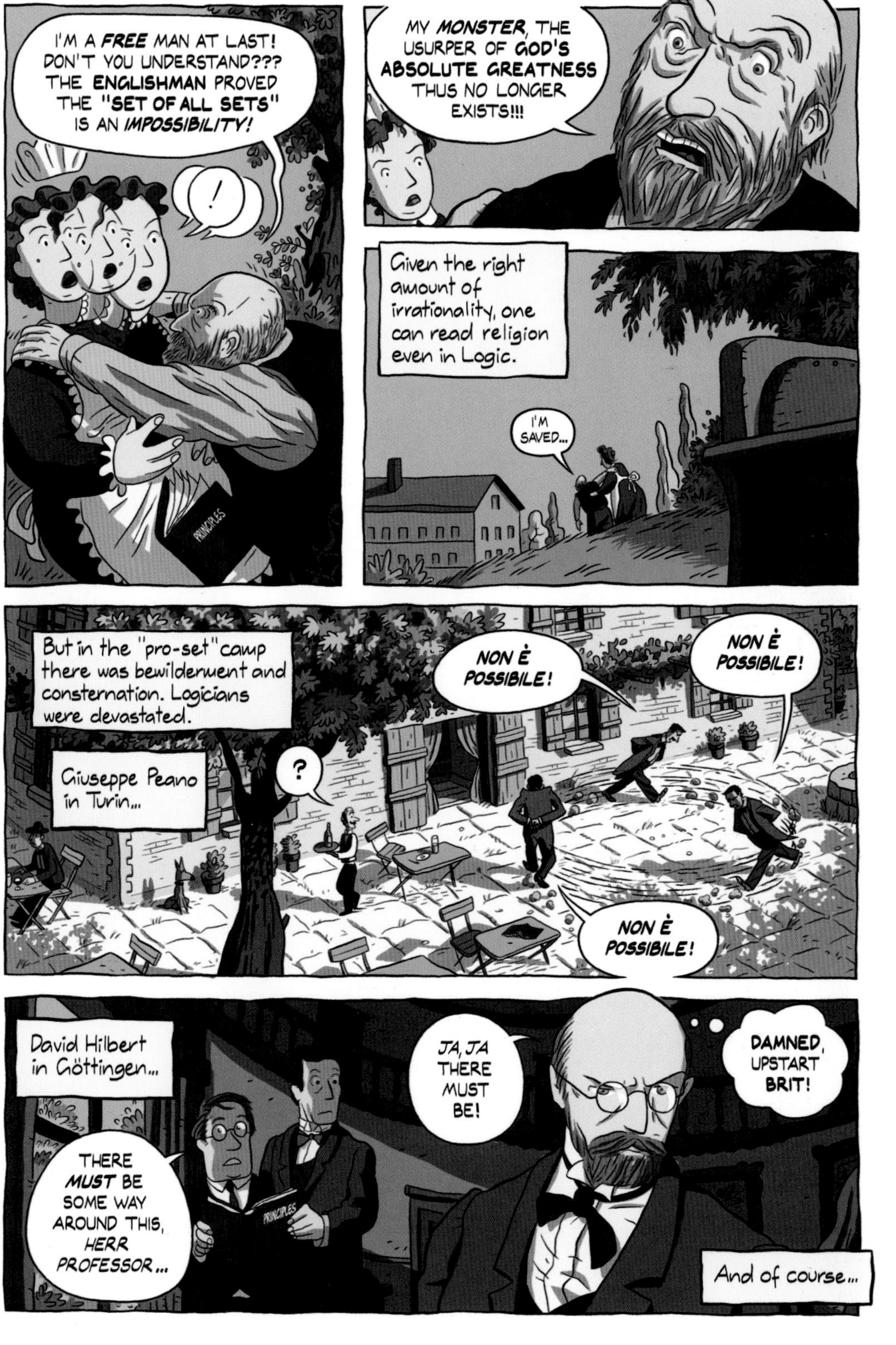
I'M A FREE MAN AT LAST! DON'T YOU UNDERSTAND??? THE ENGLISHMAN PROVED THE "SET OF ALL SETS" IS AN IMPOSSIBILITY!
!
MY MONSTER, THE USURPER OF GOD'S ABSOLUTE GREATNESS THUS NO LONGER EXISTS!!!
Given the right amount of irrationality, one can read religion even in Logic.
I'M SAVED...
PRINCIPLES
But in the "pro-set" camp there was bewilderment and consternation. Logicians were devastated.
Giuseppe Peano in Turin...
?
NON È POSSIBILE!
NON È POSSIBILE!
NON È POSSIBILE!
David Hilbert in Göttingen...
THERE MUST BE SOME WAY AROUND THIS, HERR PROFESSOR...
PRINCIPLES
JA, JA THERE MUST BE!
DAMNED, UPSTART BRIT!
And of course...

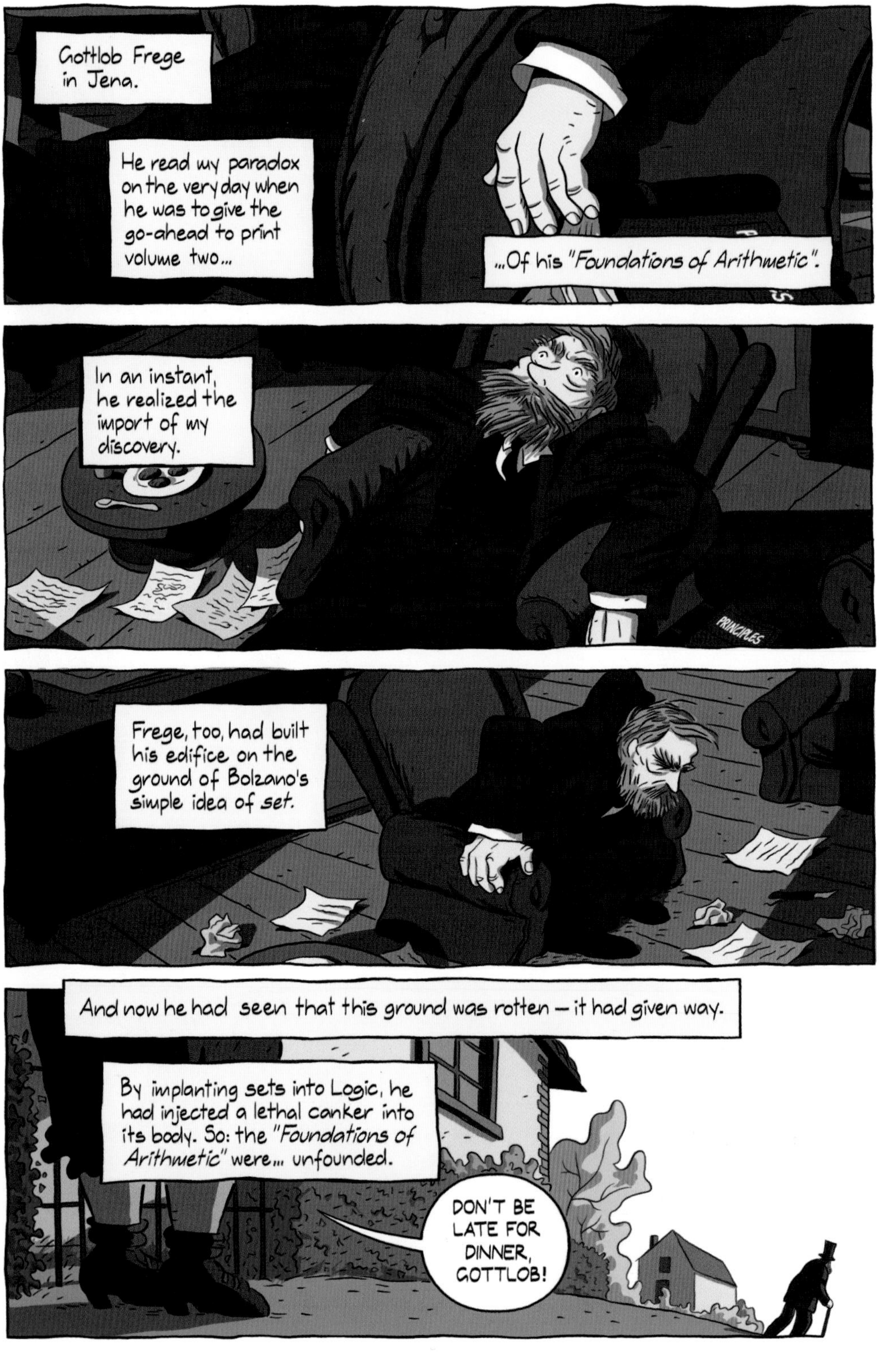
Gottlob Frege in Jena.
He read my paradox on the very day when he was to give the go-ahead to print volume two...
...Of his "Foundations of Arithmetic".
In an instant, he realized the import of my discovery.
PRINCIPLES
Frege, too, had built his edifice on the ground of Bolzano's simple idea of *set*.
And now he had seen that this ground was rotten — it had given way.
By implanting sets into Logic, he had injected a lethal canker into its body. So: the "Foundations of Arithmetic" were... unfounded.
DON'T BE LATE FOR DINNER, GOTTLOB!

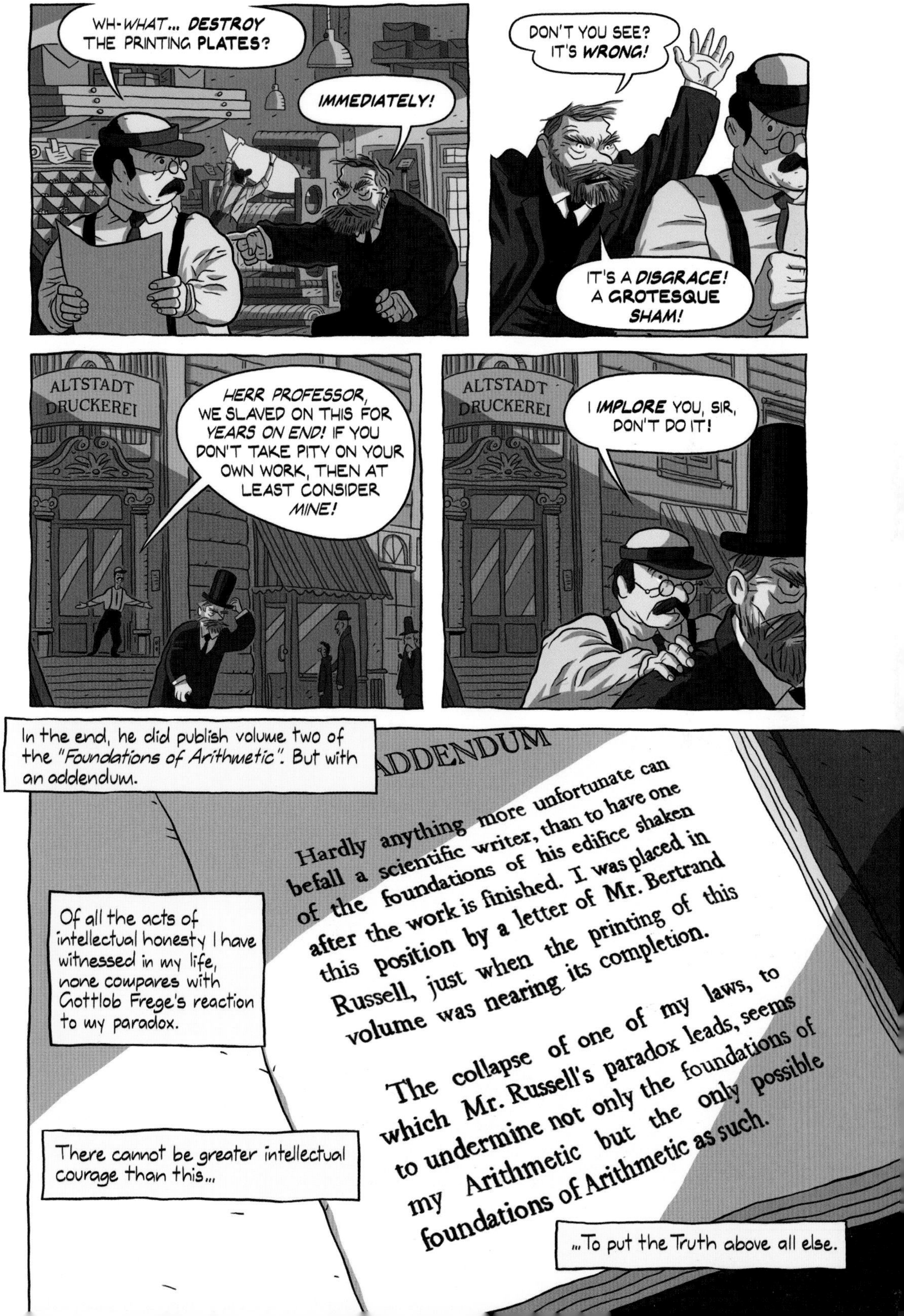

WH-WHAT... DESTROY THE PRINTING PLATES?
IMMEDIATELY!
DON'T YOU SEE? IT'S WRONG!
IT'S A DISGRACE! A GROTESQUE SHAM!
ALTSTADT DRUCKEREI
HERR PROFESSOR, WE SLAVED ON THIS FOR YEARS ON END! IF YOU DON'T TAKE PITY ON YOUR OWN WORK, THEN AT LEAST CONSIDER MINE!
ALTSTADT DRUCKEREI
I IMPLORE YOU, SIR, DON'T DO IT!
In the end, he did publish volume two of the "Foundations of Arithmetic". But with an addendum.
ADDENDUM
Hardly anything more unfortunate can befall a scientific writer, than to have one of the foundations of his edifice shaken after the work is finished. I was placed in this position by a letter of Mr. Bertrand Russell, just when the printing of this volume was nearing its completion.
The collapse of one of my laws, to which Mr. Russell's paradox leads, seems to undermine not only the foundations of my Arithmetic but the only possible foundations of Arithmetic as such.
Of all the acts of intellectual honesty I have witnessed in my life, none compares with Gottlob Frege's reaction to my paradox.
There cannot be greater intellectual courage than this...
...To put the Truth above all else.

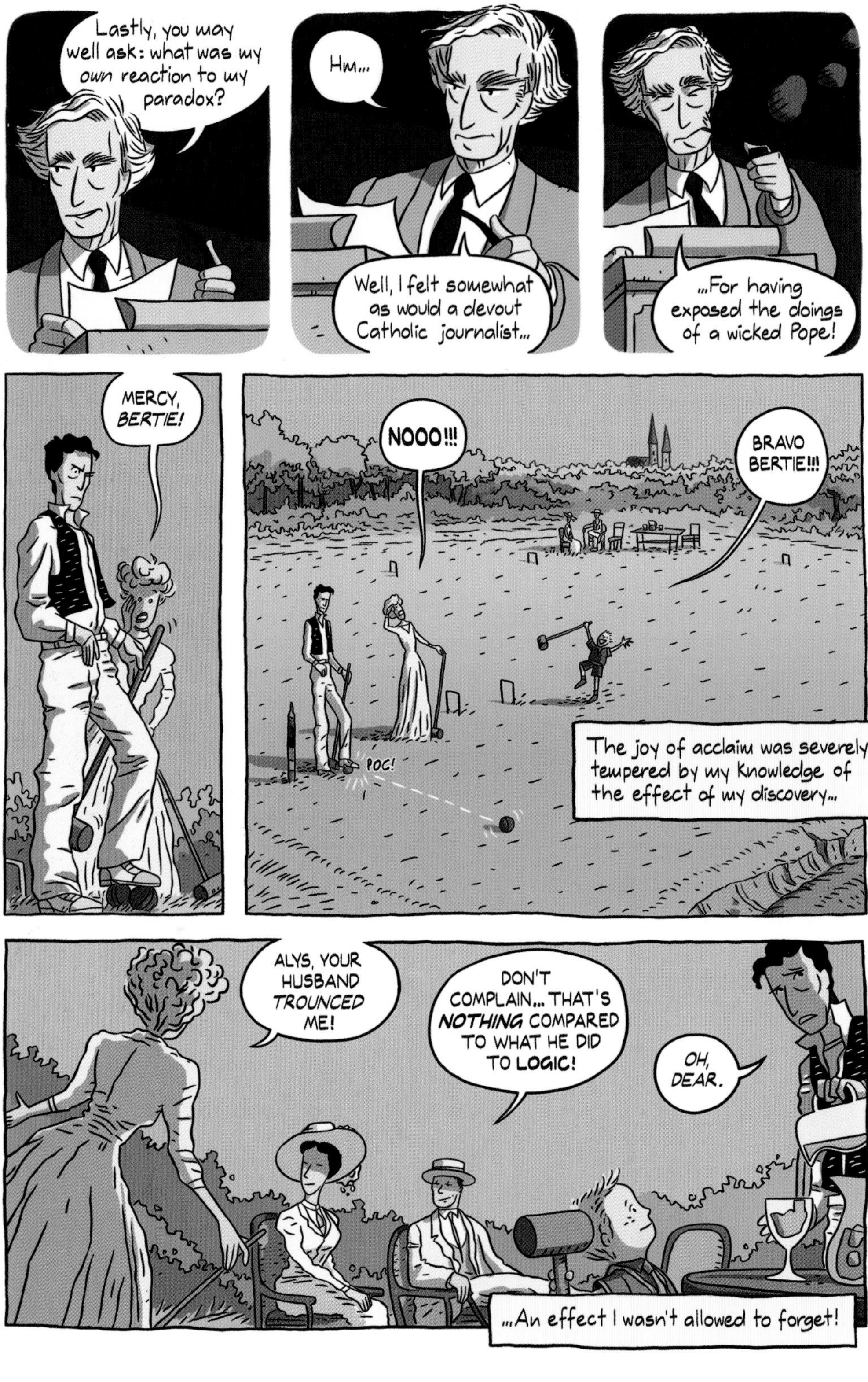

Lastly, you may well ask: what was my own reaction to my paradox?
Hm...
Well, I felt somewhat as would a devout Catholic journalist...
...For having exposed the doings of a wicked Pope!
MERCY, BERTIE!
NOOO!!!
BRAVO BERTIE!!!
POC!
The joy of acclaim was severely tempered by my knowledge of the effect of my discovery...
ALYS, YOUR HUSBAND TROUNCED ME!
DON'T COMPLAIN... THAT'S NOTHING COMPARED TO WHAT HE DID TO LOGIC!
OH, DEAR.
...An effect I wasn't allowed to forget!

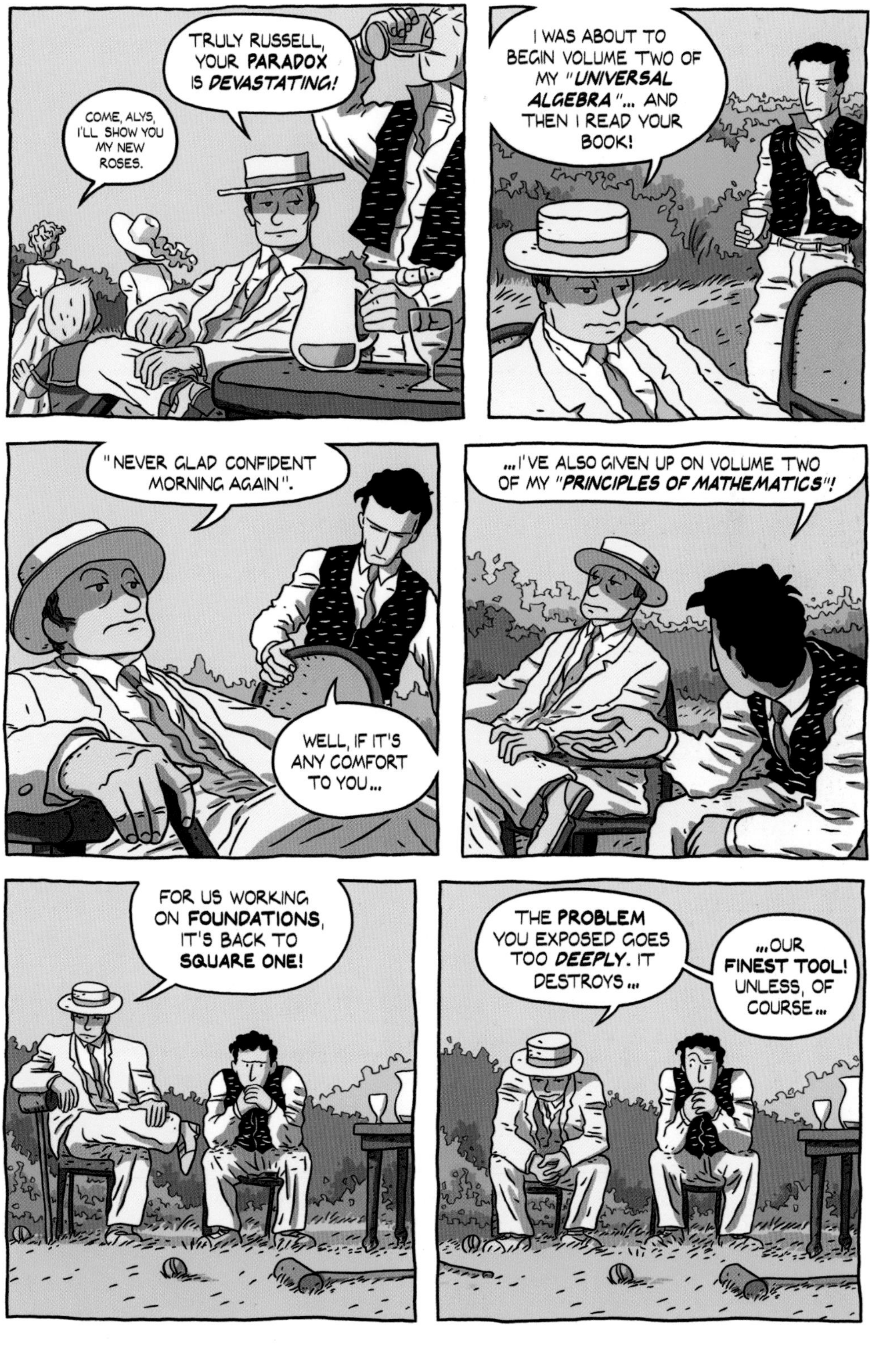
COME, ALYS, I'LL SHOW YOU MY NEW ROSES.
TRULY RUSSELL, YOUR **PARADOX** IS ***DEVASTATING!***
I WAS ABOUT TO BEGIN VOLUME TWO OF MY "***UNIVERSAL ALGEBRA***"... AND THEN I READ YOUR BOOK!
"NEVER GLAD CONFIDENT MORNING AGAIN".
WELL, IF IT'S ANY COMFORT TO YOU...
...I'VE ALSO GIVEN UP ON VOLUME TWO OF MY "***PRINCIPLES OF MATHEMATICS***"!
FOR US WORKING ON **FOUNDATIONS**, IT'S BACK TO **SQUARE ONE!**
THE **PROBLEM** YOU EXPOSED GOES TOO ***DEEPLY***. IT DESTROYS...
...OUR **FINEST TOOL!** UNLESS, OF COURSE...

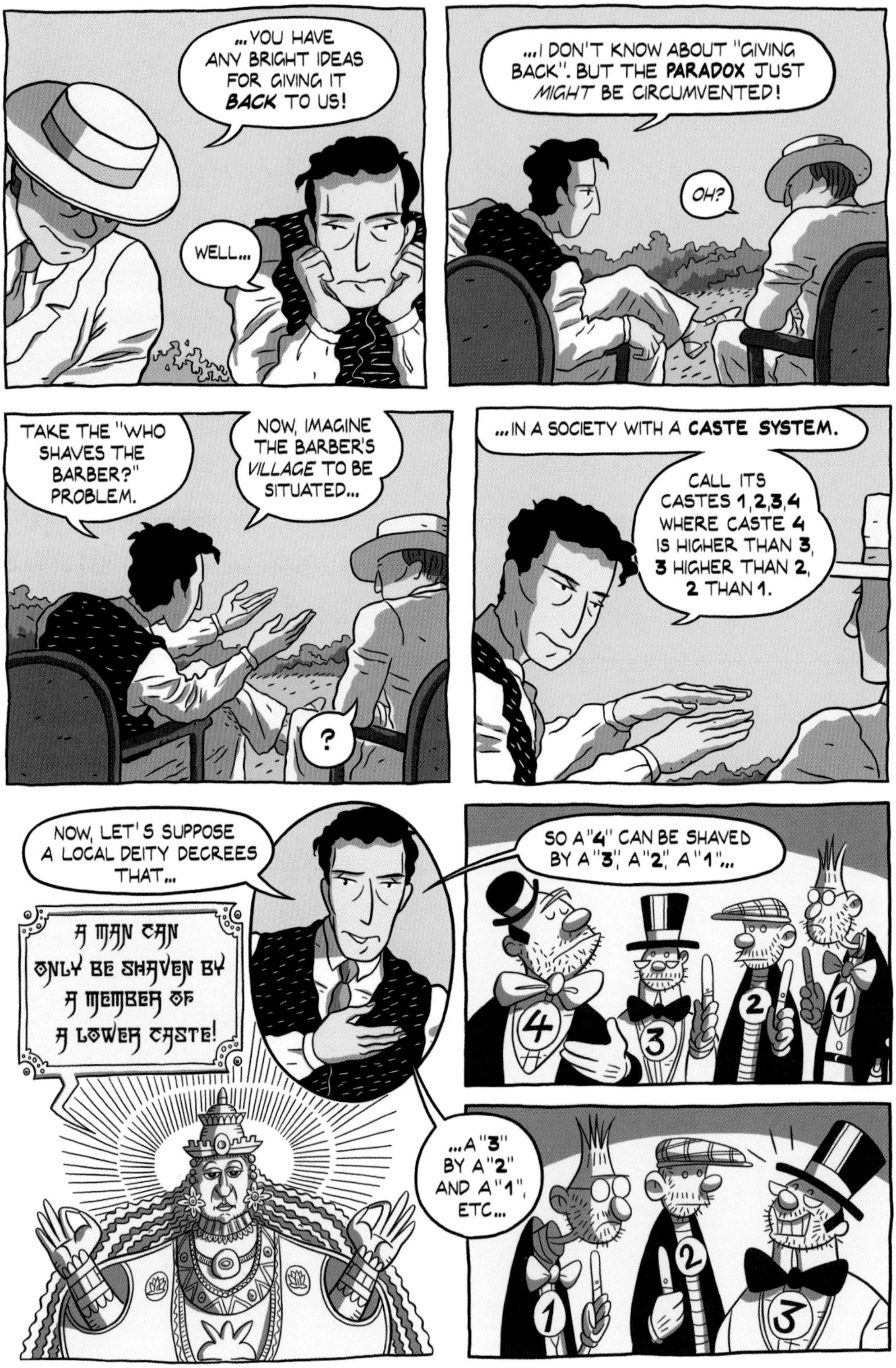
...YOU HAVE ANY BRIGHT IDEAS FOR GIVING IT BACK TO US!
WELL...
...I DON'T KNOW ABOUT "GIVING BACK". BUT THE PARADOX JUST MIGHT BE CIRCUMVENTED!
OH?
TAKE THE "WHO SHAVES THE BARBER?" PROBLEM.
NOW, IMAGINE THE BARBER'S VILLAGE TO BE SITUATED...
?
...IN A SOCIETY WITH A CASTE SYSTEM.
CALL ITS CASTES 1, 2, 3, 4 WHERE CASTE 4 IS HIGHER THAN 3, 3 HIGHER THAN 2, 2 THAN 1.
NOW, LET'S SUPPOSE A LOCAL DEITY DECREES THAT...
A MAN CAN ONLY BE SHAVEN BY A MEMBER OF A LOWER CASTE!
SO A "4" CAN BE SHAVED BY A "3", A "2", A "1"...
...A "3" BY A "2" AND A "1", ETC...

YOU SEE? BY FORBIDDING **INTRA-CASTE** SHAVING YOU ALSO RULE OUT **SELF**-SHAVING!

HM.

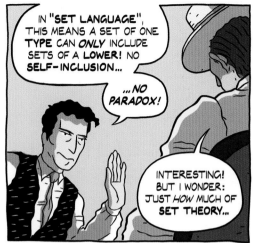

IN "SET LANGUAGE", THIS MEANS A SET OF ONE **TYPE** CAN **ONLY** INCLUDE SETS OF A **LOWER**! NO SELF-INCLUSION...

...NO PARADOX!

INTERESTING! BUT I WONDER: JUST **HOW** MUCH OF **SET THEORY**...

...YOU **THROW OUT** TOGETHER WITH THE **PARADOX** ITSELF?

ALSO, IN THIS NICE VILLAGE OF YOURS...

...YOU GET **VERY HAIRY** "1"S!

BUT THAT'S THE **IDEA**: SOME SETS **CAN'T** CONTAIN OTHERS!

HM... WELL... YOUR "**TYPES**" ARE WORTH EXPLORING. AND ANYWAY...

...THEY'RE **ALL** WE'VE GOT, FOR THE MOMENT!

I DON'T KNOW...

...EVERY MORNING, I WAKE UP AN OPTIMIST, BUT AFTER A DAY'S WORK, I DESPAIR.

THE VERY **SIZE** OF THE PROBLEM MAKES ME LOSE HEART.

RUSSELL?

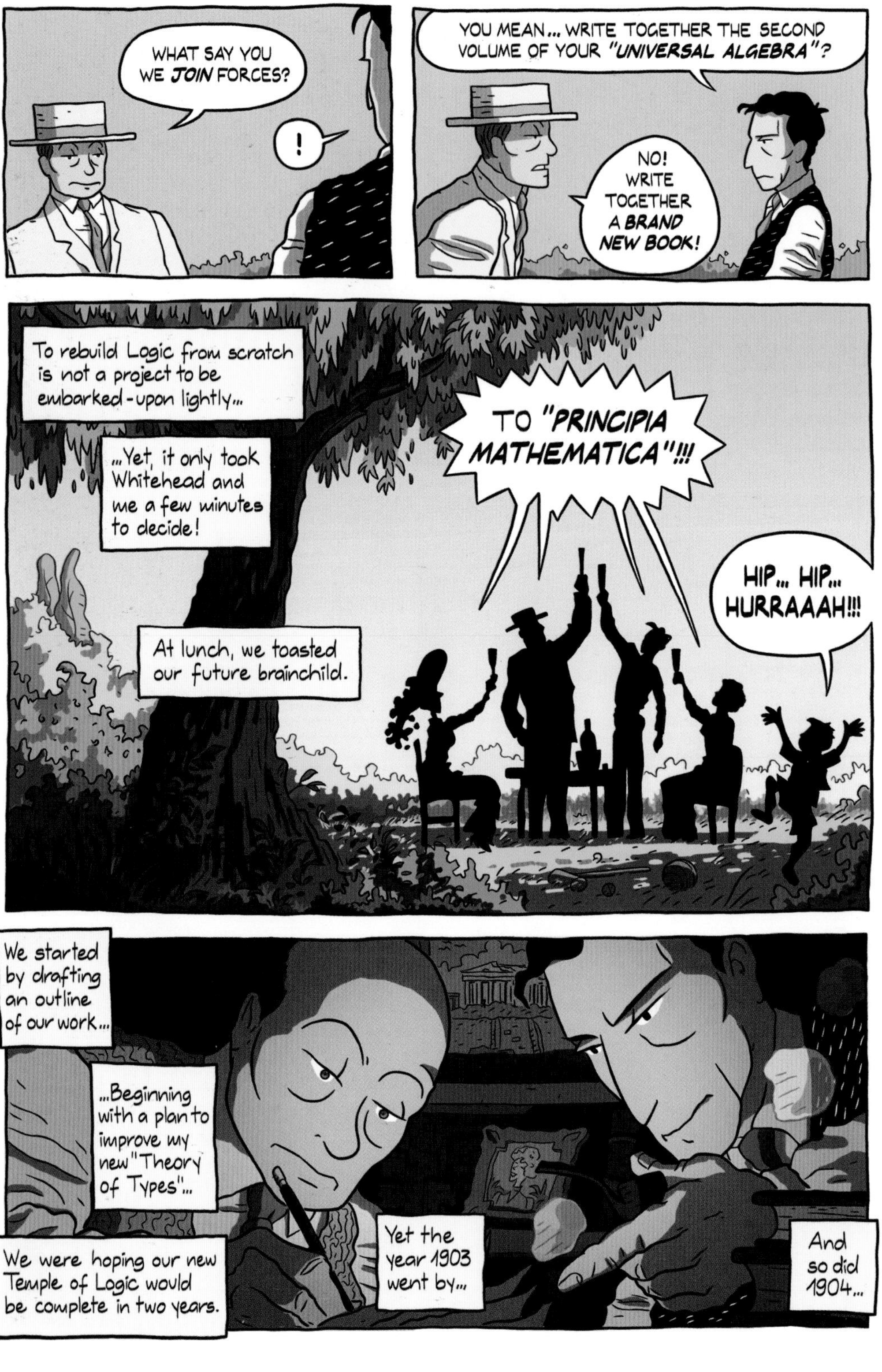
WHAT SAY YOU WE *JOIN* FORCES?
!
YOU MEAN... WRITE TOGETHER THE SECOND VOLUME OF YOUR *"UNIVERSAL ALGEBRA"*?
NO! WRITE TOGETHER A *BRAND NEW BOOK*!
To rebuild Logic from scratch is not a project to be embarked-upon lightly...
...Yet, it only took Whitehead and me a few minutes to decide!
TO *"PRINCIPIA MATHEMATICA"*!!!
At lunch, we toasted our future brainchild.
HIP... HIP... HURRAAAH!!!
We started by drafting an outline of our work...
...Beginning with a plan to improve my new "Theory of Types"...
We were hoping our new Temple of Logic would be complete in two years.
Yet the year 1903 went by...
And so did 1904...

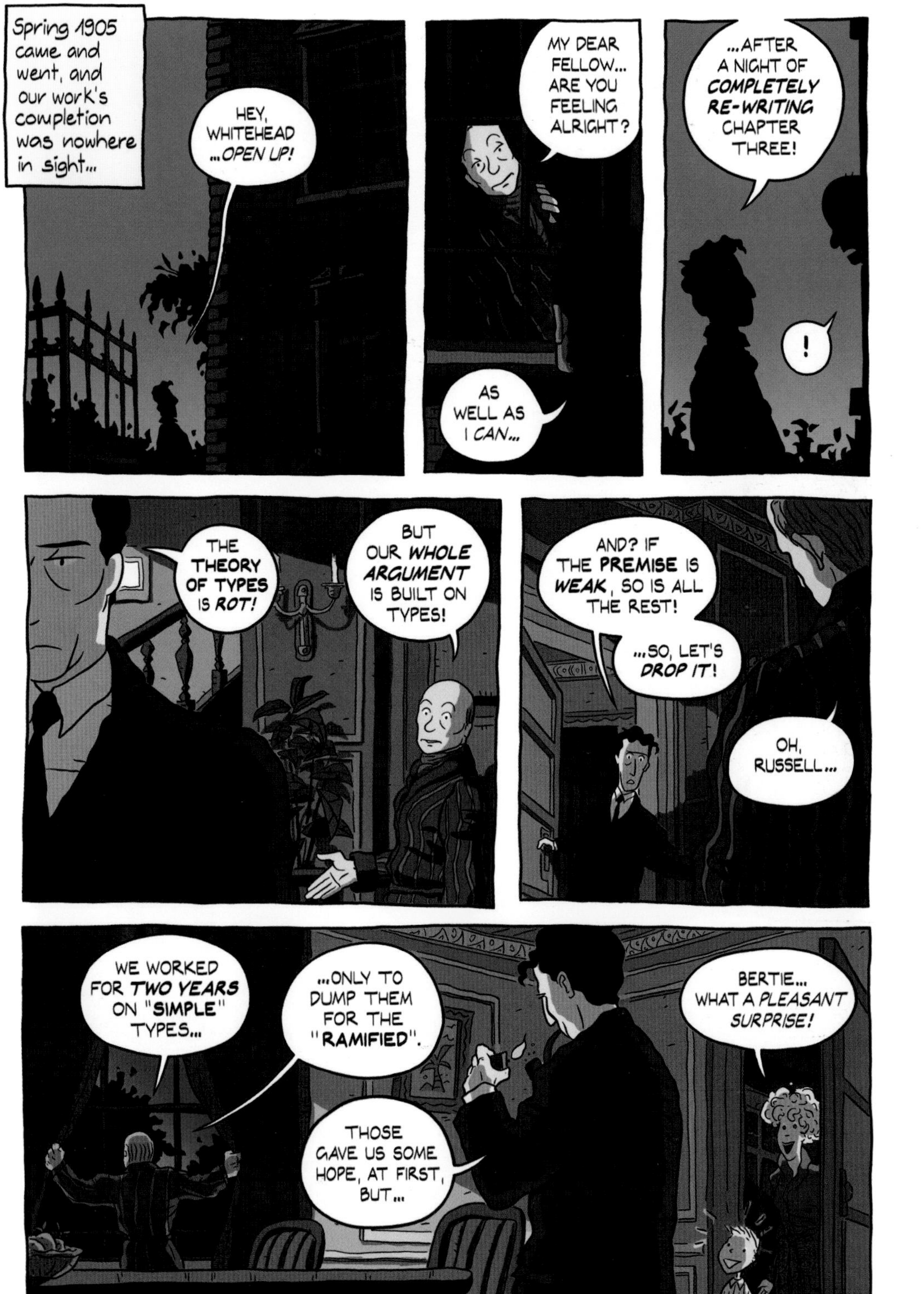
Spring 1905 came and went, and our work's completion was nowhere in sight...
HEY, WHITEHEAD ...OPEN UP!
MY DEAR FELLOW... ARE YOU FEELING ALRIGHT?
AS WELL AS I CAN...
...AFTER A NIGHT OF COMPLETELY RE-WRITING CHAPTER THREE!
!
THE THEORY OF TYPES IS ROT!
BUT OUR WHOLE ARGUMENT IS BUILT ON TYPES!
AND? IF THE PREMISE IS WEAK, SO IS ALL THE REST!
...SO, LET'S DROP IT!
OH, RUSSELL...
WE WORKED FOR TWO YEARS ON "SIMPLE" TYPES...
...ONLY TO DUMP THEM FOR THE "RAMIFIED".
THOSE GAVE US SOME HOPE, AT FIRST, BUT...
BERTIE... WHAT A PLEASANT SURPRISE!

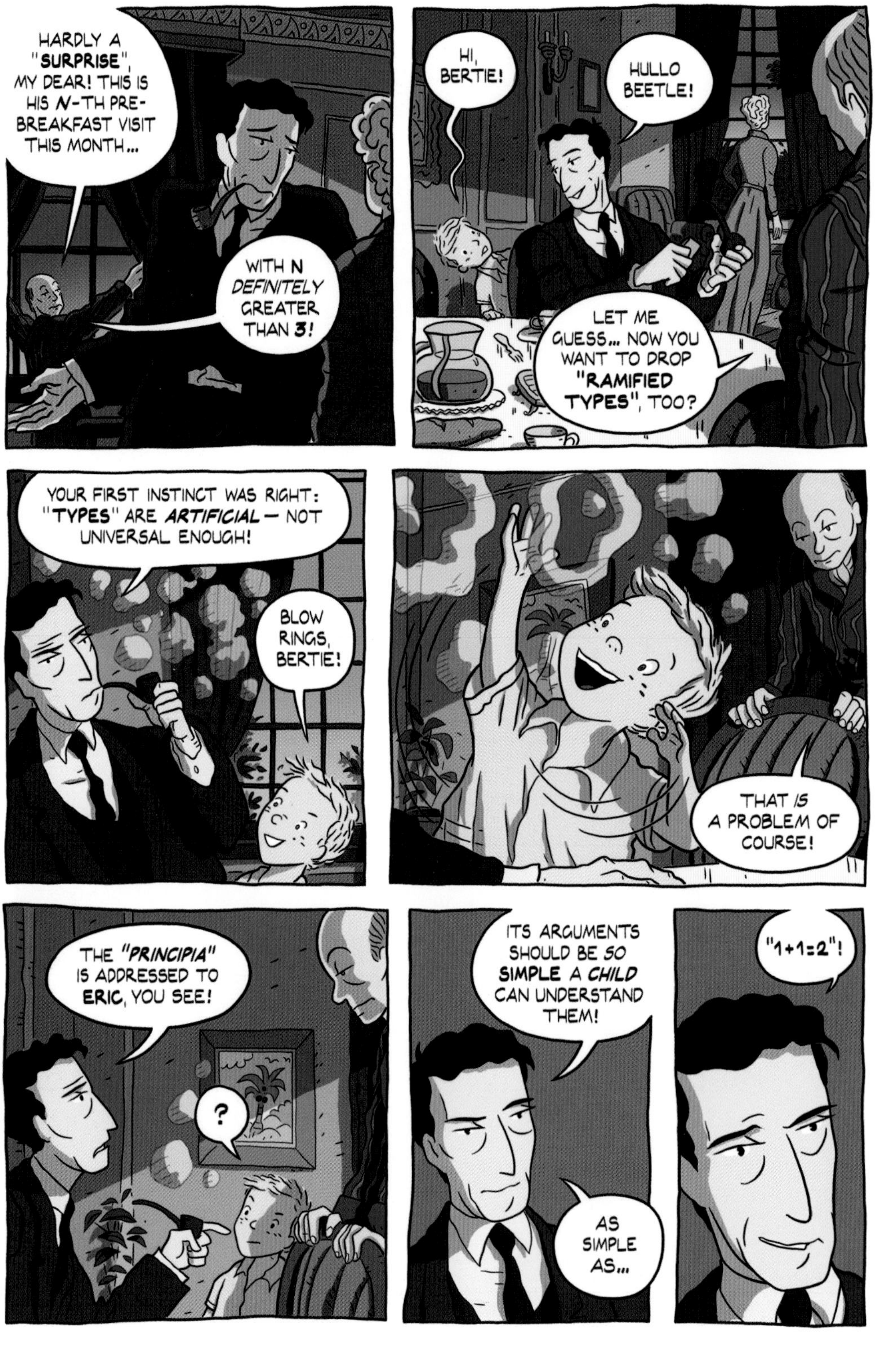
HARDLY A "SURPRISE", MY DEAR! THIS IS HIS N-TH PRE-BREAKFAST VISIT THIS MONTH...
WITH N DEFINITELY GREATER THAN 3!
HI, BERTIE!
HULLO BEETLE!
LET ME GUESS... NOW YOU WANT TO DROP "RAMIFIED TYPES", TOO?
YOUR FIRST INSTINCT WAS RIGHT: "TYPES" ARE ARTIFICIAL — NOT UNIVERSAL ENOUGH!
BLOW RINGS, BERTIE!
THAT IS A PROBLEM OF COURSE!
THE "PRINCIPIA" IS ADDRESSED TO ERIC, YOU SEE!
?
ITS ARGUMENTS SHOULD BE SO SIMPLE A CHILD CAN UNDERSTAND THEM!
AS SIMPLE AS...
"1+1=2"!

Even before I knew this was called "Occam's Razor", I felt that the simpler a theory was, the greater was its value.

The *"Principia"* could not diverge from this principle. Yet, as you can see from some lines of the book, randomly picked, our understanding of simplicity was a wee bit idiosyncratic.

$$\therefore z, w \epsilon \iota'x \cup \iota'y . \supset_{zw} . \phi(z,w) \colon \equiv . \phi(x,$$
$$\vdash . *51\cdot234 . *11\cdot62 . \supset \vdash \colon . s, w \epsilon \iota'x \cup \iota'y .$$
$$z \epsilon \iota'x \cup \iota'y . \supset . \phi$$
$$[*51\cdot234 . *10\cdot29] \equiv \colon \phi(x,x) . \phi(x,y) . \phi(y,x)$$

Whitehead and I spent that summer re-examining our premises.

...SO IF WE TAKE PREDICATE "P" TO STAND FOR...

And by the time autumn came, we had changed course again.

AND SO REDUCES TO...

Our new trick was as old as Euclid: a new set of axioms!

Once again, we rebuilt from the bottom.

To gain more time for work, my wife and I moved in with the Whiteheads.

We found the change most pleasing — well, at least half of us did!

Yet, living under the same roof was not a cure for our problems...

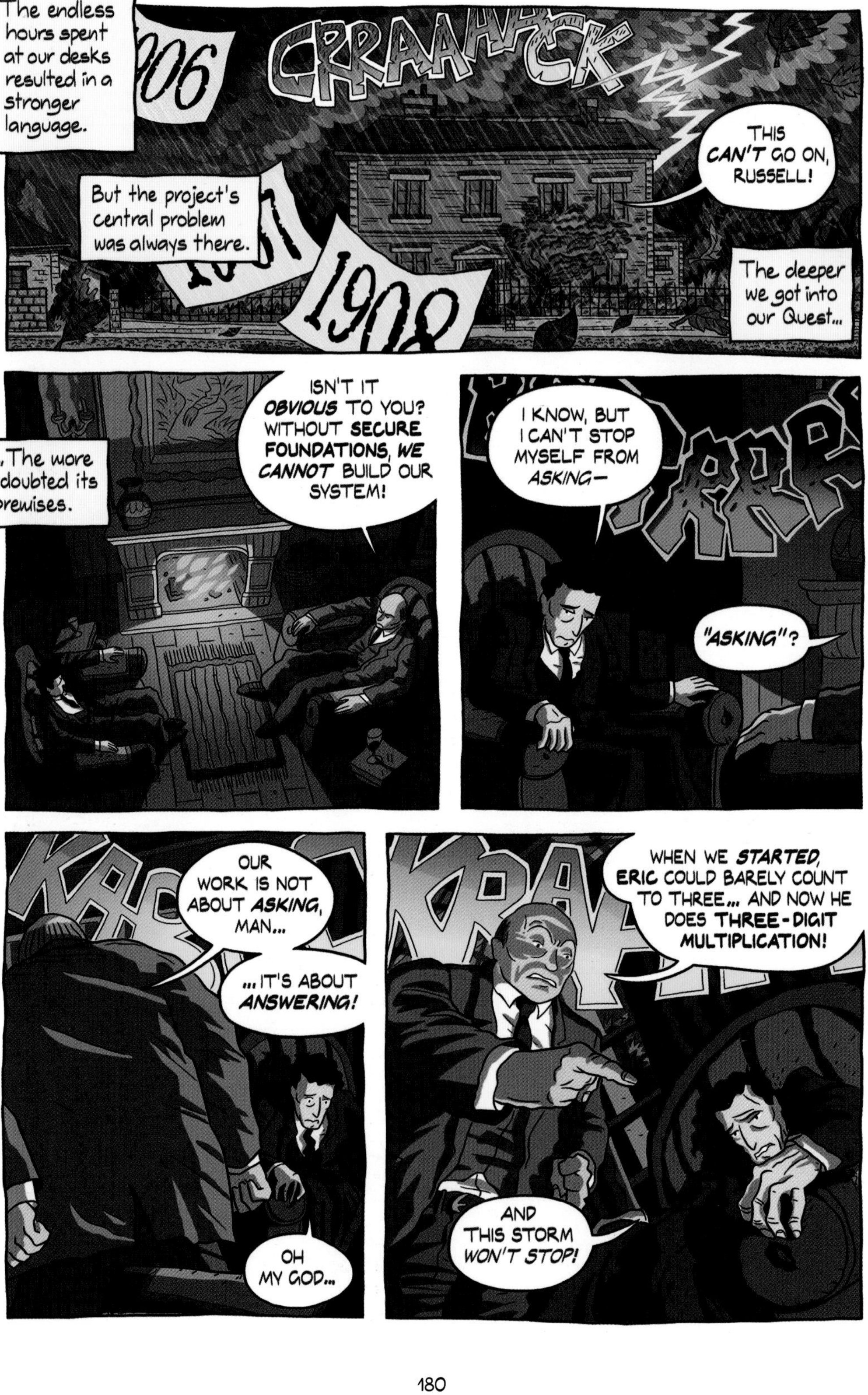
The endless hours spent at our desks resulted in a stronger language.
1908
But the project's central problem was always there.
THIS CAN'T GO ON, RUSSELL!
The deeper we got into our Quest...
...The more I doubted its premises.
ISN'T IT OBVIOUS TO YOU? WITHOUT SECURE FOUNDATIONS, WE CANNOT BUILD OUR SYSTEM!
I KNOW, BUT I CAN'T STOP MYSELF FROM ASKING—
"ASKING"?
OUR WORK IS NOT ABOUT ASKING, MAN...
...IT'S ABOUT ANSWERING!
OH MY GOD...
WHEN WE STARTED, ERIC COULD BARELY COUNT TO THREE... AND NOW HE DOES THREE-DIGIT MULTIPLICATION!
AND THIS STORM WON'T STOP!

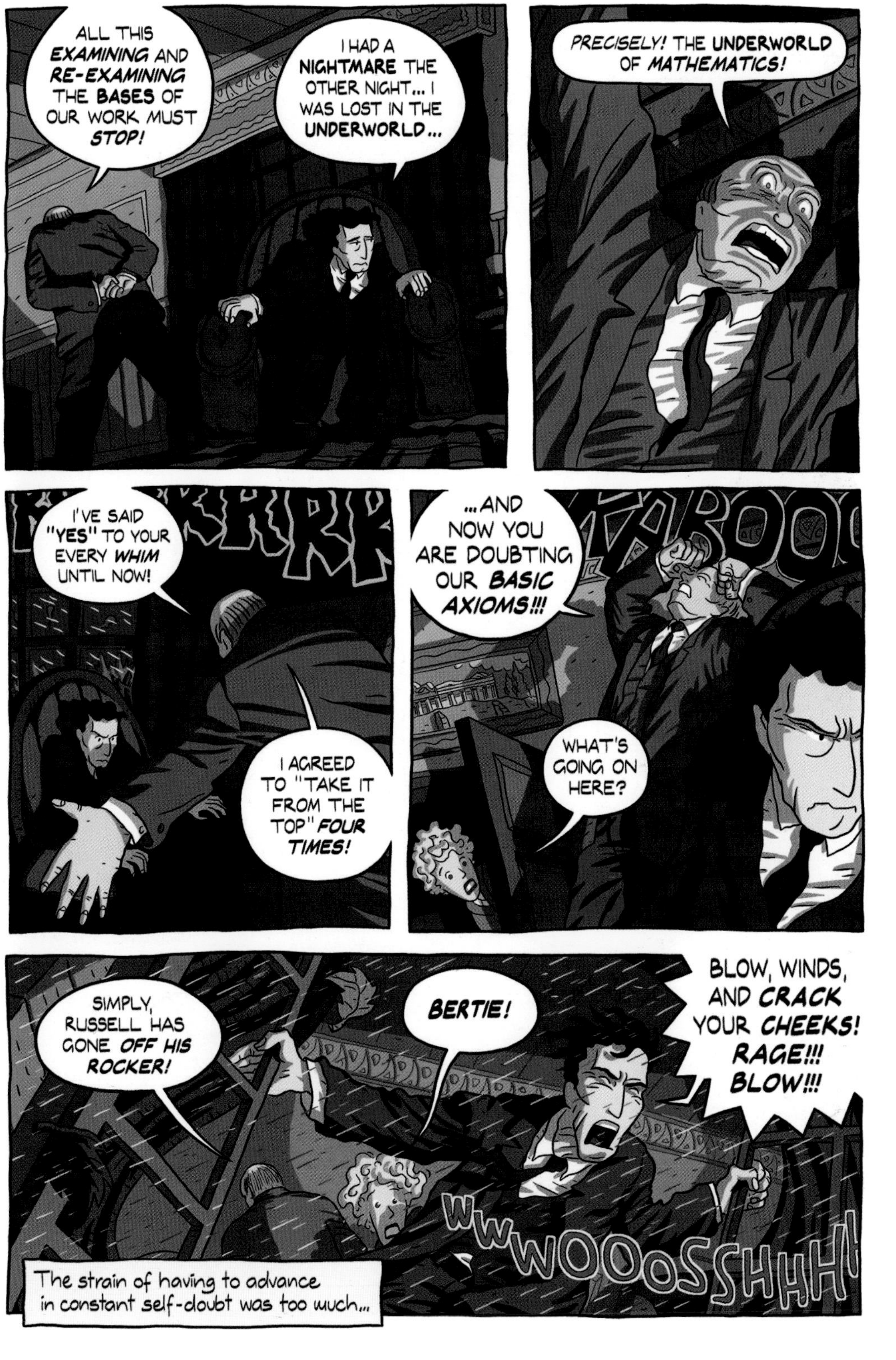
ALL THIS EXAMINING AND RE-EXAMINING THE BASES OF OUR WORK MUST STOP!
I HAD A NIGHTMARE THE OTHER NIGHT... I WAS LOST IN THE UNDERWORLD...
PRECISELY! THE UNDERWORLD OF MATHEMATICS!
I'VE SAID "YES" TO YOUR EVERY WHIM UNTIL NOW!
I AGREED TO "TAKE IT FROM THE TOP" FOUR TIMES!
...AND NOW YOU ARE DOUBTING OUR BASIC AXIOMS!!!
WHAT'S GOING ON HERE?
SIMPLY, RUSSELL HAS GONE OFF HIS ROCKER!
BERTIE!
BLOW, WINDS, AND CRACK YOUR CHEEKS! RAGE!!! BLOW!!!
WWWOOOSSHHH
The strain of having to advance in constant self-doubt was too much...

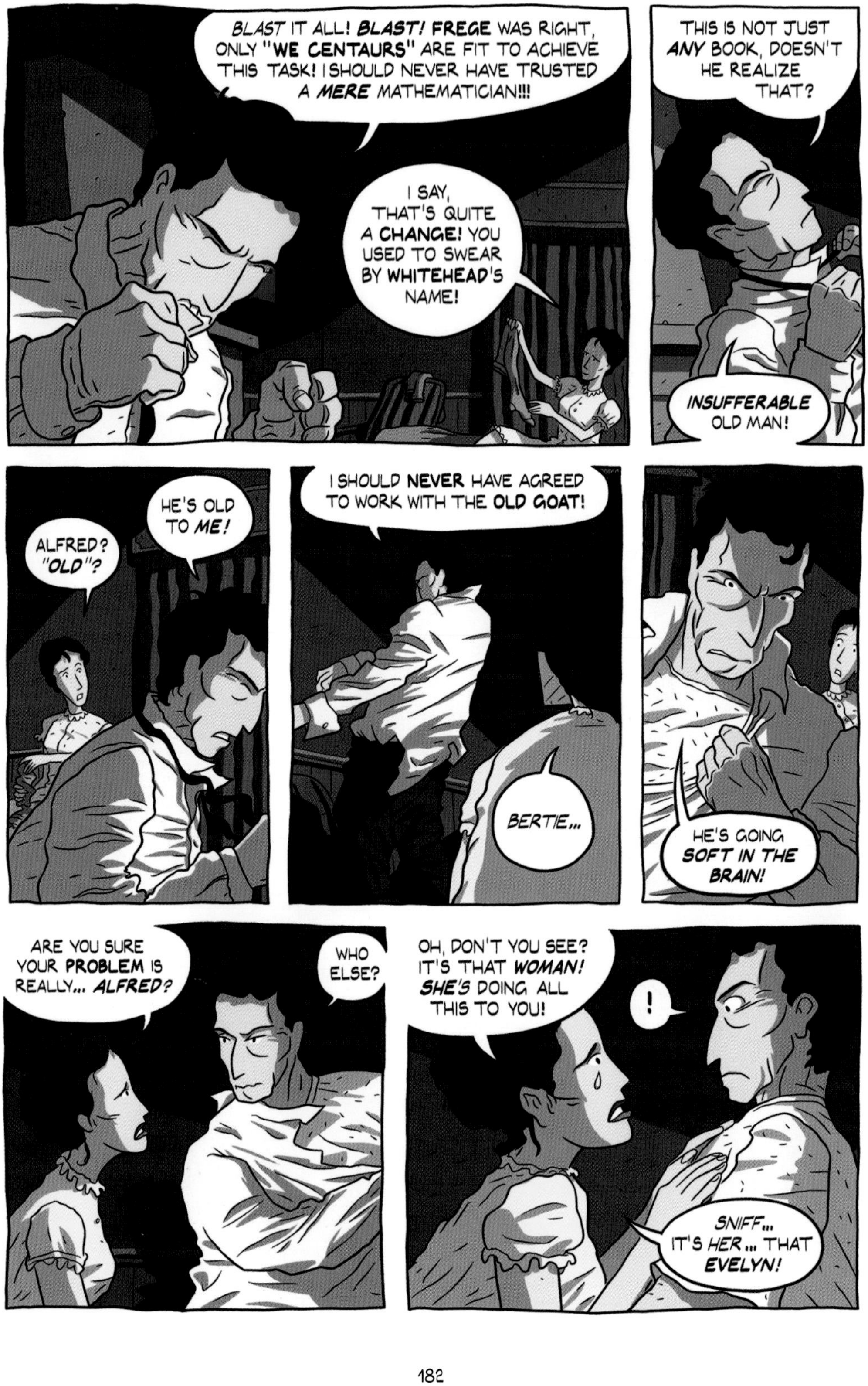
BLAST IT ALL! BLAST! FREGE WAS RIGHT, ONLY "WE CENTAURS" ARE FIT TO ACHIEVE THIS TASK! I SHOULD NEVER HAVE TRUSTED A MERE MATHEMATICIAN!!!
I SAY, THAT'S QUITE A CHANGE! YOU USED TO SWEAR BY WHITEHEAD'S NAME!
THIS IS NOT JUST ANY BOOK, DOESN'T HE REALIZE THAT?
INSUFFERABLE OLD MAN!
ALFRED? "OLD"?
HE'S OLD TO ME!
I SHOULD NEVER HAVE AGREED TO WORK WITH THE OLD GOAT!
BERTIE...
HE'S GOING SOFT IN THE BRAIN!
ARE YOU SURE YOUR PROBLEM IS REALLY... ALFRED?
WHO ELSE?
OH, DON'T YOU SEE? IT'S THAT WOMAN! SHE'S DOING ALL THIS TO YOU!
!
SNIFF... IT'S HER... THAT EVELYN!

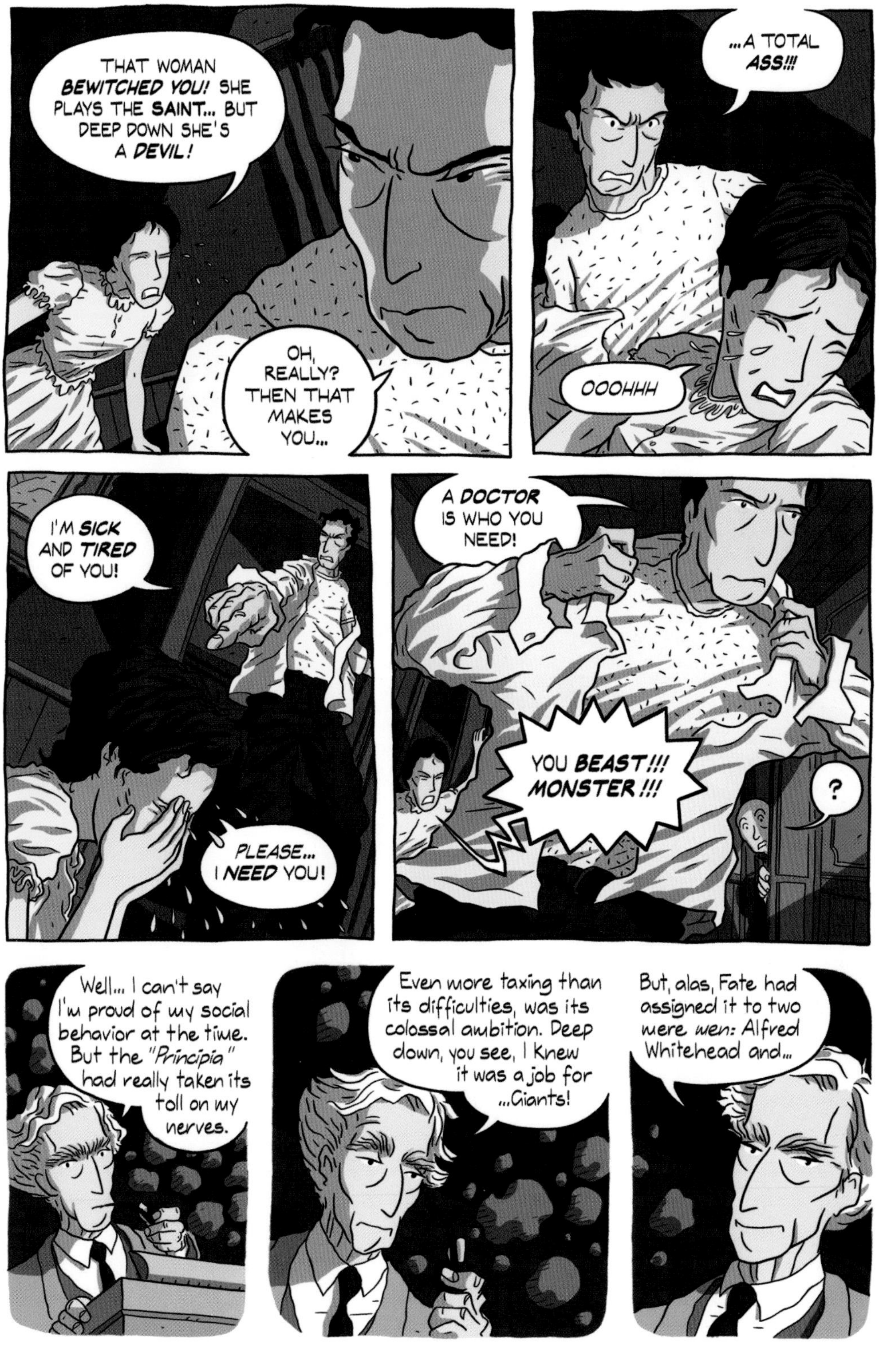
THAT WOMAN BEWITCHED YOU! SHE PLAYS THE SAINT... BUT DEEP DOWN SHE'S A DEVIL!
OH, REALLY? THEN THAT MAKES YOU...
...A TOTAL ASS!!!
OOOHHH
I'M SICK AND TIRED OF YOU!
PLEASE... I NEED YOU!
A DOCTOR IS WHO YOU NEED!
YOU BEAST!!! MONSTER!!!
?
Well... I can't say I'm proud of my social behavior at the time. But the "Principia" had really taken its toll on my nerves.
Even more taxing than its difficulties, was its colossal ambition. Deep down, you see, I knew it was a job for ...Giants!
But, alas, Fate had assigned it to two mere men: Alfred Whitehead and...

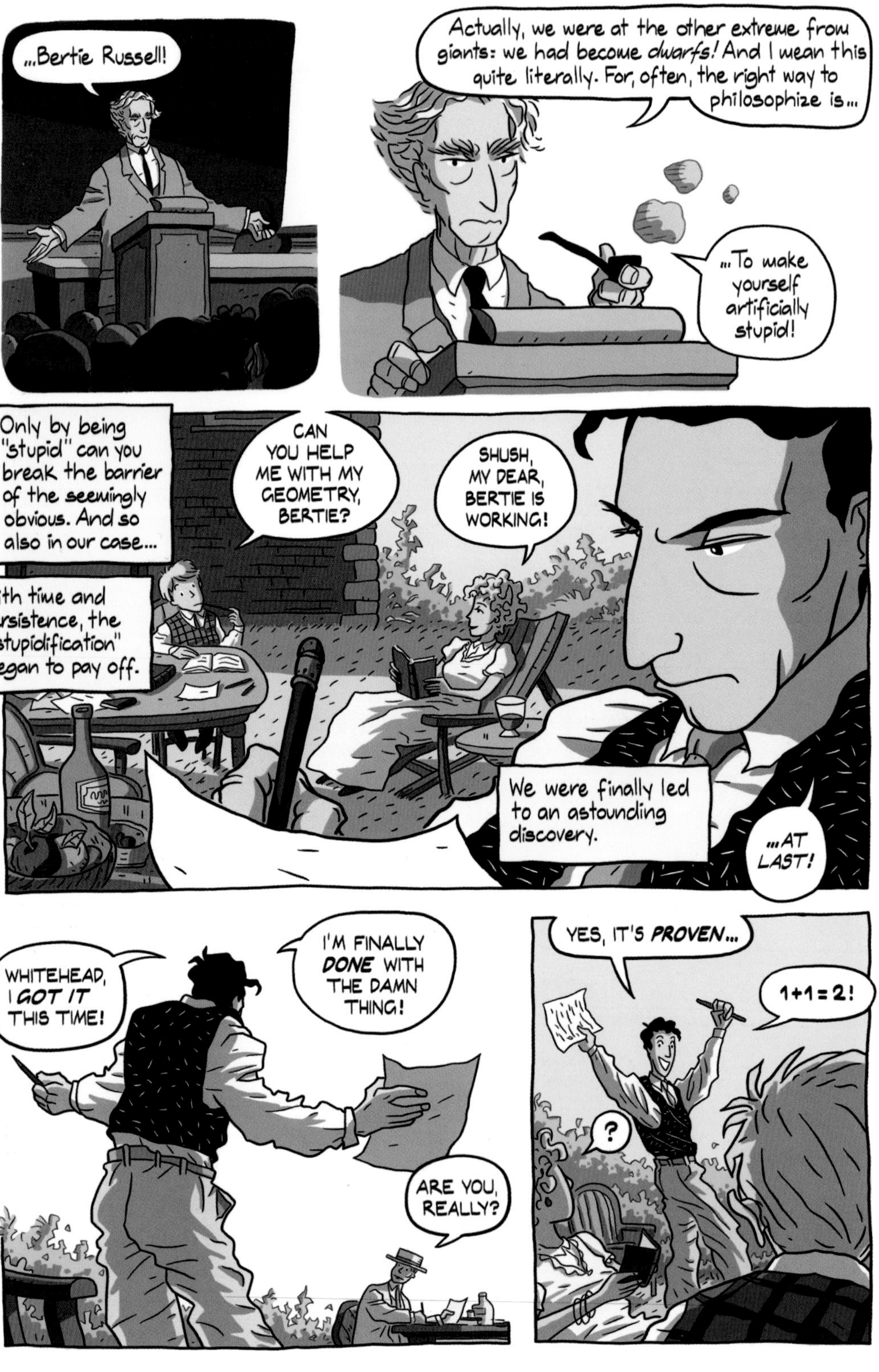
...Bertie Russell!
Actually, we were at the other extreme from giants: we had become *dwarfs!* And I mean this quite literally. For, often, the right way to philosophize is...
...To make yourself artificially stupid!
Only by being "stupid" can you break the barrier of the seemingly obvious. And so also in our case...
CAN YOU HELP ME WITH MY GEOMETRY, BERTIE?
SHUSH, MY DEAR, BERTIE IS WORKING!
With time and persistence, the "stupidification" began to pay off.
We were finally led to an astounding discovery.
...AT LAST!
WHITEHEAD, I **GOT IT** THIS TIME!
I'M FINALLY **DONE** WITH THE DAMN THING!
ARE YOU, REALLY?
YES, IT'S **PROVEN...**
1+1=2!
?

To achieve this monumental task, took us a mere... 362 pages! Think of that: 362 pages to prove what every child knows.

I DON'T *GET* IT, BERTIE.

WHY 362 PAGES?

LET ME PLAY, BEETLE OLD CHAP!

HURRAH! MISSED!

I'LL REPORT YOU TO THE NATIONAL CROQUET BOARD!

BUT WHY *ALL* THESE PAGES...

...TO PROVE 1+1=2?

HM... HOW SHALL I PUT IT?

IT'S THE PRICE YOU PAY FOR BEING *TRULY* CERTAIN.

LOOK!

THAT'S US, YOUR *DAD* AND I. WE ARE DOING ANTS' WORK...

NOT GLAMOROUS, BUT *VERY* CRUCIAL!

During the last phase of our work, my wife spent most of her time in rest homes.

ANY NEWS OF ALYS?

NO, NOT THIS WEEK...

I often think back at my stay with the Whiteheads...

Halcyon days.

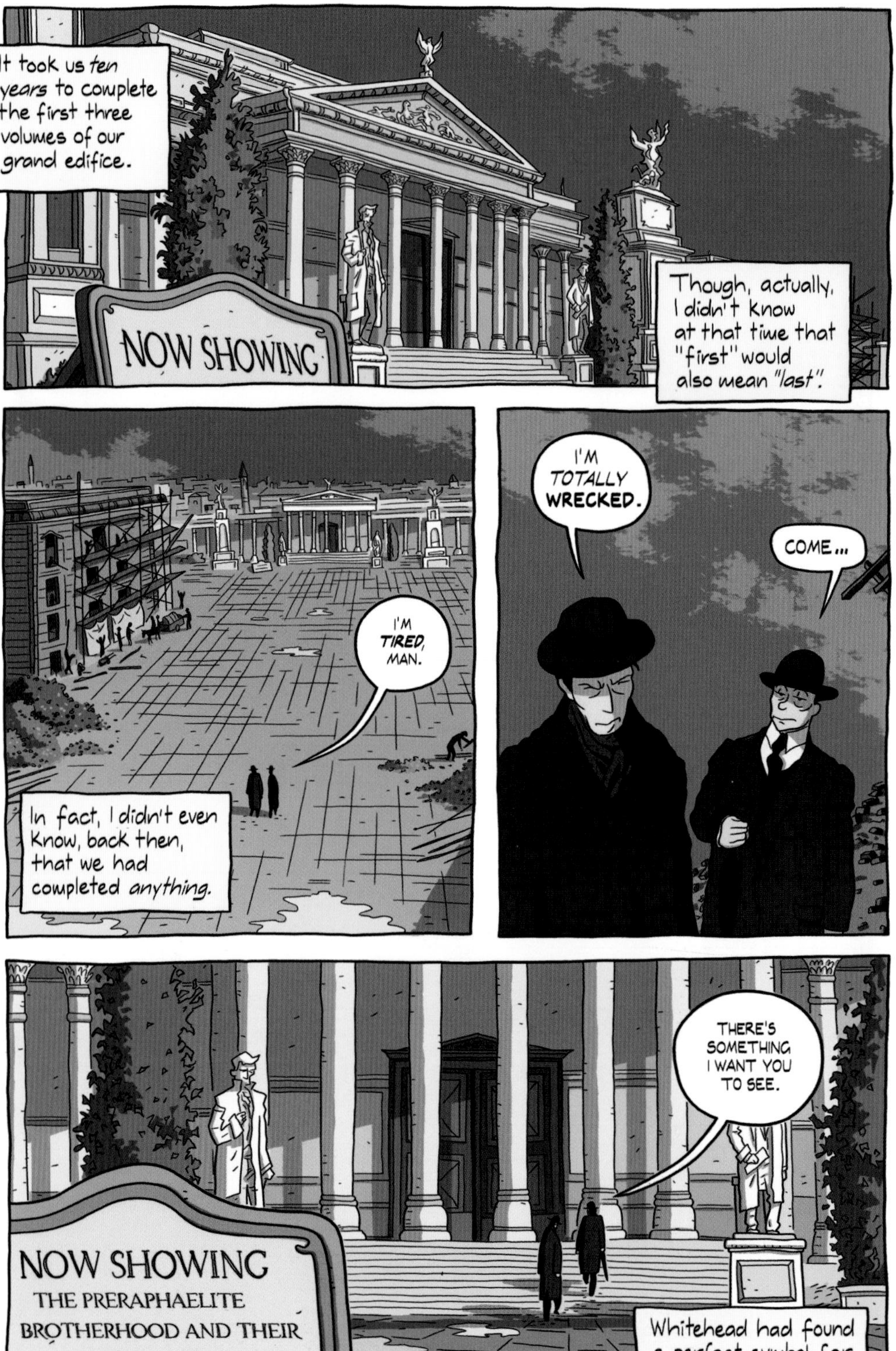
It took us *ten* years to complete the first three volumes of our grand edifice.
NOW SHOWING
Though, actually, I didn't know at that time that "first" would also mean *"last"*.
I'M *TIRED*, MAN.
In fact, I didn't even know, back then, that we had completed *anything*.
I'M *TOTALLY* **WRECKED.**
COME...
NOW SHOWING
THE PRERAPHAELITE BROTHERHOOD AND THEIR
THERE'S SOMETHING I WANT YOU TO SEE.
Whitehead had found a perfect symbol for his argument.

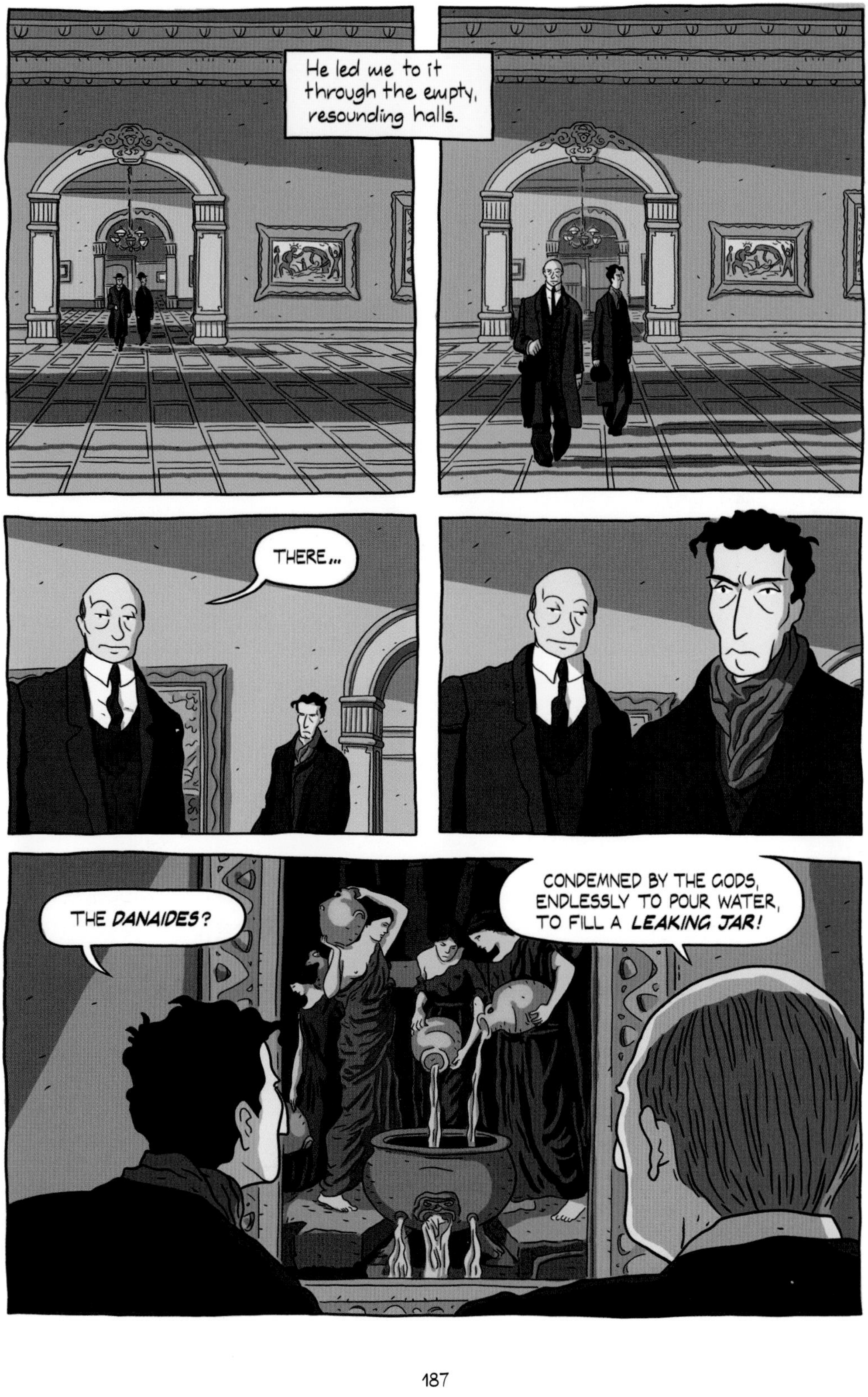
He led me to it through the empty, resounding halls.
THERE...
THE DANAIDES?
CONDEMNED BY THE GODS, ENDLESSLY TO POUR WATER, TO FILL A LEAKING JAR!

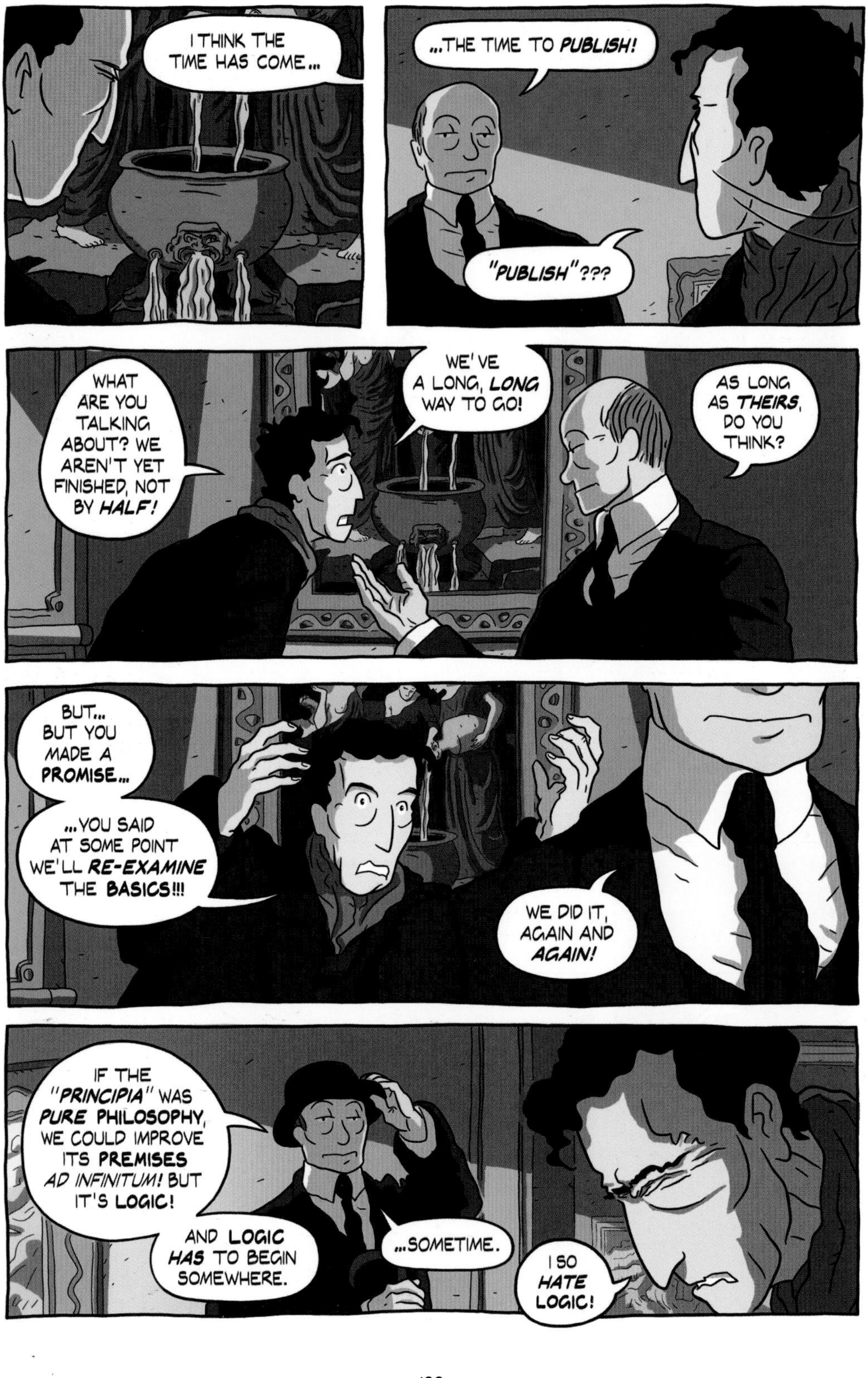
I THINK THE TIME HAS COME...
...THE TIME TO PUBLISH!
"PUBLISH"???
WHAT ARE YOU TALKING ABOUT? WE AREN'T YET FINISHED, NOT BY HALF!
WE'VE A LONG, LONG WAY TO GO!
AS LONG AS THEIRS, DO YOU THINK?
BUT... BUT YOU MADE A PROMISE...
...YOU SAID AT SOME POINT WE'LL RE-EXAMINE THE BASICS!!!
WE DID IT, AGAIN AND AGAIN!
IF THE "PRINCIPIA" WAS PURE PHILOSOPHY, WE COULD IMPROVE ITS PREMISES AD INFINITUM! BUT IT'S LOGIC!
AND LOGIC HAS TO BEGIN SOMEWHERE.
...SOMETIME.
I SO HATE LOGIC!

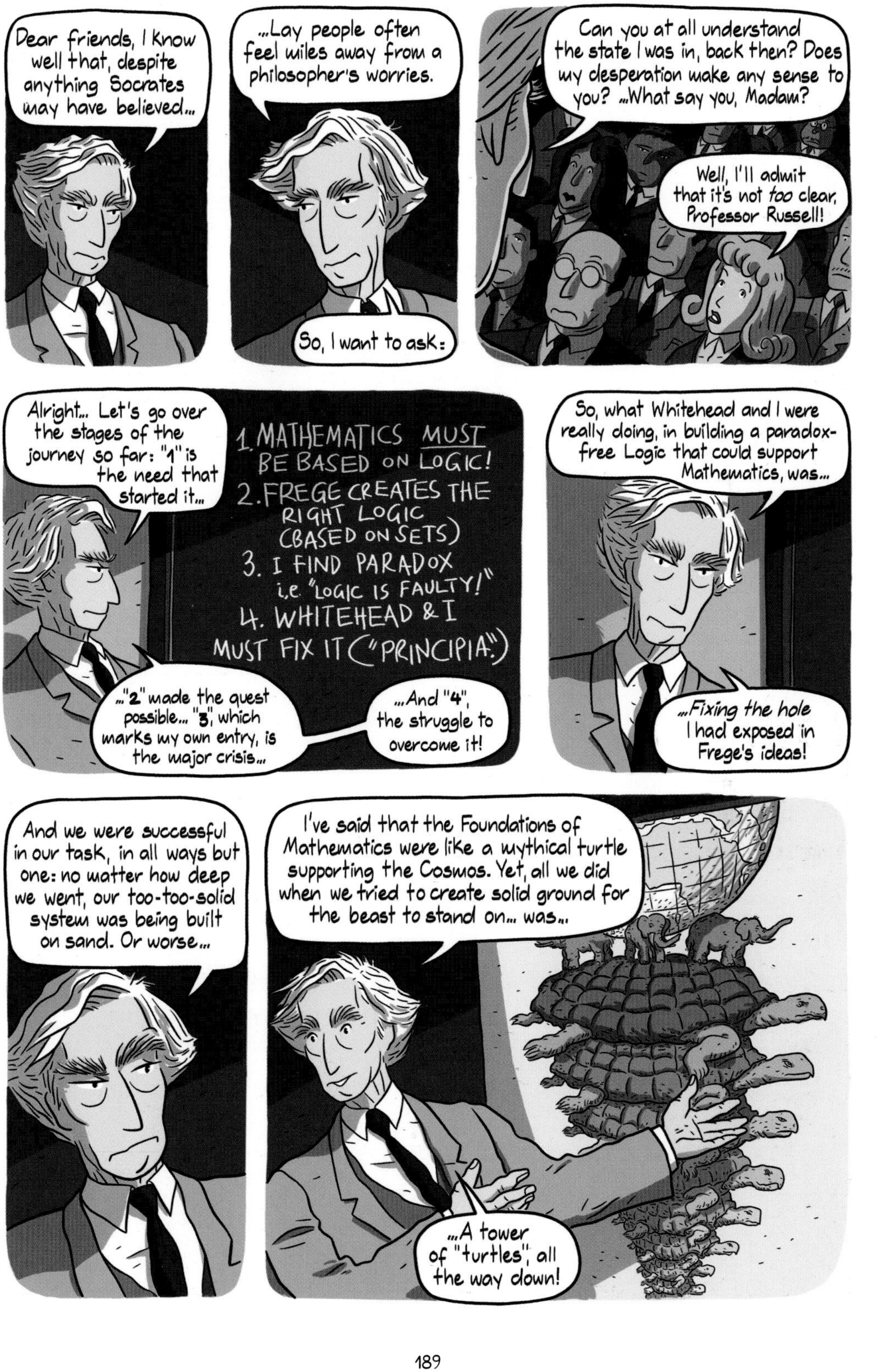
Dear friends, I know well that, despite anything Socrates may have believed...
...Lay people often feel miles away from a philosopher's worries.
So, I want to ask:
Can you at all understand the state I was in, back then? Does my desperation make any sense to you? ...What say you, Madam?
Well, I'll admit that it's not *too* clear, Professor Russell!
Alright... Let's go over the stages of the journey so far: "1" is the need that started it...
1. MATHEMATICS MUST BE BASED ON LOGIC!
2. FREGE CREATES THE RIGHT LOGIC (BASED ON SETS)
3. I FIND PARADOX i.e. "LOGIC IS FAULTY!"
4. WHITEHEAD & I MUST FIX IT ("PRINCIPIA")
..."**2**" made the quest possible... "**3**", which marks my own entry, is the major crisis...
...And "**4**", the struggle to overcome it!
So, what Whitehead and I were really doing, in building a paradox-free Logic that could support Mathematics, was...
...Fixing the hole I had exposed in Frege's ideas!
And we were successful in our task, in all ways but one: no matter how deep we went, our too-too-solid system was being built on sand. Or worse...
I've said that the Foundations of Mathematics were like a mythical turtle supporting the Cosmos. Yet, all we did when we tried to create solid ground for the beast to stand on... was...
...A tower of "turtles", all the way down!

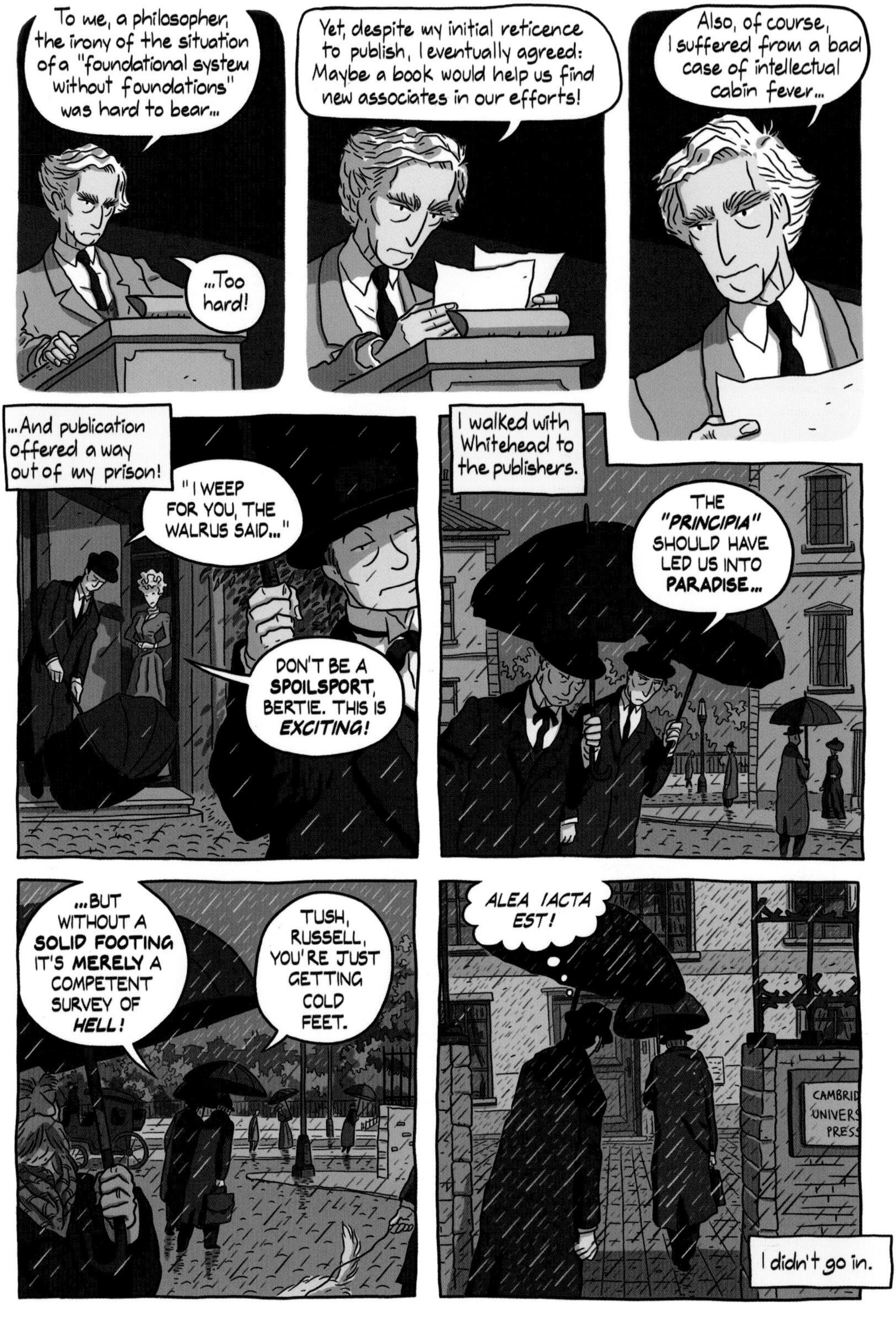
To me, a philosopher, the irony of the situation of a "foundational system without foundations" was hard to bear...
...Too hard!
Yet, despite my initial reticence to publish, I eventually agreed: Maybe a book would help us find new associates in our efforts!
Also, of course, I suffered from a bad case of intellectual cabin fever...
...And publication offered a way out of my prison!
"I WEEP FOR YOU, THE WALRUS SAID..."
DON'T BE A **SPOILSPORT**, BERTIE. THIS IS ***EXCITING!***
I walked with Whitehead to the publishers.
THE ***"PRINCIPIA"*** SHOULD HAVE LED US INTO **PARADISE...**
...BUT WITHOUT A **SOLID FOOTING** IT'S **MERELY** A COMPETENT SURVEY OF ***HELL!***
TUSH, RUSSELL, YOU'RE JUST GETTING COLD FEET.
ALEA IACTA EST!
CAMBRID
UNIVERS
PRESS
I didn't go in.

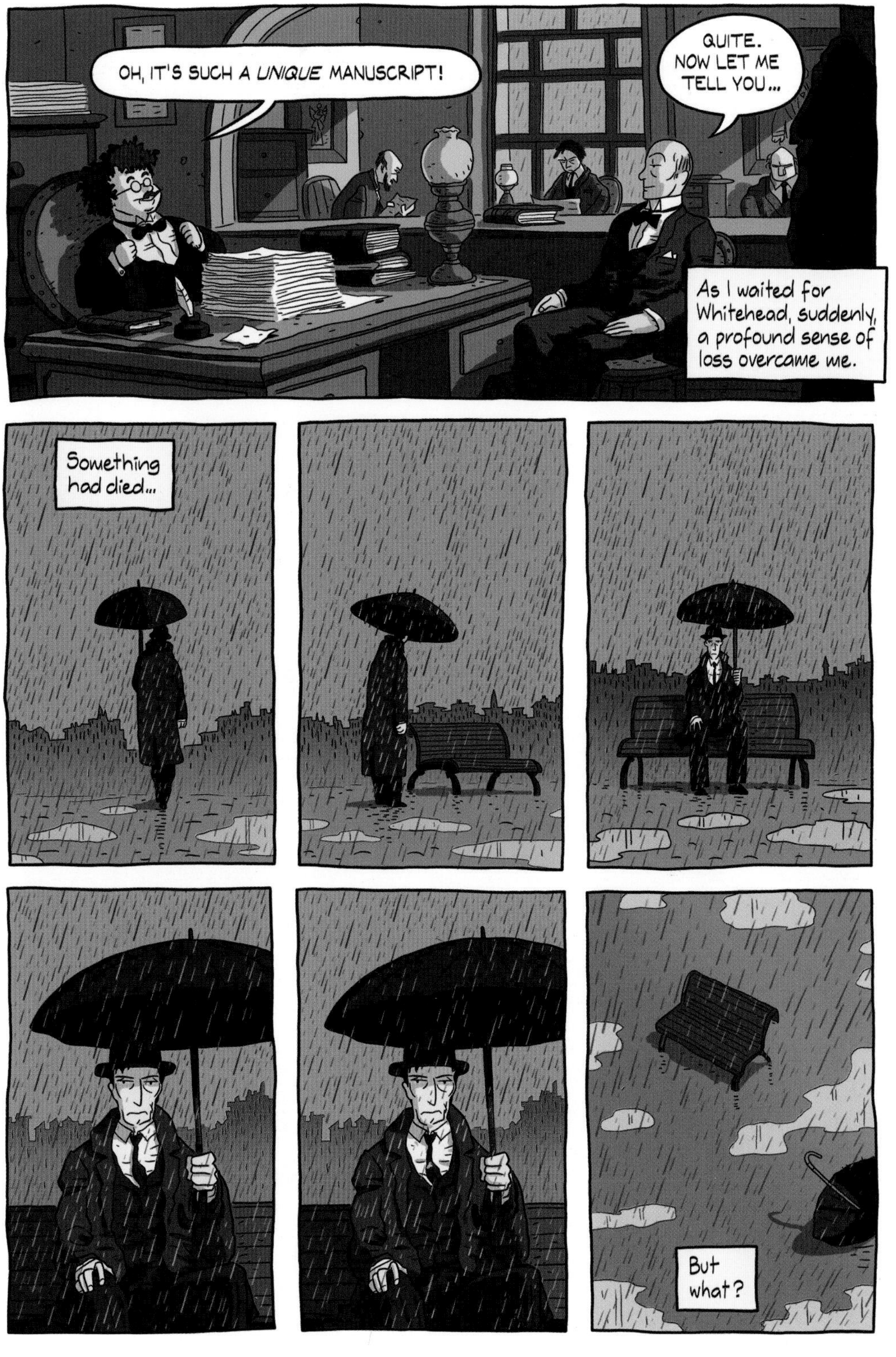
OH, IT'S SUCH A *UNIQUE* MANUSCRIPT!
QUITE. NOW LET ME TELL YOU...
As I waited for Whitehead, suddenly, a profound sense of loss overcame me.
Something had died...
But what?

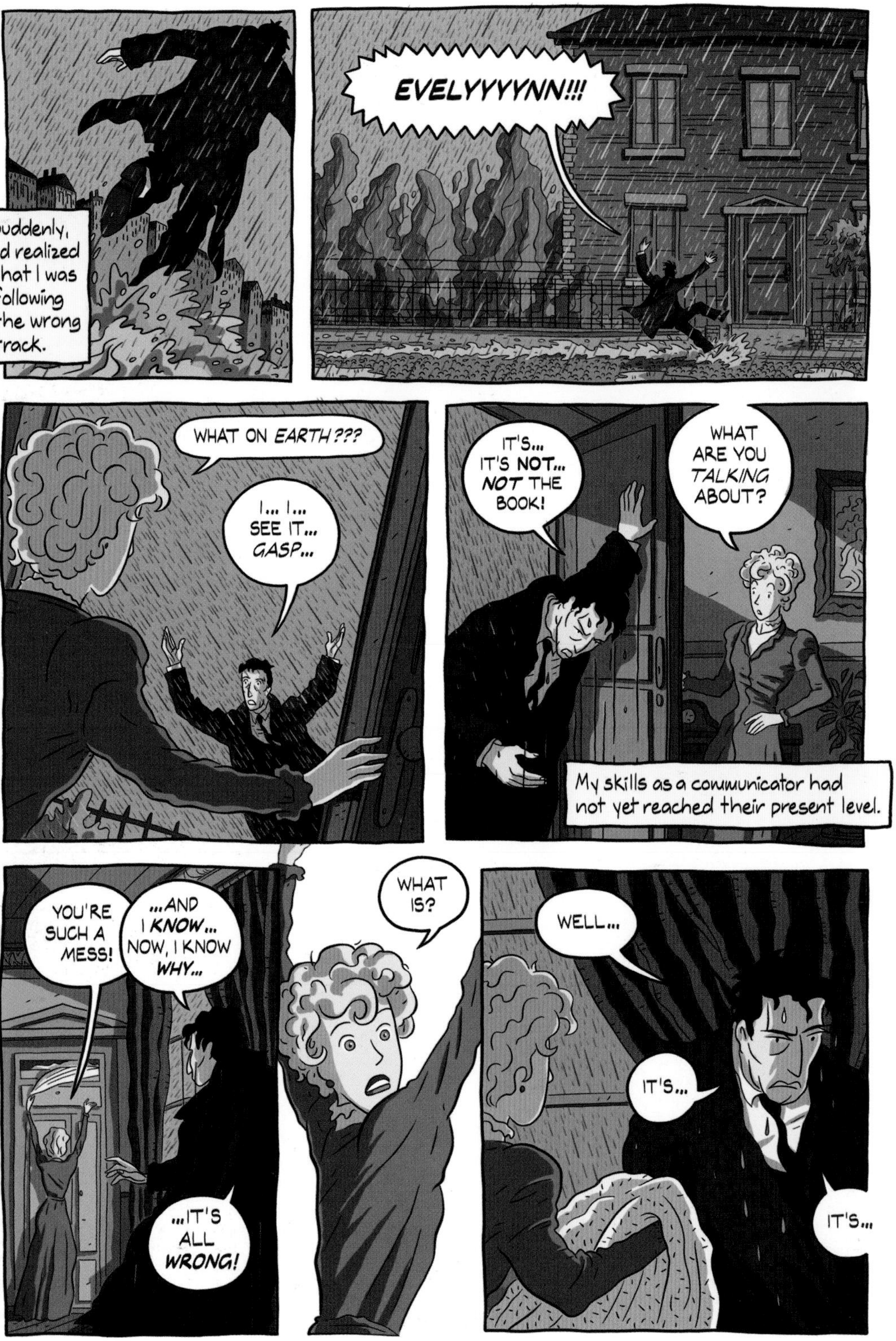
Suddenly, I'd realized that I was following the wrong track.
EVELYYYYNN!!!
WHAT ON EARTH???
I... I... SEE IT... GASP...
IT'S... IT'S NOT... NOT THE BOOK!
WHAT ARE YOU TALKING ABOUT?
My skills as a communicator had not yet reached their present level.
YOU'RE SUCH A MESS!
...AND I KNOW... NOW, I KNOW WHY...
...IT'S ALL WRONG!
WHAT IS?
WELL...
IT'S...
IT'S...

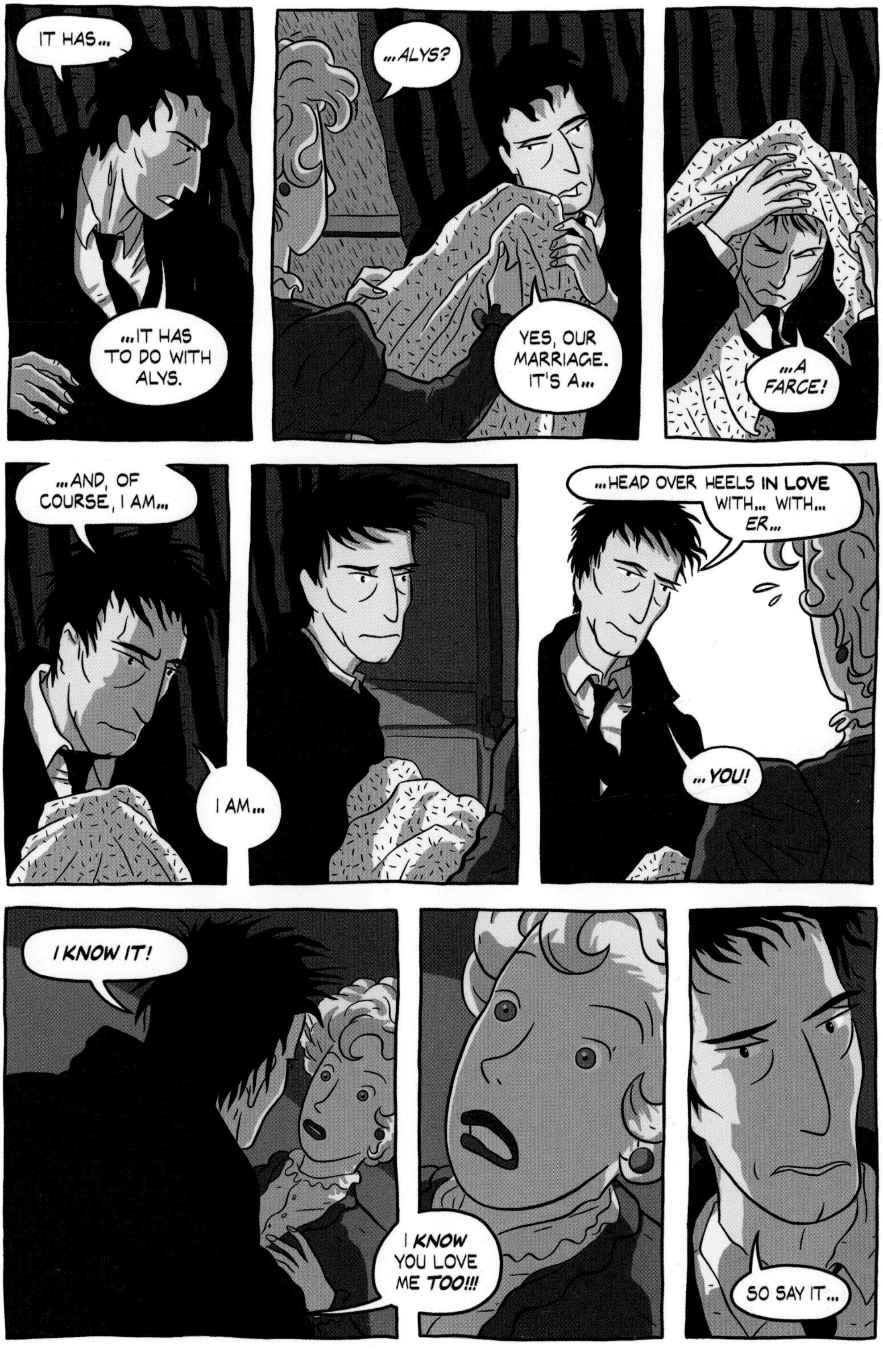
IT HAS...
...IT HAS TO DO WITH ALYS.
...ALYS?
YES, OUR MARRIAGE. IT'S A...
...A FARCE!
...AND, OF COURSE, I AM...
I AM...
...HEAD OVER HEELS IN LOVE WITH... WITH... ER...
...YOU!
I KNOW IT!
I KNOW YOU LOVE ME TOO!!!
SO SAY IT...

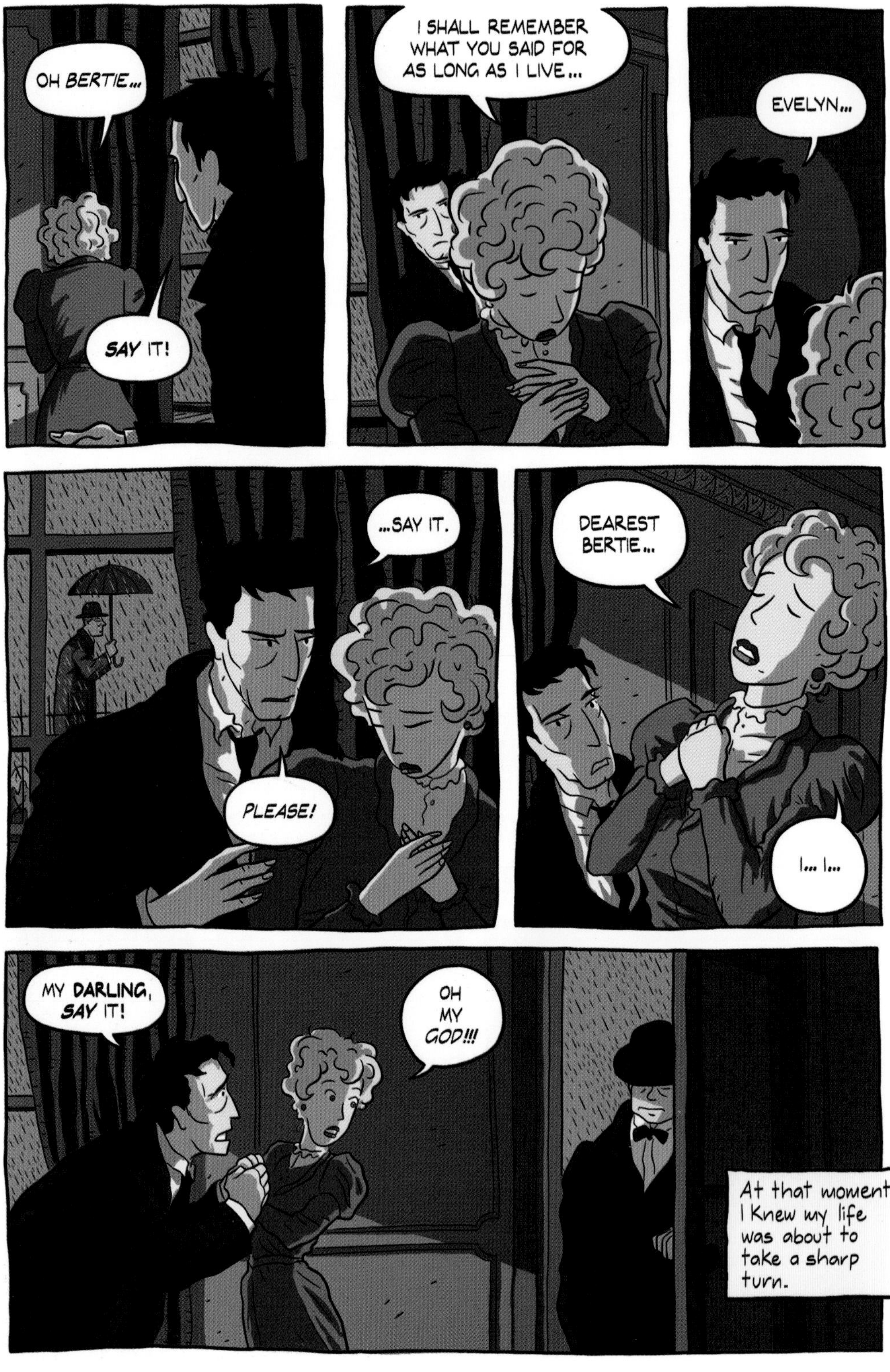
OH BERTIE...
SAY IT!
I SHALL REMEMBER WHAT YOU SAID FOR AS LONG AS I LIVE...
EVELYN...
...SAY IT.
PLEASE!
DEAREST BERTIE...
I... I...
MY DARLING, SAY IT!
OH MY GOD!!!
At that moment, I knew my life was about to take a sharp turn.

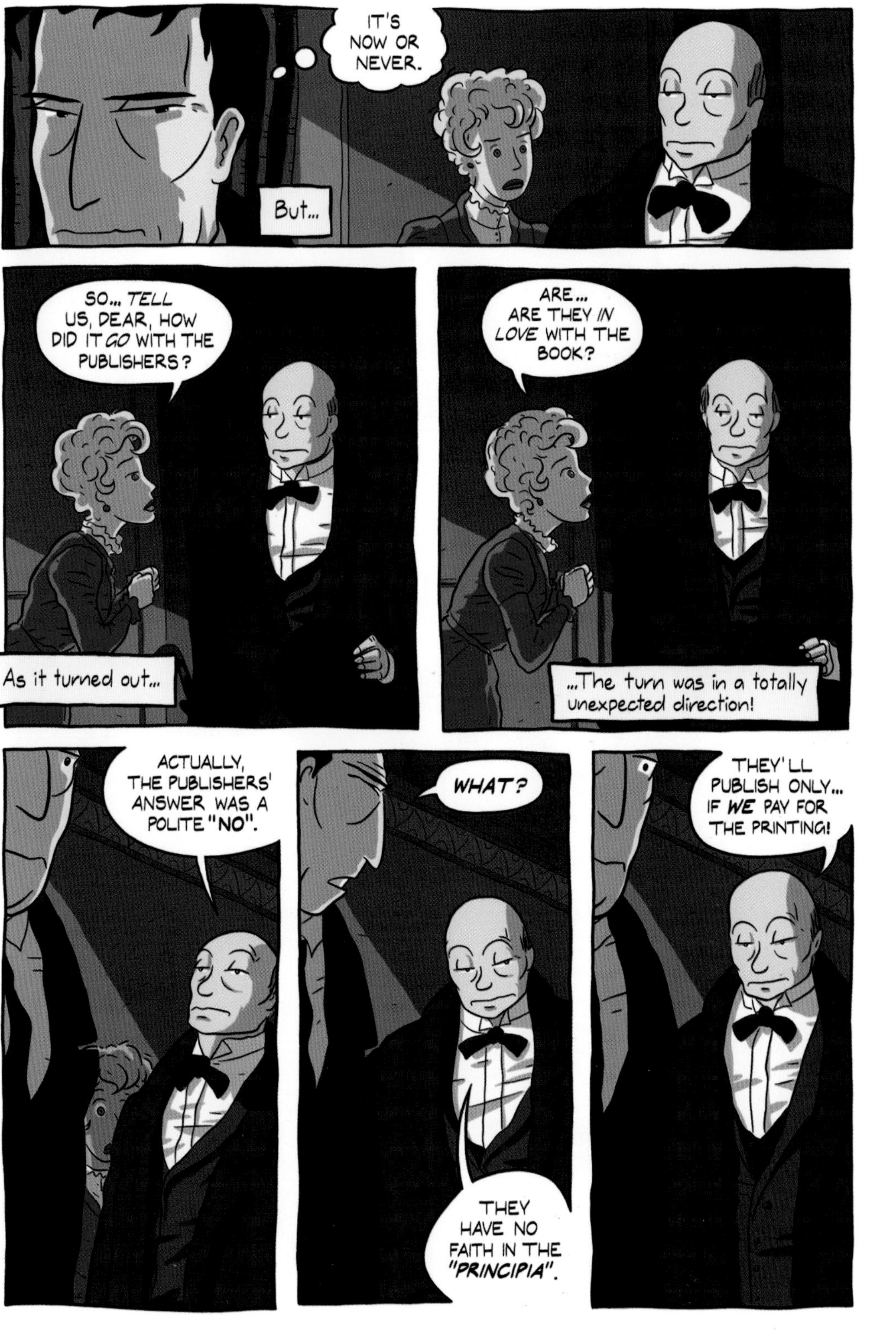
IT'S NOW OR NEVER.
But...
SO... TELL US, DEAR, HOW DID IT GO WITH THE PUBLISHERS?
As it turned out...
ARE... ARE THEY IN LOVE WITH THE BOOK?
...The turn was in a totally unexpected direction!
ACTUALLY, THE PUBLISHERS' ANSWER WAS A POLITE "NO".
WHAT?
THEY HAVE NO FAITH IN THE "PRINCIPIA".
THEY'LL PUBLISH ONLY... IF WE PAY FOR THE PRINTING!

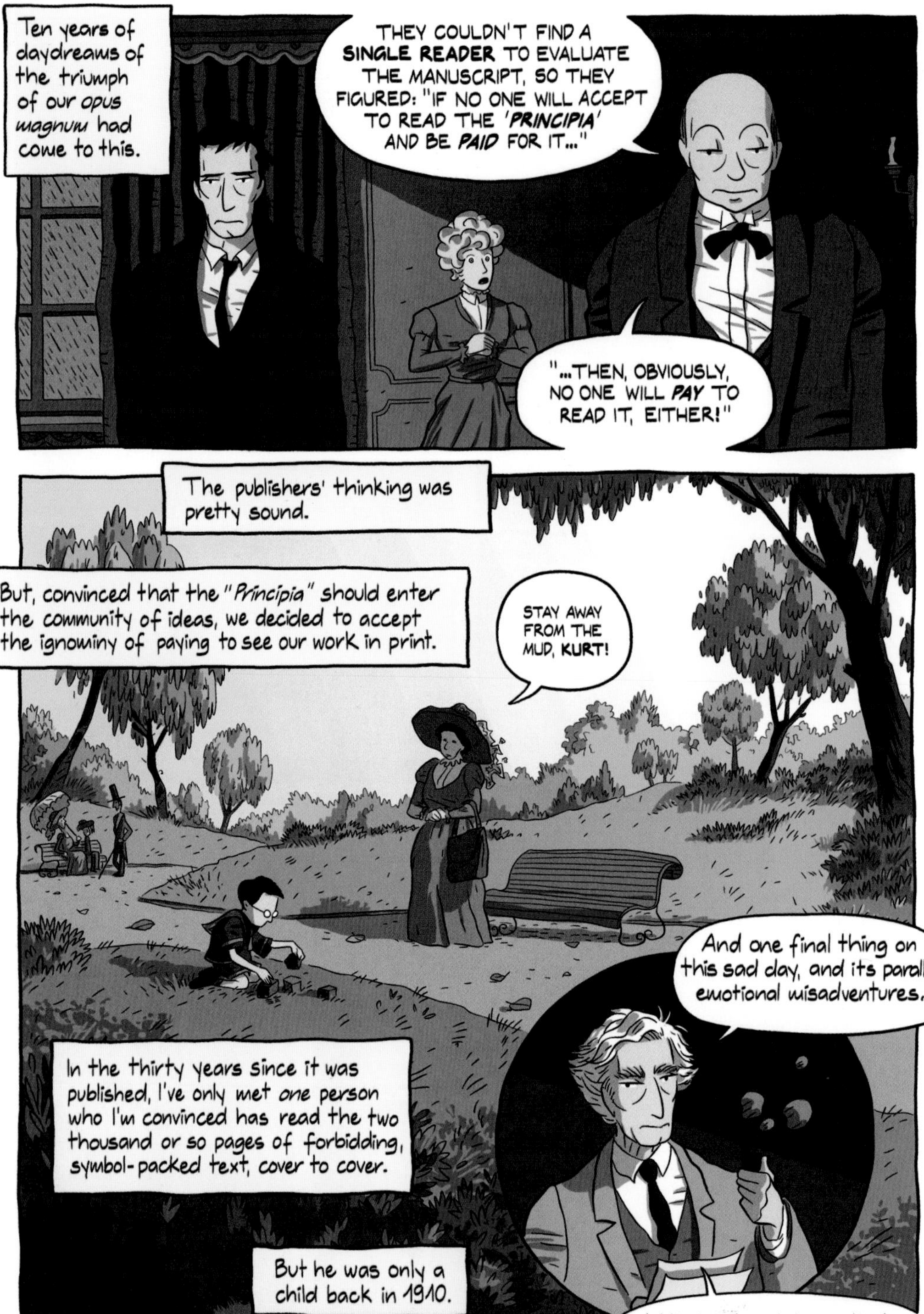
Ten years of daydreams of the triumph of our *opus magnum* had come to this.
THEY COULDN'T FIND A **SINGLE READER** TO EVALUATE THE MANUSCRIPT, SO THEY FIGURED: "IF NO ONE WILL ACCEPT TO READ THE '***PRINCIPIA***' AND BE ***PAID*** FOR IT..."
"...THEN, OBVIOUSLY, NO ONE WILL ***PAY*** TO READ IT, EITHER!"
The publishers' thinking was pretty sound.
But, convinced that the "*Principia*" should enter the community of ideas, we decided to accept the ignominy of paying to see our work in print.
STAY AWAY FROM THE MUD, **KURT!**
And one final thing on this sad day, and its parallel emotional misadventures...
In the thirty years since it was published, I've only met *one* person who I'm convinced has read the two thousand or so pages of forbidding, symbol-packed text, cover to cover.
But he was only a child back in 1910.
With hindsight, I say that I was wrong in my self-analysis: my problem *was* the book!

ENTRACTE

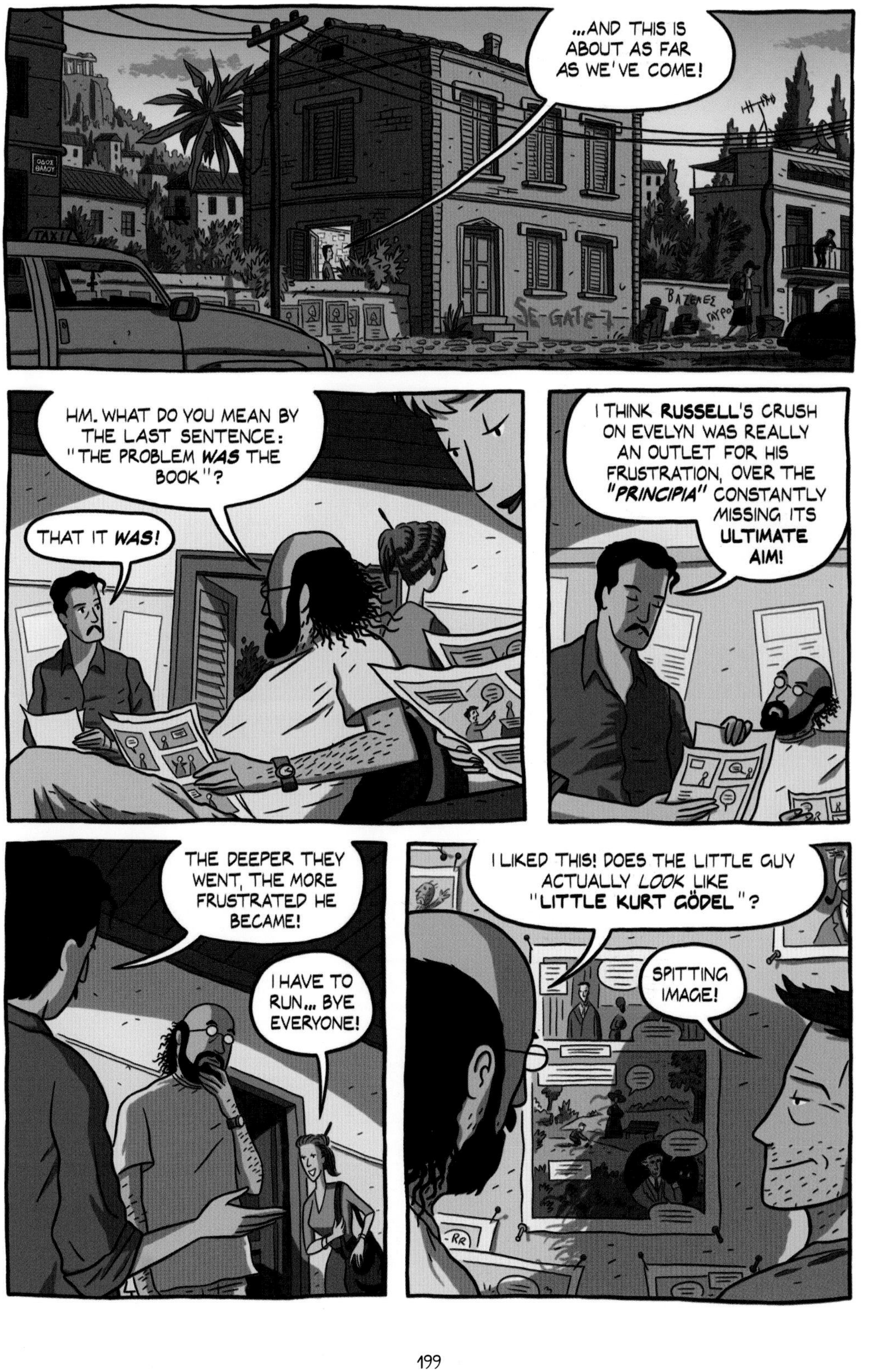
...AND THIS IS ABOUT AS FAR AS WE'VE COME!
TAXI
HM. WHAT DO YOU MEAN BY THE LAST SENTENCE: "THE PROBLEM *WAS* THE BOOK"?
THAT IT *WAS*!
I THINK **RUSSELL**'S CRUSH ON EVELYN WAS REALLY AN OUTLET FOR HIS FRUSTRATION, OVER THE "*PRINCIPIA*" CONSTANTLY MISSING ITS **ULTIMATE AIM**!
THE DEEPER THEY WENT, THE MORE FRUSTRATED HE BECAME!
I HAVE TO RUN... BYE EVERYONE!
I LIKED THIS! DOES THE LITTLE GUY ACTUALLY *LOOK* LIKE "**LITTLE KURT GÖDEL**"?
SPITTING IMAGE!

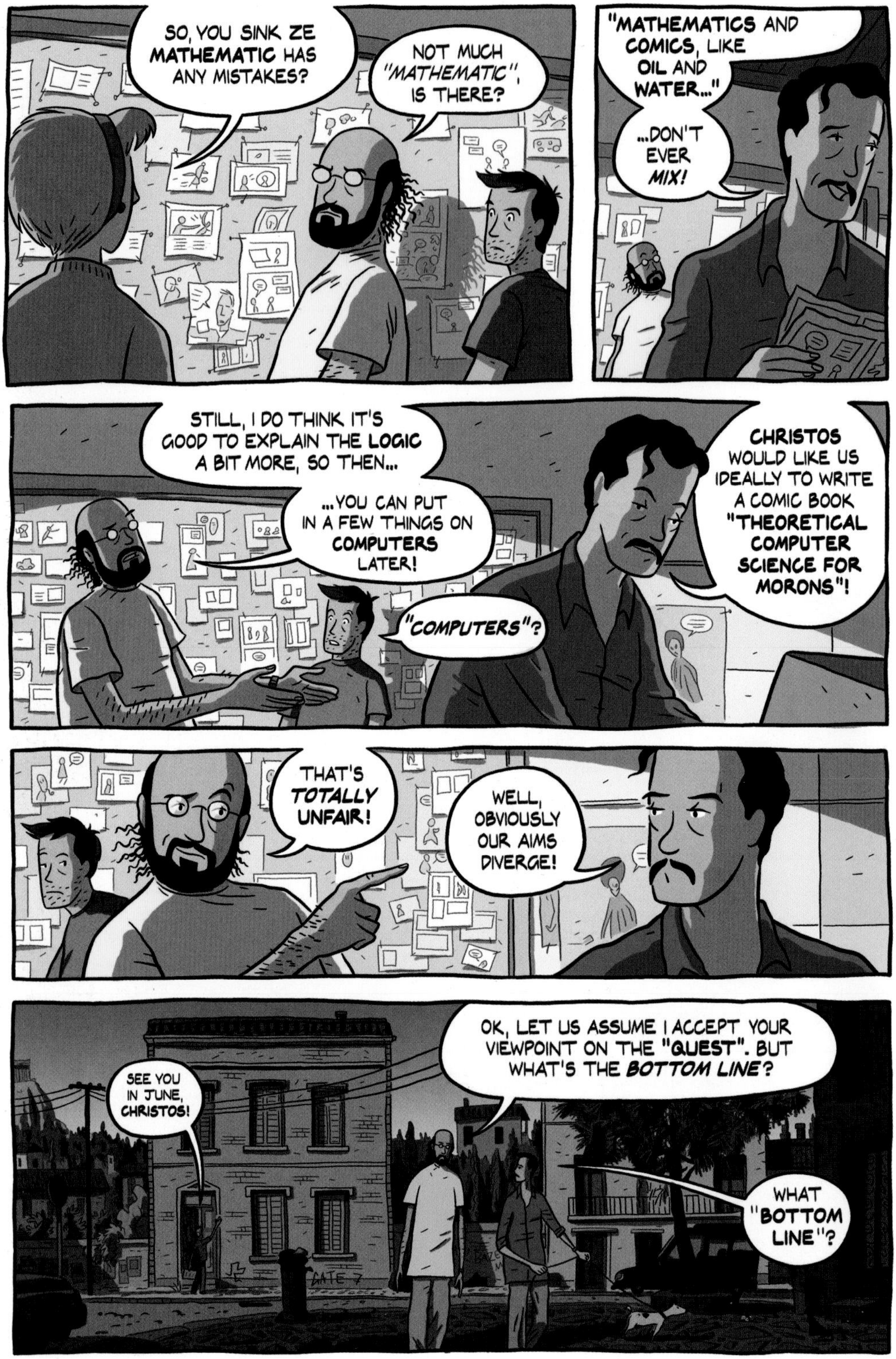
SO, YOU SINK ZE MATHEMATIC HAS ANY MISTAKES?
NOT MUCH "MATHEMATIC", IS THERE?
"MATHEMATICS AND COMICS, LIKE OIL AND WATER..."
...DON'T EVER MIX!
STILL, I DO THINK IT'S GOOD TO EXPLAIN THE LOGIC A BIT MORE, SO THEN...
...YOU CAN PUT IN A FEW THINGS ON COMPUTERS LATER!
"COMPUTERS"?
CHRISTOS WOULD LIKE US IDEALLY TO WRITE A COMIC BOOK "THEORETICAL COMPUTER SCIENCE FOR MORONS"!
THAT'S TOTALLY UNFAIR!
WELL, OBVIOUSLY OUR AIMS DIVERGE!
SEE YOU IN JUNE, CHRISTOS!
OK, LET US ASSUME I ACCEPT YOUR VIEWPOINT ON THE "QUEST". BUT WHAT'S THE BOTTOM LINE?
WHAT "BOTTOM LINE"?

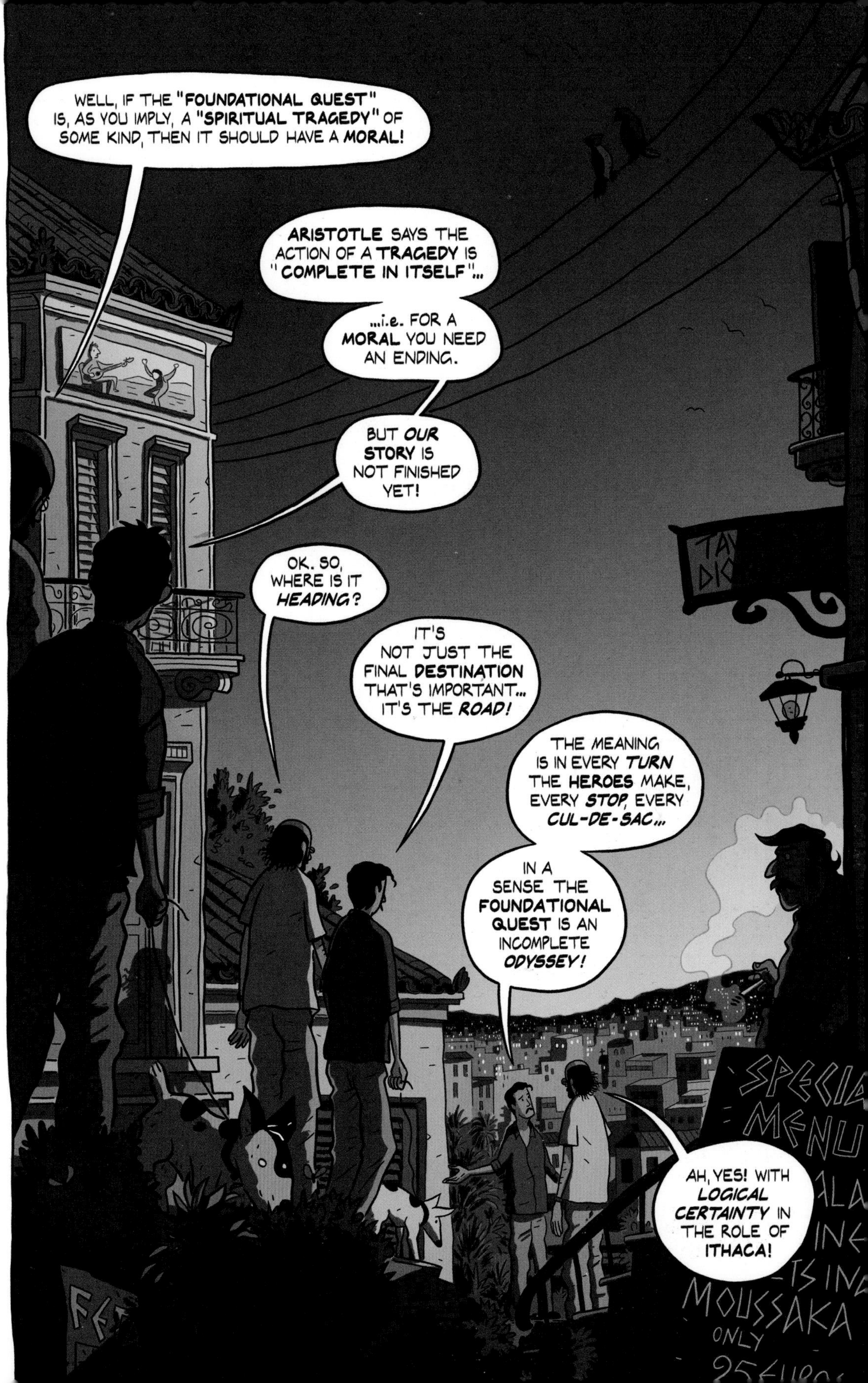
WELL, IF THE "FOUNDATIONAL QUEST" IS, AS YOU IMPLY, A "SPIRITUAL TRAGEDY" OF SOME KIND, THEN IT SHOULD HAVE A MORAL!
ARISTOTLE SAYS THE ACTION OF A TRAGEDY IS "COMPLETE IN ITSELF"...
...i.e. FOR A MORAL YOU NEED AN ENDING.
BUT OUR STORY IS NOT FINISHED YET!
OK.. SO, WHERE IS IT HEADING?
IT'S NOT JUST THE FINAL DESTINATION THAT'S IMPORTANT... IT'S THE ROAD!
THE MEANING IS IN EVERY TURN THE HEROES MAKE, EVERY STOP, EVERY CUL-DE-SAC...
IN A SENSE THE FOUNDATIONAL QUEST IS AN INCOMPLETE ODYSSEY!
AH, YES! WITH LOGICAL CERTAINTY IN THE ROLE OF ITHACA!
SPECIA
MENU
MOUSSAKA
ONLY

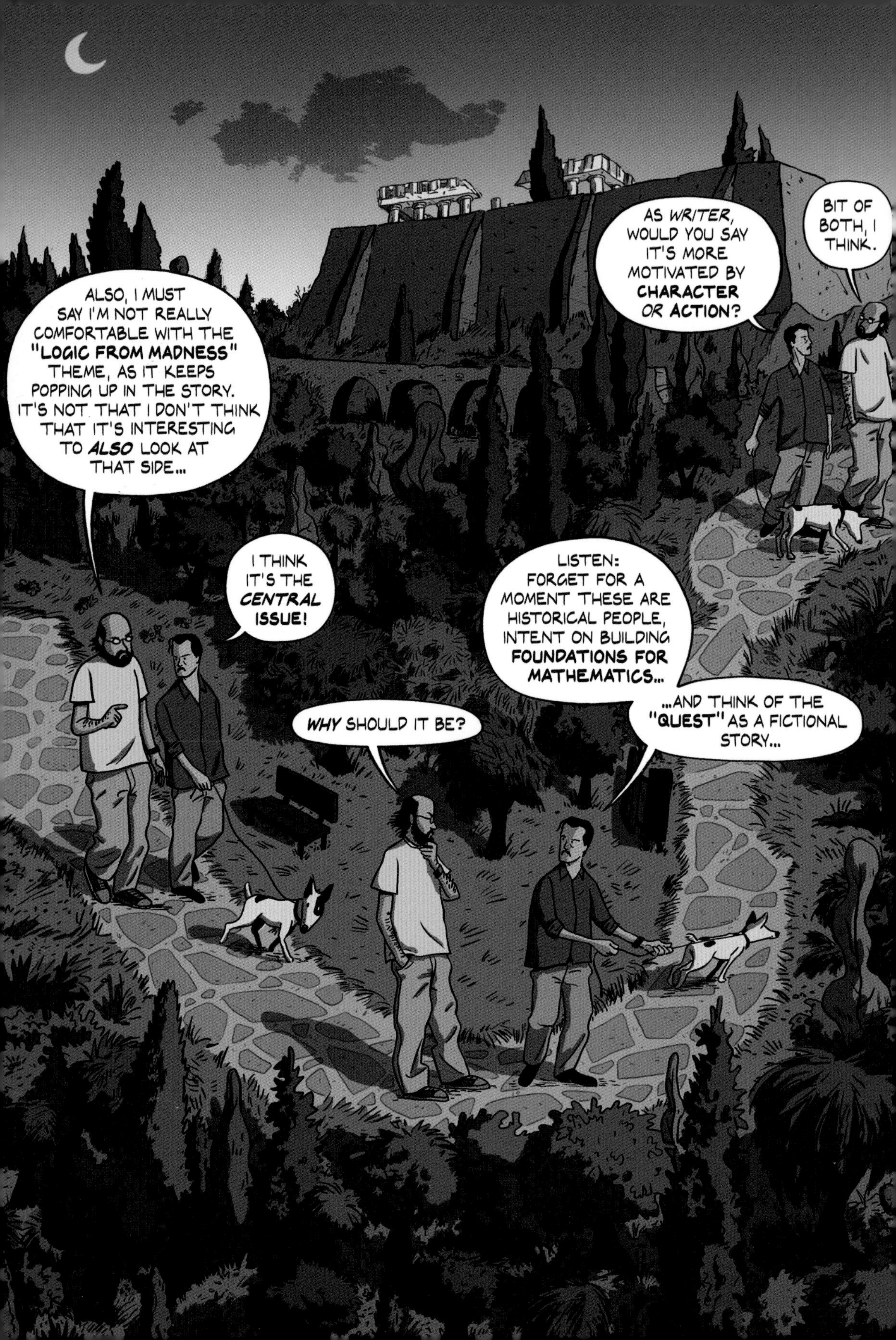
AS *WRITER*, WOULD YOU SAY IT'S MORE MOTIVATED BY **CHARACTER** *OR* **ACTION**?
BIT OF BOTH, I THINK.
ALSO, I MUST SAY I'M NOT REALLY COMFORTABLE WITH THE **"LOGIC FROM MADNESS"** THEME, AS IT KEEPS POPPING UP IN THE STORY. IT'S NOT THAT I DON'T THINK THAT IT'S INTERESTING TO ***ALSO*** LOOK AT THAT SIDE...
I THINK IT'S THE ***CENTRAL*** **ISSUE!**
WHY SHOULD IT BE?
LISTEN: FORGET FOR A MOMENT THESE ARE HISTORICAL PEOPLE, INTENT ON BUILDING **FOUNDATIONS FOR MATHEMATICS...**
...AND THINK OF THE **"QUEST"** AS A FICTIONAL STORY...

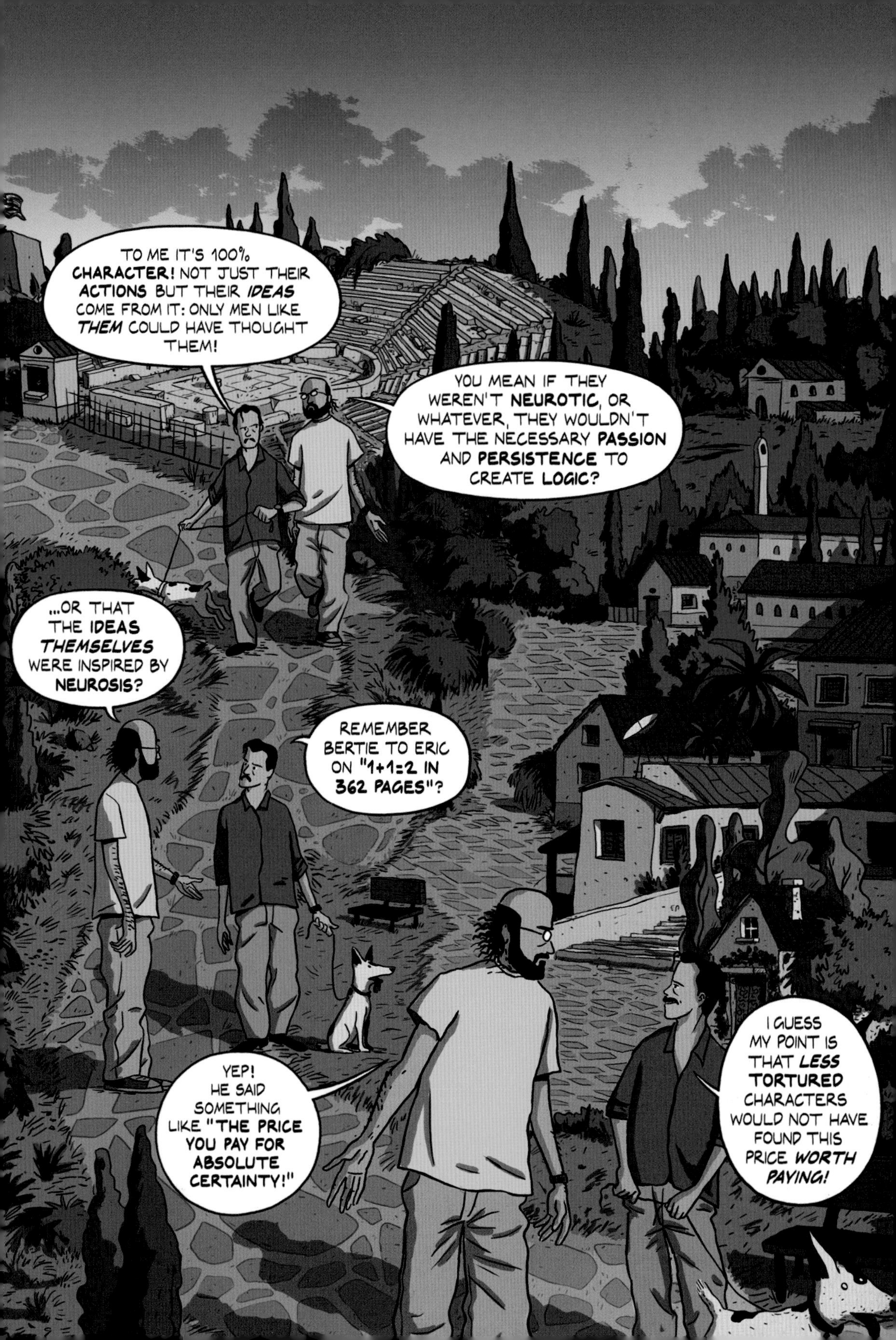
TO ME IT'S 100% CHARACTER! NOT JUST THEIR ACTIONS BUT THEIR IDEAS COME FROM IT: ONLY MEN LIKE THEM COULD HAVE THOUGHT THEM!
YOU MEAN IF THEY WEREN'T NEUROTIC, OR WHATEVER, THEY WOULDN'T HAVE THE NECESSARY PASSION AND PERSISTENCE TO CREATE LOGIC?
...OR THAT THE IDEAS THEMSELVES WERE INSPIRED BY NEUROSIS?
REMEMBER BERTIE TO ERIC ON "1+1=2 IN 362 PAGES"?
YEP! HE SAID SOMETHING LIKE "THE PRICE YOU PAY FOR ABSOLUTE CERTAINTY!"
I GUESS MY POINT IS THAT LESS TORTURED CHARACTERS WOULD NOT HAVE FOUND THIS PRICE WORTH PAYING!

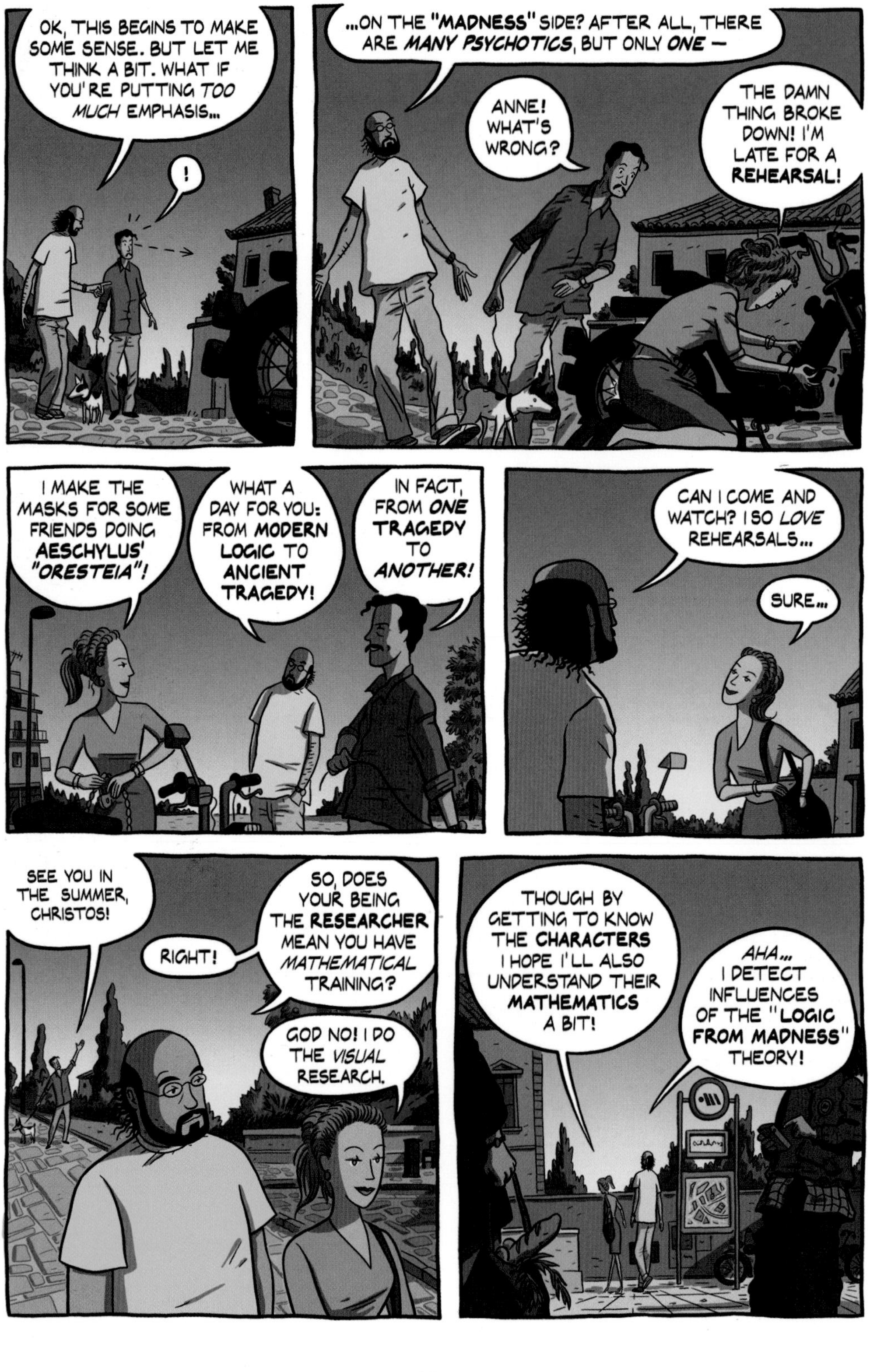
OK, THIS BEGINS TO MAKE SOME SENSE. BUT LET ME THINK A BIT. WHAT IF YOU'RE PUTTING *TOO MUCH* EMPHASIS...
!
...ON THE **"MADNESS"** SIDE? AFTER ALL, THERE ARE ***MANY PSYCHOTICS***, BUT ONLY ***ONE*** —
ANNE! WHAT'S WRONG?
THE DAMN THING BROKE DOWN! I'M LATE FOR A **REHEARSAL!**
I MAKE THE MASKS FOR SOME FRIENDS DOING **AESCHYLUS'** ***"ORESTEIA"***!
WHAT A DAY FOR YOU: FROM **MODERN LOGIC** TO **ANCIENT TRAGEDY!**
IN FACT, FROM **ONE TRAGEDY** TO ***ANOTHER!***
CAN I COME AND WATCH? I SO *LOVE* REHEARSALS...
SURE...
SEE YOU IN THE SUMMER, CHRISTOS!
RIGHT!
SO, DOES YOUR BEING THE **RESEARCHER** MEAN YOU HAVE *MATHEMATICAL* TRAINING?
GOD NO! I DO THE *VISUAL* RESEARCH.
THOUGH BY GETTING TO KNOW THE **CHARACTERS** I HOPE I'LL ALSO UNDERSTAND THEIR **MATHEMATICS** A BIT!
AHA... I DETECT INFLUENCES OF THE "**LOGIC FROM MADNESS**" THEORY!

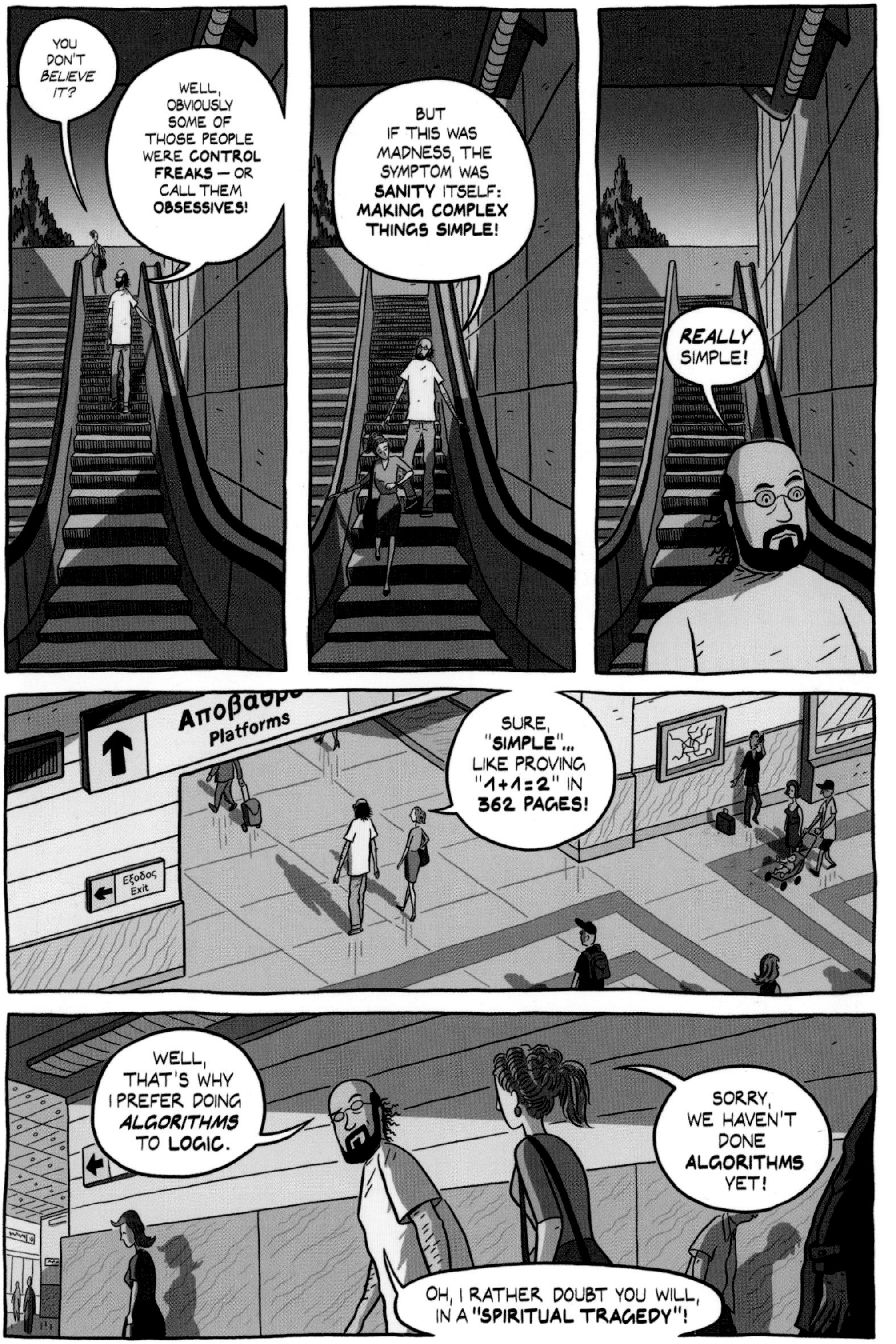
YOU DON'T *BELIEVE IT?*
WELL, OBVIOUSLY SOME OF THOSE PEOPLE WERE **CONTROL FREAKS** — OR CALL THEM **OBSESSIVES!**
BUT IF THIS WAS MADNESS, THE SYMPTOM WAS **SANITY** ITSELF: **MAKING COMPLEX THINGS SIMPLE!**
REALLY SIMPLE!
Platforms
Εξοδος
Exit
SURE, "**SIMPLE**"... LIKE PROVING "**1+1=2**" IN **362 PAGES!**
WELL, THAT'S WHY I PREFER DOING ***ALGORITHMS*** TO **LOGIC.**
SORRY, WE HAVEN'T DONE **ALGORITHMS** YET!
OH, I RATHER DOUBT YOU WILL, IN A "**SPIRITUAL TRAGEDY**"!

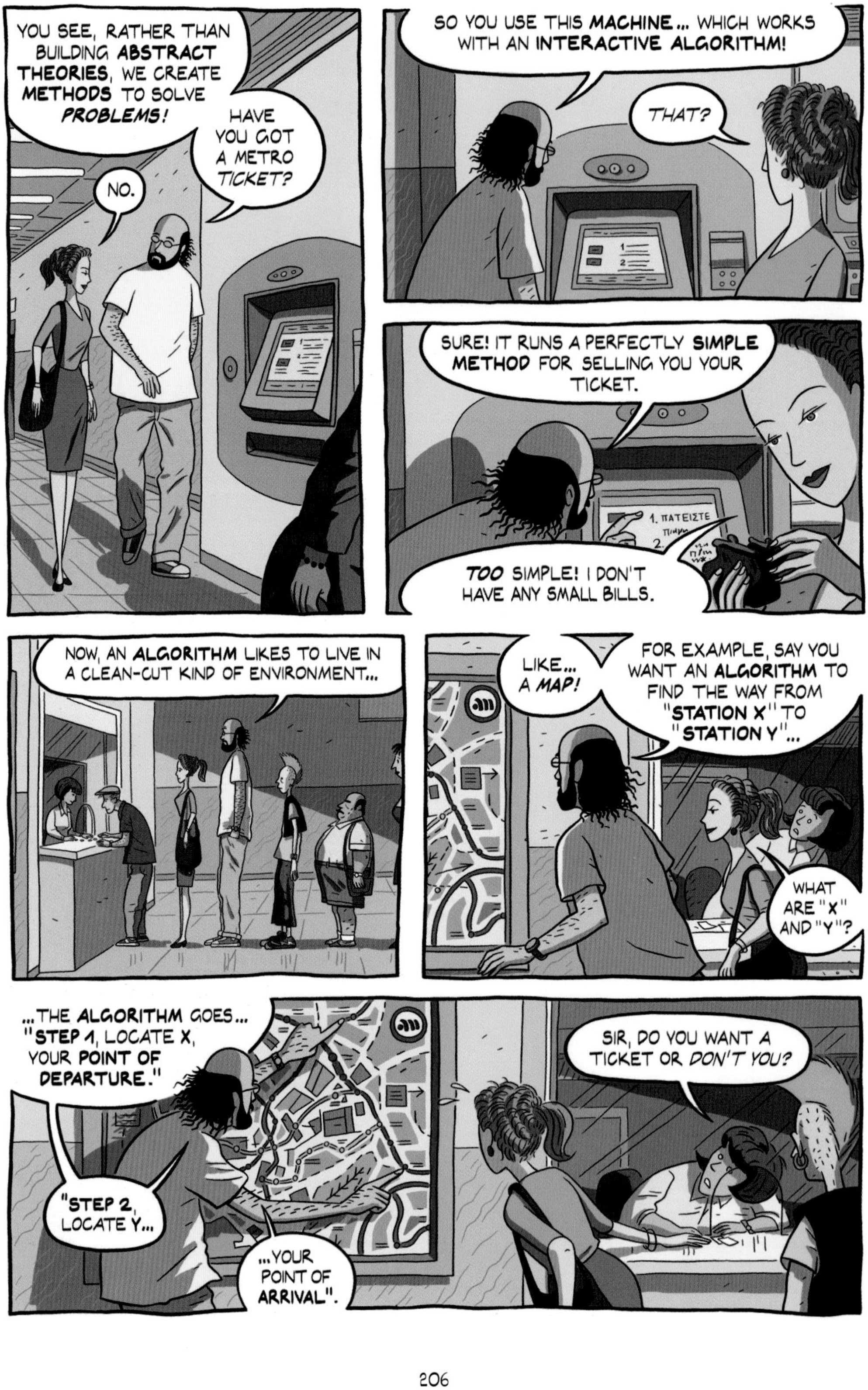
YOU SEE, RATHER THAN BUILDING **ABSTRACT THEORIES**, WE CREATE **METHODS** TO SOLVE ***PROBLEMS!***
HAVE YOU GOT A METRO *TICKET?*
NO.
SO YOU USE THIS **MACHINE...** WHICH WORKS WITH AN **INTERACTIVE ALGORITHM!**
THAT?
SURE! IT RUNS A PERFECTLY **SIMPLE METHOD** FOR SELLING YOU YOUR TICKET.
1. ΠΑΤΕΙΣΤΕ
TOO SIMPLE! I DON'T HAVE ANY SMALL BILLS.
NOW, AN **ALGORITHM** LIKES TO LIVE IN A CLEAN-CUT KIND OF ENVIRONMENT...
LIKE... A *MAP!*
FOR EXAMPLE, SAY YOU WANT AN **ALGORITHM** TO FIND THE WAY FROM "**STATION X**" TO "**STATION Y**"...
WHAT ARE "**X**" AND "**Y**"?
...THE **ALGORITHM** GOES... "**STEP 1**, LOCATE **X**, YOUR **POINT OF DEPARTURE**."
"**STEP 2**, LOCATE **Y**...
...YOUR POINT OF **ARRIVAL**".
SIR, DO YOU WANT A TICKET OR *DON'T YOU?*

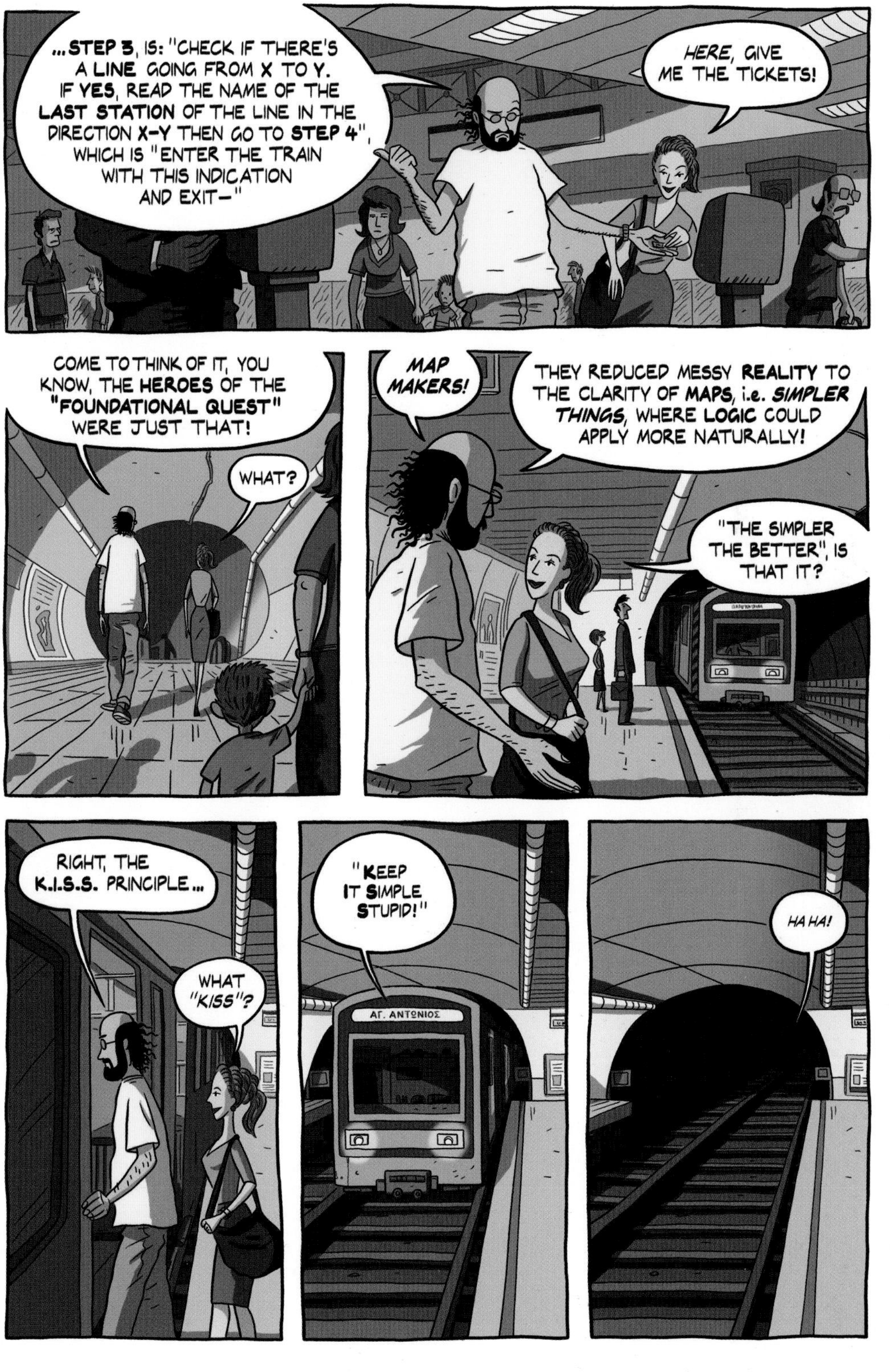
...STEP 3, IS: "CHECK IF THERE'S A LINE GOING FROM X TO Y. IF YES, READ THE NAME OF THE LAST STATION OF THE LINE IN THE DIRECTION X-Y THEN GO TO STEP 4", WHICH IS "ENTER THE TRAIN WITH THIS INDICATION AND EXIT—"
HERE, GIVE ME THE TICKETS!
COME TO THINK OF IT, YOU KNOW, THE HEROES OF THE "FOUNDATIONAL QUEST" WERE JUST THAT!
WHAT?
MAP MAKERS!
THEY REDUCED MESSY REALITY TO THE CLARITY OF MAPS, i.e. SIMPLER THINGS, WHERE LOGIC COULD APPLY MORE NATURALLY!
"THE SIMPLER THE BETTER", IS THAT IT?
RIGHT, THE K.I.S.S. PRINCIPLE...
WHAT "KISS"?
"KEEP IT SIMPLE STUPID!"
ΑΓ. ΑΝΤΩΝΙΟΣ
ΑΓ. ΑΝΤΩΝΙΟΣ
HA HA!

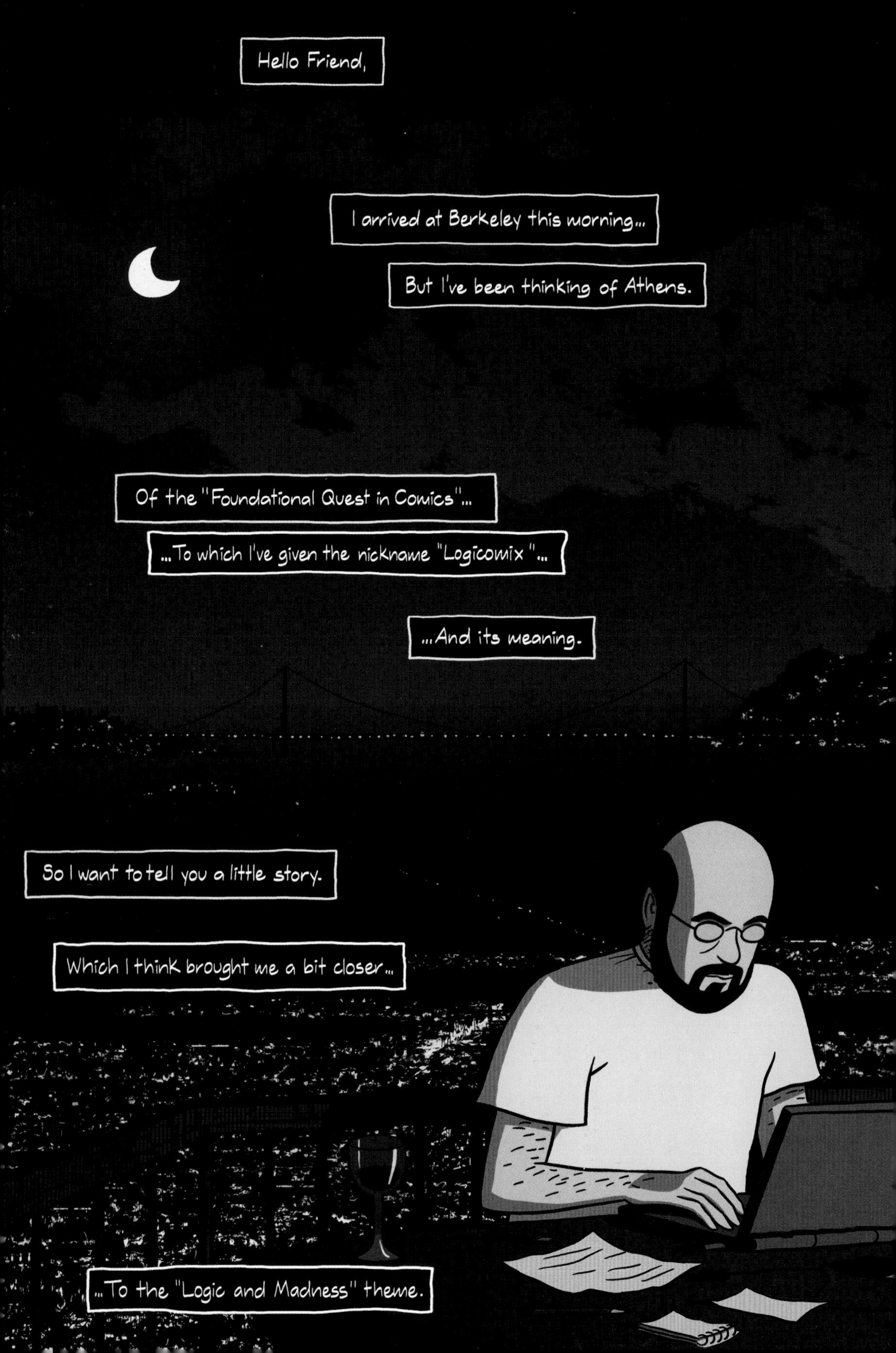
Hello Friend,
I arrived at Berkeley this morning...
But I've been thinking of Athens.
Of the "Foundational Quest in Comics"...
...To which I've given the nickname "Logicomix"...
...And its meaning.
So I want to tell you a little story.
Which I think brought me a bit closer...
...To the "Logic and Madness" theme.

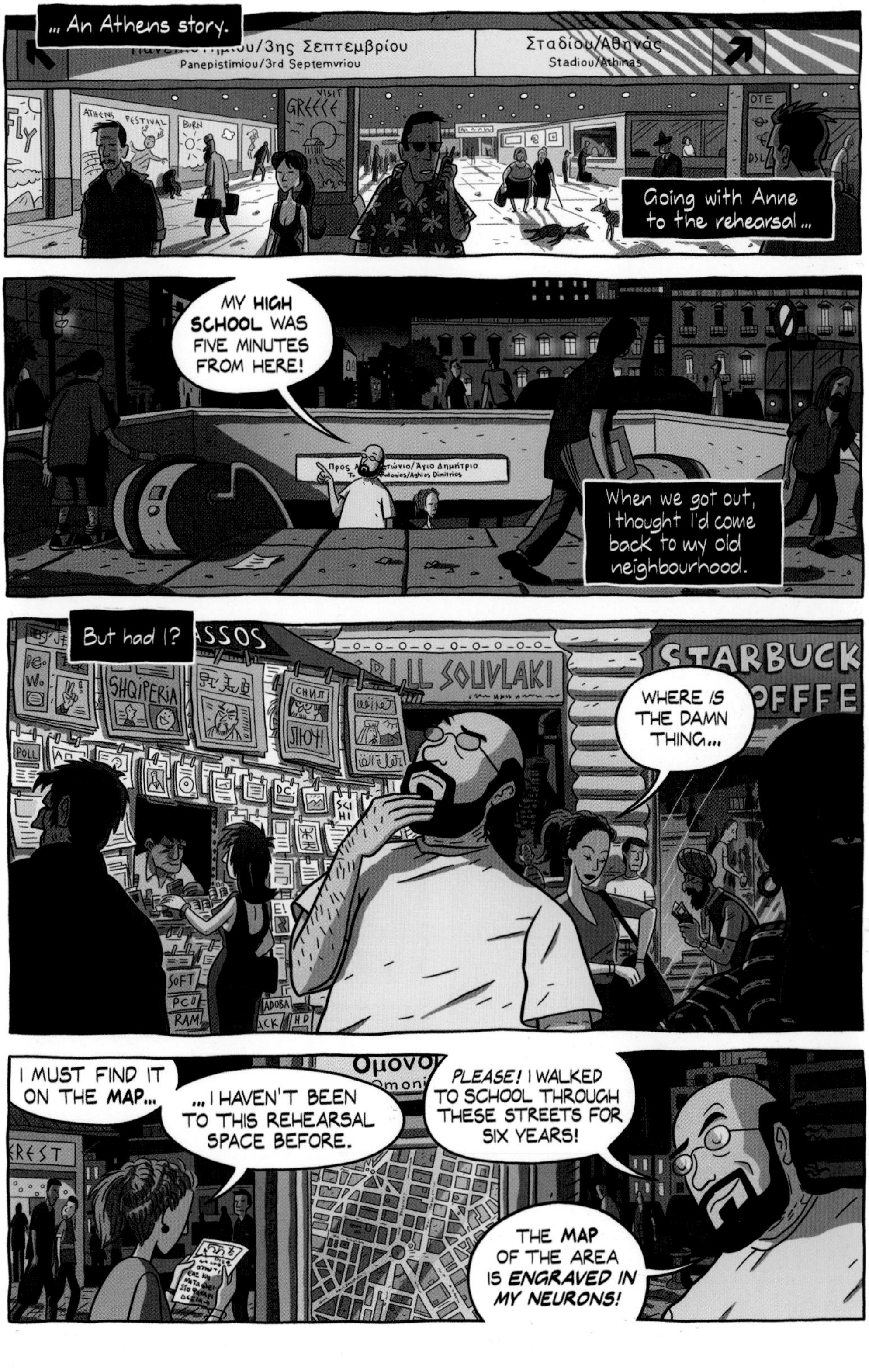
... An Athens story.
Σταδίου/Αθηνάς
Stadiou/Athinas
Panepistimiou/3rd Septemvriou
VISIT GREECE
ATHENS FESTIVAL
OTE
Going with Anne to the rehearsal...
MY HIGH SCHOOL WAS FIVE MINUTES FROM HERE!
When we got out, I thought I'd come back to my old neighbourhood.
But had I?
SHQIPERIA
GRILL SOUVLAKI
STARBUCK
OFFEE
WHERE IS THE DAMN THING...
I MUST FIND IT ON THE MAP...
... I HAVEN'T BEEN TO THIS REHEARSAL SPACE BEFORE.
PLEASE! I WALKED TO SCHOOL THROUGH THESE STREETS FOR SIX YEARS!
THE MAP OF THE AREA IS ENGRAVED IN MY NEURONS!

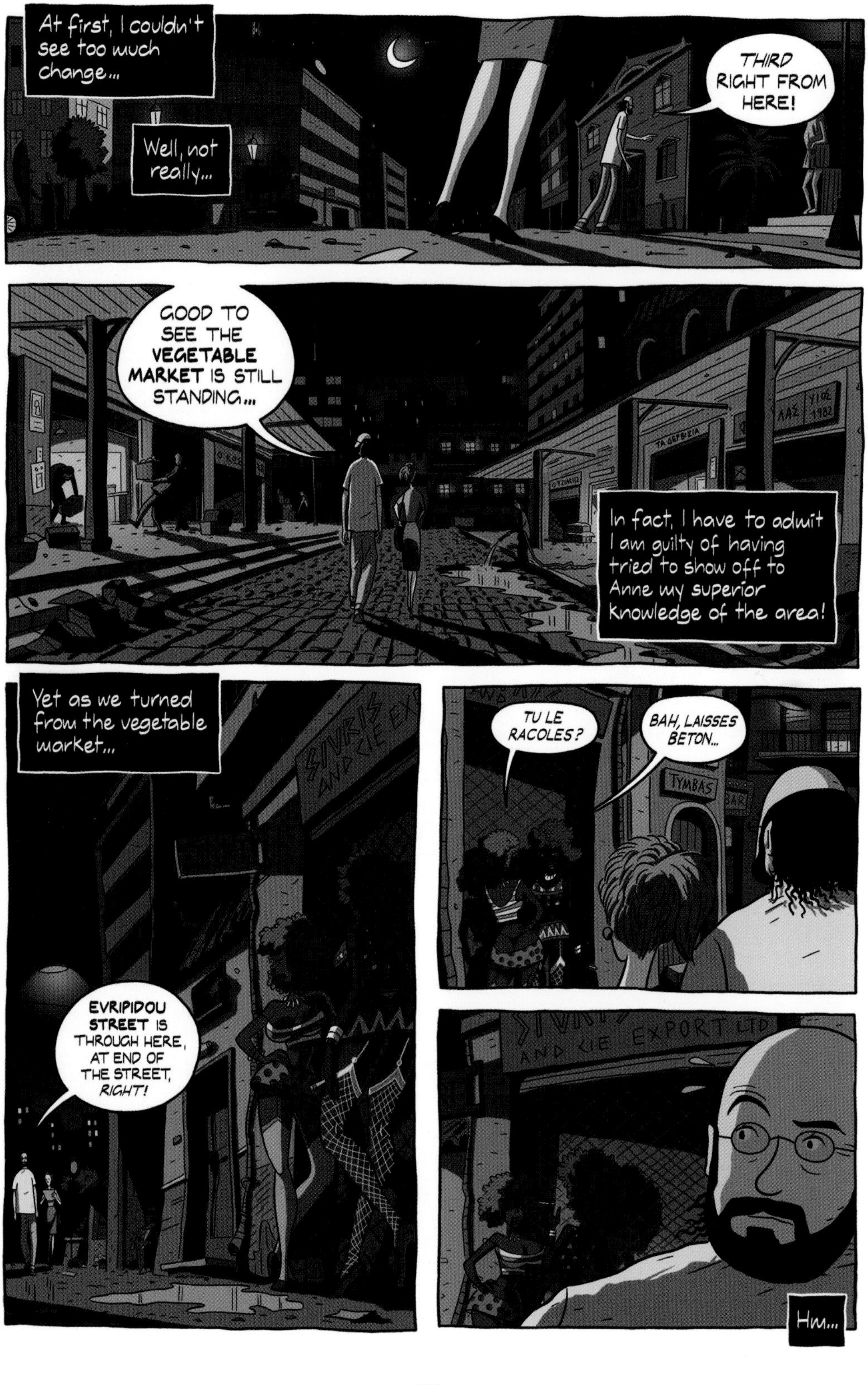
At first, I couldn't see too much change...
Well, not really...
THIRD RIGHT FROM HERE!
GOOD TO SEE THE **VEGETABLE MARKET** IS STILL STANDING...
In fact, I have to admit I am guilty of having tried to show off to Anne my superior knowledge of the area!
Yet as we turned from the vegetable market...
EVRIPIDOU STREET IS THROUGH HERE, AT END OF THE STREET, *RIGHT!*
TYMBAS
BAR
TU LE RACOLES?
BAH, LAISSES BETON...
AND CIE EXPORT LTD
Hm...

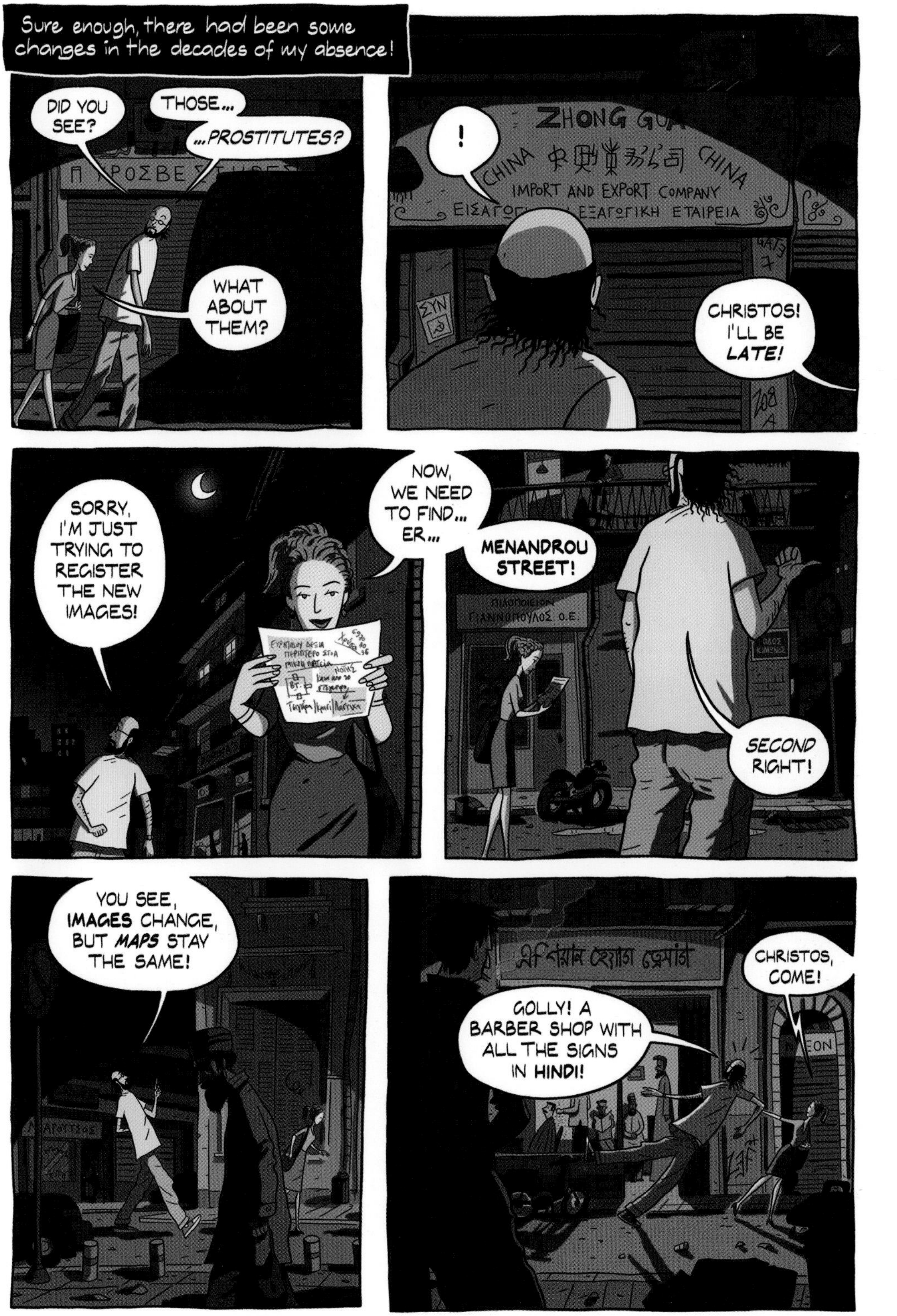
Sure enough, there had been some changes in the decades of my absence!
DID YOU SEE?
THOSE...
...PROSTITUTES?
WHAT ABOUT THEM?
!
ZHONG GUA
CHINA
CHINA
IMPORT AND EXPORT COMPANY
CHRISTOS! I'LL BE LATE!
SORRY, I'M JUST TRYING TO REGISTER THE NEW IMAGES!
NOW, WE NEED TO FIND... ER...
MENANDROU STREET!
SECOND RIGHT!
YOU SEE, IMAGES CHANGE, BUT MAPS STAY THE SAME!
CHRISTOS, COME!
GOLLY! A BARBER SHOP WITH ALL THE SIGNS IN HINDI!

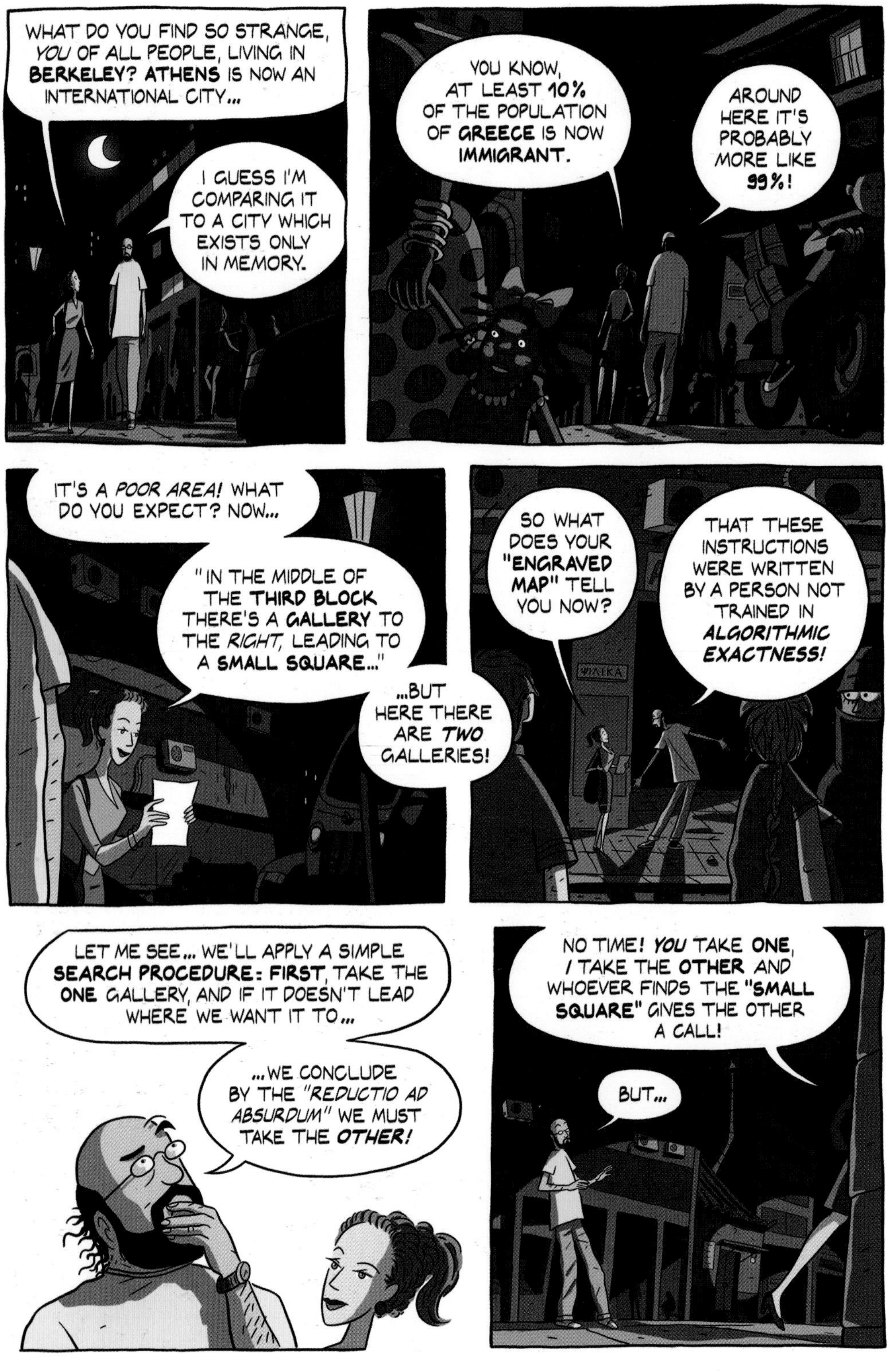
WHAT DO YOU FIND SO STRANGE, YOU OF ALL PEOPLE, LIVING IN BERKELEY? ATHENS IS NOW AN INTERNATIONAL CITY...
I GUESS I'M COMPARING IT TO A CITY WHICH EXISTS ONLY IN MEMORY.
YOU KNOW, AT LEAST 10% OF THE POPULATION OF GREECE IS NOW IMMIGRANT.
AROUND HERE IT'S PROBABLY MORE LIKE 99%!
IT'S A POOR AREA! WHAT DO YOU EXPECT? NOW...
"IN THE MIDDLE OF THE THIRD BLOCK THERE'S A GALLERY TO THE RIGHT, LEADING TO A SMALL SQUARE..."
...BUT HERE THERE ARE TWO GALLERIES!
SO WHAT DOES YOUR "ENGRAVED MAP" TELL YOU NOW?
ΨΙΛΙΚΑ
THAT THESE INSTRUCTIONS WERE WRITTEN BY A PERSON NOT TRAINED IN ALGORITHMIC EXACTNESS!
LET ME SEE... WE'LL APPLY A SIMPLE SEARCH PROCEDURE: FIRST, TAKE THE ONE GALLERY, AND IF IT DOESN'T LEAD WHERE WE WANT IT TO...
...WE CONCLUDE BY THE "REDUCTIO AD ABSURDUM" WE MUST TAKE THE OTHER!
NO TIME! YOU TAKE ONE, I TAKE THE OTHER AND WHOEVER FINDS THE "SMALL SQUARE" GIVES THE OTHER A CALL!
BUT...

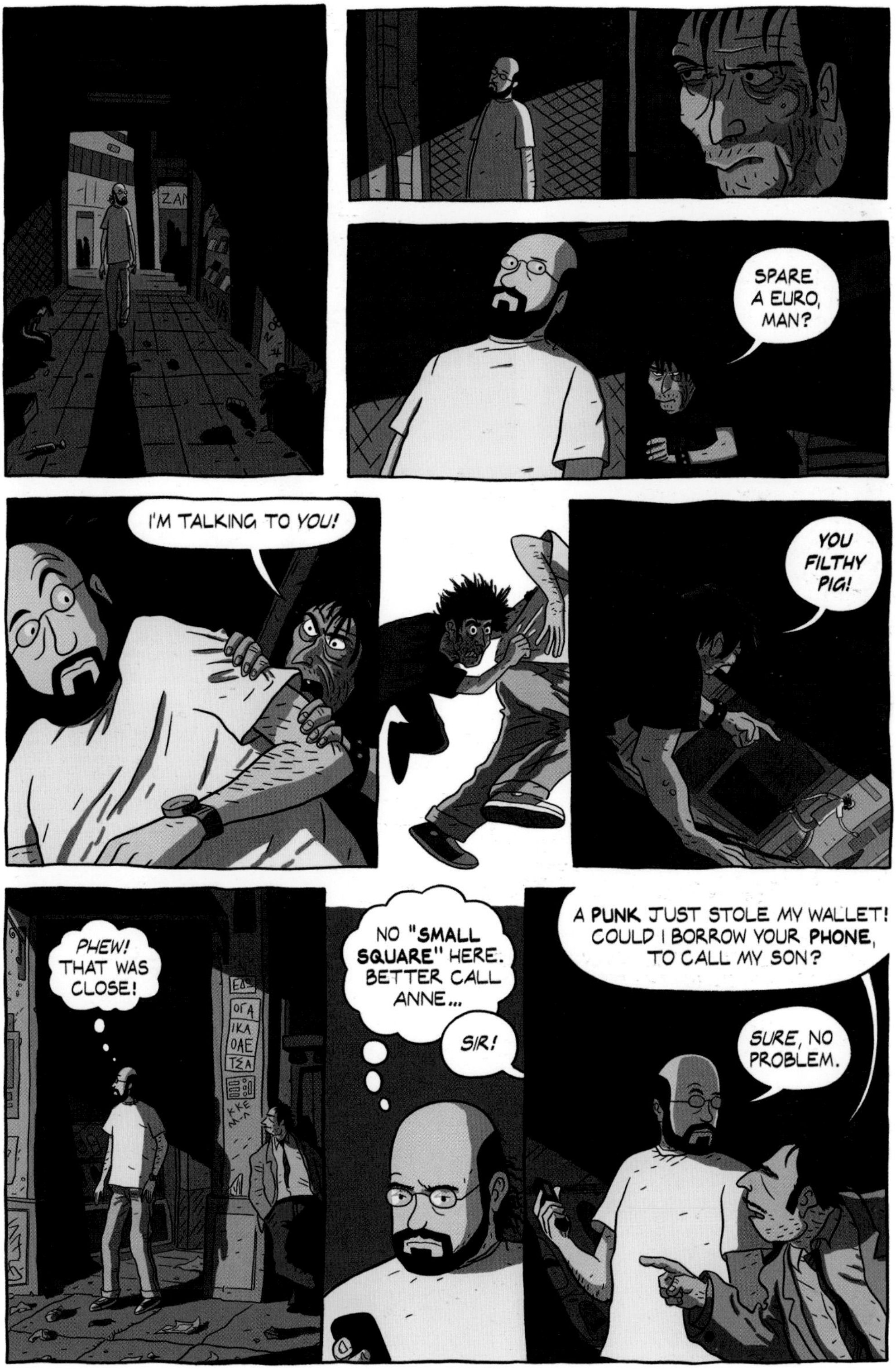

SPARE A EURO, MAN?
I'M TALKING TO *YOU!*
YOU FILTHY PIG!
PHEW! THAT WAS CLOSE!
NO "**SMALL SQUARE**" HERE. BETTER CALL ANNE...
SIR!
A **PUNK** JUST STOLE MY WALLET! COULD I BORROW YOUR **PHONE**, TO CALL MY SON?
SURE, NO PROBLEM.

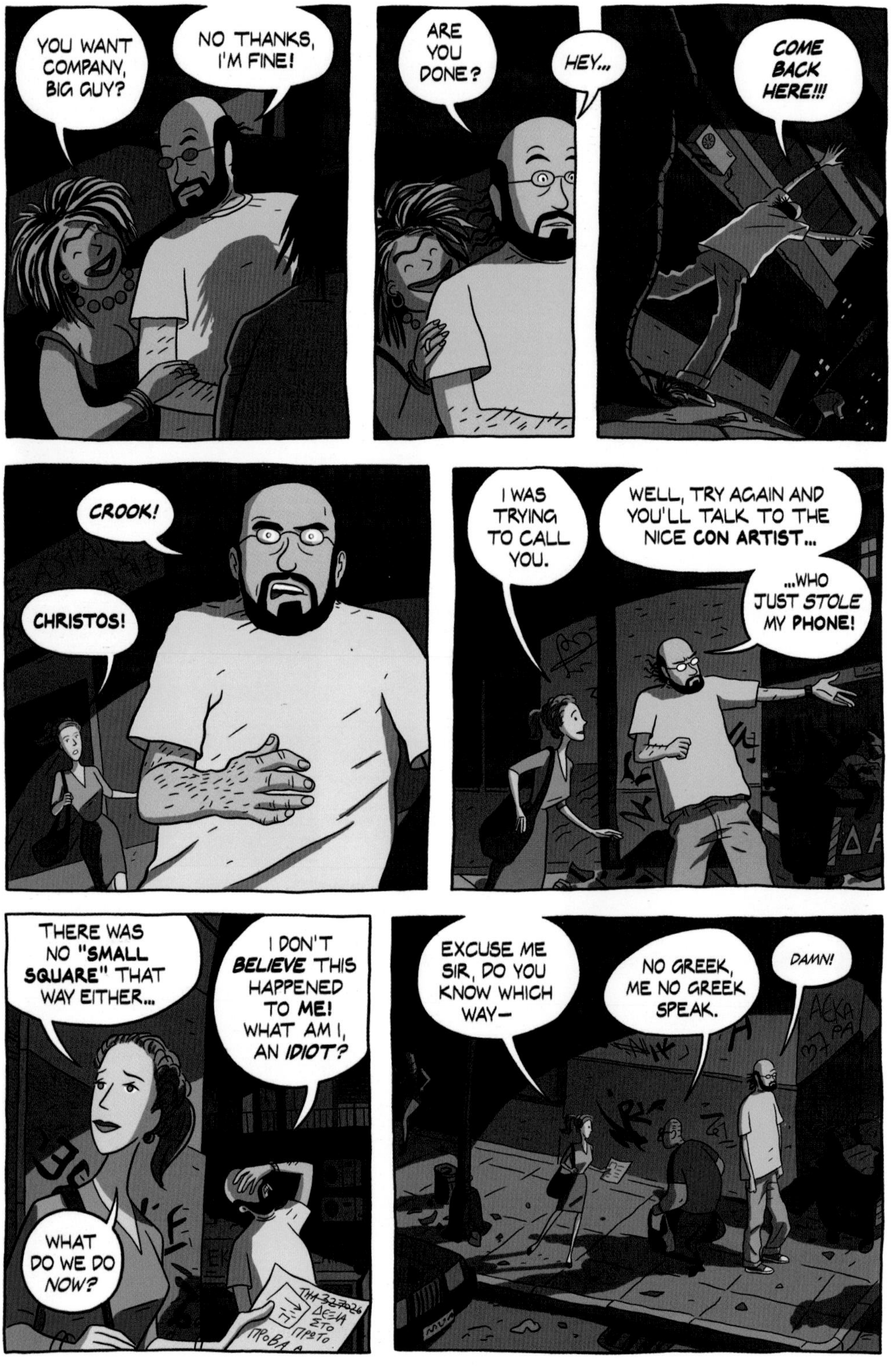
YOU WANT COMPANY, BIG GUY?
NO THANKS, I'M FINE!
ARE YOU DONE?
HEY...
COME BACK HERE!!!
CROOK!
CHRISTOS!
I WAS TRYING TO CALL YOU.
WELL, TRY AGAIN AND YOU'LL TALK TO THE NICE CON ARTIST...
...WHO JUST STOLE MY PHONE!
THERE WAS NO "SMALL SQUARE" THAT WAY EITHER...
I DON'T BELIEVE THIS HAPPENED TO ME! WHAT AM I, AN IDIOT?
WHAT DO WE DO NOW?
EXCUSE ME SIR, DO YOU KNOW WHICH WAY—
NO GREEK, ME NO GREEK SPEAK.
DAMN!

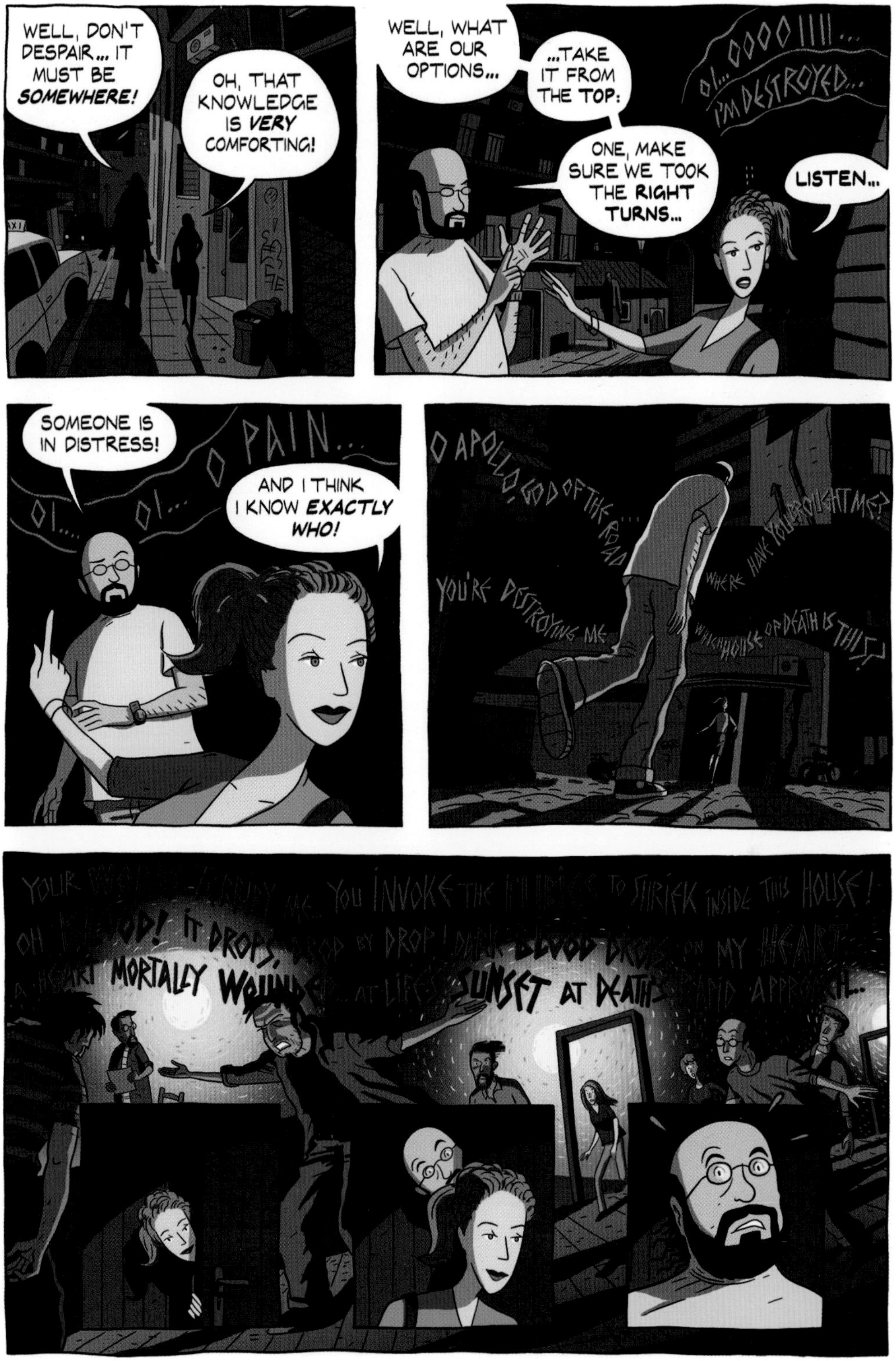

WELL, DON'T DESPAIR... IT MUST BE *SOMEWHERE!*
OH, THAT KNOWLEDGE IS *VERY* COMFORTING!
ΤΑΞΙ
WELL, WHAT ARE OUR OPTIONS...
...TAKE IT FROM THE **TOP:**
ONE, MAKE SURE WE TOOK THE **RIGHT TURNS...**
OI... OOOO IIII... I'M DESTROYED...
LISTEN...
SOMEONE IS IN DISTRESS!
OI... OI... O PAIN...
AND I THINK I KNOW ***EXACTLY WHO!***
O APOLLO, GOD OF THE ROAD...
YOU'RE DESTROYING ME
WHERE HAVE YOU BROUGHT ME?
WHICH HOUSE OF DEATH IS THIS?!
YOUR WORDS TERRIFY ME... YOU INVOKE THE FURIES TO SHRIEK INSIDE THIS HOUSE!
OH BLOOD! IT DROPS, DROP BY DROP! DARK BLOOD DROPS ON MY HEART
A HEART MORTALLY WOUNDED... AT LIFE'S SUNSET AT DEATH'S RAPID APPROACH...

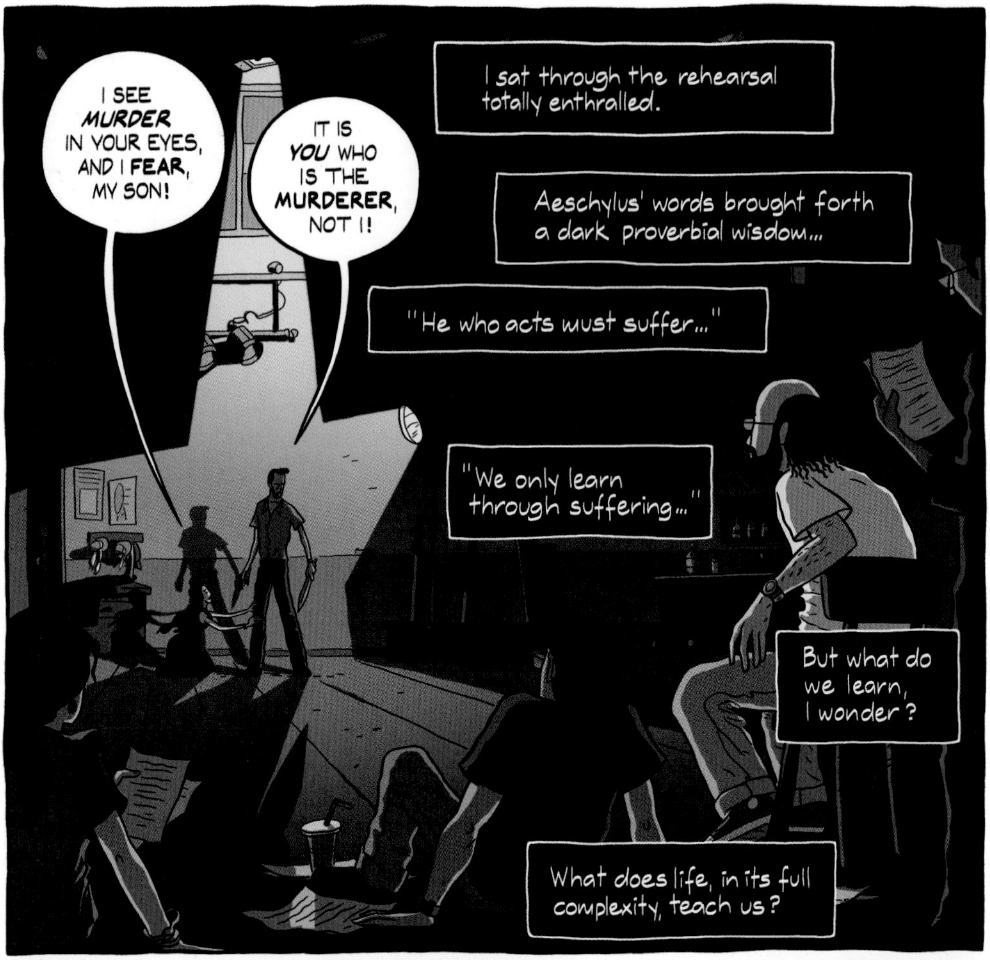
I sat through the rehearsal totally enthralled.
I SEE MURDER IN YOUR EYES, AND I FEAR, MY SON!
IT IS YOU WHO IS THE MURDERER, NOT I!
Aeschylus' words brought forth a dark proverbial wisdom...
"He who acts must suffer..."
"We only learn through suffering..."
But what do we learn, I wonder?
What does life, in its full complexity, teach us?

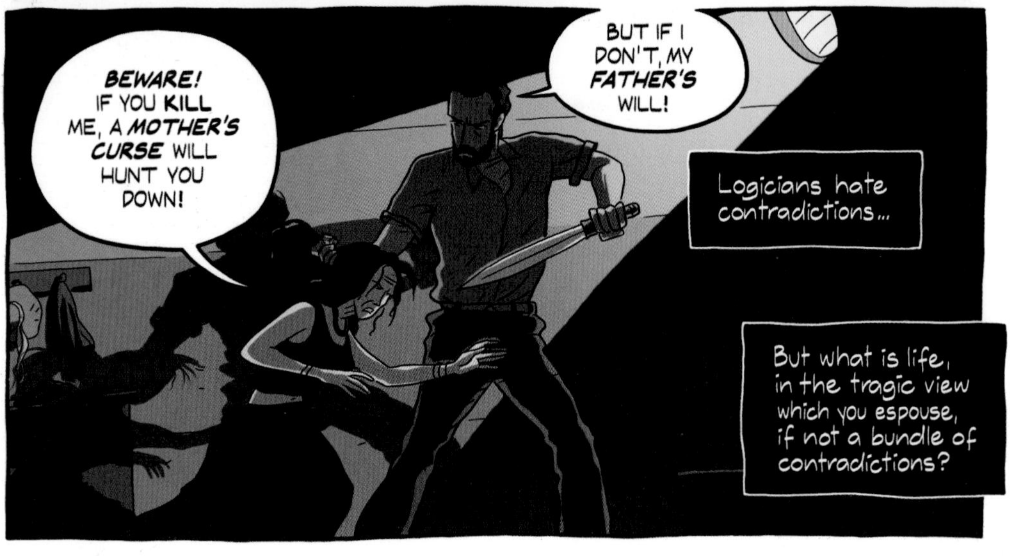
BEWARE! IF YOU KILL ME, A MOTHER'S CURSE WILL HUNT YOU DOWN!
BUT IF I DON'T, MY FATHER'S WILL!
Logicians hate contradictions...
But what is life, in the tragic view which you espouse, if not a bundle of contradictions?

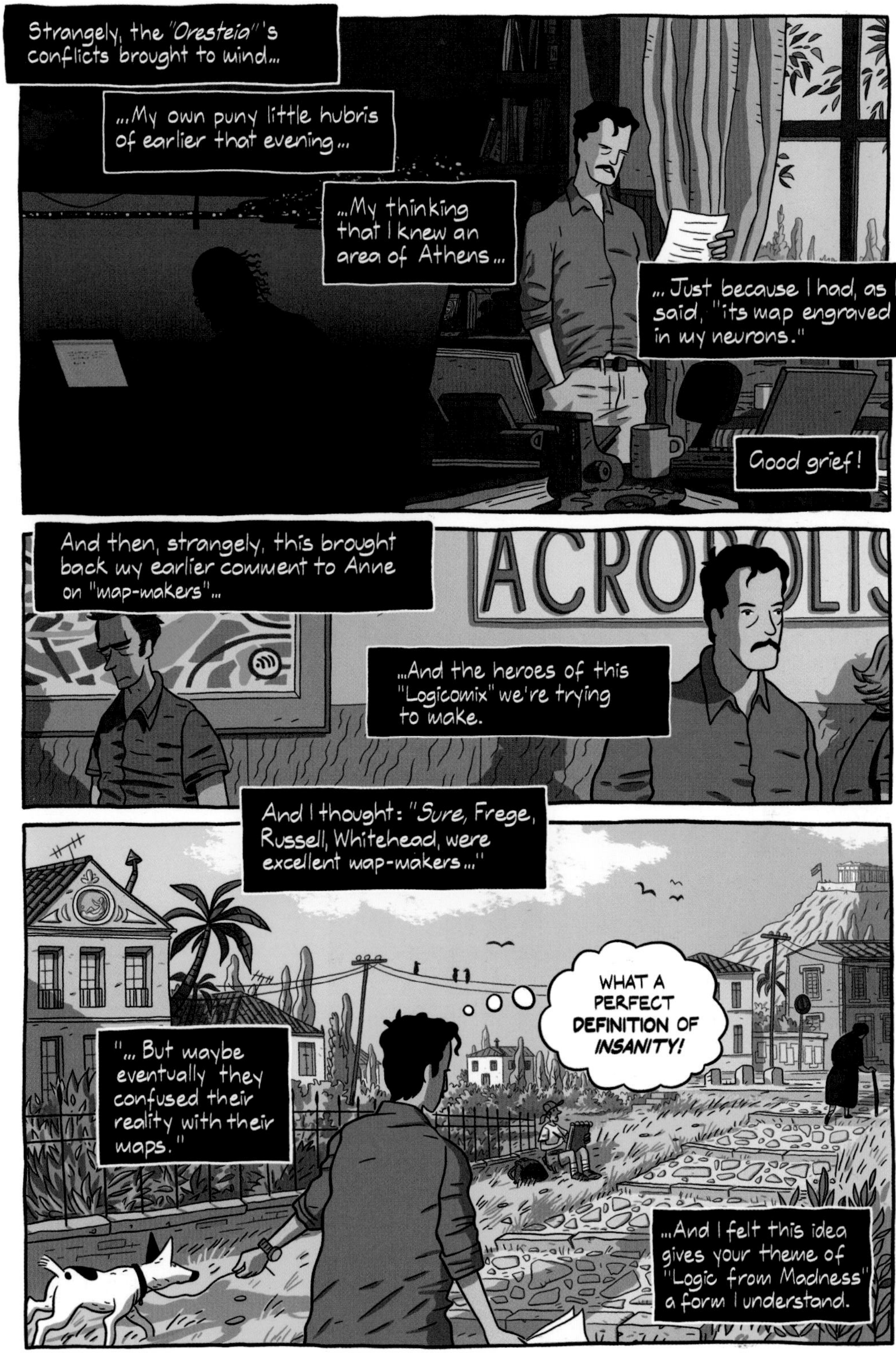

Strangely, the "Oresteia"'s conflicts brought to mind...
...My own puny little hubris of earlier that evening...
...My thinking that I knew an area of Athens...
...Just because I had, as I said, "its map engraved in my neurons."
Good grief!
And then, strangely, this brought back my earlier comment to Anne on "map-makers"...
ACROPOLIS
...And the heroes of this "Logicomix" we're trying to make.
And I thought: "Sure, Frege, Russell, Whitehead, were excellent map-makers..."
"...But maybe eventually they confused their reality with their maps."
WHAT A PERFECT DEFINITION OF INSANITY!
...And I felt this idea gives your theme of "Logic from Madness" a form I understand.

5. LOGICO-PHILOSOPHICAL WARS

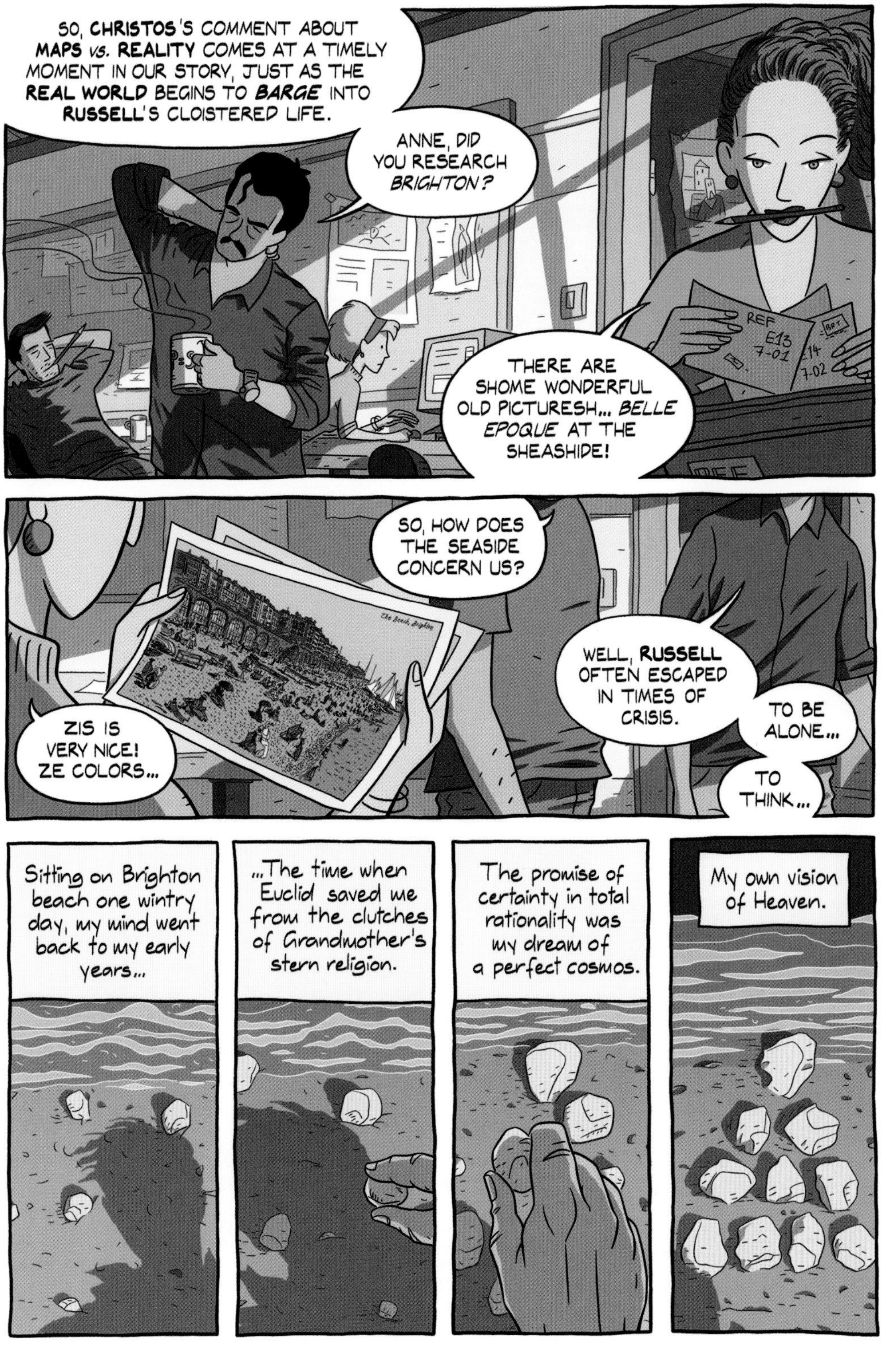
SO, CHRISTOS'S COMMENT ABOUT MAPS vs. REALITY COMES AT A TIMELY MOMENT IN OUR STORY, JUST AS THE REAL WORLD BEGINS TO BARGE INTO RUSSELL'S CLOISTERED LIFE.
ANNE, DID YOU RESEARCH BRIGHTON?
REF E13 7-01
E14 7-02
THERE ARE SHOME WONDERFUL OLD PICTURESH... BELLE EPOQUE AT THE SHEASHIDE!
The Beach, Brighton
SO, HOW DOES THE SEASIDE CONCERN US?
ZIS IS VERY NICE! ZE COLORS...
WELL, RUSSELL OFTEN ESCAPED IN TIMES OF CRISIS.
TO BE ALONE...
TO THINK...
Sitting on Brighton beach one wintry day, my mind went back to my early years...
...The time when Euclid saved me from the clutches of Grandmother's stern religion.
The promise of certainty in total rationality was my dream of a perfect cosmos.
My own vision of Heaven.

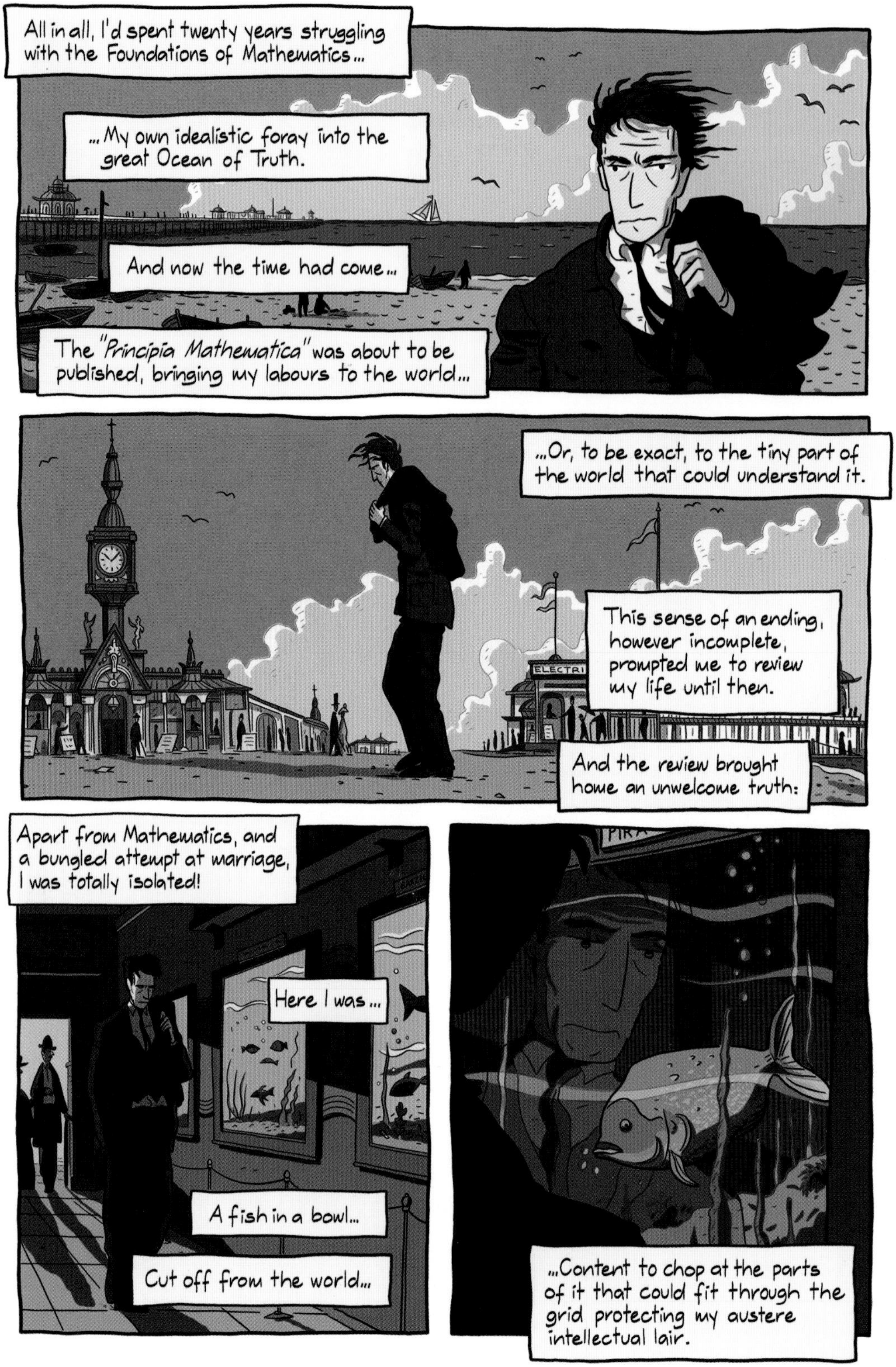
All in all, I'd spent twenty years struggling with the Foundations of Mathematics...
...My own idealistic foray into the great Ocean of Truth.
And now the time had come...
The "Principia Mathematica" was about to be published, bringing my labours to the world...
...Or, to be exact, to the tiny part of the world that could understand it.
This sense of an ending, however incomplete, prompted me to review my life until then.
And the review brought home an unwelcome truth:
Apart from Mathematics, and a bungled attempt at marriage, I was totally isolated!
Here I was...
A fish in a bowl...
Cut off from the world...
...Content to chop at the parts of it that could fit through the grid protecting my austere intellectual lair.

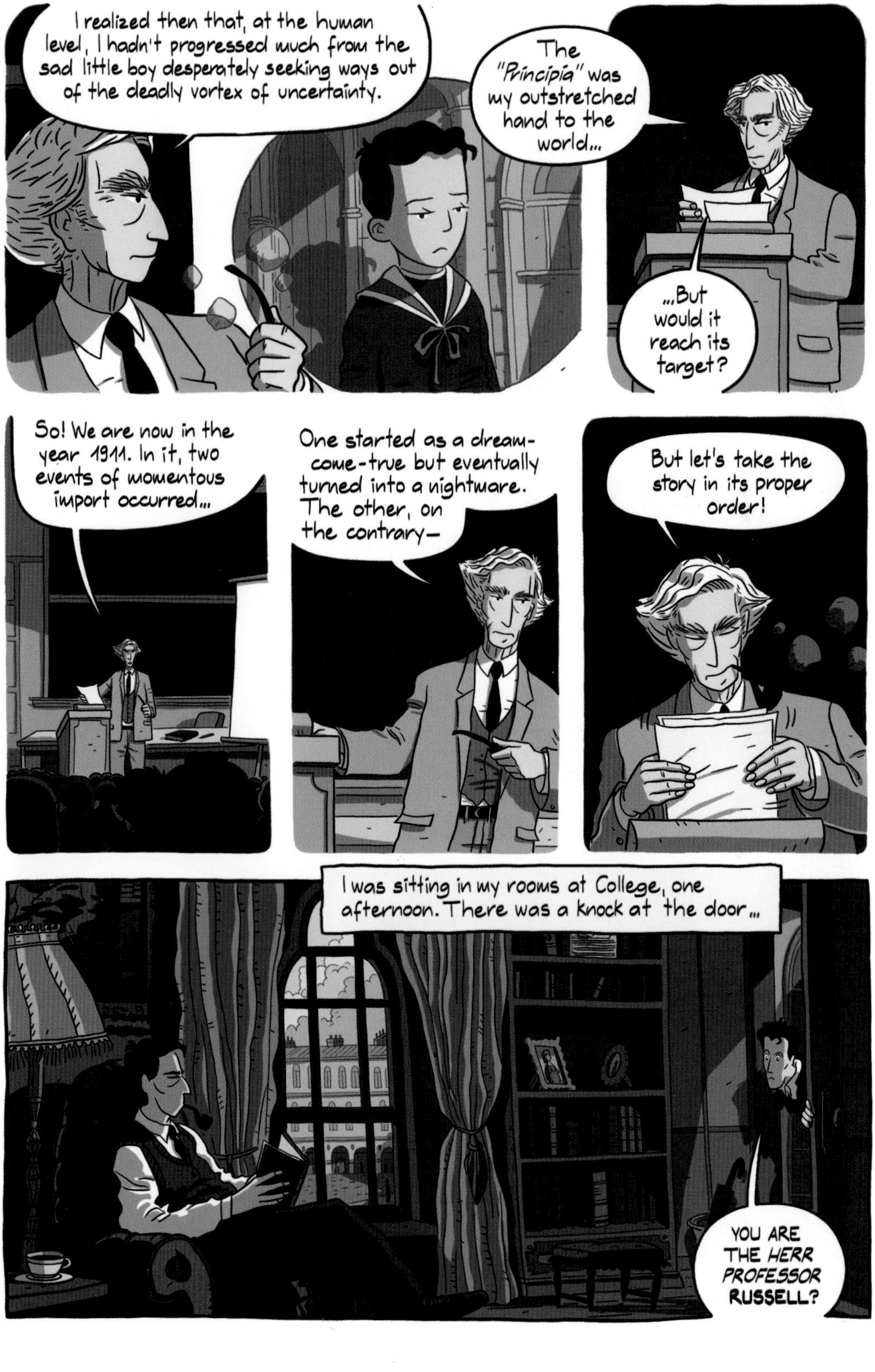
I realized then that, at the human level, I hadn't progressed much from the sad little boy desperately seeking ways out of the deadly vortex of uncertainty.
The "Principia" was my outstretched hand to the world...
...But would it reach its target?
So! We are now in the year 1911. In it, two events of momentous import occurred...
One started as a dream-come-true but eventually turned into a nightmare. The other, on the contrary—
But let's take the story in its proper order!
I was sitting in my rooms at College, one afternoon. There was a knock at the door...
YOU ARE THE HERR PROFESSOR RUSSELL?

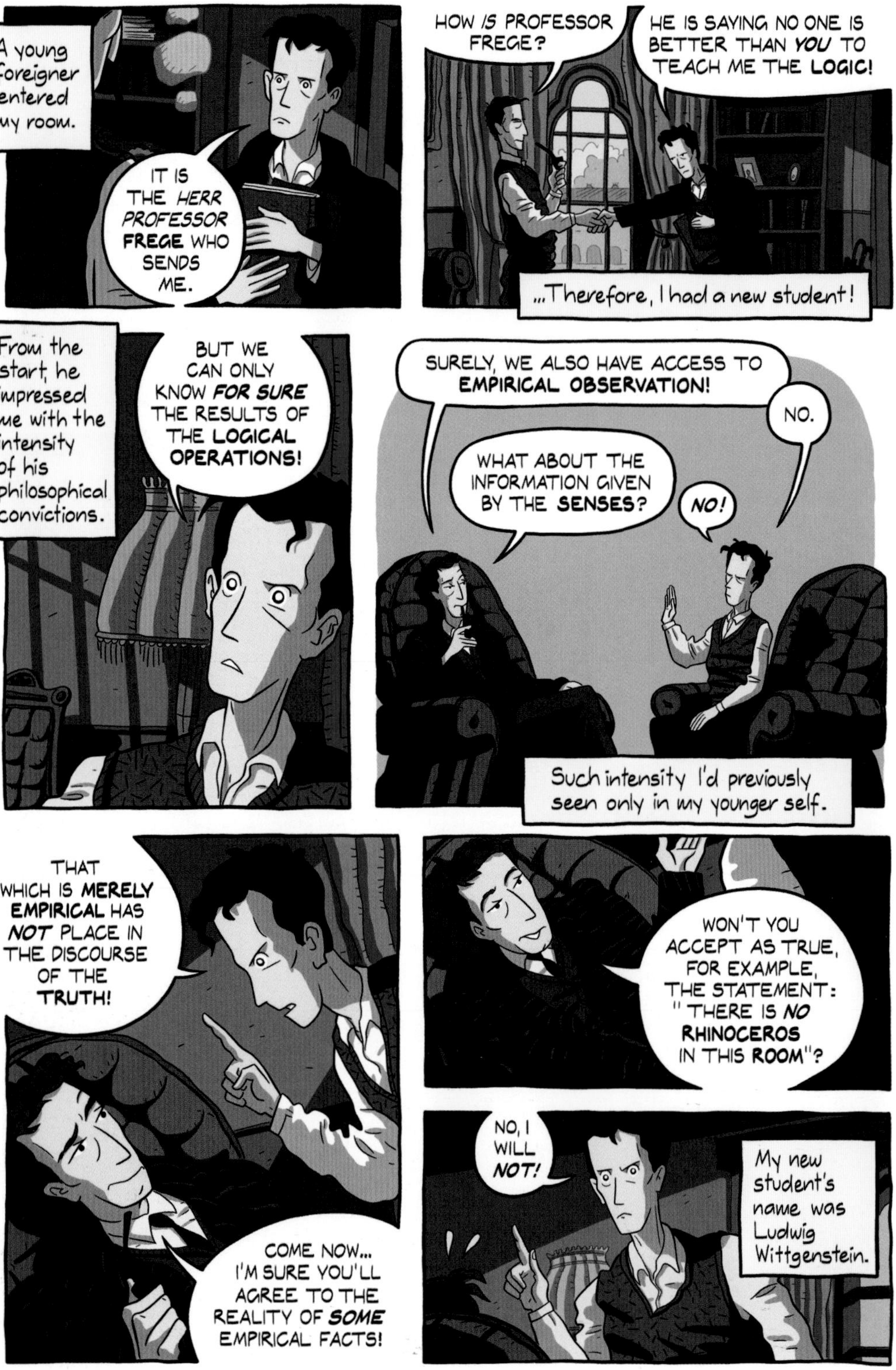
A young foreigner entered my room.
IT IS THE *HERR PROFESSOR* **FREGE** WHO SENDS ME.
HOW *IS* PROFESSOR FREGE?
HE IS SAYING NO ONE IS BETTER THAN ***YOU*** TO TEACH ME THE **LOGIC!**
...Therefore, I had a new student!
From the start, he impressed me with the intensity of his philosophical convictions.
BUT WE CAN ONLY KNOW ***FOR SURE*** THE RESULTS OF THE **LOGICAL OPERATIONS!**
SURELY, WE ALSO HAVE ACCESS TO **EMPIRICAL OBSERVATION!**
NO.
WHAT ABOUT THE INFORMATION GIVEN BY THE **SENSES?**
NO!
Such intensity I'd previously seen only in my younger self.
THAT WHICH IS **MERELY EMPIRICAL** HAS ***NOT*** PLACE IN THE DISCOURSE OF THE **TRUTH!**
COME NOW... I'M SURE YOU'LL AGREE TO THE REALITY OF ***SOME*** EMPIRICAL FACTS!
WON'T YOU ACCEPT AS TRUE, FOR EXAMPLE, THE STATEMENT: "THERE IS ***NO*** **RHINOCEROS** IN THIS **ROOM**"?
NO, I WILL ***NOT!***
My new student's name was Ludwig Wittgenstein.

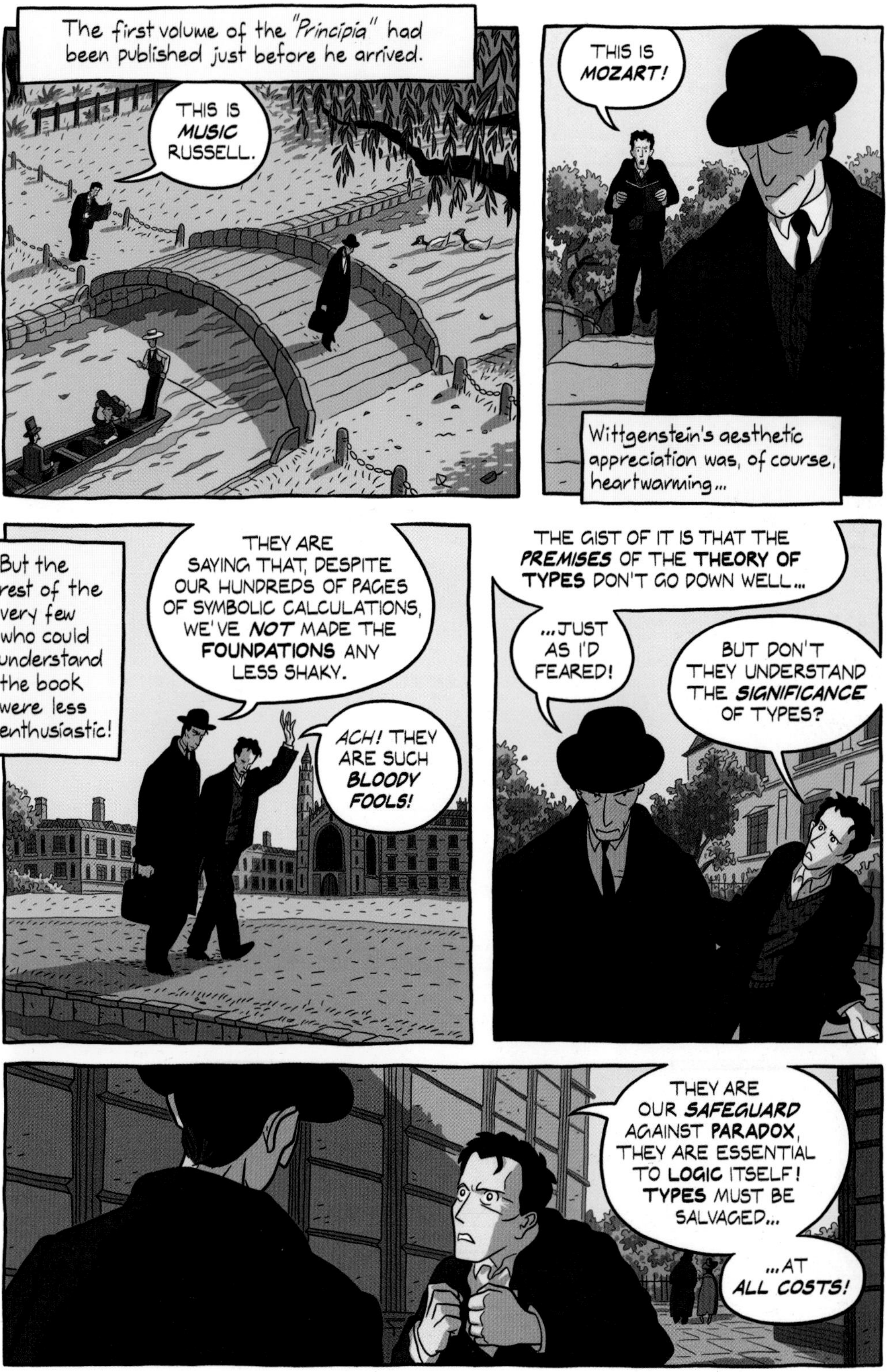
The first volume of the "Principia" had been published just before he arrived.
THIS IS MUSIC RUSSELL.
THIS IS MOZART!
Wittgenstein's aesthetic appreciation was, of course, heartwarming...
But the rest of the very few who could understand the book were less enthusiastic!
THEY ARE SAYING THAT, DESPITE OUR HUNDREDS OF PAGES OF SYMBOLIC CALCULATIONS, WE'VE NOT MADE THE FOUNDATIONS ANY LESS SHAKY.
ACH! THEY ARE SUCH BLOODY FOOLS!
THE GIST OF IT IS THAT THE PREMISES OF THE THEORY OF TYPES DON'T GO DOWN WELL...
...JUST AS I'D FEARED!
BUT DON'T THEY UNDERSTAND THE SIGNIFICANCE OF TYPES?
THEY ARE OUR SAFEGUARD AGAINST PARADOX, THEY ARE ESSENTIAL TO LOGIC ITSELF! TYPES MUST BE SALVAGED...
...AT ALL COSTS!

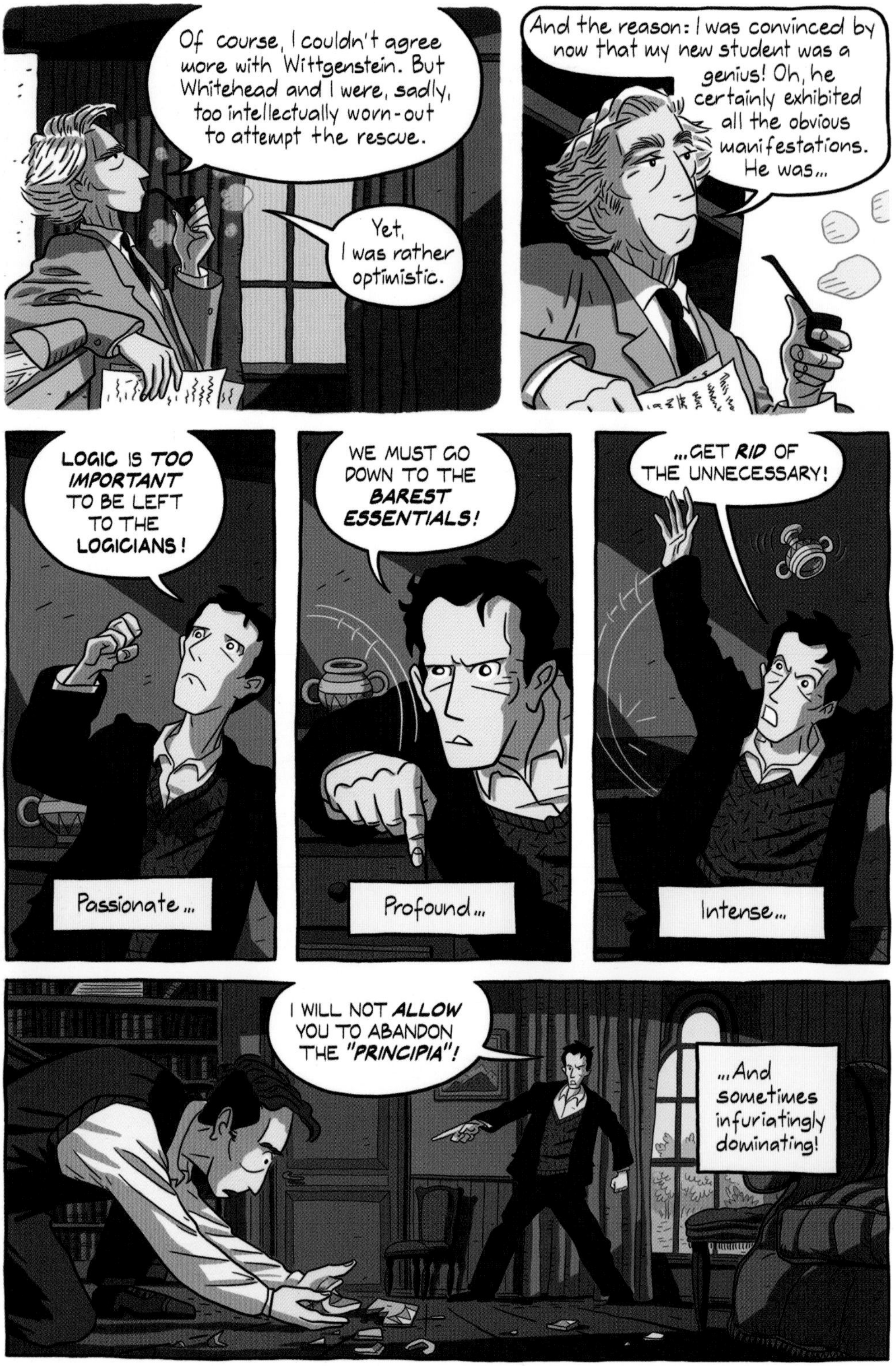
Of course, I couldn't agree more with Wittgenstein. But Whitehead and I were, sadly, too intellectually worn-out to attempt the rescue.
Yet, I was rather optimistic.
And the reason: I was convinced by now that my new student was a genius! Oh, he certainly exhibited all the obvious manifestations. He was...
LOGIC IS TOO IMPORTANT TO BE LEFT TO THE LOGICIANS!
Passionate...
WE MUST GO DOWN TO THE BAREST ESSENTIALS!
Profound...
...GET RID OF THE UNNECESSARY!
Intense...
I WILL NOT ALLOW YOU TO ABANDON THE "PRINCIPIA"!
...And sometimes infuriatingly dominating!

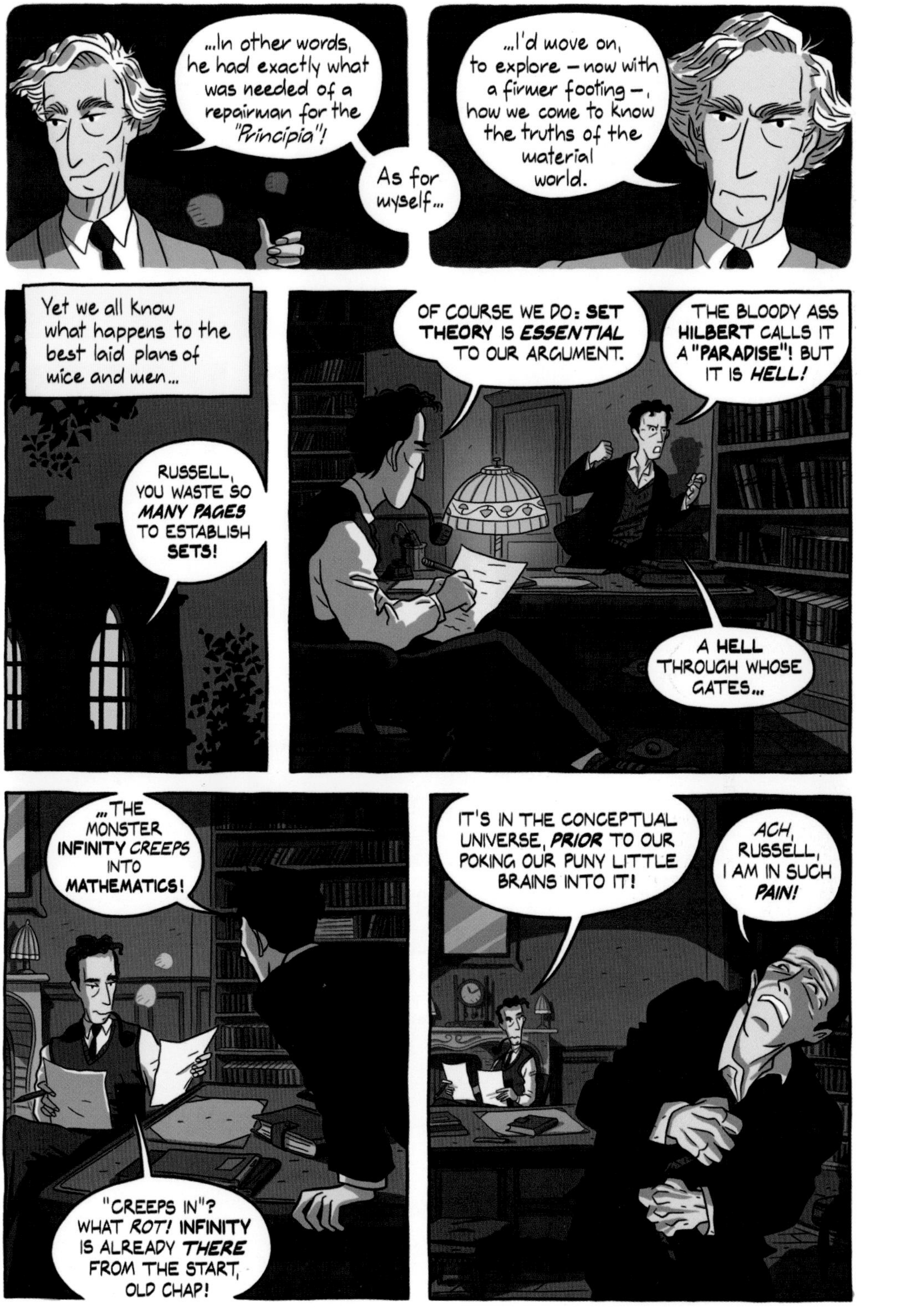
...In other words, he had exactly what was needed of a repairman for the "Principia"!
As for myself...
...I'd move on, to explore – now with a firmer footing –, how we come to know the truths of the material world.
Yet we all know what happens to the best laid plans of mice and men...
RUSSELL, YOU WASTE SO MANY PAGES TO ESTABLISH SETS!
OF COURSE WE DO: SET THEORY IS ESSENTIAL TO OUR ARGUMENT.
THE BLOODY ASS HILBERT CALLS IT A "PARADISE"! BUT IT IS HELL!
A HELL THROUGH WHOSE GATES...
...THE MONSTER INFINITY CREEPS INTO MATHEMATICS!
"CREEPS IN"? WHAT ROT! INFINITY IS ALREADY THERE FROM THE START, OLD CHAP!
IT'S IN THE CONCEPTUAL UNIVERSE, PRIOR TO OUR POKING OUR PUNY LITTLE BRAINS INTO IT!
ACH, RUSSELL, I AM IN SUCH PAIN!

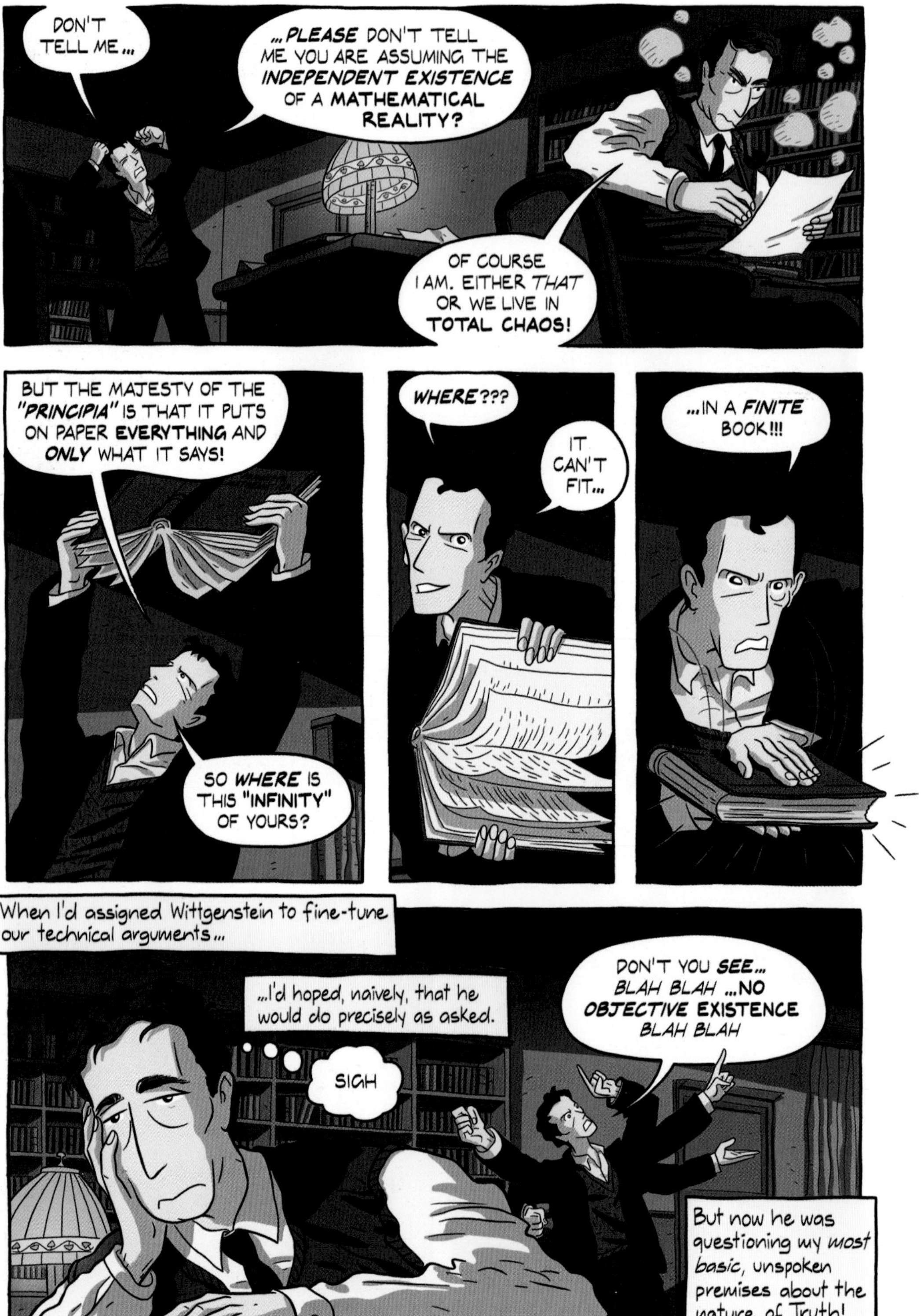
DON'T TELL ME...
...PLEASE DON'T TELL ME YOU ARE ASSUMING THE INDEPENDENT EXISTENCE OF A MATHEMATICAL REALITY?
OF COURSE I AM. EITHER THAT OR WE LIVE IN TOTAL CHAOS!
BUT THE MAJESTY OF THE "PRINCIPIA" IS THAT IT PUTS ON PAPER EVERYTHING AND ONLY WHAT IT SAYS!
SO WHERE IS THIS "INFINITY" OF YOURS?
WHERE???
IT CAN'T FIT...
...IN A FINITE BOOK!!!
When I'd assigned Wittgenstein to fine-tune our technical arguments...
...I'd hoped, naively, that he would do precisely as asked.
SIGH
DON'T YOU SEE... BLAH BLAH ...NO OBJECTIVE EXISTENCE BLAH BLAH
But now he was questioning my most basic, unspoken premises about the nature of Truth!

THIS QUESTIONING CREATED IN **RUSSELL** EXTREME ANXIETY, ATTACKING ONE OF HIS **STRONGEST** DEFENCES, HIS BELIEF IN **OBJECTIVE REALITY!**

...AND THIS AT A TIME WHEN **OTHER** REFUGES WERE NO MORE AVAILABLE!

HE'D ALREADY LEFT **ALYS.** AND DEAR **EVELYN** WAS NOT WILLING TO SUCCUMB TO HIS CHARM.

AT LEAST ONE OF ZEM WAS REMEMBERING SHE WAS MARRIED TO HIS CLOSEST FRIEND!

I THINK THAT **RUSSELL** SAW **WITTGENSTEIN** AS A *MIRROR:* HE HAD SO MANY ELEMENTS OF HIMSELF!

BUT ABOVE ALL THIS "**INTENSITY**"

WHICH HE GRADUALLY REALIZED WAS BUT AN EXPRESSION OF...

...AN UNDERLYING **INSTABILITY!**

Wittgenstein barged into my rooms at 3 AM one night, in extreme agony about some fine logical point.

I warned him that he should beware: the way he was driving himself he could well go insane.

But he said...

GOD **PREVENT** ME FROM **SANITY!**

GOD CERTAINLY **WILL!**

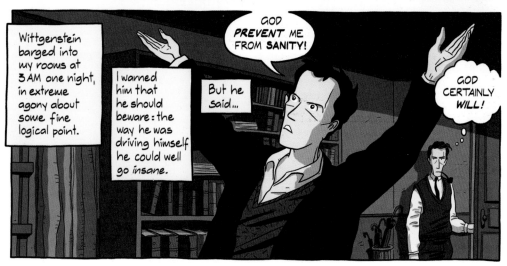

229

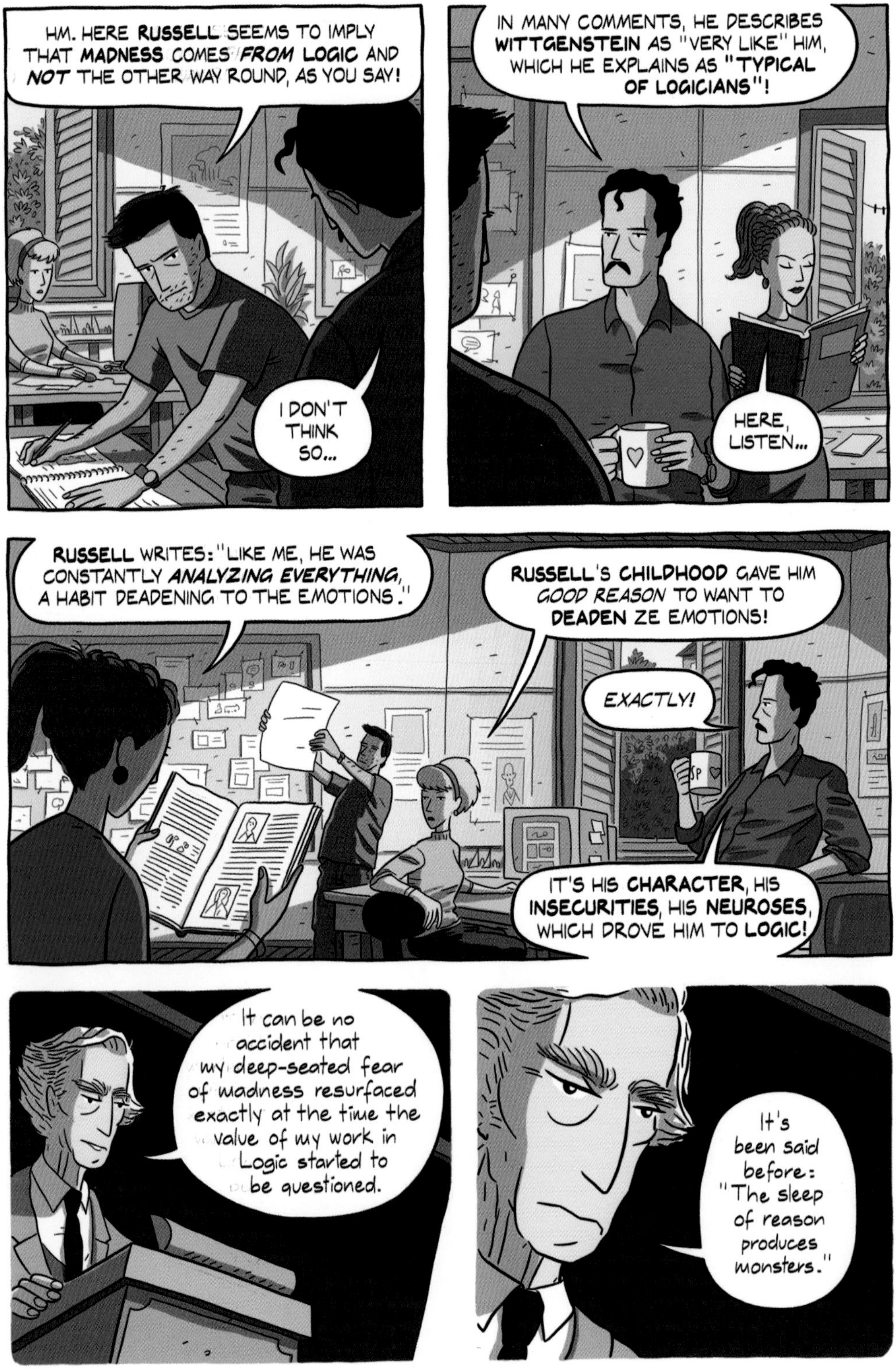
HM. HERE RUSSELL SEEMS TO IMPLY THAT MADNESS COMES FROM LOGIC AND NOT THE OTHER WAY ROUND, AS YOU SAY!
I DON'T THINK SO...
IN MANY COMMENTS, HE DESCRIBES WITTGENSTEIN AS "VERY LIKE" HIM, WHICH HE EXPLAINS AS "TYPICAL OF LOGICIANS"!
HERE, LISTEN...
RUSSELL WRITES: "LIKE ME, HE WAS CONSTANTLY ANALYZING EVERYTHING, A HABIT DEADENING TO THE EMOTIONS."
RUSSELL'S CHILDHOOD GAVE HIM GOOD REASON TO WANT TO DEADEN ZE EMOTIONS!
EXACTLY!
IT'S HIS CHARACTER, HIS INSECURITIES, HIS NEUROSES, WHICH DROVE HIM TO LOGIC!
It can be no accident that my deep-seated fear of madness resurfaced exactly at the time the value of my work in Logic started to be questioned.
It's been said before: "The sleep of reason produces monsters."

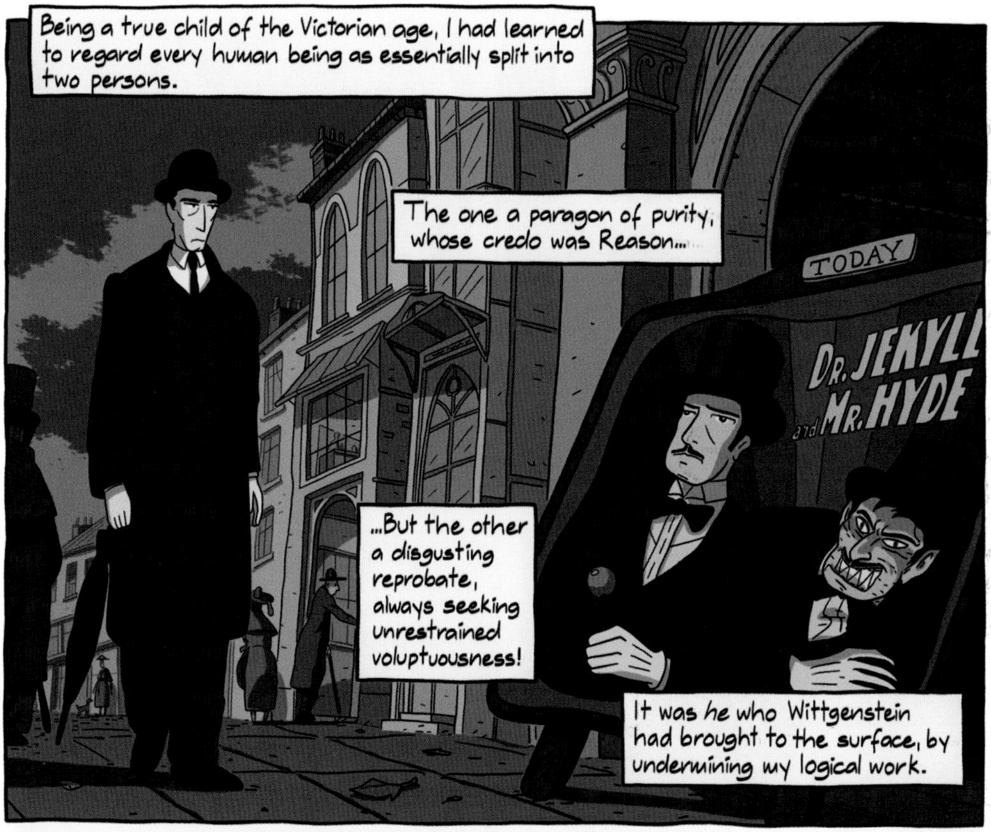
Being a true child of the Victorian age, I had learned to regard every human being as essentially split into two persons.
The one a paragon of purity, whose credo was Reason...
TODAY
DR. JEKYLL and MR. HYDE
...But the other a disgusting reprobate, always seeking unrestrained voluptuousness!
It was *he* who Wittgenstein had brought to the surface, by undermining my logical work.

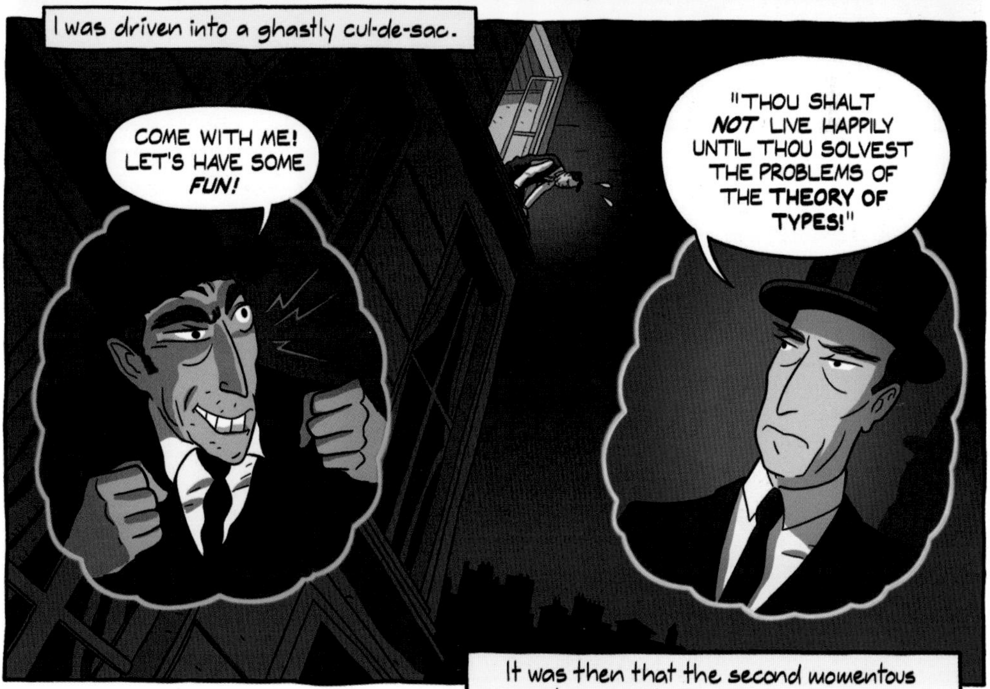
I was driven into a ghastly cul-de-sac.
COME WITH ME! LET'S HAVE SOME ***FUN!***
"THOU SHALT ***NOT*** LIVE HAPPILY UNTIL THOU SOLVEST THE PROBLEMS OF THE **THEORY OF TYPES!**"
It was then that the second momentous event occurred.

During this crisis, I was clinging for some kind of support to the external routines of academic life.

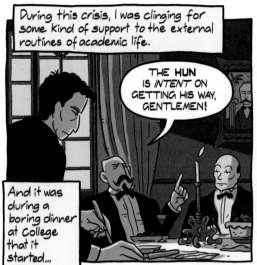

THE HUN IS *INTENT* ON GETTING HIS WAY, GENTLEMEN!

And it was during a boring dinner at College that it started...

AND IF HE CAN'T GET IT PEACEFULLY, HE WILL RESORT TO *ARMS!*

I BEG YOUR PARDON, GENTLEMEN!

A short sequence of events which...

?

YOU ARE *URGENTLY* REQUESTED AT YOUR *HOME,* SIR!

...Showed me, totally unexpectedly, a brand-new way to face life.

WAIT FOR ME, RUSSELL!

IT'S AN EMERGENCY, MAN! *RUN!!!*

AH, WHITEHEAD, AT LAST!

WE HAVE *SEVERE* CHEST PAIN...

...AND PALPITATIONS...

...INDICATING HEART *FAILURE!*

HEART... FAILURE?

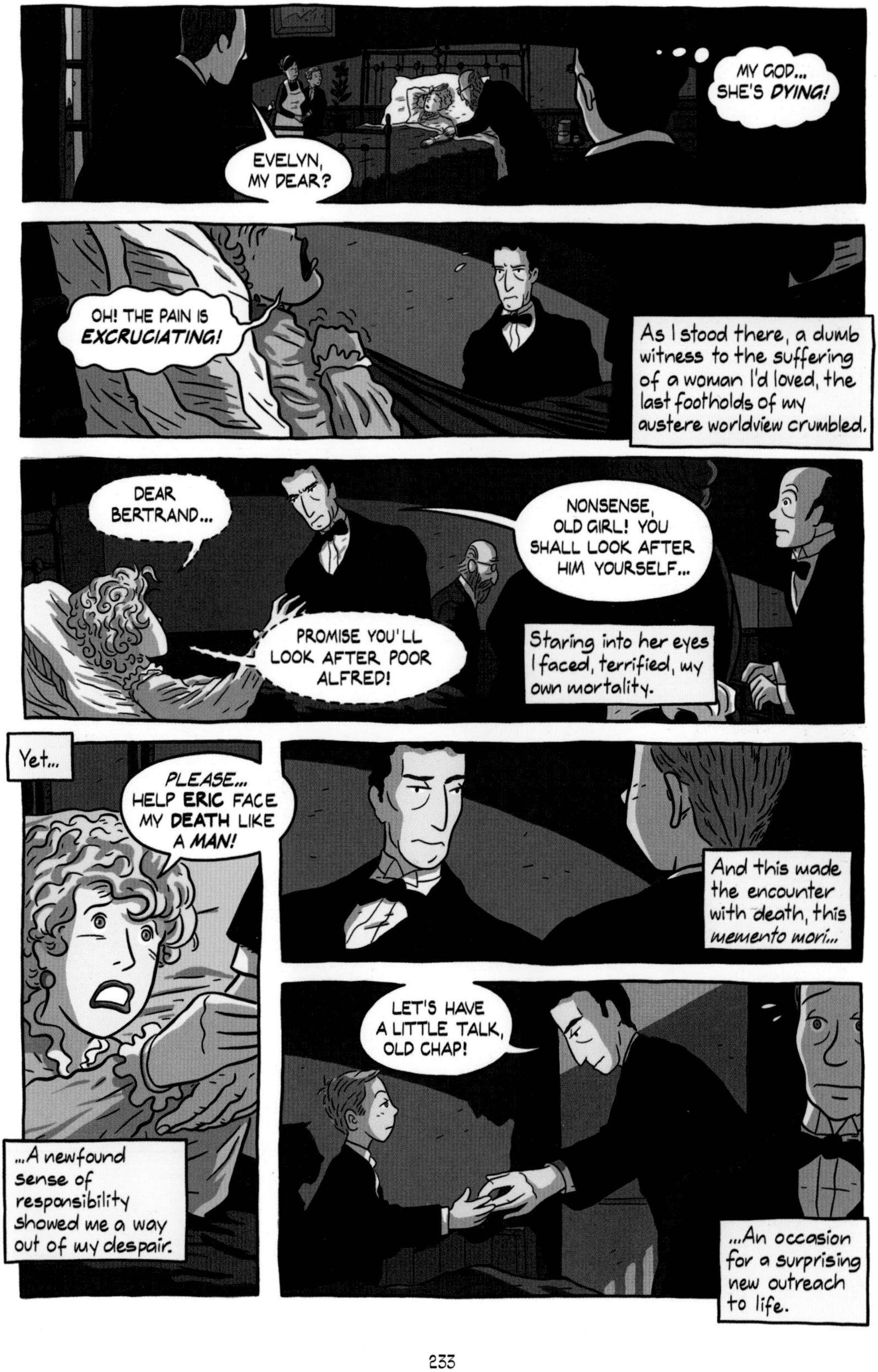
EVELYN, MY DEAR?
MY GOD... SHE'S DYING!
OH! THE PAIN IS EXCRUCIATING!
As I stood there, a dumb witness to the suffering of a woman I'd loved, the last footholds of my austere worldview crumbled.
DEAR BERTRAND...
PROMISE YOU'LL LOOK AFTER POOR ALFRED!
NONSENSE, OLD GIRL! YOU SHALL LOOK AFTER HIM YOURSELF...
Staring into her eyes I faced, terrified, my own mortality.
Yet...
PLEASE... HELP ERIC FACE MY DEATH LIKE A MAN!
...A newfound sense of responsibility showed me a way out of my despair.
And this made the encounter with death, this memento mori...
LET'S HAVE A LITTLE TALK, OLD CHAP!
...An occasion for a surprising new outreach to life.

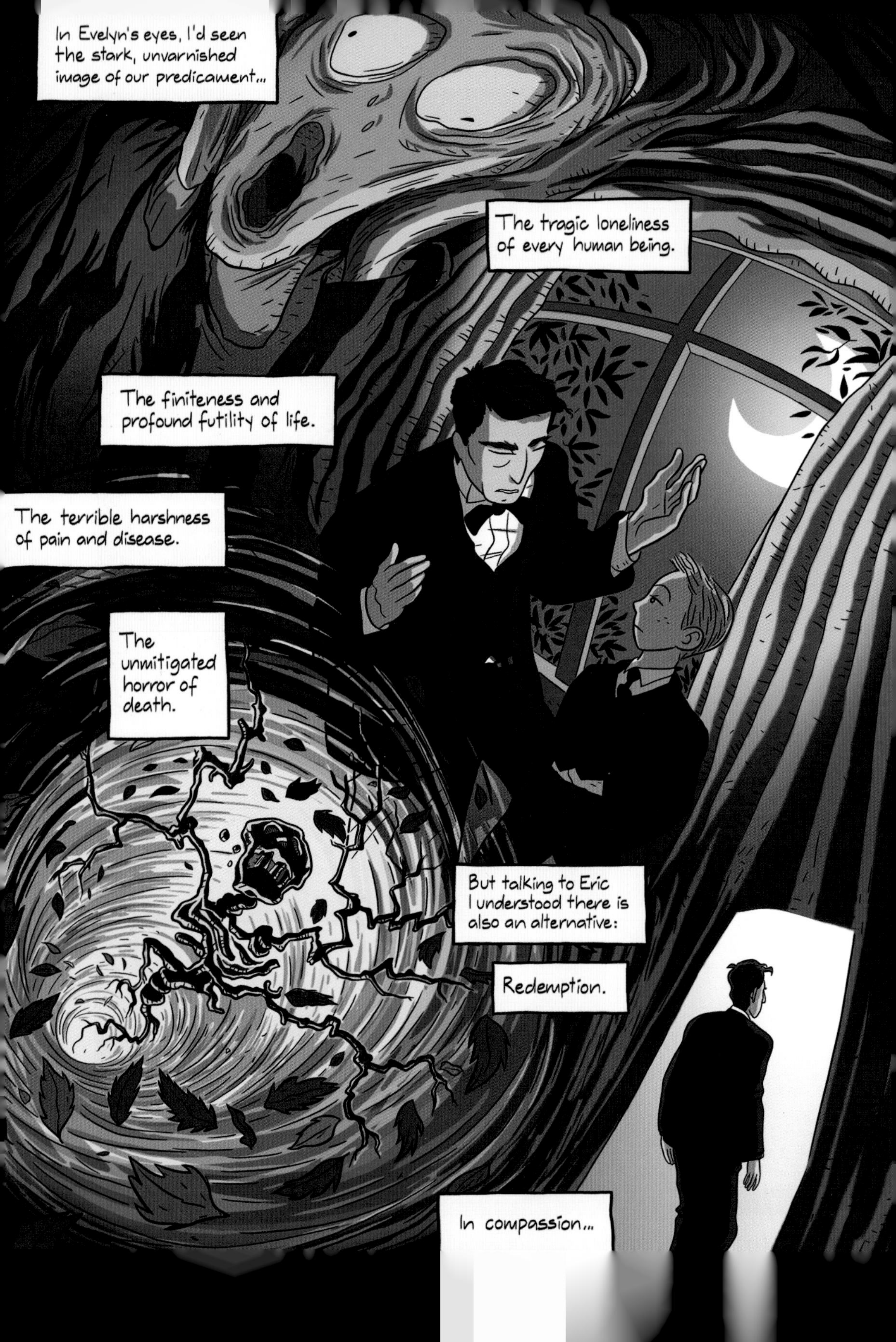
In Evelyn's eyes, I'd seen the stark, unvarnished image of our predicament...
The tragic loneliness of every human being.
The finiteness and profound futility of life.
The terrible harshness of pain and disease.
The unmitigated horror of death.
But talking to Eric I understood there is also an alternative:
Redemption.
In compassion...

...I must tell you that Mrs. Whitehead, soon after...
...Totally recovered!
...And is still alive and well today!
Apparently, her "heart failure" was indigestion, slightly aggravated by a nervous disposition!
Yet, though the report of her imminent demise had been rather grossly exaggerated, the transformation it caused in me was totally real.
So real, in fact that when I received a letter from Wittgenstein, who had gone to a God-forsaken Norwegian fjord "to think about the meaning of logical propositions"...
...I was not so affected by his doubts, or so perturbed by his criticisms.
my house
Dear Russell,
I'm thinking about your BEASTLY Theory of Types. Can't make heads or tails of it! Do we really need this ABERRATION???
My new concern with the welfare of my fellow human beings had tempered my passion for the Foundations of Mathematics.
...SO, IN ORDER TO ACT *REASONABLY*...
In this new spirit, I also started giving lectures — like the one I'm giving to you right now — trying to apply the Higher Logic to human affairs.
ickshire
ation Society
nd Russell
WHAT IS THE NATURE OF LOGIC?
This, mind you, at a time when any kind of Logic was in very short supply!

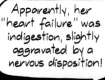

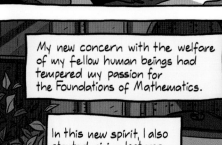

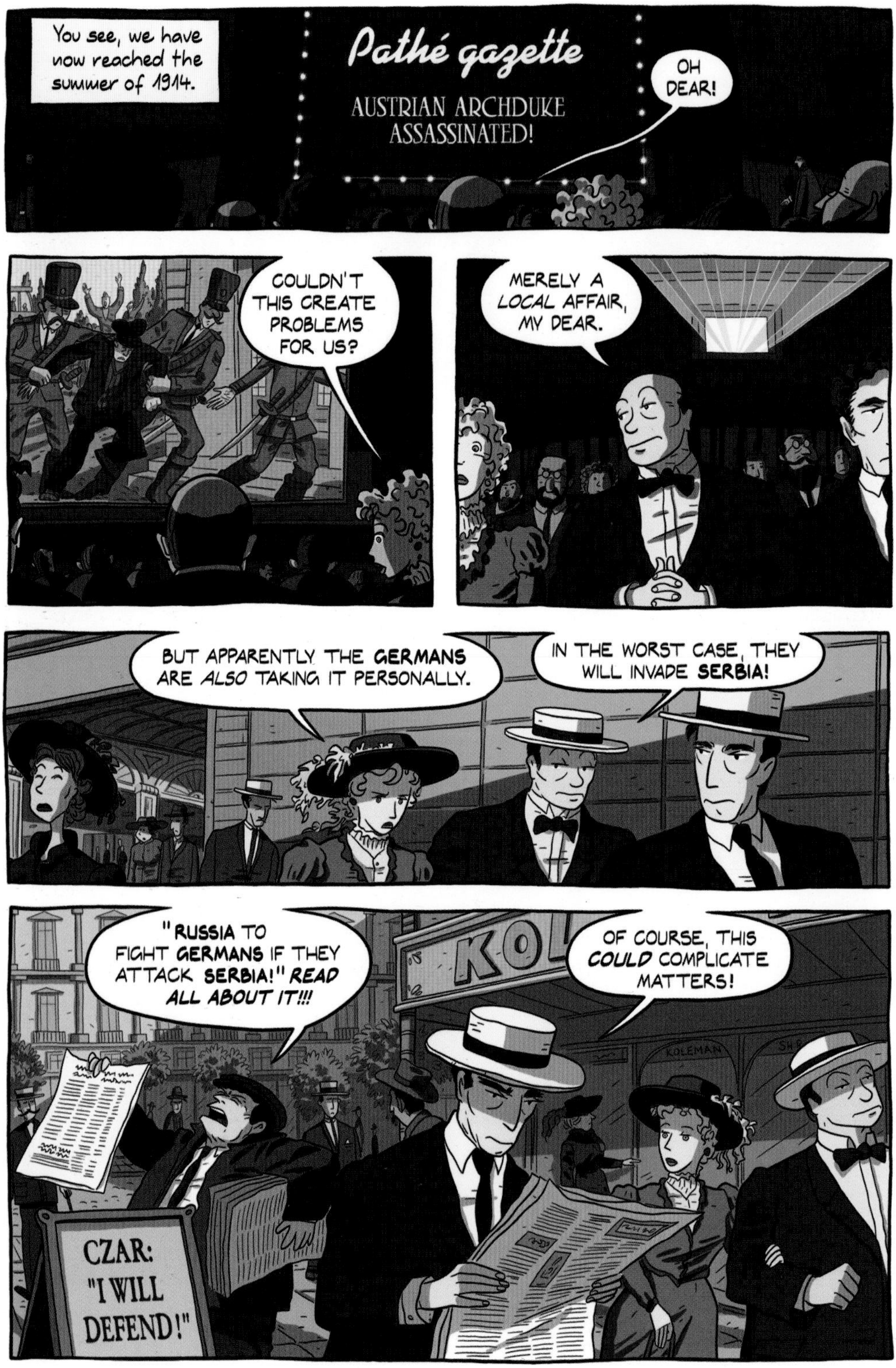
You see, we have now reached the summer of 1914.
Pathé gazette
AUSTRIAN ARCHDUKE ASSASSINATED!
OH DEAR!
COULDN'T THIS CREATE PROBLEMS FOR US?
MERELY A *LOCAL* AFFAIR, MY DEAR.
BUT APPARENTLY THE **GERMANS** ARE *ALSO* TAKING IT PERSONALLY.
IN THE WORST CASE, THEY WILL INVADE **SERBIA!**
"**RUSSIA** TO FIGHT **GERMANS** IF THEY ATTACK **SERBIA!**" ***READ ALL ABOUT IT!!!***
OF COURSE, THIS ***COULD*** COMPLICATE MATTERS!
KOLEMAN
CZAR: "I WILL DEFEND!"

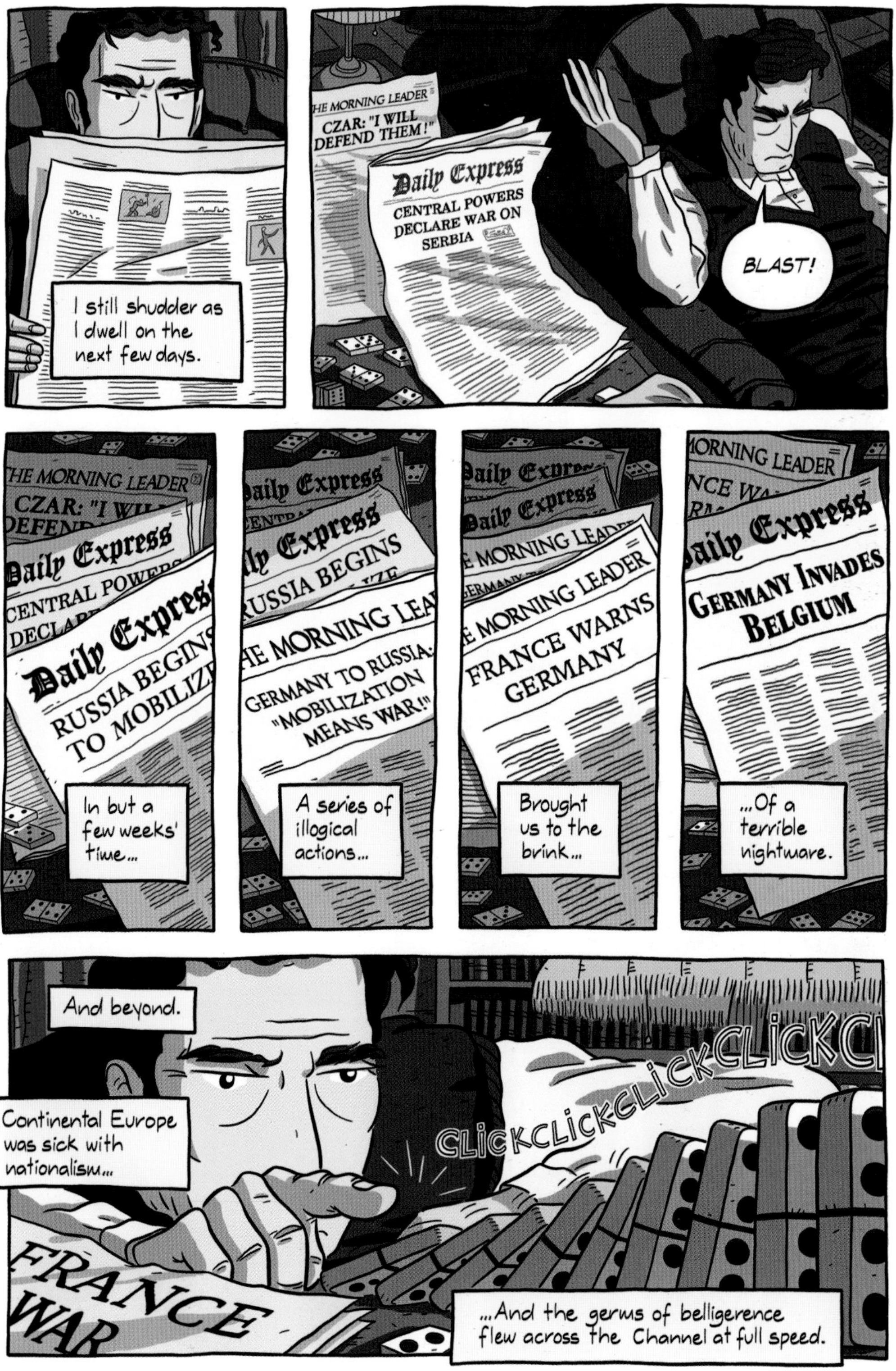
I still shudder as I dwell on the next few days.
THE MORNING LEADER
CZAR: "I WILL DEFEND THEM!"
Daily Express
CENTRAL POWERS DECLARE WAR ON SERBIA
BLAST!
Daily Express
In but a few weeks' time...
Daily Express
RUSSIA BEGINS
GERMANY TO RUSSIA: "MOBILIZATION MEANS WAR!"
A series of illogical actions...
FRANCE WARNS GERMANY
Brought us to the brink...
Daily Express
GERMANY INVADES BELGIUM
...Of a terrible nightmare.
And beyond.
Continental Europe was sick with nationalism...
CLICKCLICKCLICKCLICKCL
FRANCE
WAR
...And the germs of belligerence flew across the Channel at full speed.

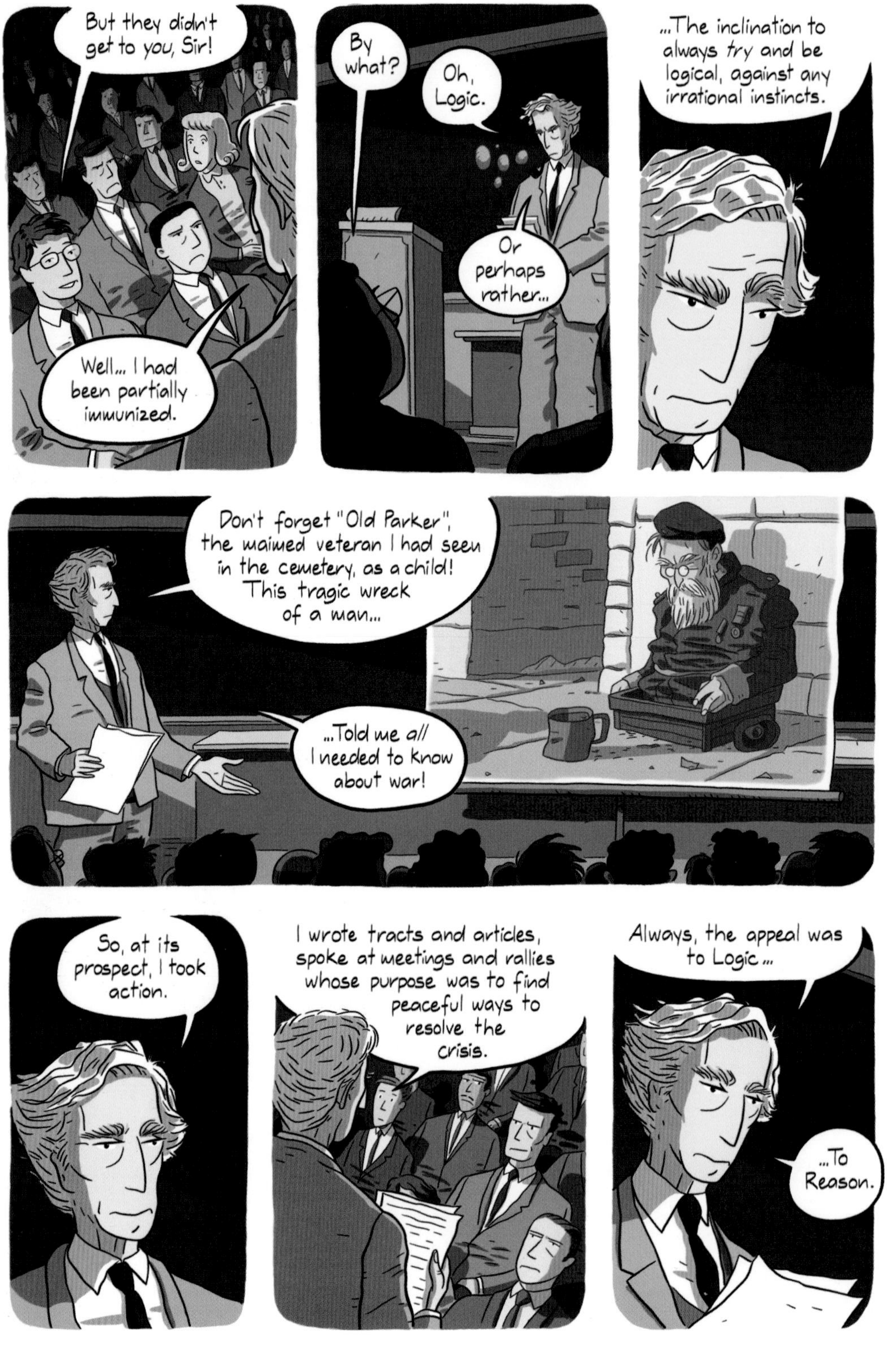
But they didn't get to you, Sir!
Well... I had been partially immunized.
By what?
Oh, Logic.
Or perhaps rather...
...The inclination to always *try* and be logical, against any irrational instincts.
Don't forget "Old Parker", the maimed veteran I had seen in the cemetery, as a child! This tragic wreck of a man...
...Told me *all* I needed to know about war!
So, at its prospect, I took action.
I wrote tracts and articles, spoke at meetings and rallies whose purpose was to find peaceful ways to resolve the crisis.
Always, the appeal was to Logic...
...To Reason.

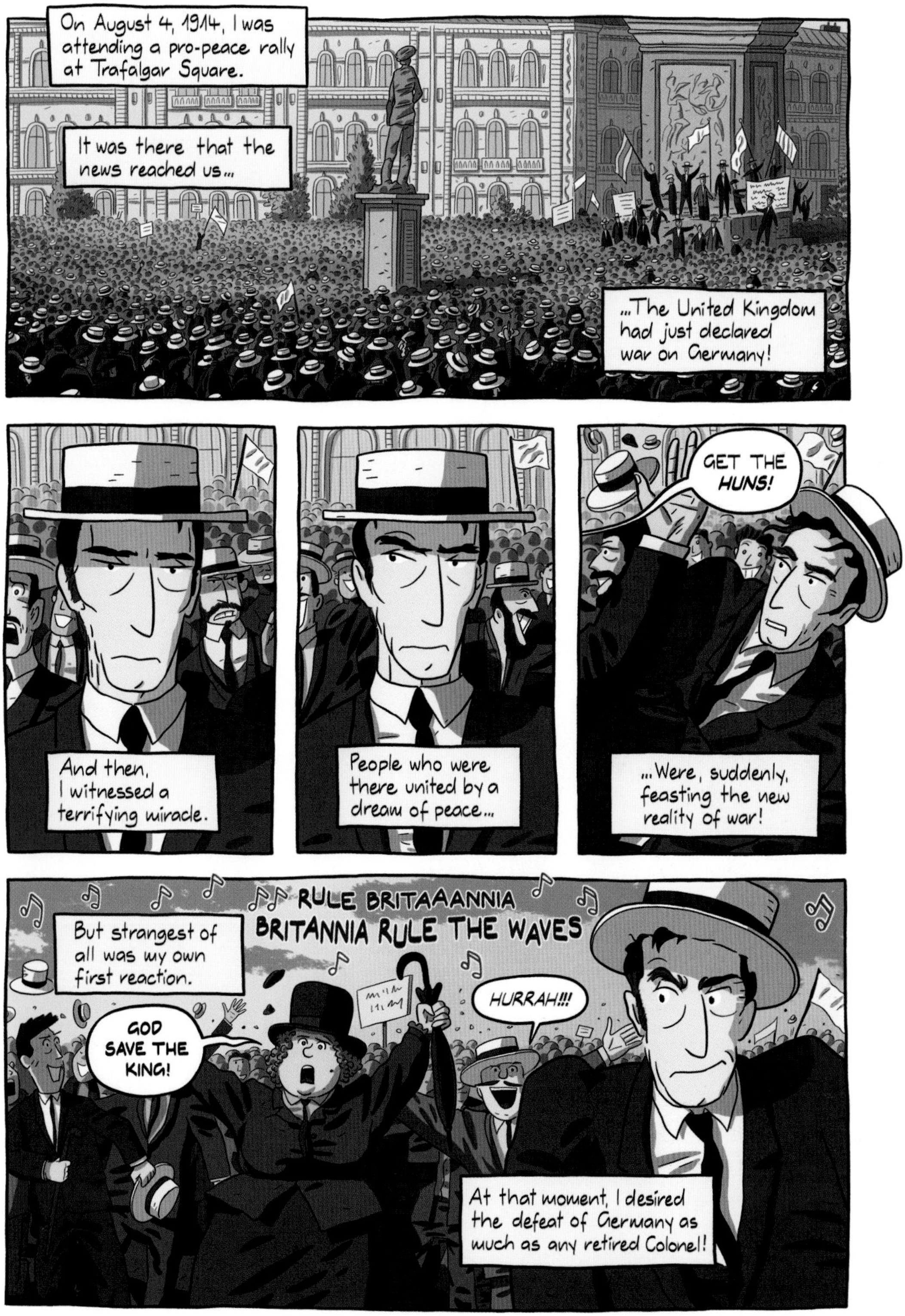
On August 4, 1914, I was attending a pro-peace rally at Trafalgar Square.
It was there that the news reached us...
...The United Kingdom had just declared war on Germany!
And then, I witnessed a terrifying miracle.
People who were there united by a dream of peace...
GET THE HUNS!
...Were, suddenly, feasting the new reality of war!
RULE BRITAAANNIA BRITANNIA RULE THE WAVES
But strangest of all was my own first reaction.
GOD SAVE THE KING!
HURRAH!!!
At that moment, I desired the defeat of Germany as much as any retired Colonel!

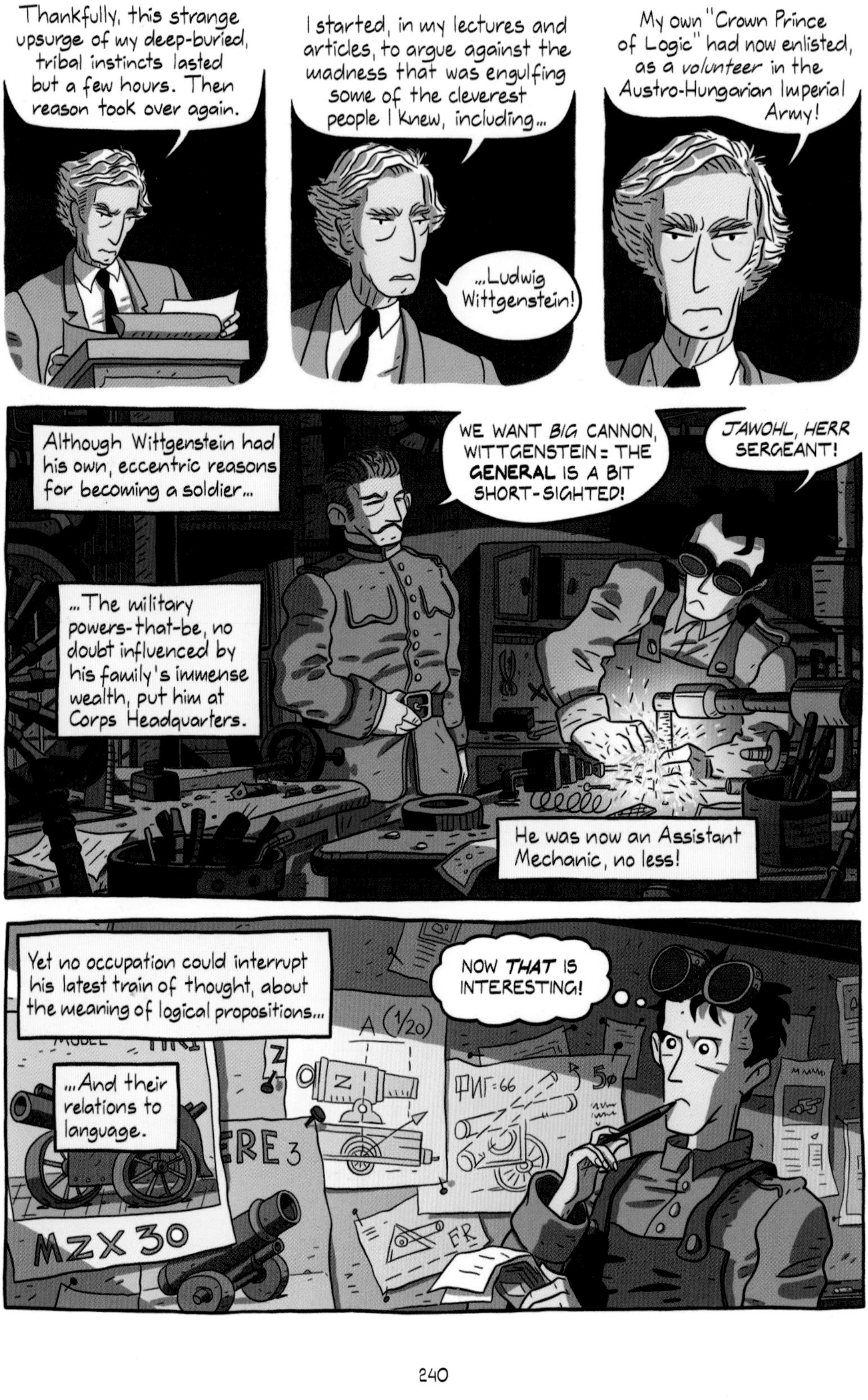
Thankfully, this strange upsurge of my deep-buried, tribal instincts lasted but a few hours. Then reason took over again.
I started, in my lectures and articles, to argue against the madness that was engulfing some of the cleverest people I knew, including...
...Ludwig Wittgenstein!
My own "Crown Prince of Logic" had now enlisted, as a *volunteer* in the Austro-Hungarian Imperial Army!
Although Wittgenstein had his own, eccentric reasons for becoming a soldier...
WE WANT *BIG* CANNON, WITTGENSTEIN: THE **GENERAL** IS A BIT SHORT-SIGHTED!
JAWOHL, HERR SERGEANT!
...The military powers-that-be, no doubt influenced by his family's immense wealth, put him at Corps Headquarters.
He was now an Assistant Mechanic, no less!
Yet no occupation could interrupt his latest train of thought, about the meaning of logical propositions...
NOW ***THAT*** IS INTERESTING!
...And their relations to language.
MZX 30

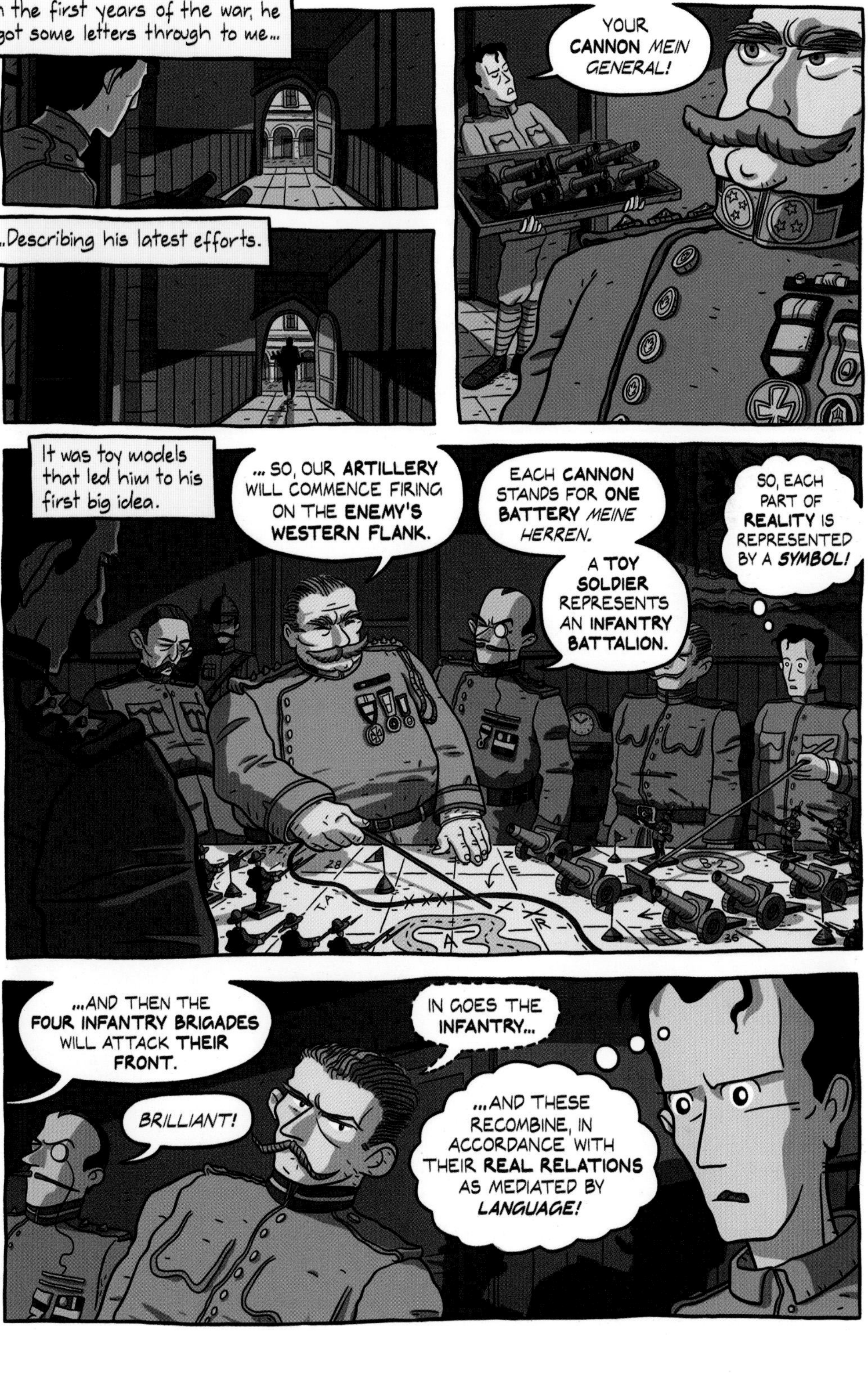
In the first years of the war, he got some letters through to me...
YOUR CANNON *MEIN GENERAL!*
...Describing his latest efforts.
It was toy models that led him to his first big idea.
... SO, OUR ARTILLERY WILL COMMENCE FIRING ON THE ENEMY'S WESTERN FLANK.
EACH CANNON STANDS FOR ONE BATTERY *MEINE HERREN.*
A TOY SOLDIER REPRESENTS AN INFANTRY BATTALION.
SO, EACH PART OF REALITY IS REPRESENTED BY A *SYMBOL!*
...AND THEN THE FOUR INFANTRY BRIGADES WILL ATTACK THEIR FRONT.
BRILLIANT!
IN GOES THE INFANTRY...
...AND THESE RECOMBINE, IN ACCORDANCE WITH THEIR REAL RELATIONS AS MEDIATED BY *LANGUAGE!*

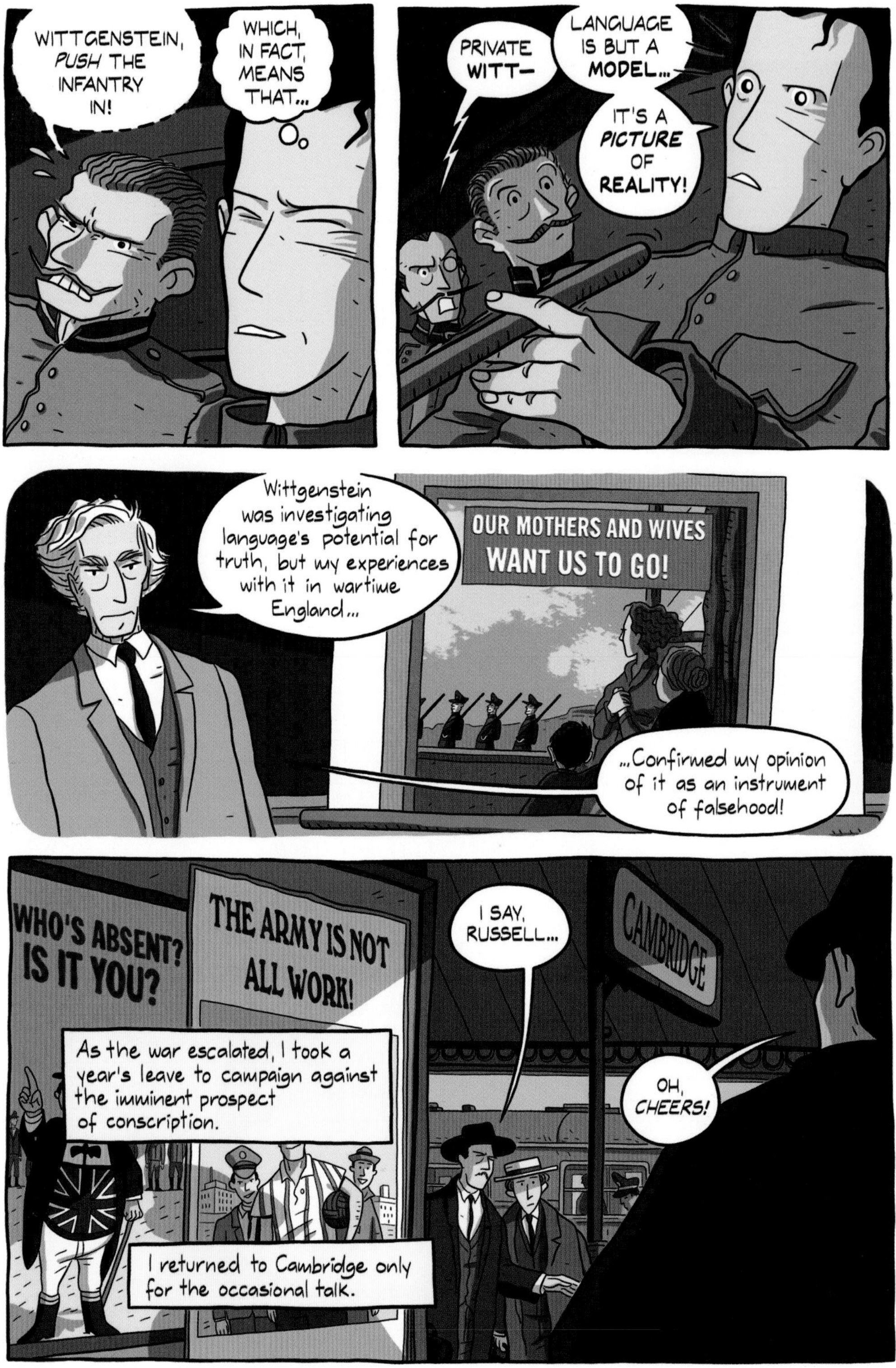
WITTGENSTEIN, *PUSH* THE INFANTRY IN!
WHICH, IN FACT, MEANS THAT...
PRIVATE WITT—
LANGUAGE IS BUT A MODEL...
IT'S A *PICTURE* OF REALITY!
Wittgenstein was investigating language's potential for truth, but my experiences with it in wartime England...
OUR MOTHERS AND WIVES WANT US TO GO!
...Confirmed my opinion of it as an instrument of falsehood!
WHO'S ABSENT? IS IT YOU?
THE ARMY IS NOT ALL WORK!
As the war escalated, I took a year's leave to campaign against the imminent prospect of conscription.
I SAY, RUSSELL...
CAMBRIDGE
OH, *CHEERS!*
I returned to Cambridge only for the occasional talk.

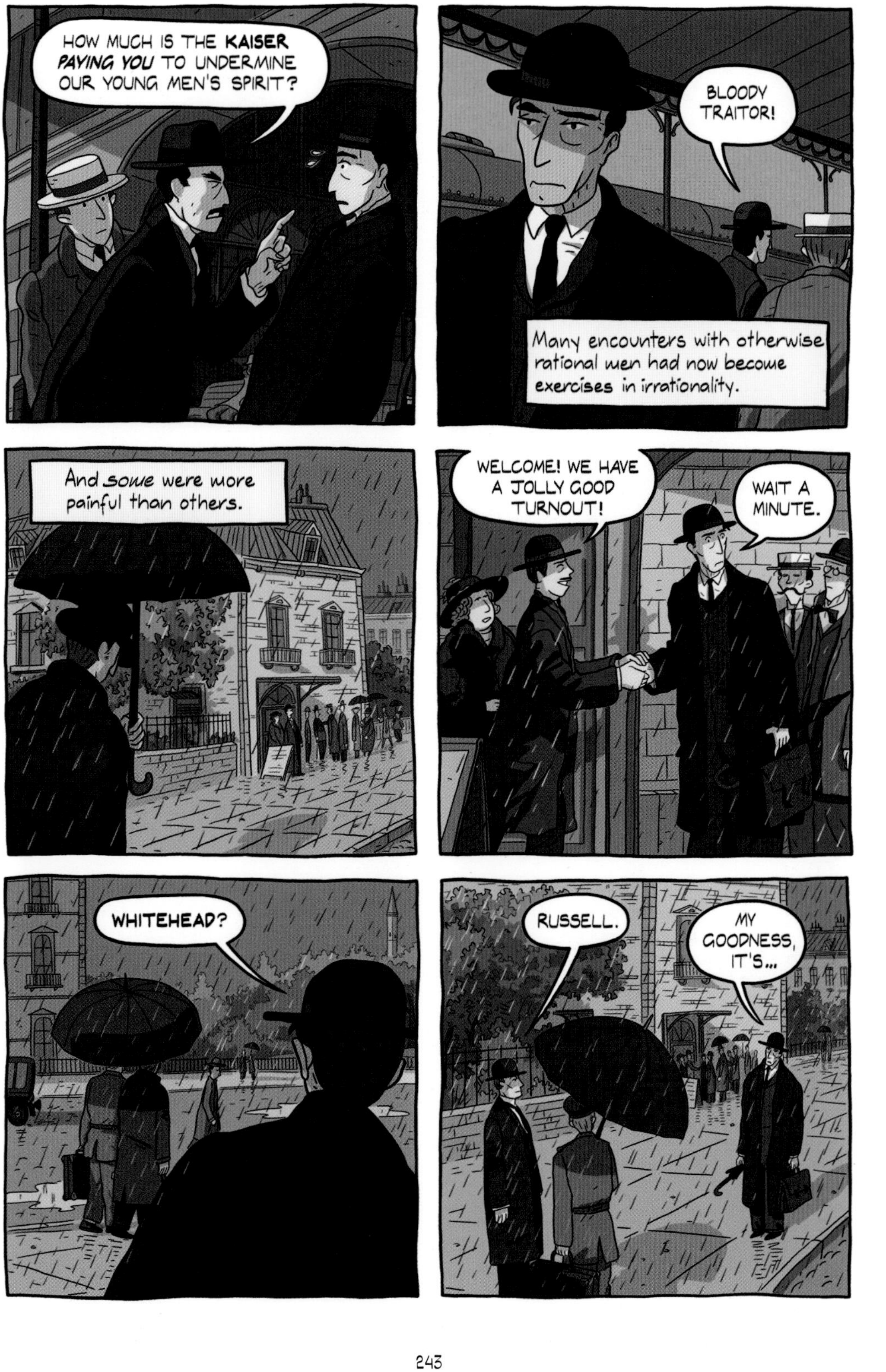
HOW MUCH IS THE **KAISER** ***PAYING YOU*** TO UNDERMINE OUR YOUNG MEN'S SPIRIT?
BLOODY TRAITOR!
Many encounters with otherwise rational men had now become exercises in irrationality.
And *some* were more painful than others.
WELCOME! WE HAVE A JOLLY GOOD TURNOUT!
WAIT A MINUTE.
WHITEHEAD?
RUSSELL.
MY GOODNESS, IT'S...

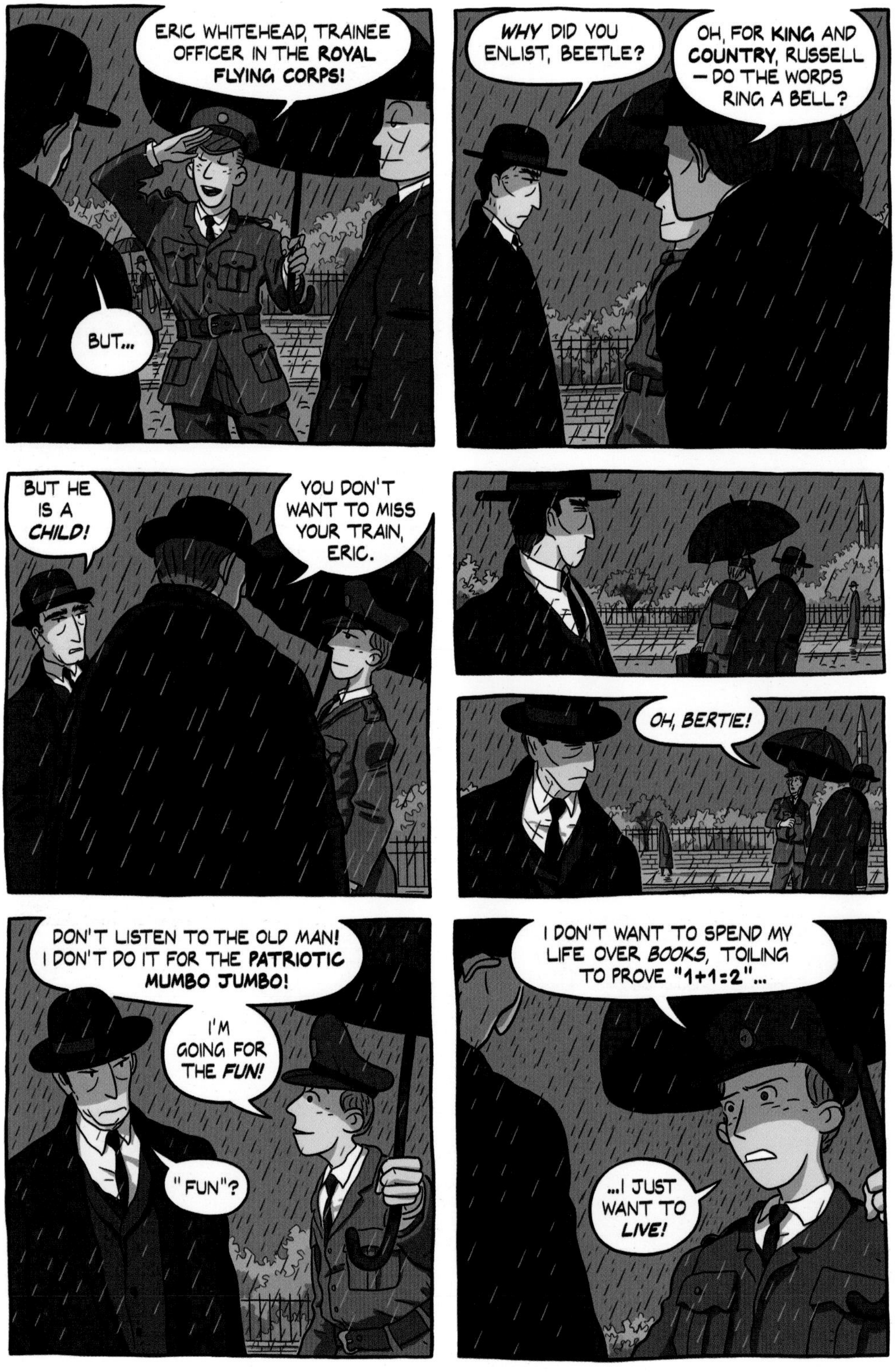
ERIC WHITEHEAD, TRAINEE OFFICER IN THE ROYAL FLYING CORPS!
BUT...
WHY DID YOU ENLIST, BEETLE?
OH, FOR KING AND COUNTRY, RUSSELL — DO THE WORDS RING A BELL?
BUT HE IS A CHILD!
YOU DON'T WANT TO MISS YOUR TRAIN, ERIC.
OH, BERTIE!
DON'T LISTEN TO THE OLD MAN! I DON'T DO IT FOR THE PATRIOTIC MUMBO JUMBO!
I'M GOING FOR THE FUN!
"FUN"?
I DON'T WANT TO SPEND MY LIFE OVER BOOKS, TOILING TO PROVE "1+1=2"...
...I JUST WANT TO LIVE!

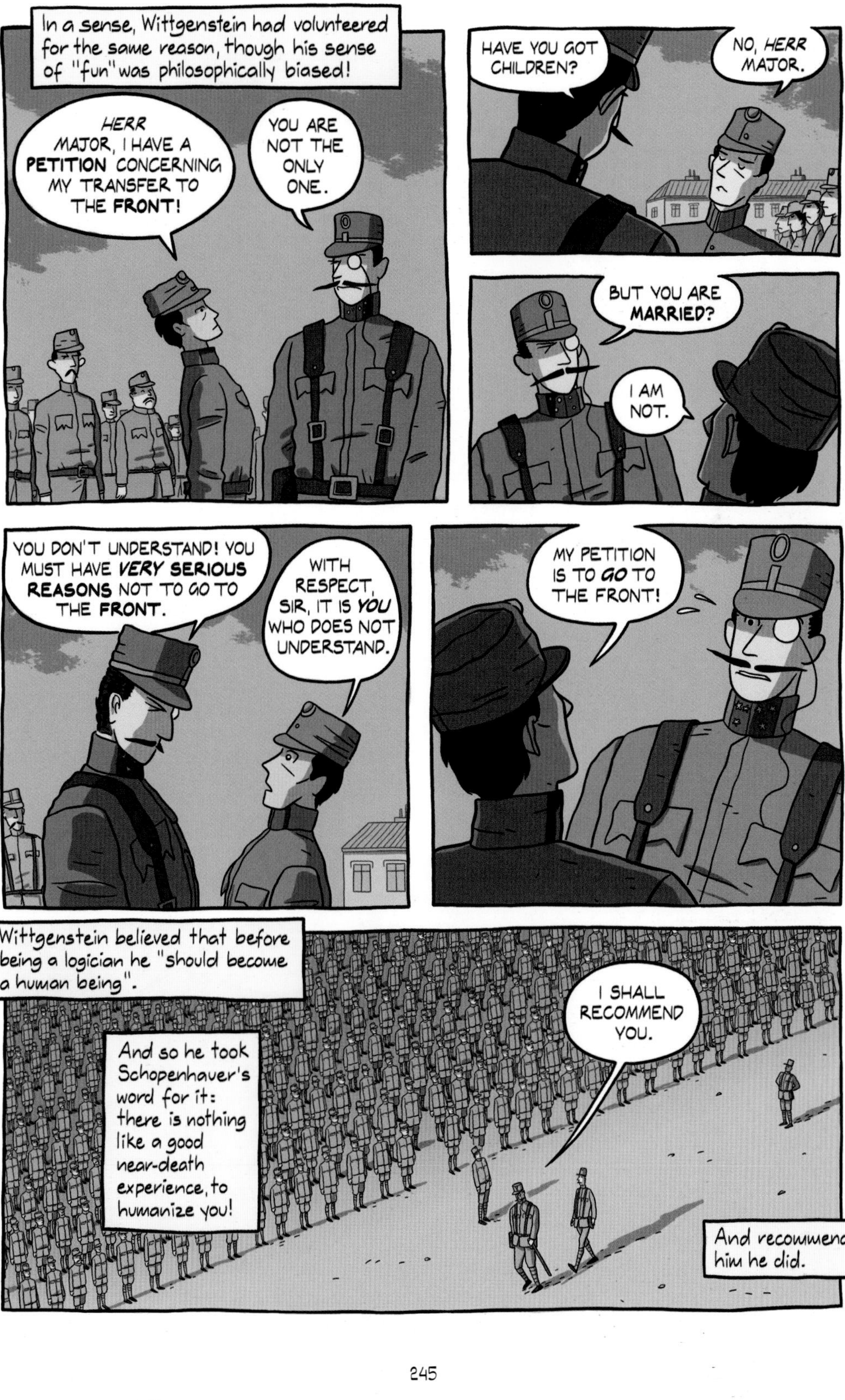
In a sense, Wittgenstein had volunteered for the same reason, though his sense of "fun" was philosophically biased!
HERR MAJOR, I HAVE A PETITION CONCERNING MY TRANSFER TO THE FRONT!
YOU ARE NOT THE ONLY ONE.
HAVE YOU GOT CHILDREN?
NO, HERR MAJOR.
BUT YOU ARE MARRIED?
I AM NOT.
YOU DON'T UNDERSTAND! YOU MUST HAVE VERY SERIOUS REASONS NOT TO GO TO THE FRONT.
WITH RESPECT, SIR, IT IS YOU WHO DOES NOT UNDERSTAND.
MY PETITION IS TO GO TO THE FRONT!
Wittgenstein believed that before being a logician he "should become a human being".
And so he took Schopenhauer's word for it: there is nothing like a good near-death experience, to humanize you!
I SHALL RECOMMEND YOU.
And recommend him he did.

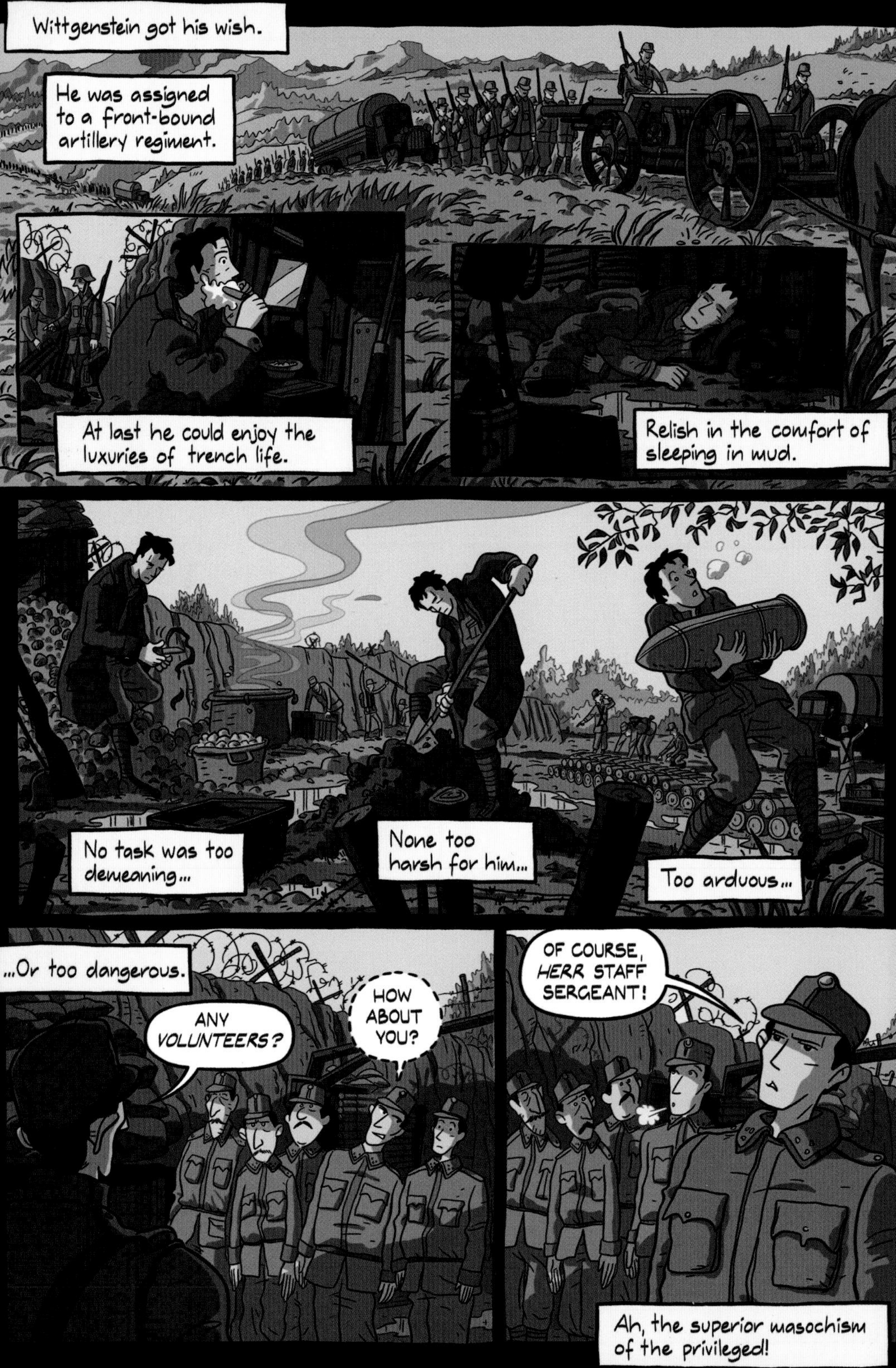
Wittgenstein got his wish.
He was assigned to a front-bound artillery regiment.
At last he could enjoy the luxuries of trench life.
Relish in the comfort of sleeping in mud.
No task was too demeaning...
None too harsh for him...
Too arduous...
...Or too dangerous.
ANY *VOLUNTEERS*?
HOW ABOUT YOU?
OF COURSE, *HERR* STAFF SERGEANT!
Ah, the superior masochism of the privileged!

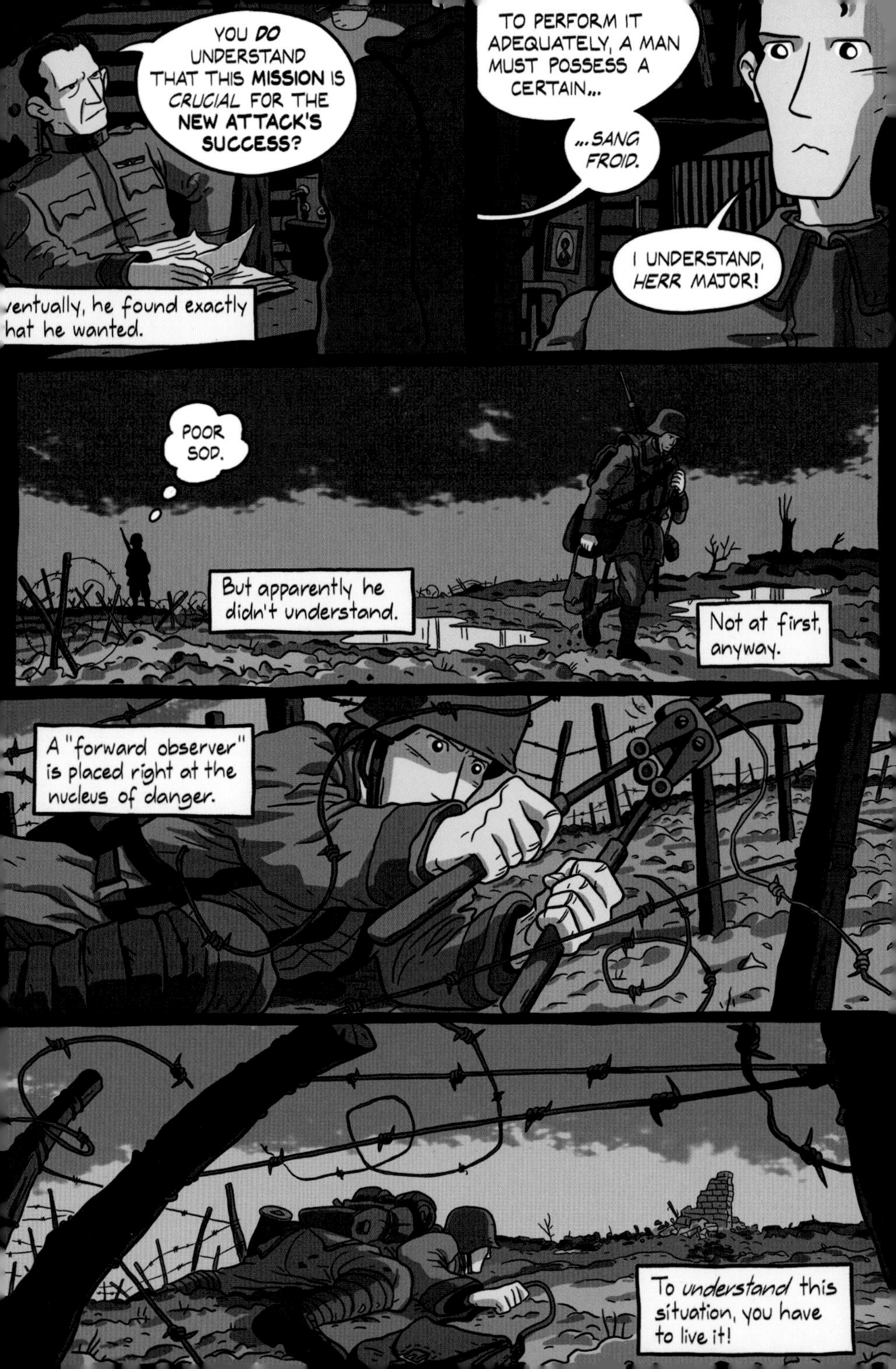

YOU DO UNDERSTAND THAT THIS MISSION IS CRUCIAL FOR THE NEW ATTACK'S SUCCESS?
ventually, he found exactly hat he wanted.
TO PERFORM IT ADEQUATELY, A MAN MUST POSSESS A CERTAIN...
...SANG FROID.
I UNDERSTAND, HERR MAJOR!
POOR SOD.
But apparently he didn't understand.
Not at first, anyway.
A "forward observer" is placed right at the nucleus of danger.
To understand this situation, you have to live it!

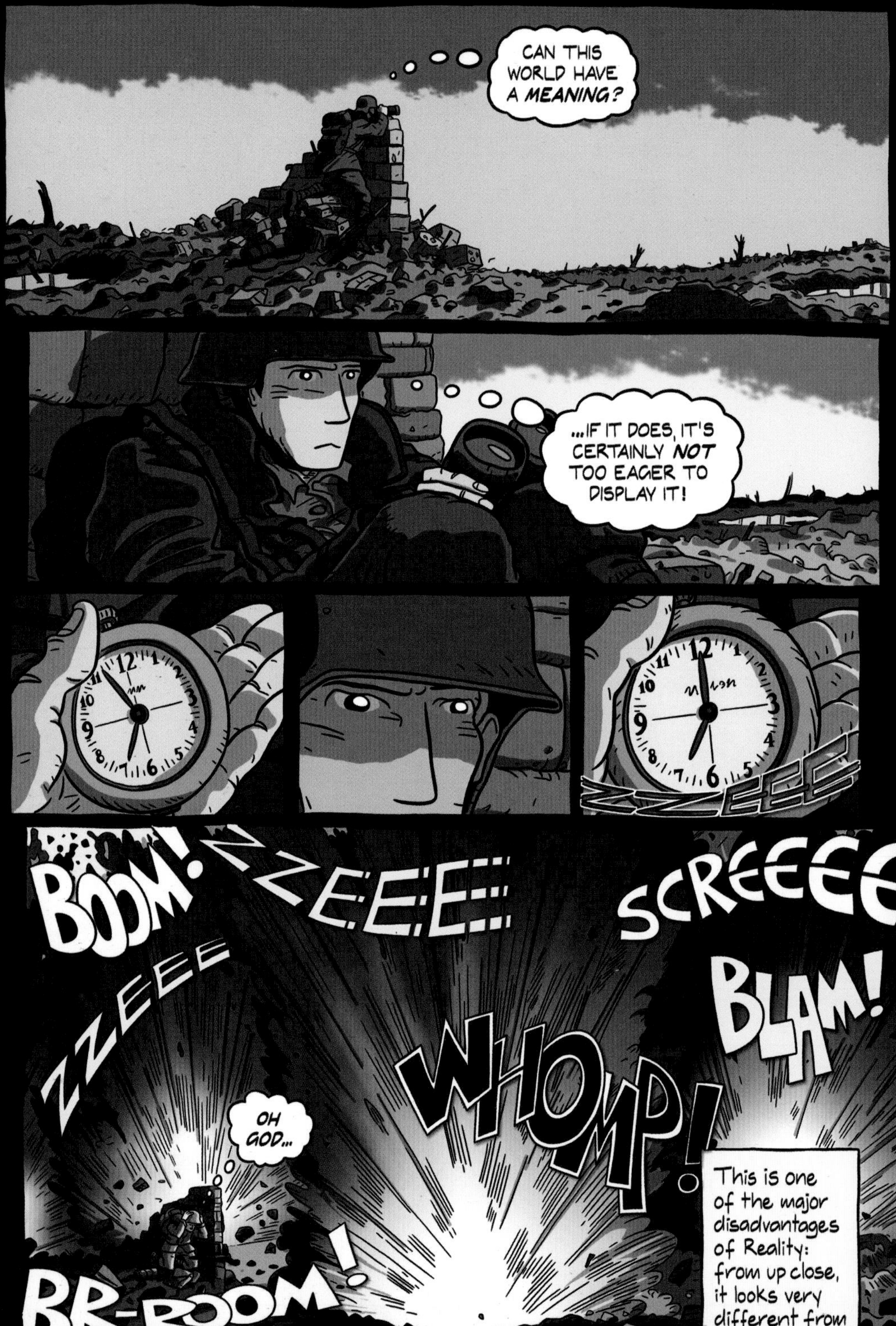

CAN THIS WORLD HAVE A *MEANING*?
...IF IT DOES, IT'S CERTAINLY *NOT* TOO EAGER TO DISPLAY IT!
ZZEEEE
BOOM!
ZZEEEE
SCREEEEE
ZZEEEE
BLAM!
WHOMP!
OH GOD...
BR-ROOM!
This is one of the major disadvantages of Reality: from up close, it looks very different from any "picture".

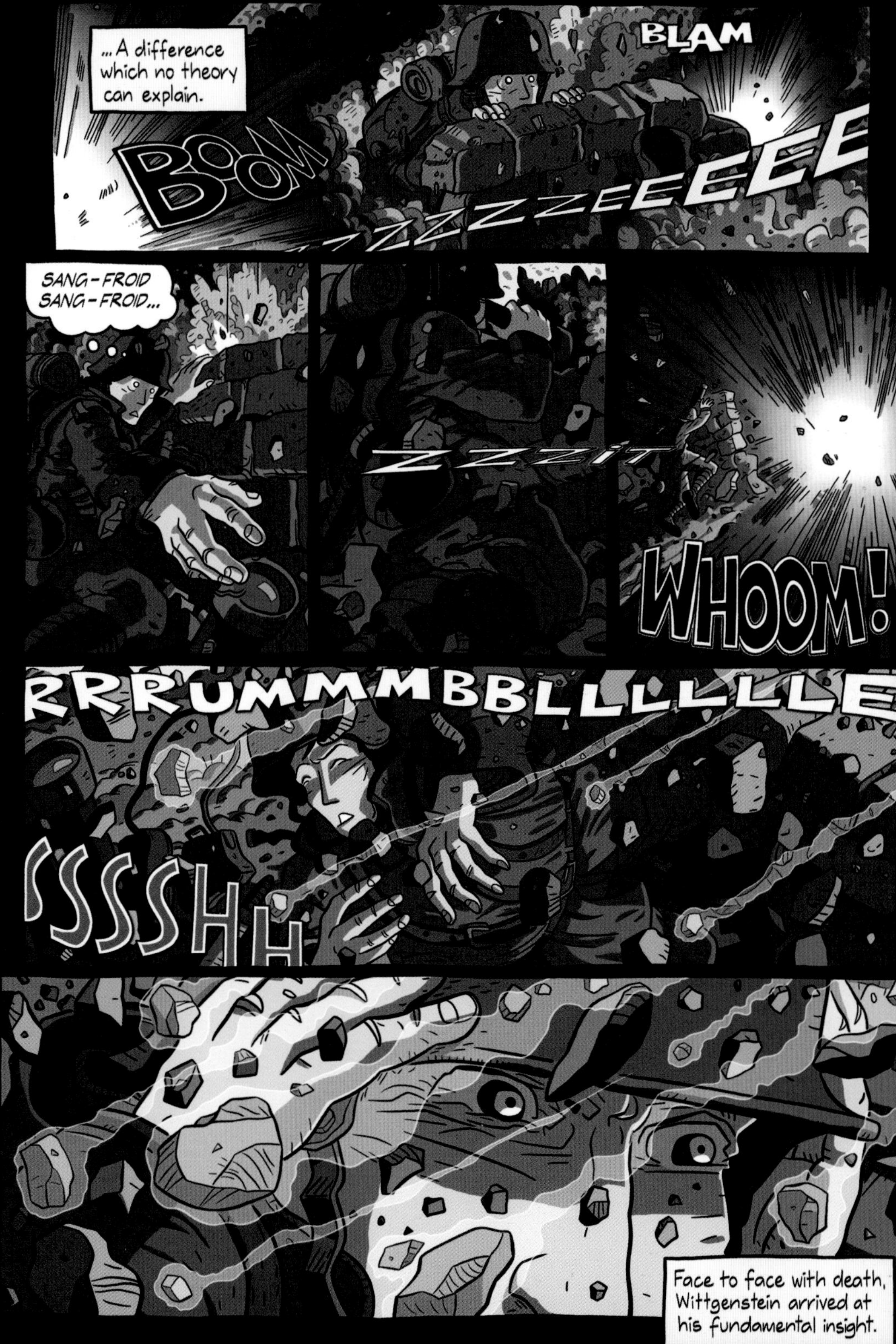
...A difference which no theory can explain.
BLAM
BOOM
ZZZZZZZZEEEEEEE
SANG-FROID SANG-FROID...
ZZZIT
WHOOM!
RRRUMMMBBLLLLLLE
SSSSHH
Face to face with death, Wittgenstein arrived at his fundamental insight.

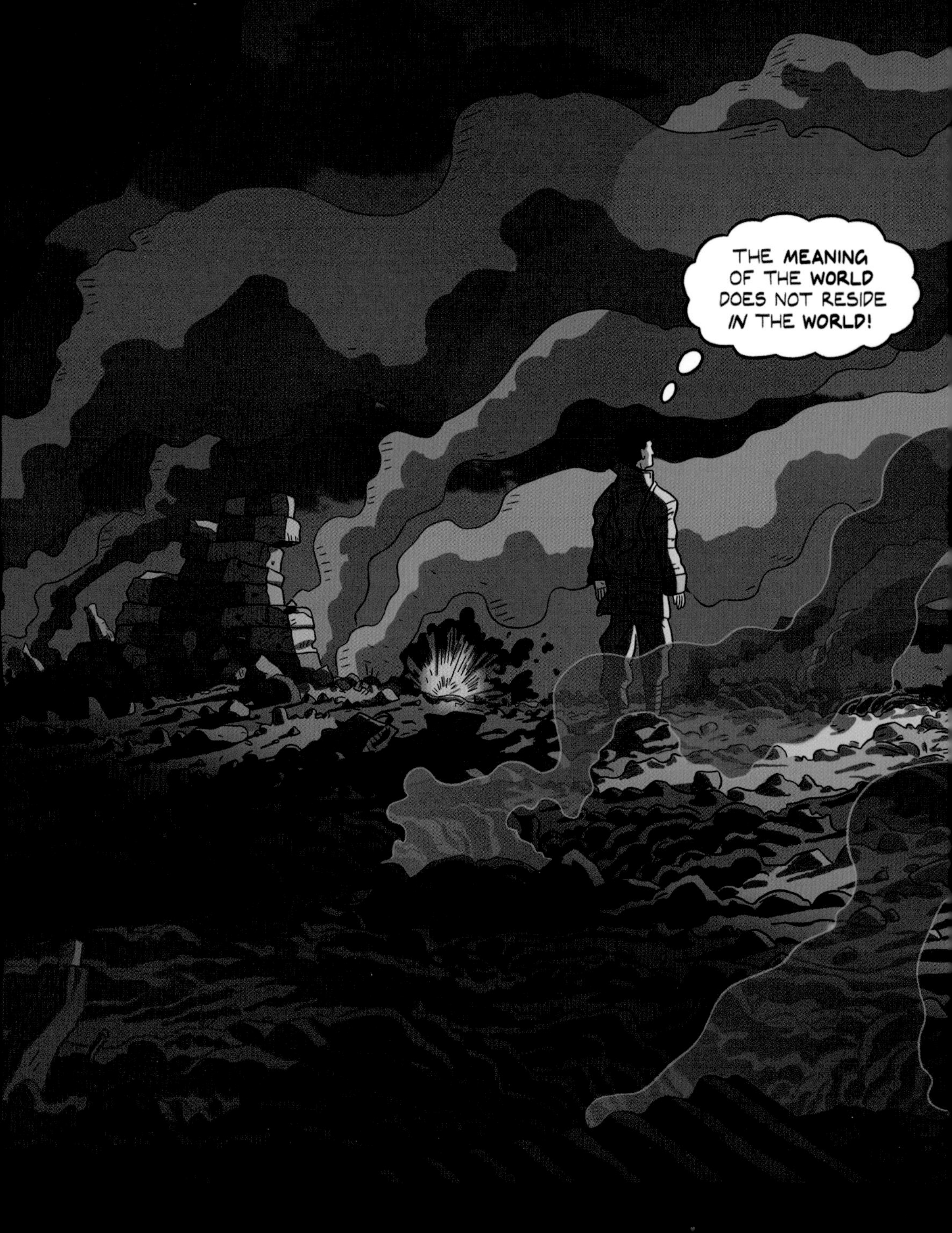
THE MEANING OF THE WORLD DOES NOT RESIDE IN THE WORLD!

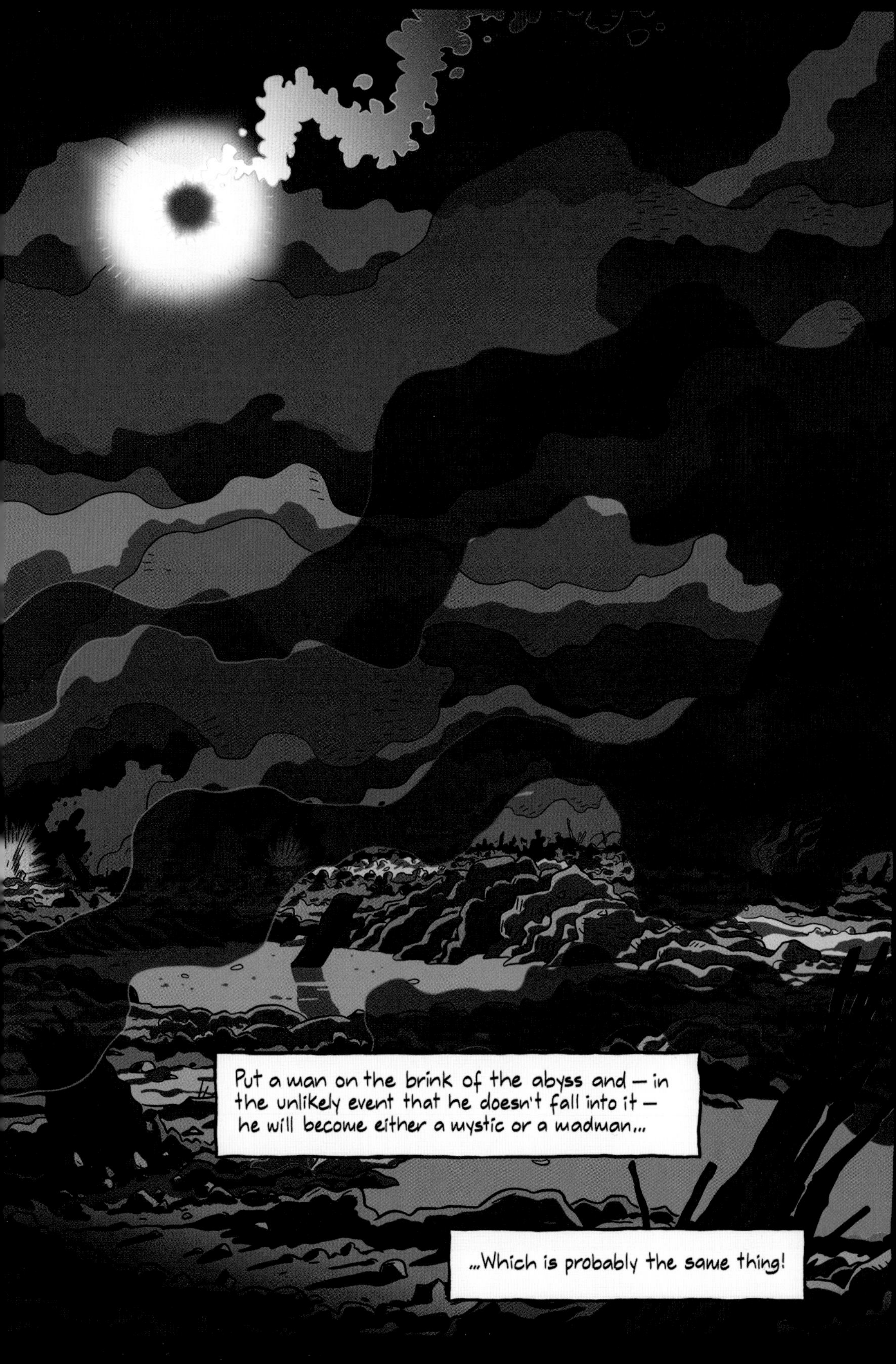
Put a man on the brink of the abyss and — in the unlikely event that he doesn't fall into it — he will become either a mystic or a madman...
...Which is probably the same thing!

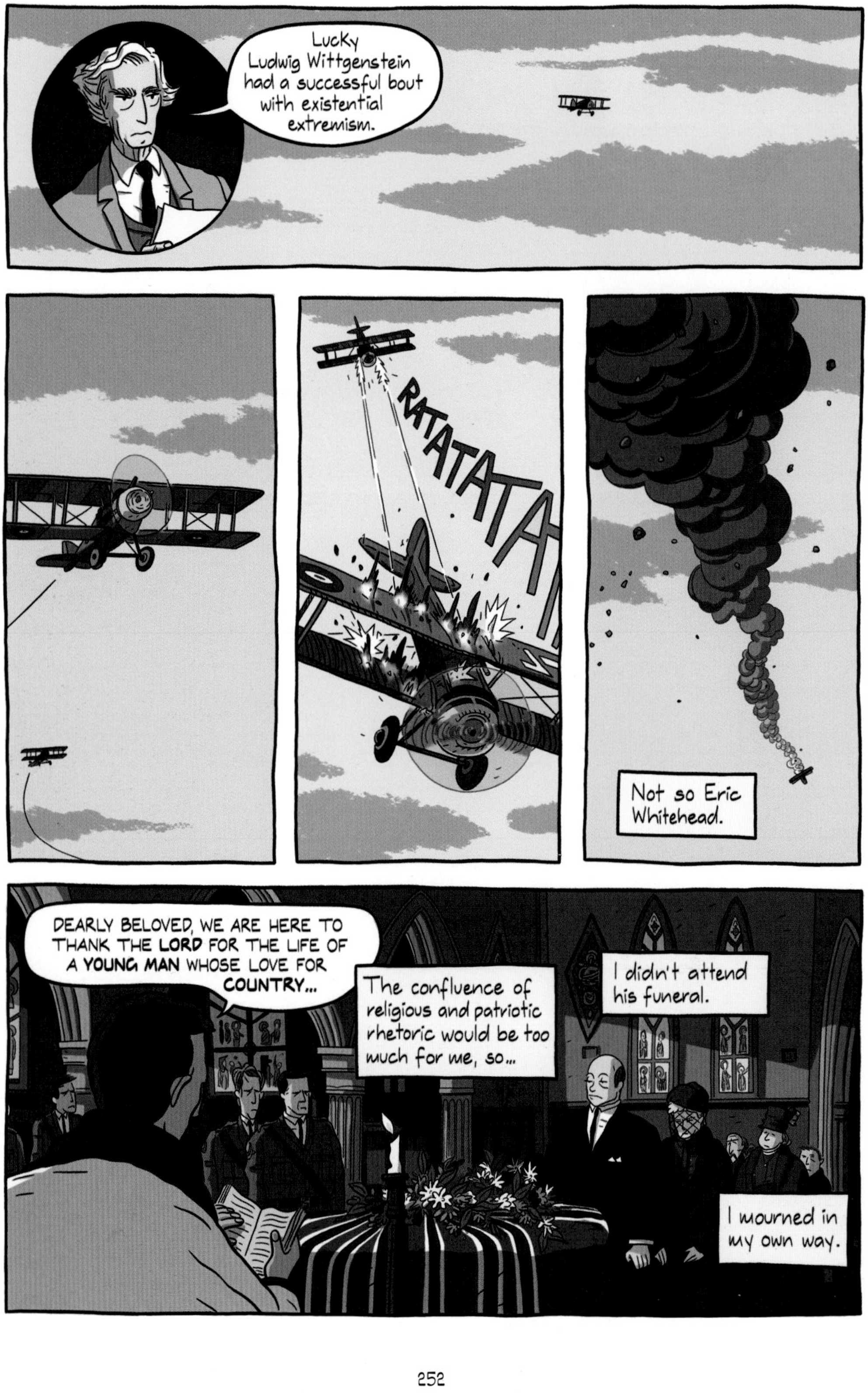
Lucky Ludwig Wittgenstein had a successful bout with existential extremism.
RATATATAT
Not so Eric Whitehead.
DEARLY BELOVED, WE ARE HERE TO THANK THE **LORD** FOR THE LIFE OF A **YOUNG MAN** WHOSE LOVE FOR **COUNTRY**...
The confluence of religious and patriotic rhetoric would be too much for me, so...
I didn't attend his funeral.
I mourned in my own way.

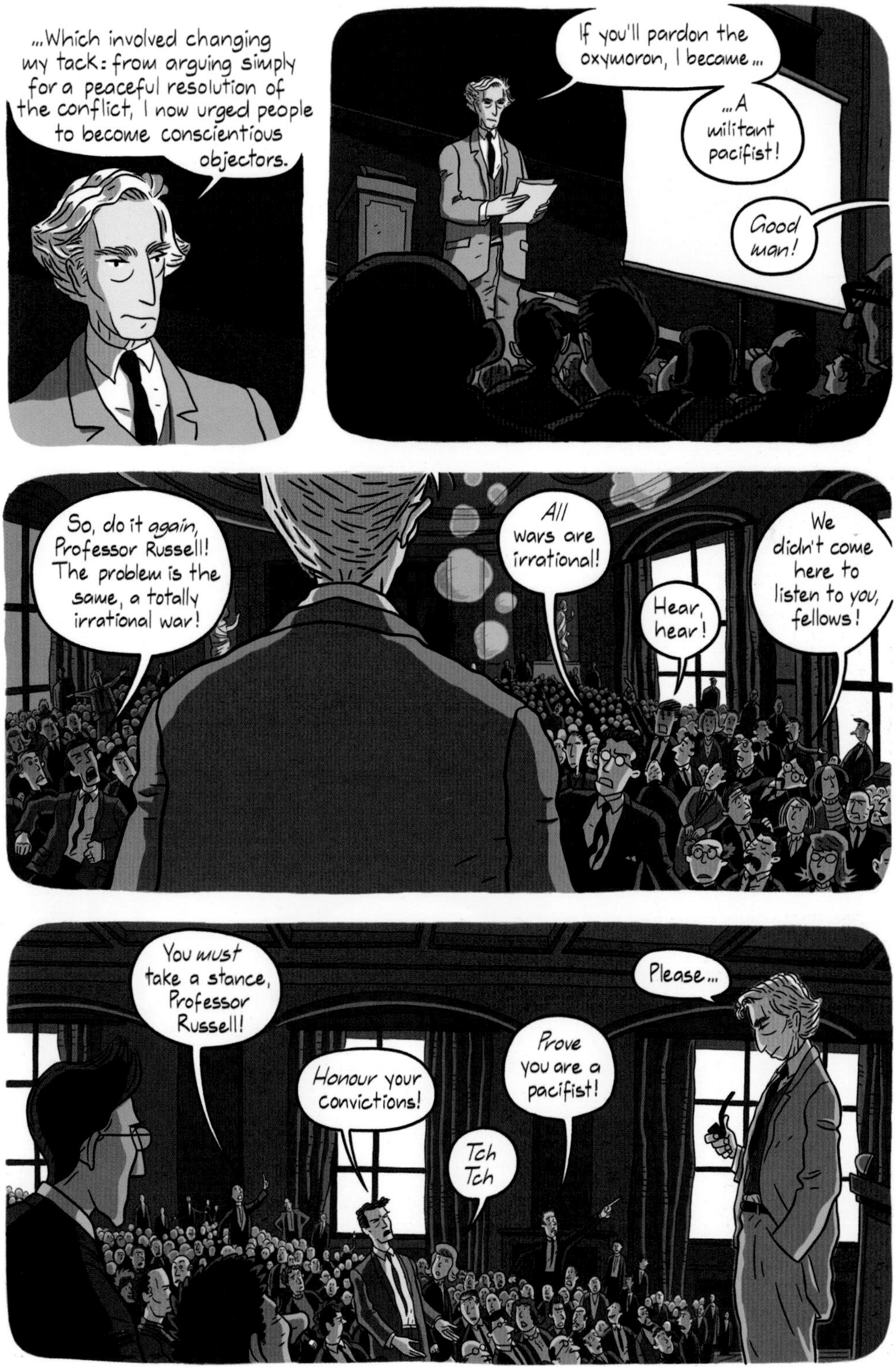

...Which involved changing my tack: from arguing simply for a peaceful resolution of the conflict, I now urged people to become conscientious objectors.
If you'll pardon the oxymoron, I became...
...A militant pacifist!
Good man!
So, do it again, Professor Russell! The problem is the same, a totally irrational war!
All wars are irrational!
We didn't come here to listen to you, fellows!
Hear, hear!
You must take a stance, Professor Russell!
Please...
Prove you are a pacifist!
Honour your convictions!
Tch Tch

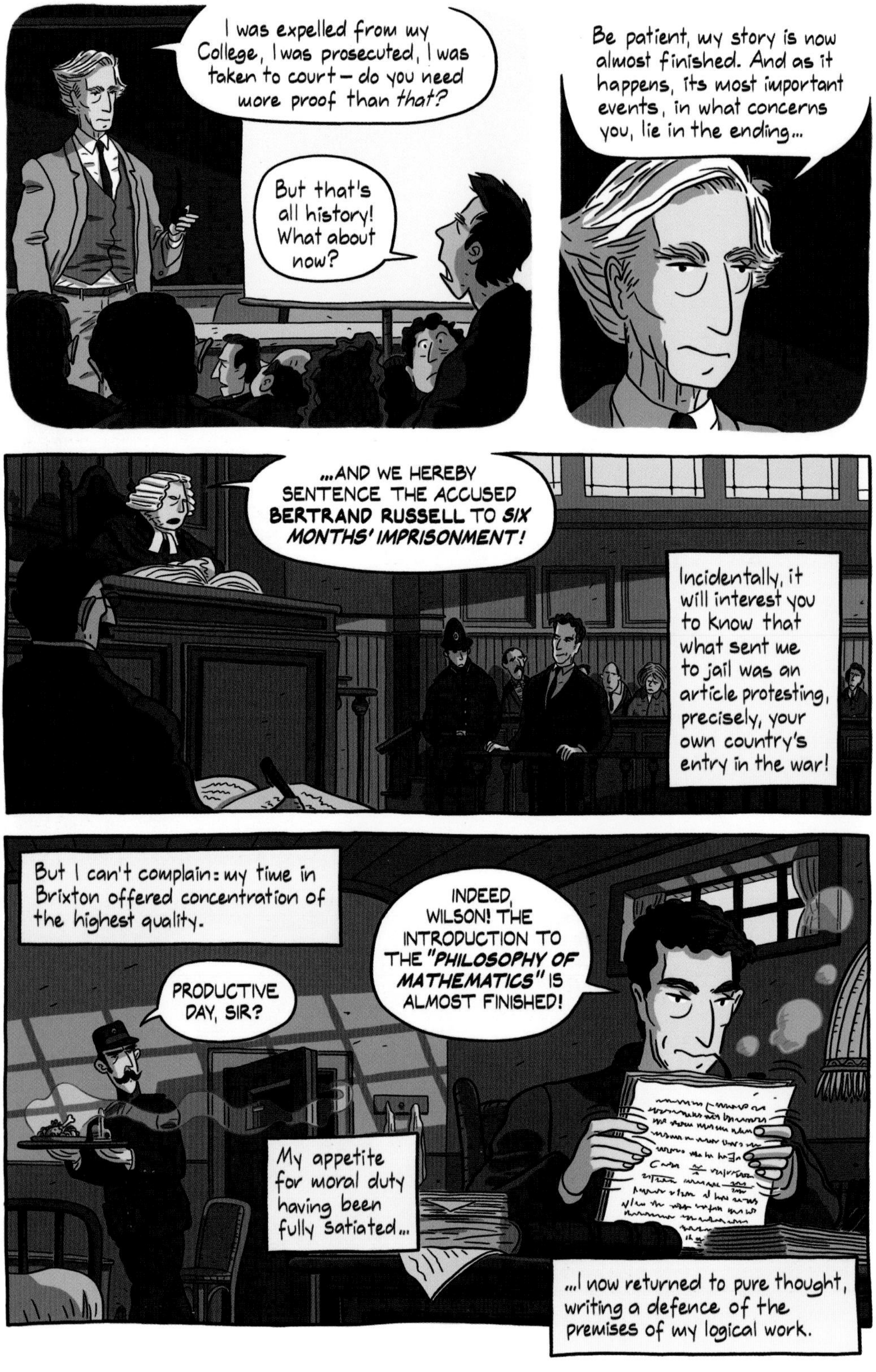
I was expelled from my College, I was prosecuted, I was taken to court – do you need more proof than *that?*
But that's all history! What about now?
Be patient, my story is now almost finished. And as it happens, its most important events, in what concerns you, lie in the ending...
...AND WE HEREBY SENTENCE THE ACCUSED **BERTRAND RUSSELL** TO ***SIX MONTHS' IMPRISONMENT!***
Incidentally, it will interest you to know that what sent me to jail was an article protesting, precisely, your own country's entry in the war!
But I can't complain: my time in Brixton offered concentration of the highest quality.
PRODUCTIVE DAY, SIR?
INDEED, WILSON! THE INTRODUCTION TO THE ***"PHILOSOPHY OF MATHEMATICS"*** IS ALMOST FINISHED!
My appetite for moral duty having been fully satiated...
...I now returned to pure thought, writing a defence of the premises of my logical work.

A few months after the end of the war, I received a totally unexpected gift.

THANK GOD... HE IS **ALIVE!**

"I HAVE **TOTALLY** SOLVED **ALL** THE PROBLEMS OF PHILOSOPHY!"

Good old Wittgenstein... Modesty was never his strong point!

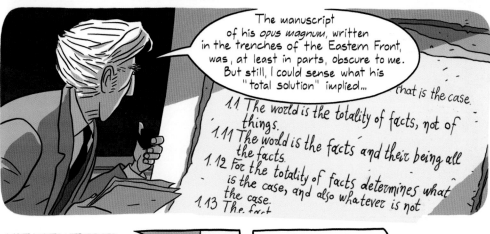

The manuscript of his *opus magnum*, written in the trenches of the Eastern Front, was, at least in parts, obscure to me. But still, I could sense what his "total solution" implied...

...that is the case.

1.1 The world is the totality of facts, not of things.
1.11 The world is the facts and their being all the facts.
1.12 For the totality of facts determines what is the case, and also whatever is not the case.
1.13 The fact

WAIT, WAIT! I TRIED TO READ THIS "**TRACTATUS LOGICO-PHILOSOPHICUS**"...

STILL CALLED, BACK THEN, "**LOGISCHE-PHILOSOPHISCHE ABHANDLUNG**".

...WHATEVER, AND I GOT **VERY** CONFUSED!

IN THAT, YOU ARE **NOT** ALONE.

WHAT IS ZE "NUTSHELL" OF IT?

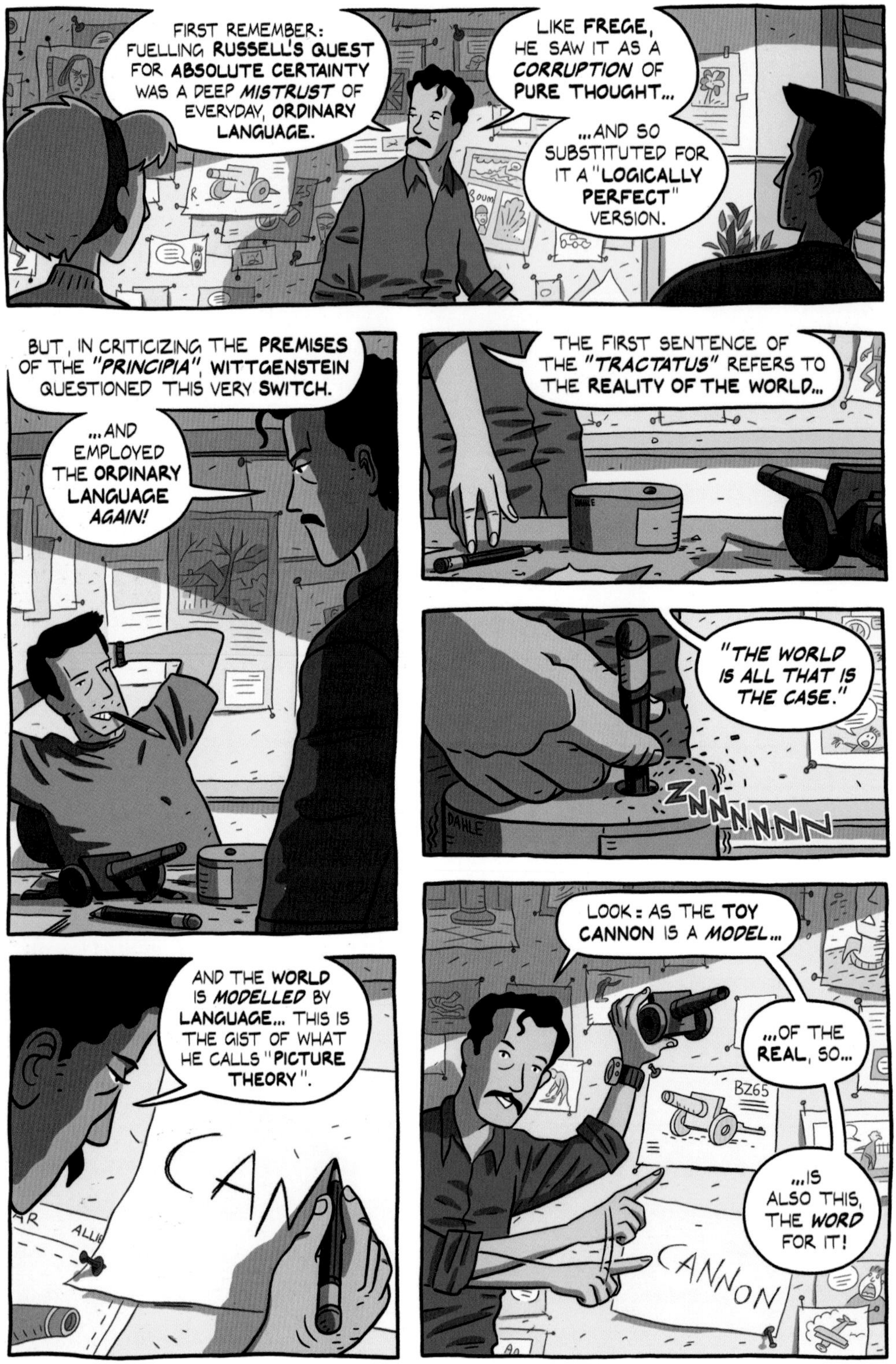
FIRST REMEMBER: FUELLING **RUSSELL'S QUEST** FOR **ABSOLUTE CERTAINTY** WAS A DEEP ***MISTRUST*** OF EVERYDAY, **ORDINARY LANGUAGE.**
LIKE **FREGE,** HE SAW IT AS A ***CORRUPTION*** OF **PURE THOUGHT...**
...AND SO SUBSTITUTED FOR IT A "**LOGICALLY PERFECT**" VERSION.
BUT, IN CRITICIZING THE **PREMISES** OF THE ***"PRINCIPIA"***, **WITTGENSTEIN** QUESTIONED THIS VERY **SWITCH.**
...AND EMPLOYED THE **ORDINARY LANGUAGE** ***AGAIN!***
THE FIRST SENTENCE OF THE ***"TRACTATUS"*** REFERS TO THE **REALITY OF THE WORLD...**
DAHLE
"THE WORLD IS ALL THAT IS THE CASE."
DAHLE
ZNNNNNN
LOOK: AS THE **TOY CANNON** IS A ***MODEL...***
...OF THE **REAL,** SO...
BZ65
...IS ALSO THIS, THE ***WORD*** FOR IT!
CANNON
AND THE **WORLD** IS ***MODELLED*** BY **LANGUAGE...** THIS IS THE GIST OF WHAT HE CALLS "**PICTURE THEORY**".
CAN

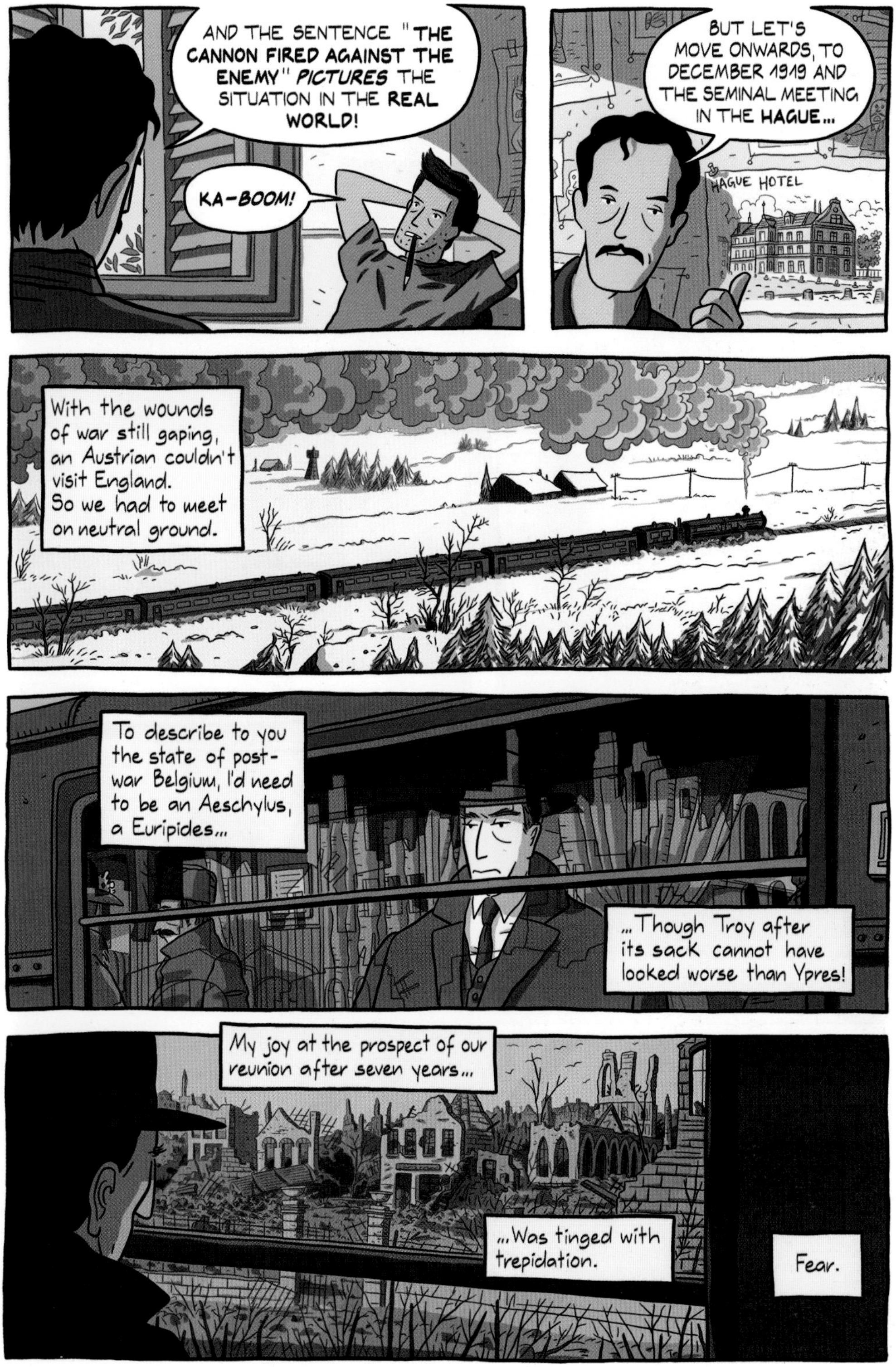
AND THE SENTENCE "THE CANNON FIRED AGAINST THE ENEMY" PICTURES THE SITUATION IN THE REAL WORLD!
KA-BOOM!
BUT LET'S MOVE ONWARDS, TO DECEMBER 1919 AND THE SEMINAL MEETING IN THE HAGUE...
HAGUE HOTEL
With the wounds of war still gaping, an Austrian couldn't visit England. So we had to meet on neutral ground.
To describe to you the state of post-war Belgium, I'd need to be an Aeschylus, a Euripides...
...Though Troy after its sack cannot have looked worse than Ypres!
My joy at the prospect of our reunion after seven years...
...Was tinged with trepidation.
Fear.

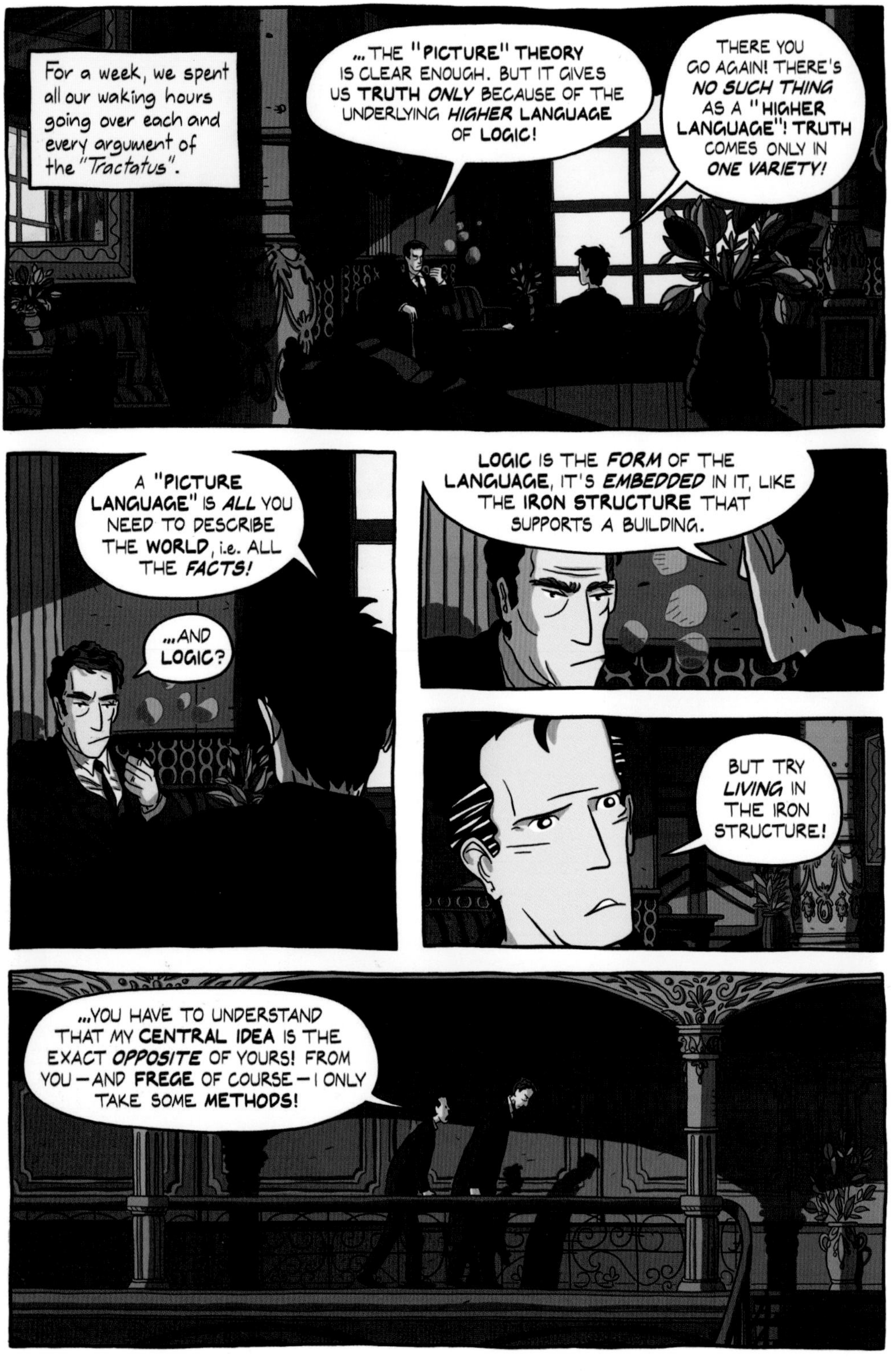
For a week, we spent all our waking hours going over each and every argument of the "Tractatus".
...THE "PICTURE" THEORY IS CLEAR ENOUGH. BUT IT GIVES US TRUTH ONLY BECAUSE OF THE UNDERLYING HIGHER LANGUAGE OF LOGIC!
THERE YOU GO AGAIN! THERE'S NO SUCH THING AS A "HIGHER LANGUAGE"! TRUTH COMES ONLY IN ONE VARIETY!
A "PICTURE LANGUAGE" IS ALL YOU NEED TO DESCRIBE THE WORLD, i.e. ALL THE FACTS!
...AND LOGIC?
LOGIC IS THE FORM OF THE LANGUAGE, IT'S EMBEDDED IN IT, LIKE THE IRON STRUCTURE THAT SUPPORTS A BUILDING.
BUT TRY LIVING IN THE IRON STRUCTURE!
...YOU HAVE TO UNDERSTAND THAT MY CENTRAL IDEA IS THE EXACT OPPOSITE OF YOURS! FROM YOU — AND FREGE OF COURSE — I ONLY TAKE SOME METHODS!

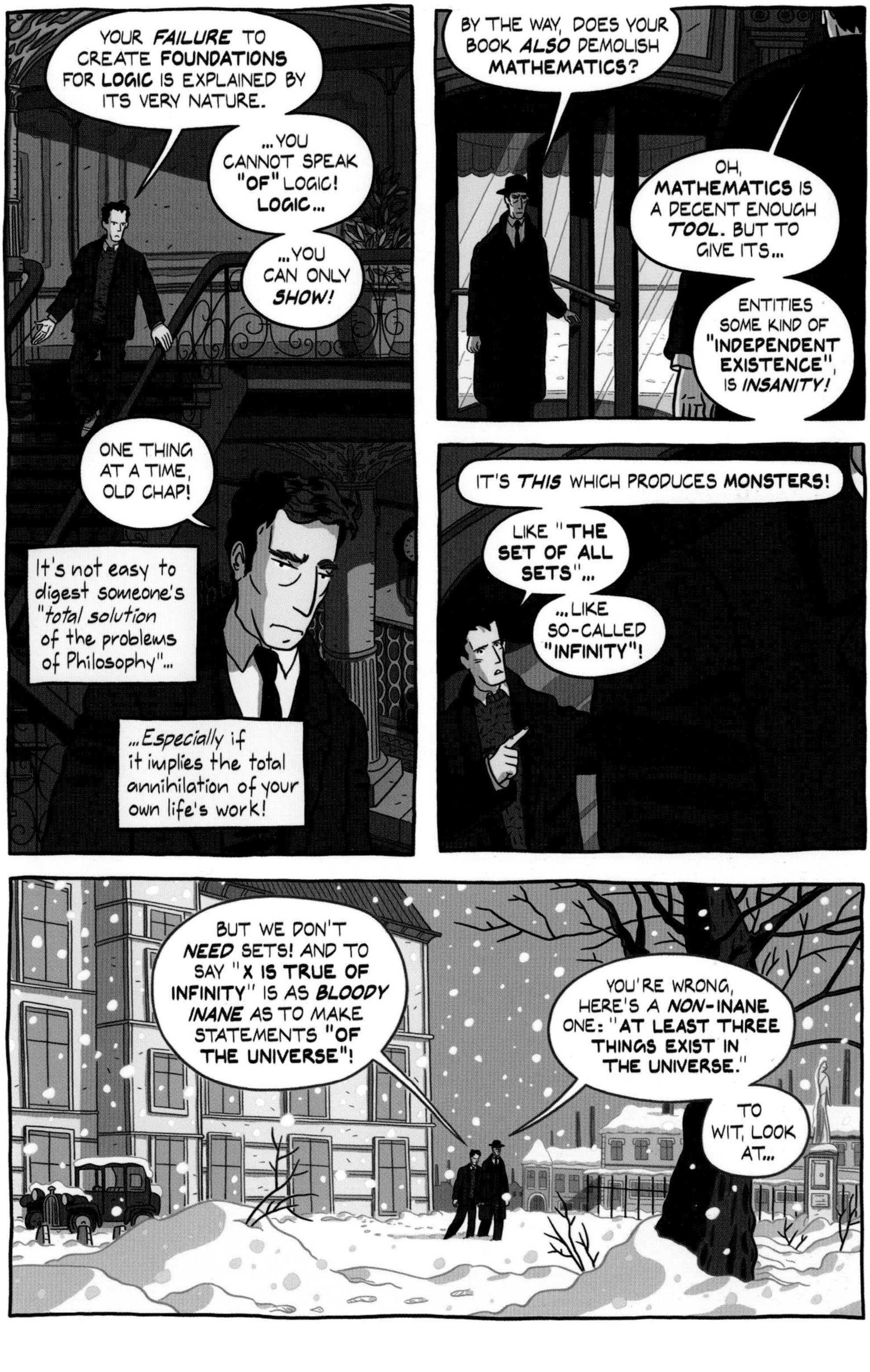

YOUR **FAILURE** TO CREATE **FOUNDATIONS** FOR **LOGIC** IS EXPLAINED BY ITS VERY NATURE.
...YOU CANNOT SPEAK **"OF"** LOGIC! **LOGIC...**
...YOU CAN ONLY ***SHOW!***
ONE THING AT A TIME, OLD CHAP!
It's not easy to digest someone's "*total solution* of the problems of Philosophy"...
...*Especially* if it implies the total annihilation of your own life's work!
BY THE WAY, DOES YOUR BOOK ***ALSO*** DEMOLISH **MATHEMATICS?**
OH, **MATHEMATICS** IS A DECENT ENOUGH ***TOOL***. BUT TO GIVE ITS...
ENTITIES SOME KIND OF **"INDEPENDENT EXISTENCE"**, IS ***INSANITY!***
IT'S ***THIS*** WHICH PRODUCES **MONSTERS!**
LIKE **"THE SET OF ALL SETS"...**
...LIKE SO-CALLED **"INFINITY"!**
BUT WE DON'T ***NEED*** SETS! AND TO SAY **"X IS TRUE OF INFINITY"** IS AS ***BLOODY INANE*** AS TO MAKE STATEMENTS **"OF THE UNIVERSE"!**
YOU'RE WRONG, HERE'S A ***NON***-INANE ONE: **"AT LEAST THREE THINGS EXIST IN THE UNIVERSE."**
TO WIT, LOOK AT...

...THIS BRANCH: THREE LEAVES EXIST ON IT, THEREFORE "AT LEAST THREE THINGS EXIST IN THE UNIVERSE!"
NO, NO, NO! YOU ARE SO WRONG! YOU CAN SAY "AT LEAST THREE LEAVES EXIST ON THIS TREE..."
...BUT YOU CANNOT SAY "IN THE UNIVERSE"! LOGIC MUST NOT ALLOW IT, SINCE YOU CANNOT PICTURE THE UNIVERSE!
THE PARADOXES SHOULD HAVE ALERTED YOU, RUSSELL!
LOGIC IS VACUOUS... IT CANNOT SPEAK REALITY!
TRY AND MAKE EMPTY FORM SPEAK OF SUBSTANCE AND YOU GET POPPYCOCK!
WHAT ABOUT THE STATEMENT "TOMORROW IT WILL EITHER SNOW OR NOT SNOW"? IT'S "EMPTY FORM" YET TOTALLY TRUE!
YES, BUT IT TELLS US NOTHING OF THE WEATHER TOMORROW!
There it was: for twenty years, I had sweated to justify the existence of a machine for manufacturing tautologies!

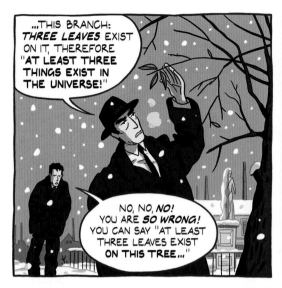

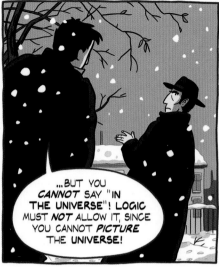

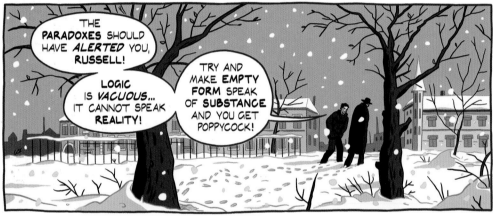

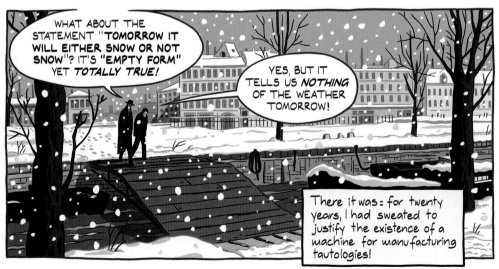

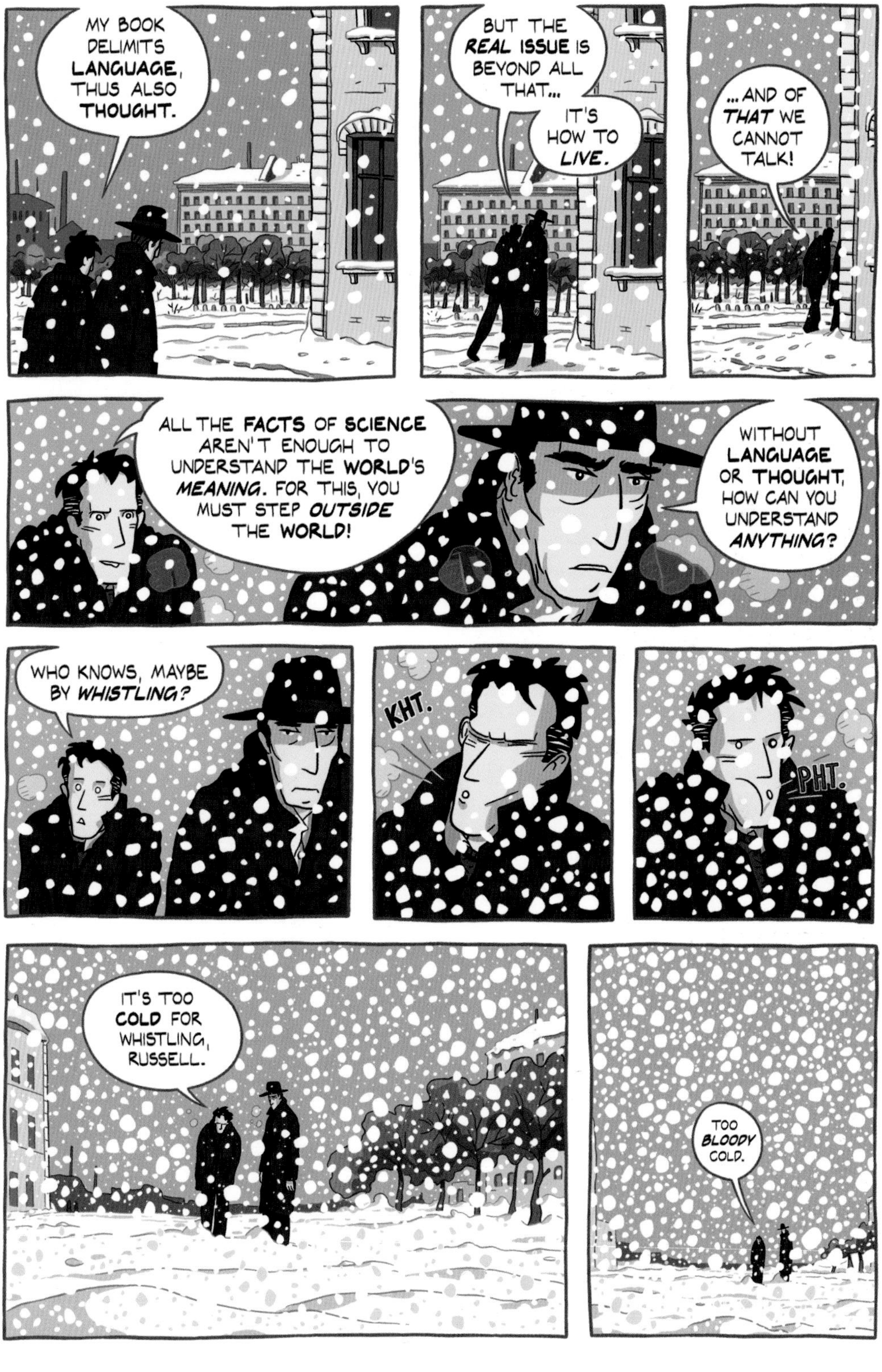
MY BOOK DELIMITS LANGUAGE, THUS ALSO THOUGHT.
BUT THE REAL ISSUE IS BEYOND ALL THAT...
IT'S HOW TO LIVE.
...AND OF THAT WE CANNOT TALK!
ALL THE FACTS OF SCIENCE AREN'T ENOUGH TO UNDERSTAND THE WORLD'S MEANING. FOR THIS, YOU MUST STEP OUTSIDE THE WORLD!
WITHOUT LANGUAGE OR THOUGHT, HOW CAN YOU UNDERSTAND ANYTHING?
WHO KNOWS, MAYBE BY WHISTLING?
KHT.
PHT.
IT'S TOO COLD FOR WHISTLING, RUSSELL.
TOO BLOODY COLD.

6. INCOMPLETENESS

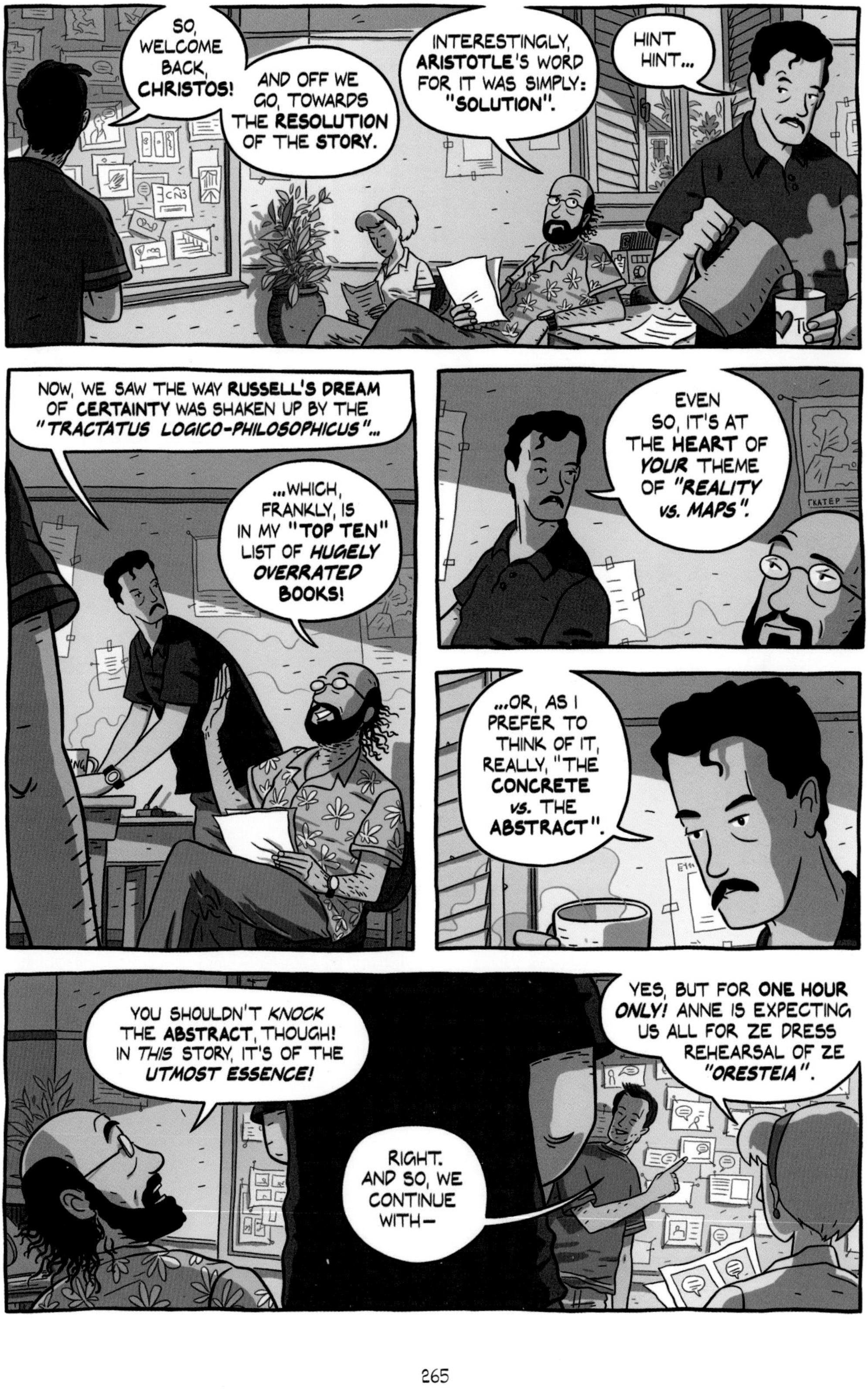
SO, WELCOME BACK, CHRISTOS!
AND OFF WE GO, TOWARDS THE RESOLUTION OF THE STORY.
INTERESTINGLY, ARISTOTLE'S WORD FOR IT WAS SIMPLY: "SOLUTION".
HINT HINT...
NOW, WE SAW THE WAY RUSSELL'S DREAM OF CERTAINTY WAS SHAKEN UP BY THE "TRACTATUS LOGICO-PHILOSOPHICUS"...
...WHICH, FRANKLY, IS IN MY "TOP TEN" LIST OF HUGELY OVERRATED BOOKS!
EVEN SO, IT'S AT THE HEART OF YOUR THEME OF "REALITY vs. MAPS".
...OR, AS I PREFER TO THINK OF IT, REALLY, "THE CONCRETE vs. THE ABSTRACT".
YOU SHOULDN'T KNOCK THE ABSTRACT, THOUGH! IN THIS STORY, IT'S OF THE UTMOST ESSENCE!
RIGHT. AND SO, WE CONTINUE WITH—
YES, BUT FOR ONE HOUR ONLY! ANNE IS EXPECTING US ALL FOR ZE DRESS REHEARSAL OF ZE "ORESTEIA".

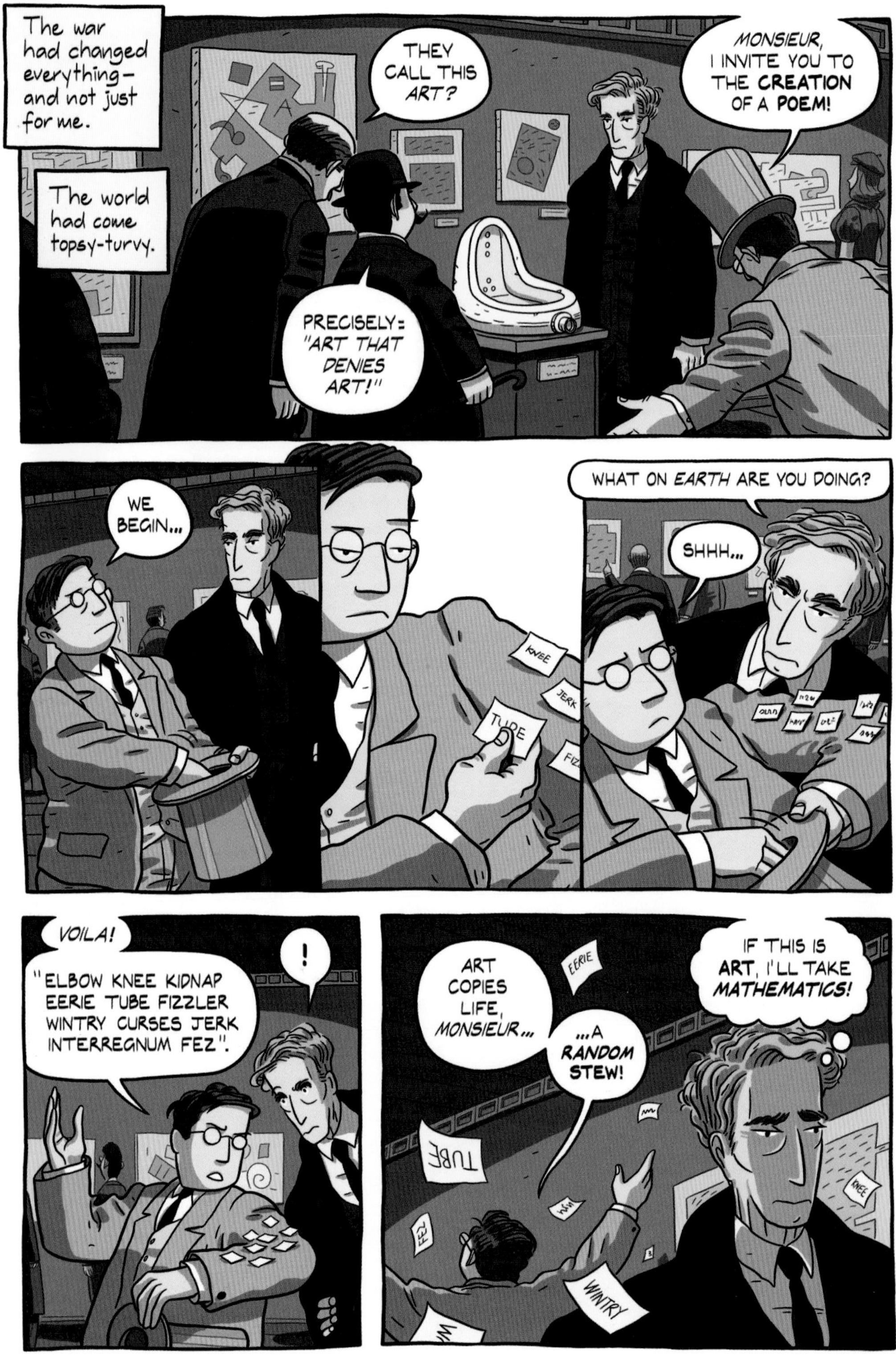
The war had changed everything— and not just for me.
The world had come topsy-turvy.
THEY CALL THIS ART?
MONSIEUR, I INVITE YOU TO THE CREATION OF A POEM!
PRECISELY: "ART THAT DENIES ART!"
WE BEGIN...
KNEE
JERK
TUBE
WHAT ON EARTH ARE YOU DOING?
SHHH...
VOILA!
"ELBOW KNEE KIDNAP EERIE TUBE FIZZLER WINTRY CURSES JERK INTERREGNUM FEZ".
!
ART COPIES LIFE, MONSIEUR...
EERIE
...A RANDOM STEW!
IF THIS IS ART, I'LL TAKE MATHEMATICS!
TUBE
KNEE
WINTRY

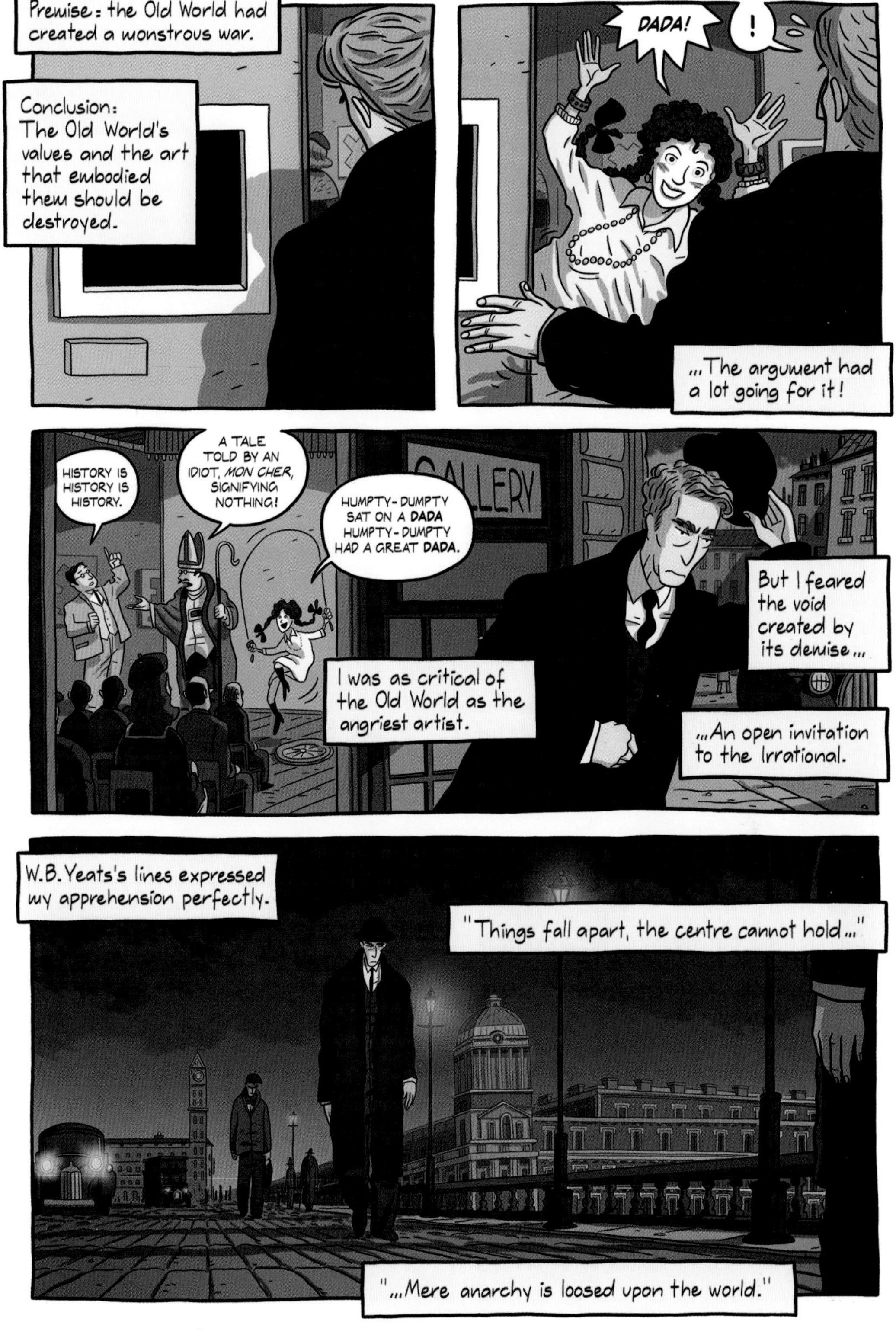
Premise: the Old World had created a monstrous war.
Conclusion: The Old World's values and the art that embodied them should be destroyed.
DADA!
!
...The argument had a lot going for it!
HISTORY IS HISTORY IS HISTORY.
A TALE TOLD BY AN IDIOT, *MON CHER*, SIGNIFYING NOTHING!
HUMPTY-DUMPTY SAT ON A **DADA** HUMPTY-DUMPTY HAD A GREAT **DADA**.
I was as critical of the Old World as the angriest artist.
But I feared the void created by its demise...
...An open invitation to the Irrational.
W.B. Yeats's lines expressed my apprehension perfectly.
"Things fall apart, the centre cannot hold..."
"...Mere anarchy is loosed upon the world."

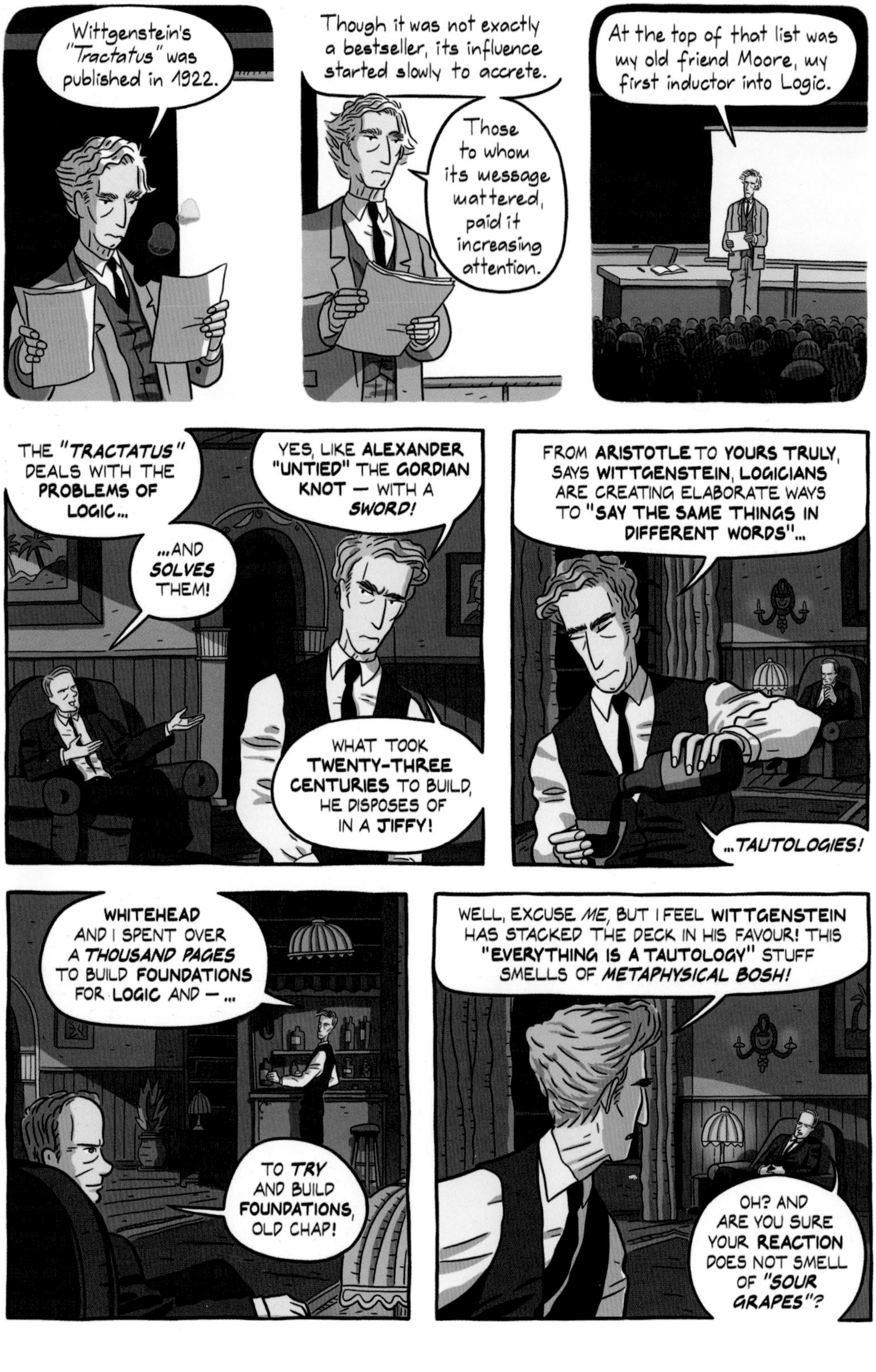

Wittgenstein's "*Tractatus*" was published in 1922.
Though it was not exactly a bestseller, its influence started slowly to accrete.
Those to whom its message mattered, paid it increasing attention.
At the top of that list was my old friend Moore, my first inductor into Logic.
THE "***TRACTATUS***" DEALS WITH THE **PROBLEMS OF LOGIC**...
...AND ***SOLVES*** THEM!
YES, LIKE **ALEXANDER "UNTIED"** THE **GORDIAN KNOT** — WITH A ***SWORD!***
WHAT TOOK **TWENTY-THREE CENTURIES** TO BUILD, HE DISPOSES OF IN A **JIFFY!**
FROM **ARISTOTLE** TO **YOURS TRULY**, SAYS **WITTGENSTEIN**, **LOGICIANS** ARE CREATING ELABORATE WAYS TO "**SAY THE SAME THINGS IN DIFFERENT WORDS**"...
...***TAUTOLOGIES!***
WHITEHEAD AND I SPENT OVER A ***THOUSAND PAGES*** TO BUILD **FOUNDATIONS** FOR **LOGIC** AND — ...
TO ***TRY*** AND BUILD **FOUNDATIONS**, OLD CHAP!
WELL, EXCUSE *ME*, BUT I FEEL **WITTGENSTEIN** HAS STACKED THE DECK IN HIS FAVOUR! THIS "**EVERYTHING IS A TAUTOLOGY**" STUFF SMELLS OF ***METAPHYSICAL BOSH!***
OH? AND ARE YOU SURE YOUR **REACTION** DOES NOT SMELL OF "***SOUR GRAPES***"?

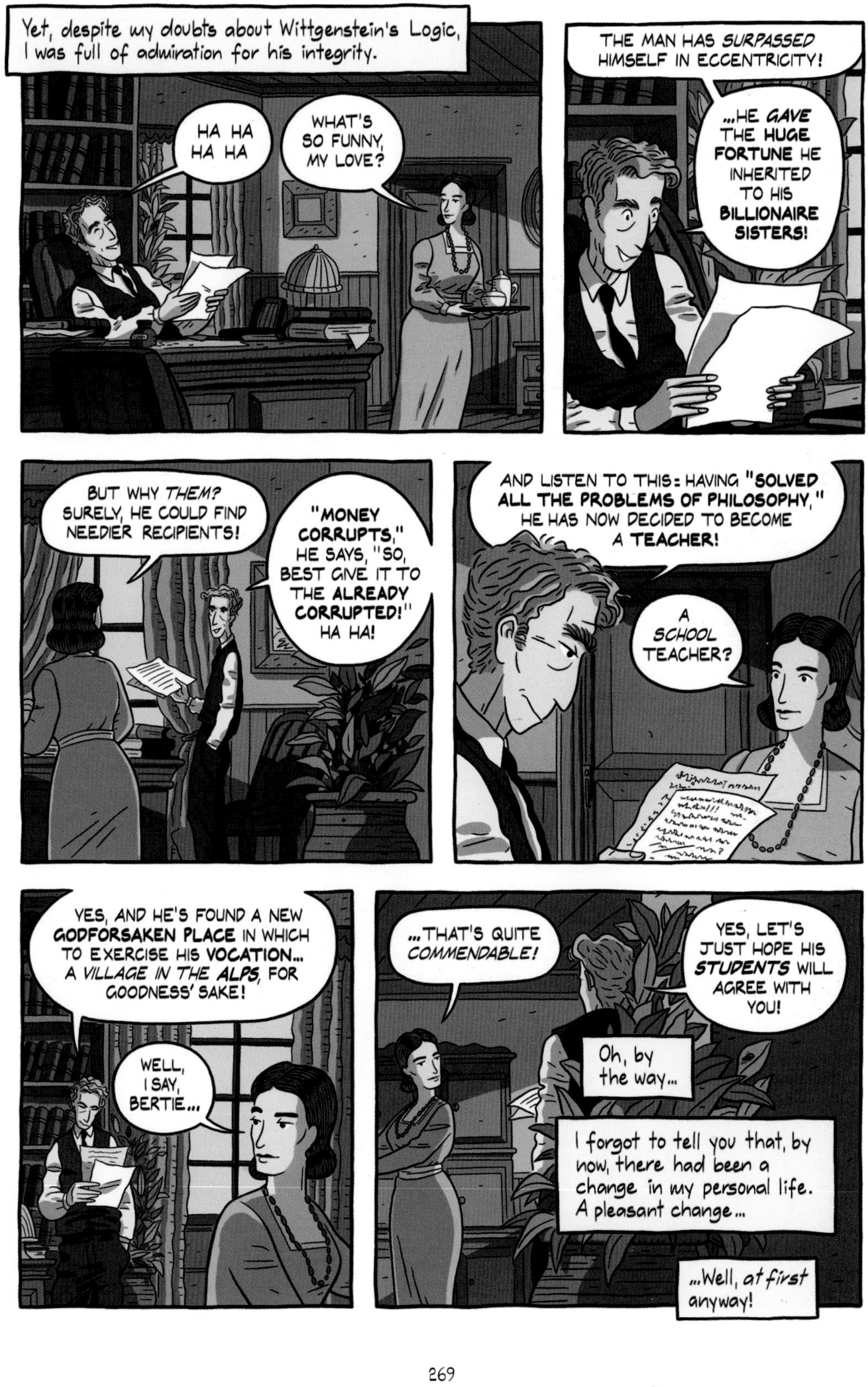
Yet, despite my doubts about Wittgenstein's Logic, I was full of admiration for his integrity.
HA HA HA HA
WHAT'S SO FUNNY, MY LOVE?
THE MAN HAS *SURPASSED* HIMSELF IN ECCENTRICITY!
...HE **GAVE** THE **HUGE FORTUNE** HE INHERITED TO HIS **BILLIONAIRE SISTERS!**
BUT WHY *THEM?* SURELY, HE COULD FIND NEEDIER RECIPIENTS!
"**MONEY CORRUPTS,**" HE SAYS, "SO, BEST GIVE IT TO THE **ALREADY CORRUPTED!**" HA HA!
AND LISTEN TO THIS: HAVING "**SOLVED ALL THE PROBLEMS OF PHILOSOPHY,**" HE HAS NOW DECIDED TO BECOME A **TEACHER!**
A *SCHOOL* TEACHER?
YES, AND HE'S FOUND A NEW **GODFORSAKEN PLACE** IN WHICH TO EXERCISE HIS **VOCATION...** A *VILLAGE IN THE* **ALPS**, FOR GOODNESS' SAKE!
WELL, I SAY, BERTIE...
...THAT'S QUITE *COMMENDABLE!*
YES, LET'S JUST HOPE HIS ***STUDENTS*** WILL AGREE WITH YOU!
Oh, by the way...
I forgot to tell you that, by now, there had been a change in my personal life. A pleasant change...
...Well, *at first* anyway!

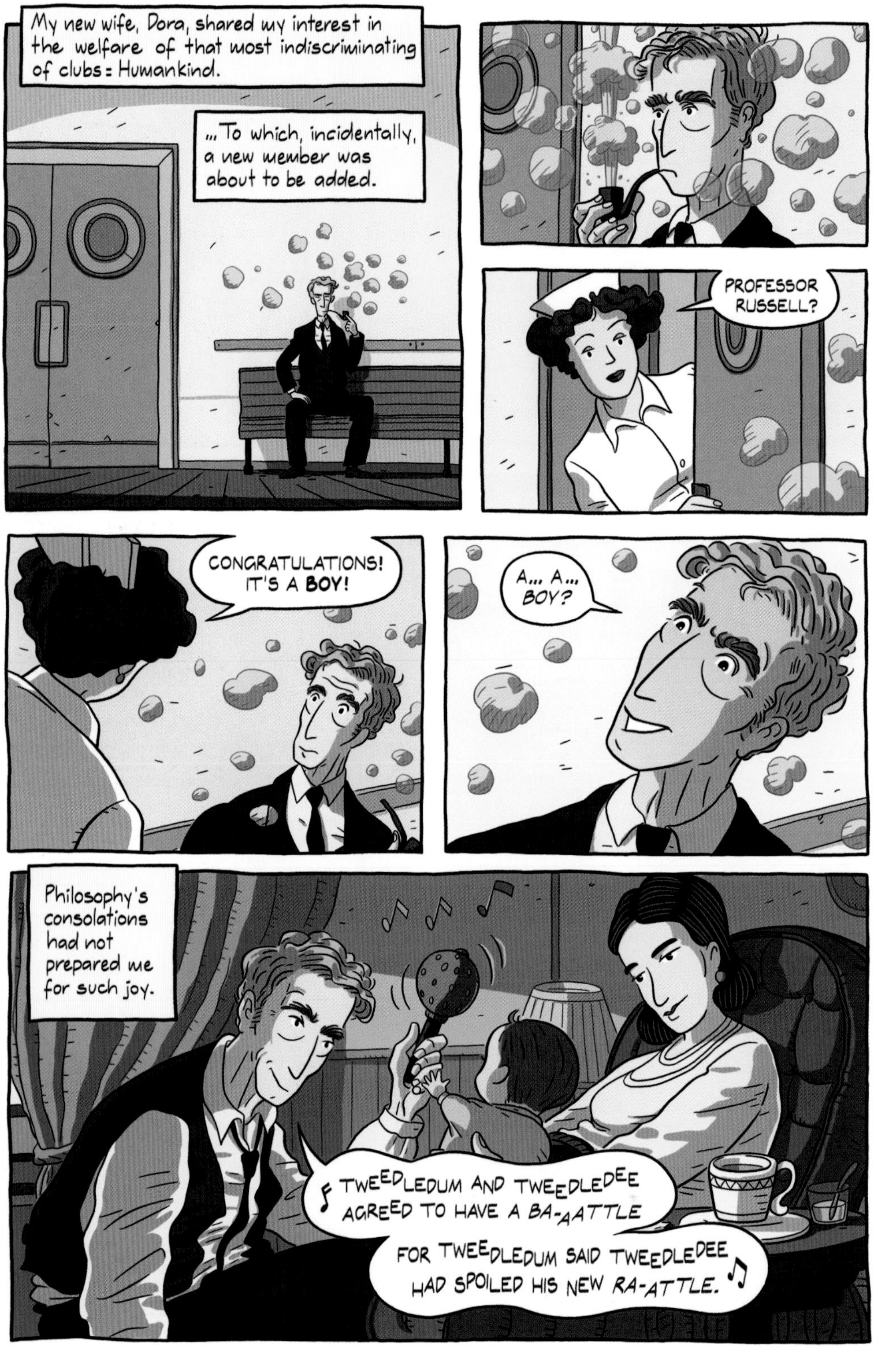
My new wife, Dora, shared my interest in the welfare of that most indiscriminating of clubs: Humankind.
...To which, incidentally, a new member was about to be added.
PROFESSOR RUSSELL?
CONGRATULATIONS! IT'S A **BOY**!
A... A... BOY?
Philosophy's consolations had not prepared me for such joy.
♪ TWEEDLEDUM AND TWEEDLEDEE AGREED TO HAVE A *BA-AATTLE*
FOR TWEEDLEDUM SAID TWEEDLEDEE HAD SPOILED HIS NEW *RA-ATTLE*. ♪

...Which, like all joys, was not unadulterated!

WAAAH WAAR! WAAAA!! WAAAAR

DARLING....

MMM?

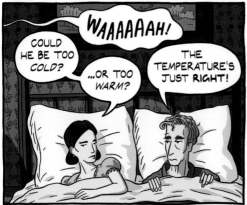

WAAAAAAH!

COULD HE BE TOO COLD?

...OR TOO WARM?

THE TEMPERATURE'S JUST **RIGHT!**

WAAAAAAAH!

IS HE HUNGRY MAYBE?

HE ATE HALF AN HOUR AGO!

WWAAAAAAAAAAAA

I'M GOING TO CHECK!

DAMN!

AAAHHH

HUSH MMM

As always in my adult life, I turned to Reason for assistance.

HMM... THAT'S INTERESTING...

HUMAN BEHAVIOUR

The new science of psychology seemed to offer a way out.

In fact, the time seemed propitious for an extension of my logicist project.

A group of visionaries in Vienna had drafted a manifesto advocating "the scientific conception of the world"...

... A project to apply the tools of Logic, Mathematics and the Physical Sciences to the study of human matters.

So, despite the failure of my own work in Logic, I did not—

WAIT!

YOU JUST CAN'T GO CALLING RUSSELL'S WORK IN LOGIC A "FAILURE"... ...NO WAY!

IT'S HIS WORDS WE ARE USING!

BUT THE "PRINCIPIA" IS THE BASIS OF EVERYTHING THAT FOLLOWED!

BUT— MMM... OK, MAYBE WE'LL PUT SOME OF THAT IN BERTIE'S TRIP TO VIENNA!

...IT IS THE GREATEST HONOUR FOR US, DEAR COLLEAGUES, TO HAVE LISTENED TO PROFESSOR RUSSELL...

MATHEMATISCHES SEMINAR
DER K.K. UNIVERSITÄT

...WHO LAID THE GROUND FOR A LOGICAL LANGUAGE MAKING POSSIBLE THE SCIENTIFIC WORLD-VIEW! IT IS HIS PIONEERING VISION WHICH INSPIRES THE WORK OF OUR CIRCLE...

TOGETHER WITH THAT OF THOSE VENERABLE OLD GENTLEMEN, FREGE AND WITTGENSTEIN!

?

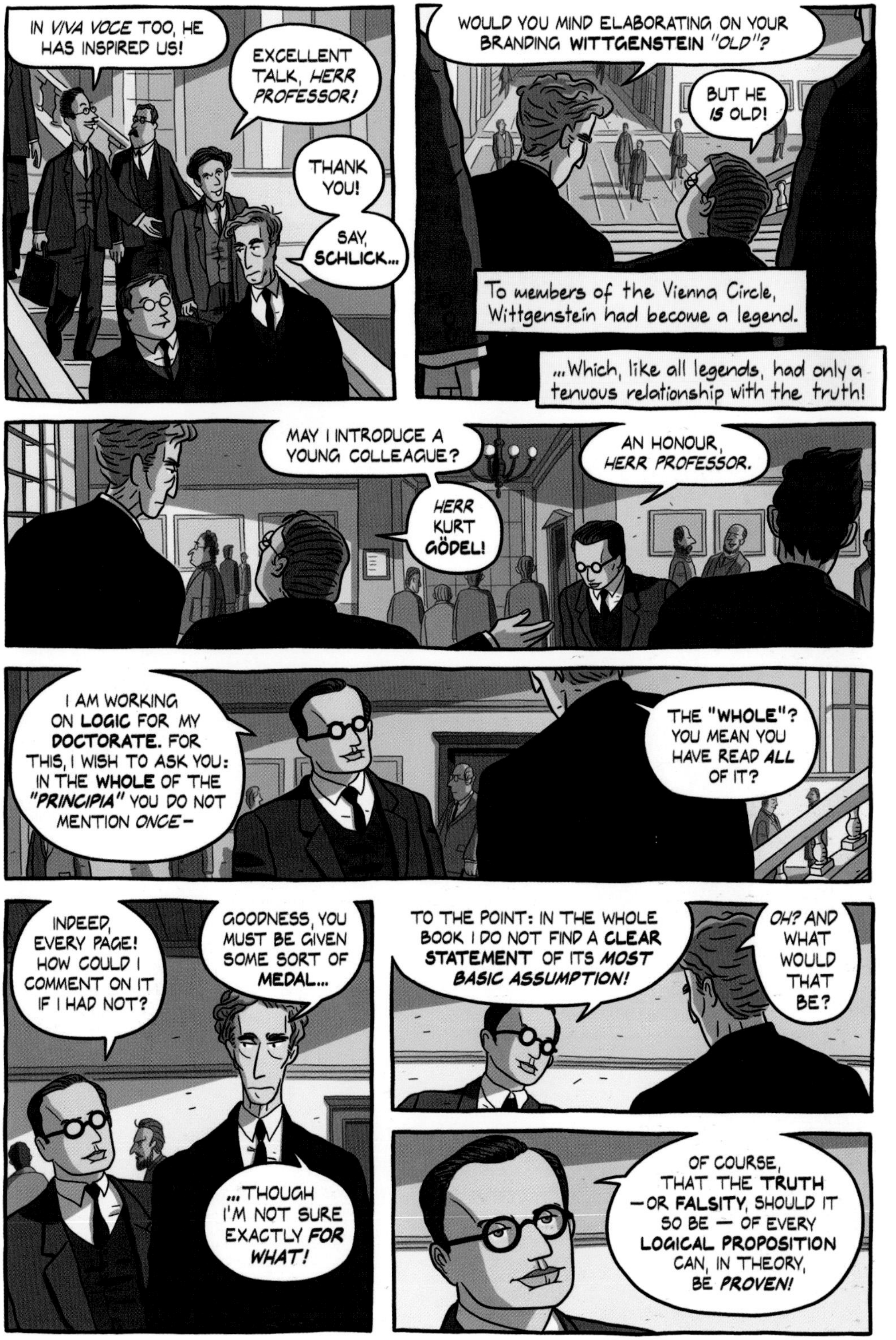
IN *VIVA VOCE* TOO, HE HAS INSPIRED US!
EXCELLENT TALK, *HERR PROFESSOR!*
THANK YOU!
SAY, **SCHLICK...**
WOULD YOU MIND ELABORATING ON YOUR BRANDING **WITTGENSTEIN** *"OLD"*?
BUT HE ***IS*** OLD!
To members of the Vienna Circle, Wittgenstein had become a legend.
...Which, like all legends, had only a tenuous relationship with the truth!
MAY I INTRODUCE A YOUNG COLLEAGUE?
HERR KURT **GÖDEL!**
AN HONOUR, *HERR PROFESSOR.*
I AM WORKING ON **LOGIC** FOR MY **DOCTORATE.** FOR THIS, I WISH TO ASK YOU: IN THE **WHOLE** OF THE ***"PRINCIPIA"*** YOU DO NOT MENTION *ONCE*—
THE **"WHOLE"**? YOU MEAN YOU HAVE READ ***ALL*** OF IT?
INDEED, EVERY PAGE! HOW COULD I COMMENT ON IT IF I HAD NOT?
GOODNESS, YOU MUST BE GIVEN SOME SORT OF **MEDAL...**
...THOUGH I'M NOT SURE EXACTLY ***FOR WHAT!***
TO THE POINT: IN THE WHOLE BOOK I DO NOT FIND A **CLEAR STATEMENT** OF ITS ***MOST BASIC ASSUMPTION!***
OH? AND WHAT WOULD THAT BE?
OF COURSE, THAT THE **TRUTH** —OR **FALSITY**, SHOULD IT SO BE — OF EVERY **LOGICAL PROPOSITION** CAN, IN THEORY, BE ***PROVEN!***

SURELY, THAT'S THE *BASIS* OF THE **LOGICAL POINT OF VIEW**, IS IT NOT?

THAT SOMETHING IS **TRUE** IS **SYNONYMOUS** TO IT BEING **PROVABLE!**

SO, YOU POSIT IT AS AN **AXIOM?**

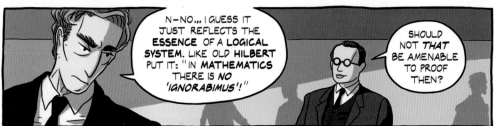

N—NO... I GUESS IT JUST REFLECTS THE **ESSENCE** OF A **LOGICAL SYSTEM.** LIKE OLD **HILBERT** PUT IT: "IN **MATHEMATICS** THERE IS **NO** '*IGNORABIMUS*'!"

SHOULD NOT *THAT* BE AMENABLE TO PROOF THEN?

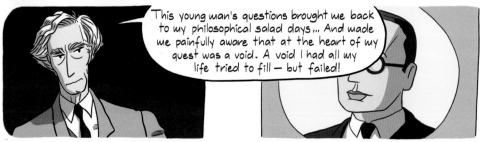

This young man's questions brought me back to my philosophical salad days... And made me painfully aware that at the heart of my quest was a void. A void I had all my life tried to fill — but failed!

SATISFIED NOW?

WITH **RUSSELL** HAVING "*FAILED*"?

ANYWAY, I'LL TAKE HIS WHINING, AS LONG AS THE POINT IS CLEAR: WITHOUT THE "*PRINCIPIA*" DOING THE DONKEYWORK, **GÖDEL** COULDN'T ASK HIS **QUESTIONS!**

WELL, HE DIDN'T GET MANY *ANSWERS* FROM **RUSSELL**... OR ANYONE ELSE, FOR THAT MATTER!

AND THAT'S WHY HE HAD TO *INVESTIGATE* ON HIS OWN!

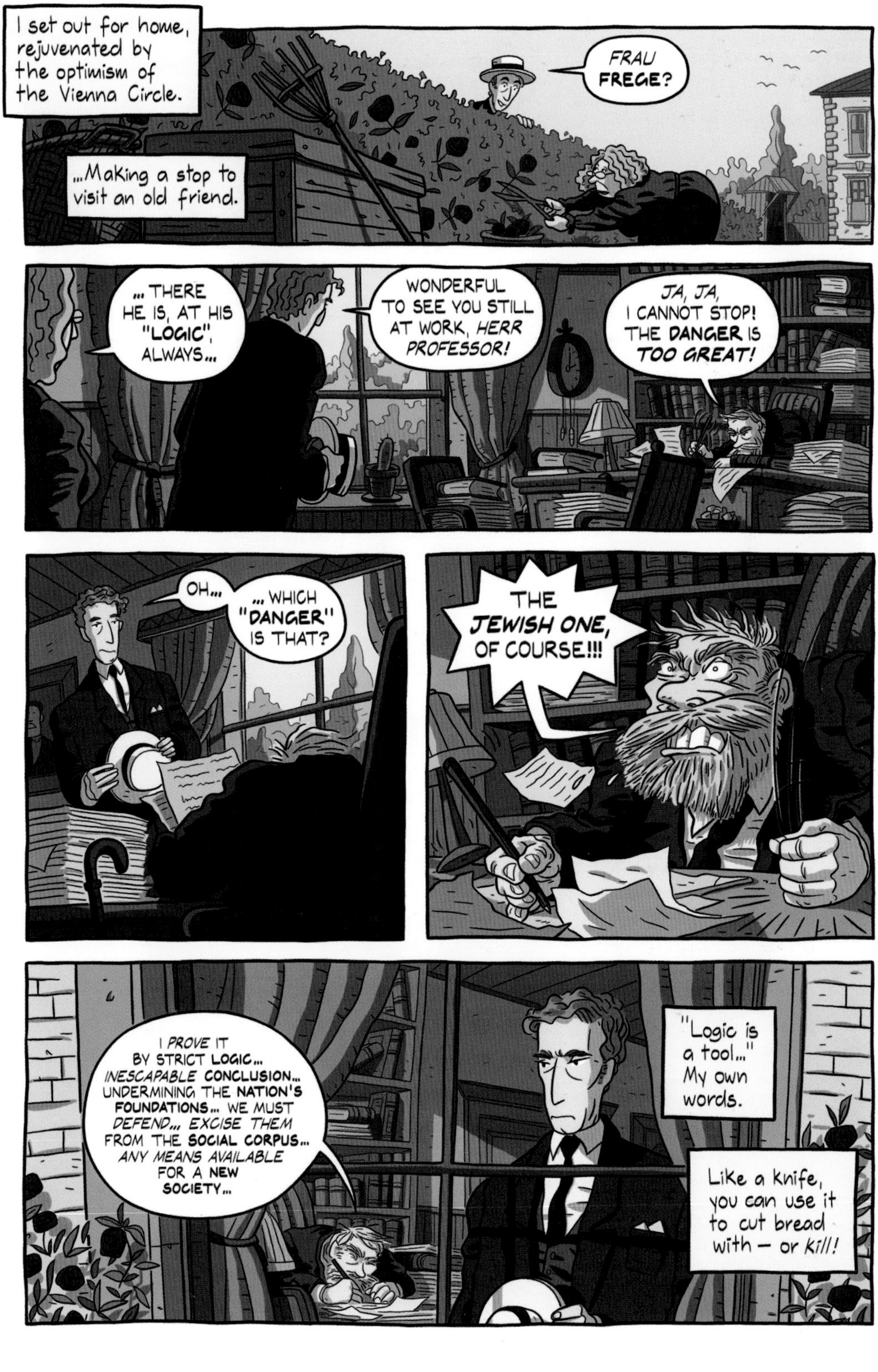
I set out for home, rejuvenated by the optimism of the Vienna Circle.
FRAU FREGE?
...Making a stop to visit an old friend.
... THERE HE IS, AT HIS "LOGIC", ALWAYS...
WONDERFUL TO SEE YOU STILL AT WORK, HERR PROFESSOR!
JA, JA, I CANNOT STOP! THE DANGER IS TOO GREAT!
OH...
... WHICH "DANGER" IS THAT?
THE JEWISH ONE, OF COURSE!!!
I PROVE IT BY STRICT LOGIC... INESCAPABLE CONCLUSION... UNDERMINING THE NATION'S FOUNDATIONS... WE MUST DEFEND... EXCISE THEM FROM THE SOCIAL CORPUS... ANY MEANS AVAILABLE FOR A NEW SOCIETY...
"Logic is a tool..." My own words.
Like a knife, you can use it to cut bread with — or kill!

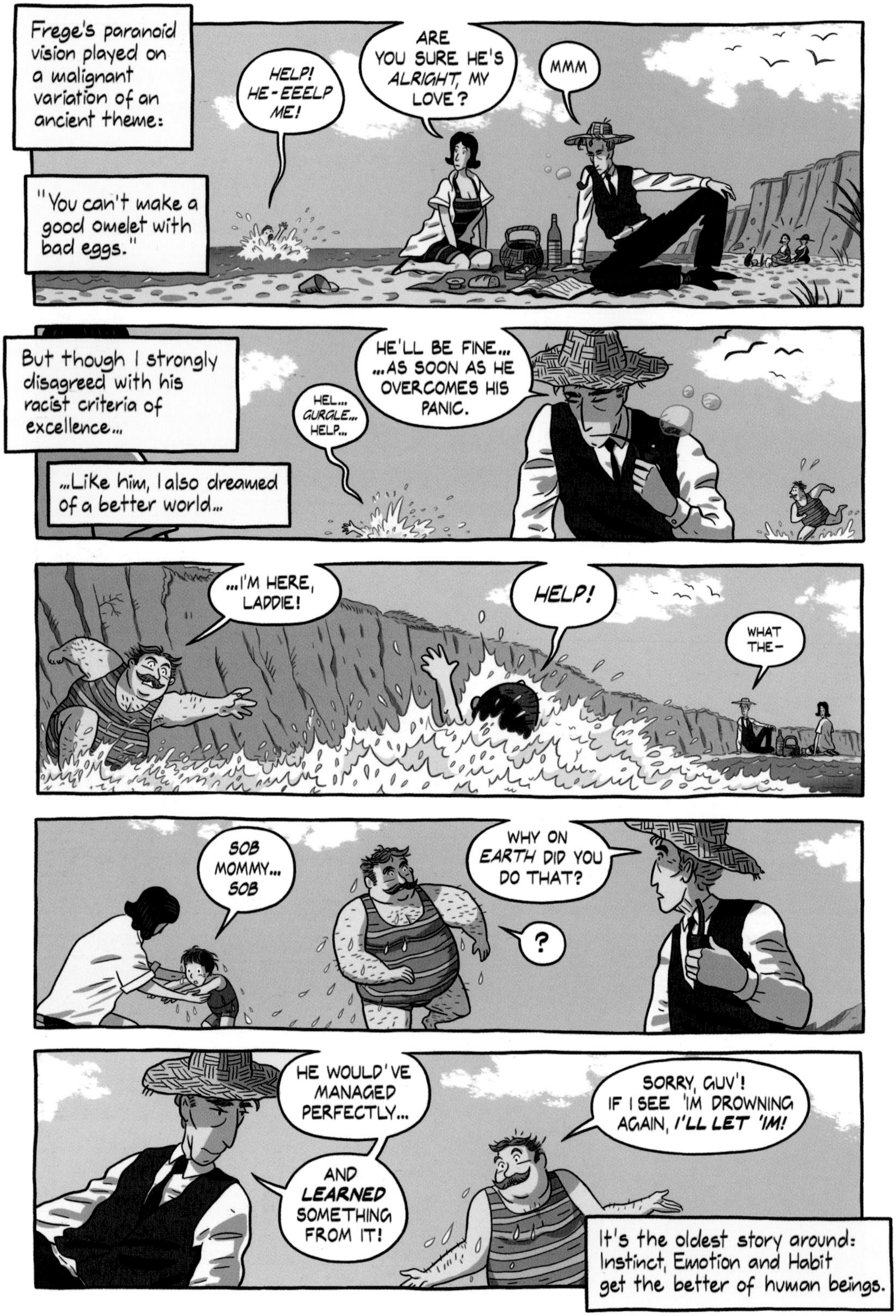
Frege's paranoid vision played on a malignant variation of an ancient theme:
"You can't make a good omelet with bad eggs."
HELP! HE-EEELP ME!
ARE YOU SURE HE'S *ALRIGHT*, MY LOVE?
MMM
But though I strongly disagreed with his racist criteria of excellence...
...Like him, I also dreamed of a better world...
HEL... GURGLE... HELP...
HE'LL BE FINE... ...AS SOON AS HE OVERCOMES HIS PANIC.
...I'M HERE, LADDIE!
HELP!
WHAT THE-
SOB MOMMY... *SOB*
WHY ON *EARTH* DID YOU DO THAT?
?
HE WOULD'VE MANAGED PERFECTLY...
AND ***LEARNED*** SOMETHING FROM IT!
SORRY, GUV'! IF I SEE 'IM DROWNING AGAIN, ***I'LL LET 'IM!***
It's the oldest story around: Instinct, Emotion and Habit get the better of human beings.

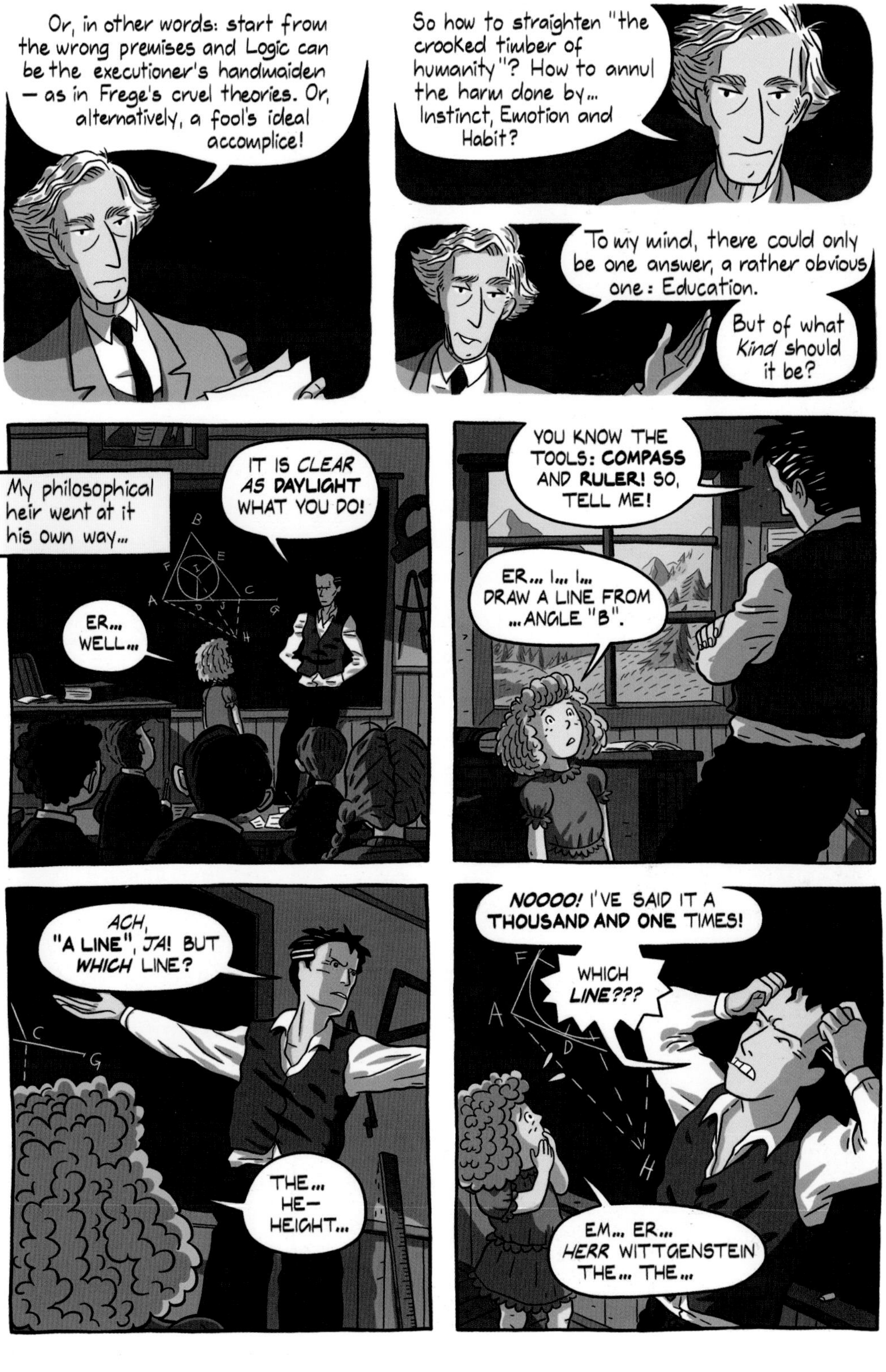
Or, in other words: start from the wrong premises and Logic can be the executioner's handmaiden — as in Frege's cruel theories. Or, alternatively, a fool's ideal accomplice!
So how to straighten "the crooked timber of humanity"? How to annul the harm done by... Instinct, Emotion and Habit?
To my mind, there could only be one answer, a rather obvious one: Education.
But of what *Kind* should it be?
My philosophical heir went at it his own way...
IT IS *CLEAR AS* **DAYLIGHT** WHAT YOU DO!
ER... WELL...
YOU KNOW THE TOOLS: **COMPASS** AND **RULER!** SO, TELL ME!
ER... I... I... DRAW A LINE FROM ...ANGLE "B".
ACH, **"A LINE"**, *JA!* BUT ***WHICH*** LINE?
THE... HE— HEIGHT...
NOOOO! I'VE SAID IT A **THOUSAND AND ONE** TIMES!
WHICH ***LINE???***
EM... ER... *HERR* WITTGENSTEIN THE... THE...

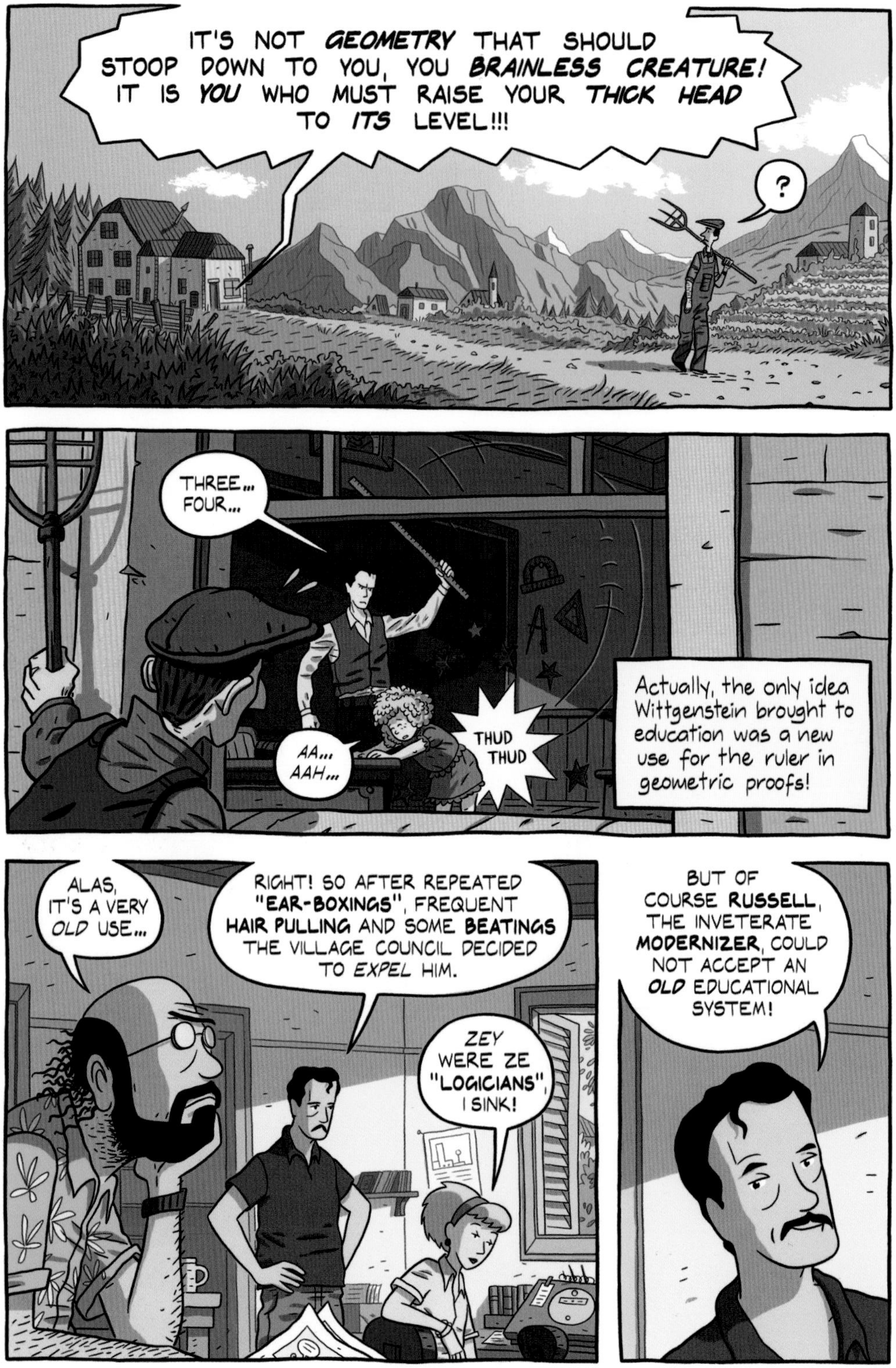
IT'S NOT GEOMETRY THAT SHOULD STOOP DOWN TO YOU, YOU BRAINLESS CREATURE! IT IS YOU WHO MUST RAISE YOUR THICK HEAD TO ITS LEVEL!!!
?
THREE... FOUR...
AA... AAH...
THUD THUD
Actually, the only idea Wittgenstein brought to education was a new use for the ruler in geometric proofs!
ALAS, IT'S A VERY OLD USE...
RIGHT! SO AFTER REPEATED "EAR-BOXINGS", FREQUENT HAIR PULLING AND SOME BEATINGS THE VILLAGE COUNCIL DECIDED TO EXPEL HIM.
ZEY WERE ZE "LOGICIANS", I SINK!
BUT OF COURSE RUSSELL, THE INVETERATE MODERNIZER, COULD NOT ACCEPT AN OLD EDUCATIONAL SYSTEM!

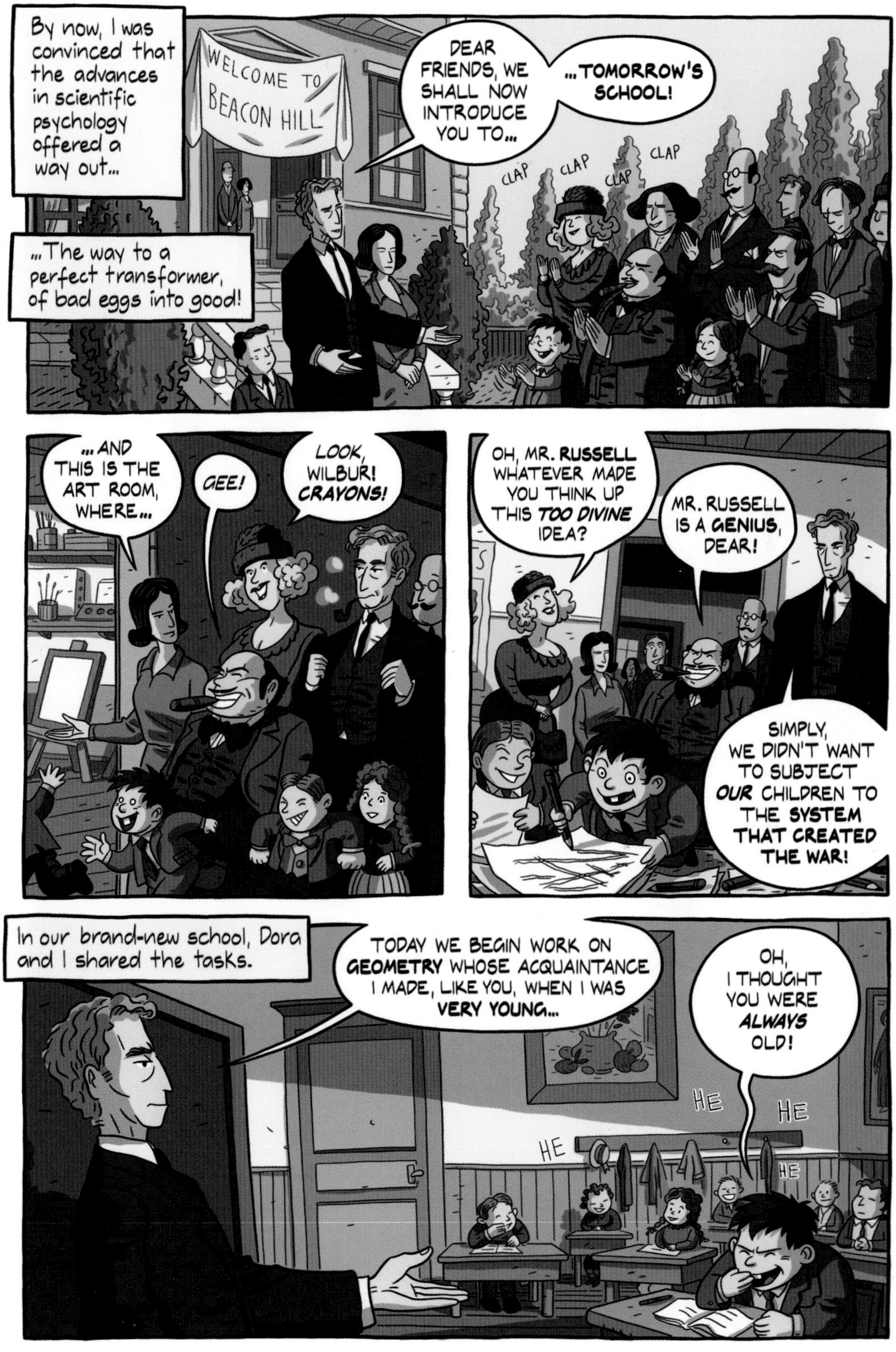

By now, I was convinced that the advances in scientific psychology offered a way out...
WELCOME TO BEACON HILL
DEAR FRIENDS, WE SHALL NOW INTRODUCE YOU TO...
...TOMORROW'S SCHOOL!
CLAP CLAP CLAP CLAP
...The way to a perfect transformer, of bad eggs into good!
...AND THIS IS THE ART ROOM, WHERE...
GEE!
LOOK, WILBUR! CRAYONS!
OH, MR. RUSSELL WHATEVER MADE YOU THINK UP THIS TOO DIVINE IDEA?
MR. RUSSELL IS A GENIUS, DEAR!
SIMPLY, WE DIDN'T WANT TO SUBJECT OUR CHILDREN TO THE SYSTEM THAT CREATED THE WAR!
In our brand-new school, Dora and I shared the tasks.
TODAY WE BEGIN WORK ON GEOMETRY WHOSE ACQUAINTANCE I MADE, LIKE YOU, WHEN I WAS VERY YOUNG...
OH, I THOUGHT YOU WERE ALWAYS OLD!
HE HE HE HE

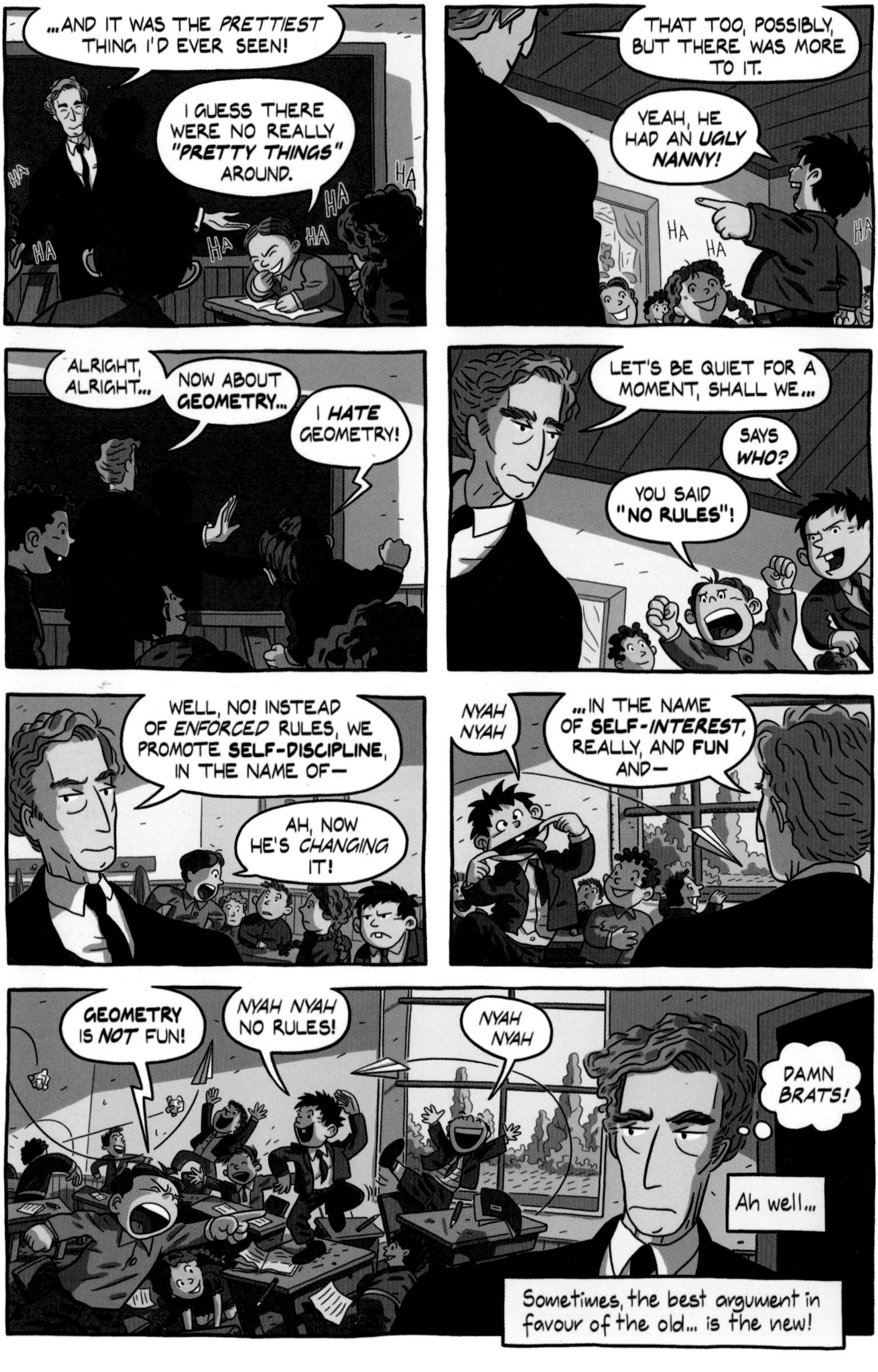
...AND IT WAS THE PRETTIEST THING I'D EVER SEEN!
I GUESS THERE WERE NO REALLY "PRETTY THINGS" AROUND.
HA HA HA HA HA HA HA
THAT TOO, POSSIBLY, BUT THERE WAS MORE TO IT.
YEAH, HE HAD AN UGLY NANNY!
HA HA HA
ALRIGHT, ALRIGHT...
NOW ABOUT GEOMETRY...
I HATE GEOMETRY!
LET'S BE QUIET FOR A MOMENT, SHALL WE...
SAYS WHO?
YOU SAID "NO RULES"!
WELL, NO! INSTEAD OF ENFORCED RULES, WE PROMOTE SELF-DISCIPLINE, IN THE NAME OF—
AH, NOW HE'S CHANGING IT!
NYAH NYAH
...IN THE NAME OF SELF-INTEREST, REALLY, AND FUN AND—
GEOMETRY IS NOT FUN!
NYAH NYAH NO RULES!
NYAH NYAH
DAMN BRATS!
Ah well...
Sometimes, the best argument in favour of the old... is the new!

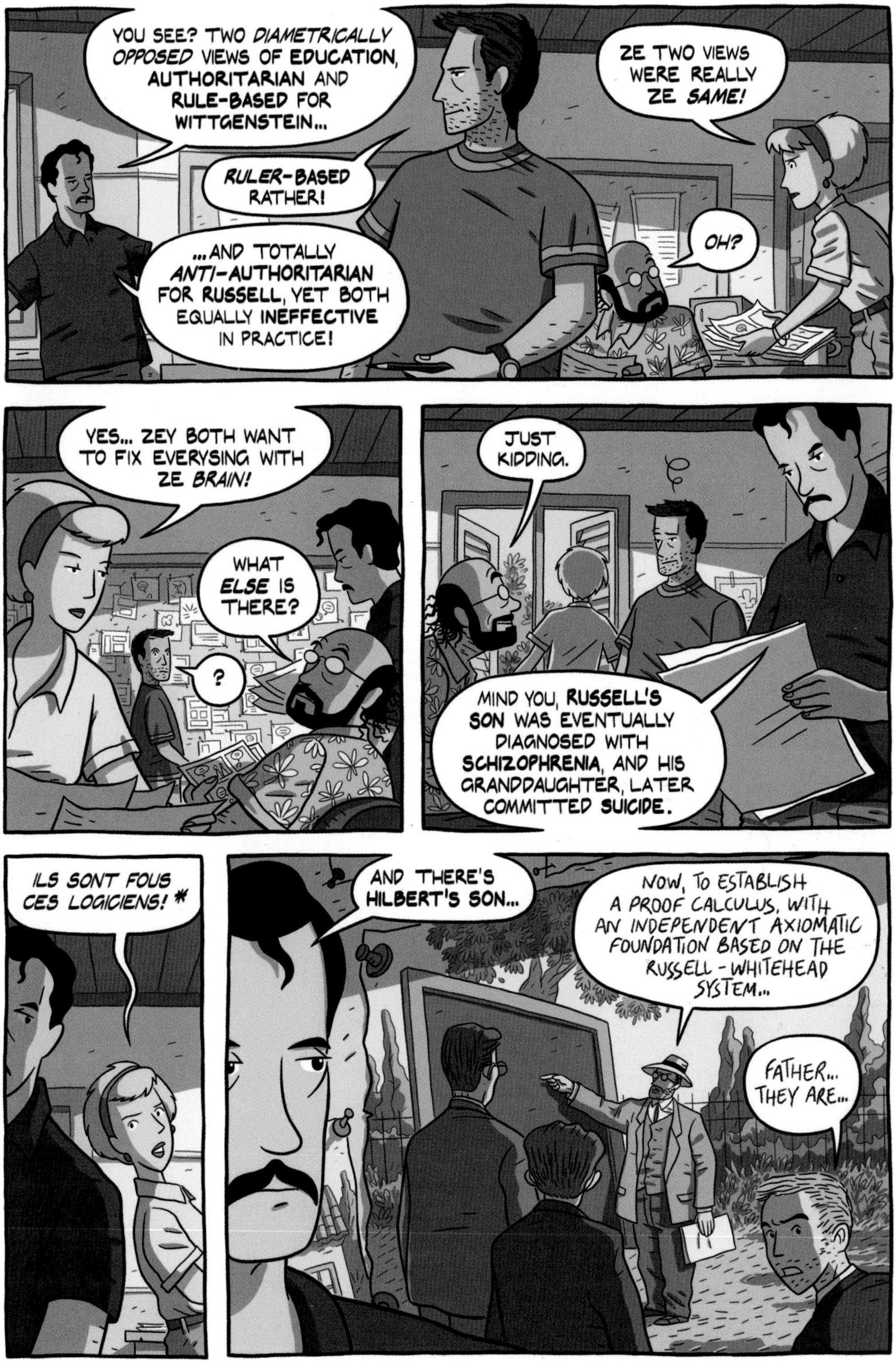
YOU SEE? TWO DIAMETRICALLY OPPOSED VIEWS OF EDUCATION, AUTHORITARIAN AND RULE-BASED FOR WITTGENSTEIN...
RULER-BASED RATHER!
...AND TOTALLY ANTI-AUTHORITARIAN FOR RUSSELL, YET BOTH EQUALLY INEFFECTIVE IN PRACTICE!
ZE TWO VIEWS WERE REALLY ZE SAME!
OH?
YES... ZEY BOTH WANT TO FIX EVERYSING WITH ZE BRAIN!
WHAT ELSE IS THERE?
?
JUST KIDDING.
MIND YOU, RUSSELL'S SON WAS EVENTUALLY DIAGNOSED WITH SCHIZOPHRENIA, AND HIS GRANDDAUGHTER, LATER COMMITTED SUICIDE.
ILS SONT FOUS CES LOGICIENS! *
AND THERE'S HILBERT'S SON...
NOW, TO ESTABLISH A PROOF CALCULUS, WITH AN INDEPENDENT AXIOMATIC FOUNDATION BASED ON THE RUSSELL-WHITEHEAD SYSTEM...
FATHER... THEY ARE...

*Zey are crazy zese logicians!

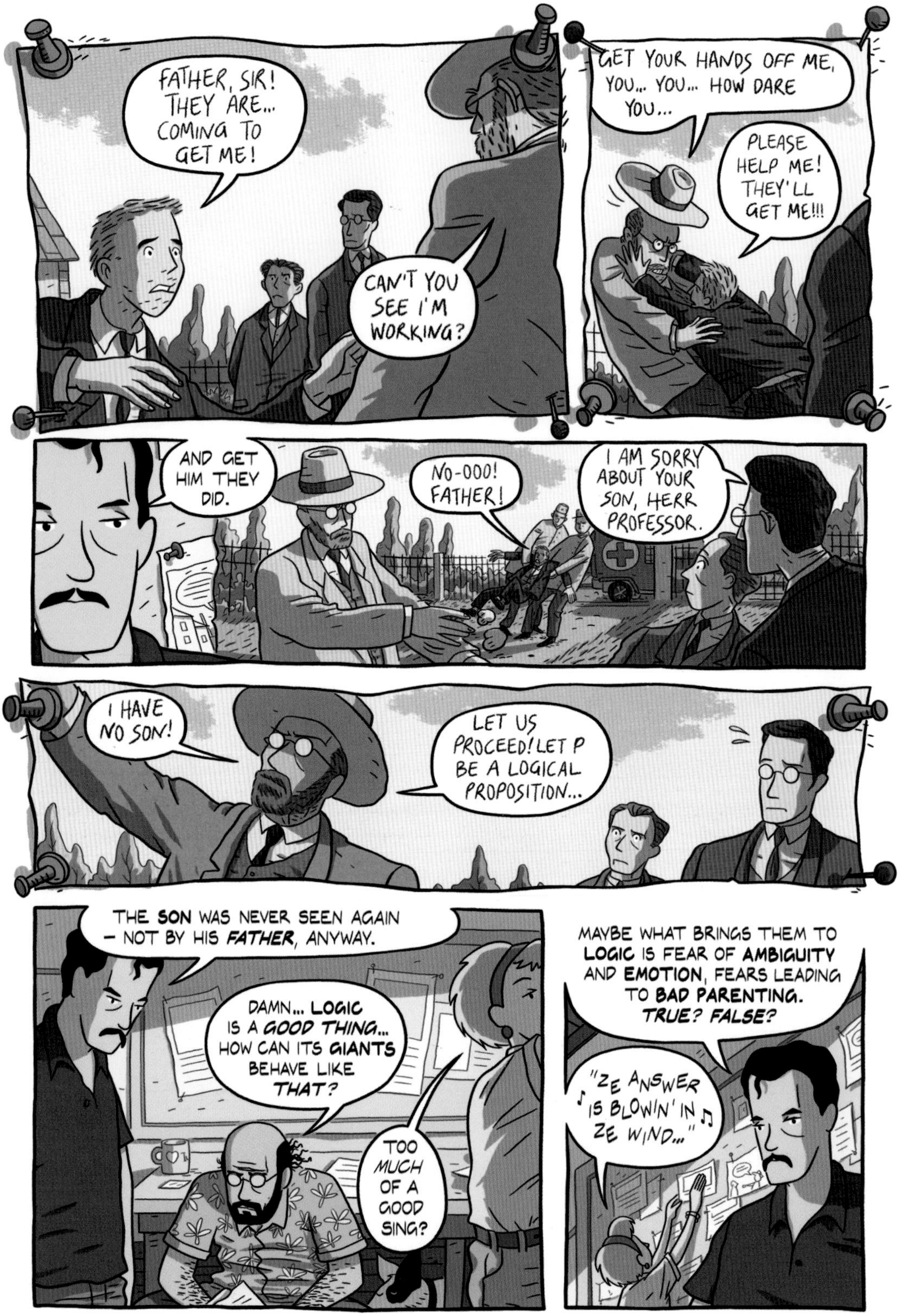
FATHER, SIR! THEY ARE... COMING TO GET ME!
CAN'T YOU SEE I'M WORKING?
GET YOUR HANDS OFF ME, YOU... YOU... HOW DARE YOU...
PLEASE HELP ME! THEY'LL GET ME!!!
AND GET HIM THEY DID.
NO-OOO! FATHER!
I AM SORRY ABOUT YOUR SON, HERR PROFESSOR.
I HAVE NO SON!
LET US PROCEED! LET P BE A LOGICAL PROPOSITION...
THE SON WAS NEVER SEEN AGAIN – NOT BY HIS FATHER, ANYWAY.
DAMN... LOGIC IS A GOOD THING... HOW CAN ITS GIANTS BEHAVE LIKE THAT?
TOO MUCH OF A GOOD SING?
MAYBE WHAT BRINGS THEM TO LOGIC IS FEAR OF AMBIGUITY AND EMOTION, FEARS LEADING TO BAD PARENTING. TRUE? FALSE?
♪ "ZE ANSWER IS BLOWIN' IN ZE WIND..." ♪

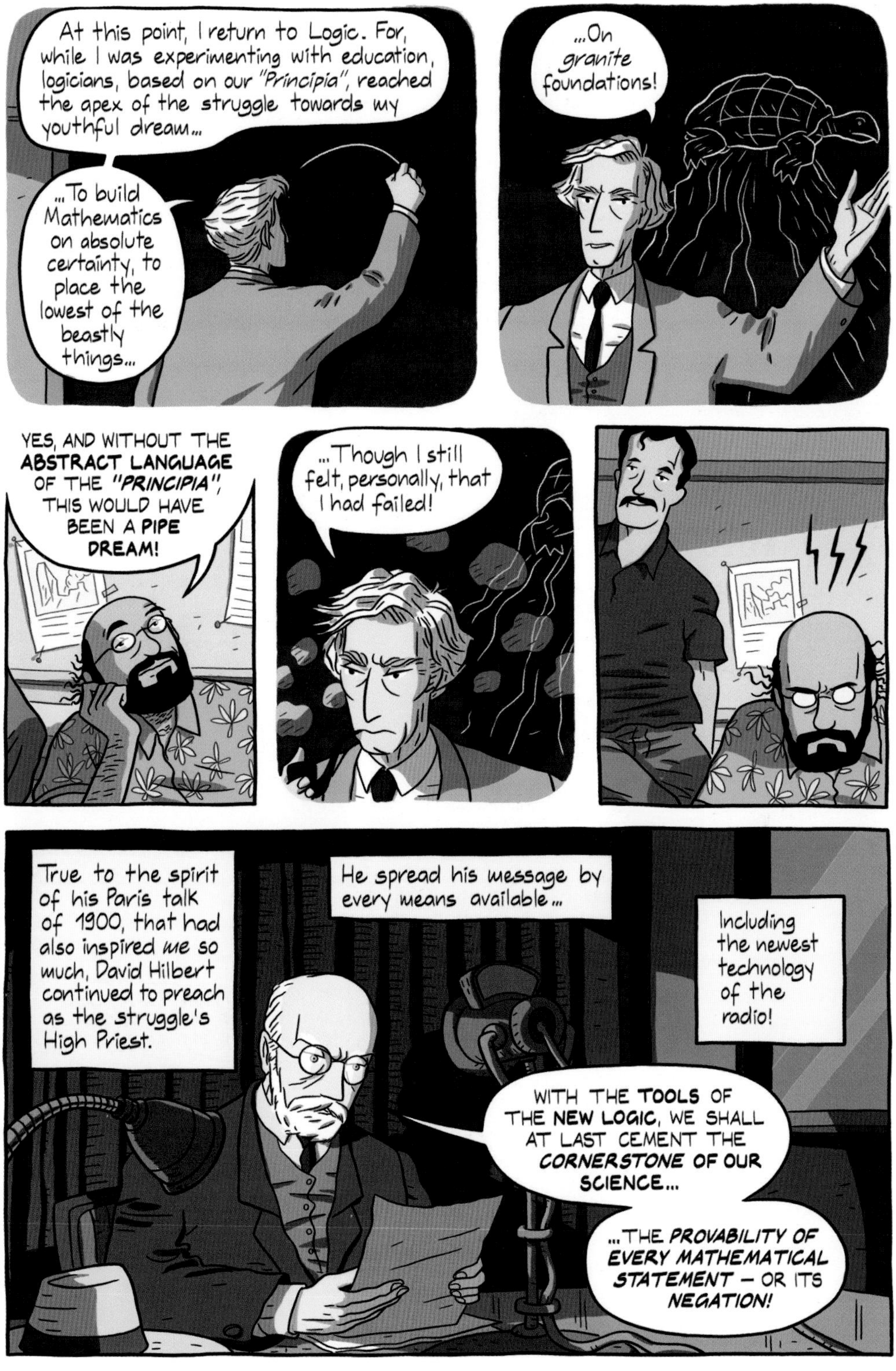
At this point, I return to Logic. For, while I was experimenting with education, logicians, based on our "Principia", reached the apex of the struggle towards my youthful dream...
...To build Mathematics on absolute certainty, to place the lowest of the beastly things...
...On granite foundations!
YES, AND WITHOUT THE **ABSTRACT LANGUAGE** OF THE ***"PRINCIPIA"***, THIS WOULD HAVE BEEN A **PIPE DREAM!**
...Though I still felt, personally, that I had failed!
True to the spirit of his Paris talk of 1900, that had also inspired *me* so much, David Hilbert continued to preach as the struggle's High Priest.
He spread his message by every means available...
Including the newest technology of the radio!
WITH THE **TOOLS** OF THE **NEW LOGIC**, WE SHALL AT LAST CEMENT THE ***CORNERSTONE* OF OUR SCIENCE...**
...THE ***PROVABILITY OF EVERY MATHEMATICAL STATEMENT*** — OR ITS ***NEGATION!***

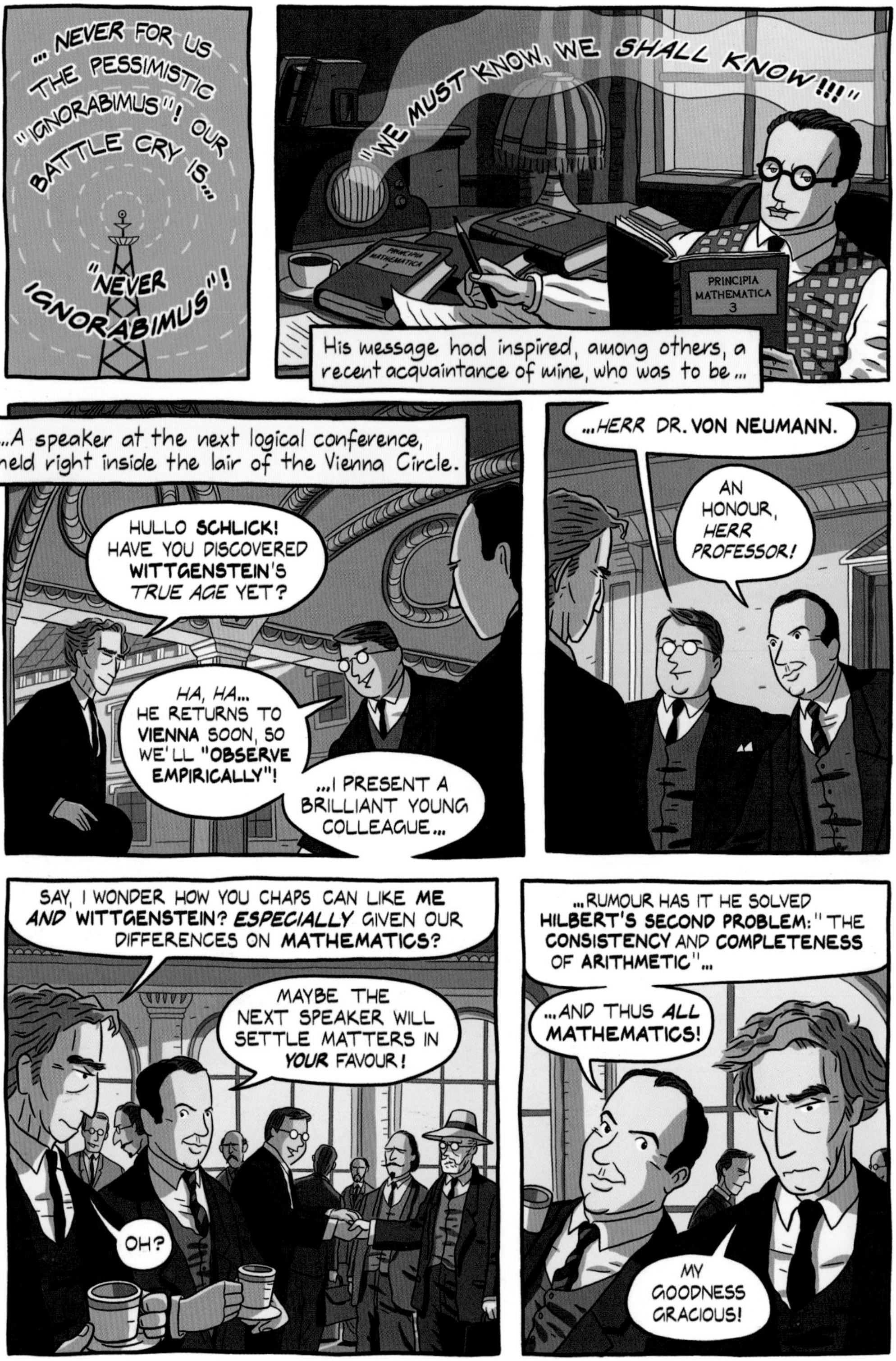
... NEVER FOR US THE PESSIMISTIC "IGNORABIMUS"! OUR BATTLE CRY IS...
"NEVER IGNORABIMUS"!
"WE MUST KNOW, WE SHALL KNOW!!!"
PRINCIPIA MATHEMATICA 3
His message had inspired, among others, a recent acquaintance of mine, who was to be...
...A speaker at the next logical conference, held right inside the lair of the Vienna Circle.
HULLO SCHLICK! HAVE YOU DISCOVERED WITTGENSTEIN'S TRUE AGE YET?
HA, HA... HE RETURNS TO VIENNA SOON, SO WE'LL "OBSERVE EMPIRICALLY"!
...I PRESENT A BRILLIANT YOUNG COLLEAGUE...
...HERR DR. VON NEUMANN.
AN HONOUR, HERR PROFESSOR!
SAY, I WONDER HOW YOU CHAPS CAN LIKE ME AND WITTGENSTEIN? ESPECIALLY GIVEN OUR DIFFERENCES ON MATHEMATICS?
MAYBE THE NEXT SPEAKER WILL SETTLE MATTERS IN YOUR FAVOUR!
OH?
...RUMOUR HAS IT HE SOLVED HILBERT'S SECOND PROBLEM: "THE CONSISTENCY AND COMPLETENESS OF ARITHMETIC"...
...AND THUS ALL MATHEMATICS!
MY GOODNESS GRACIOUS!

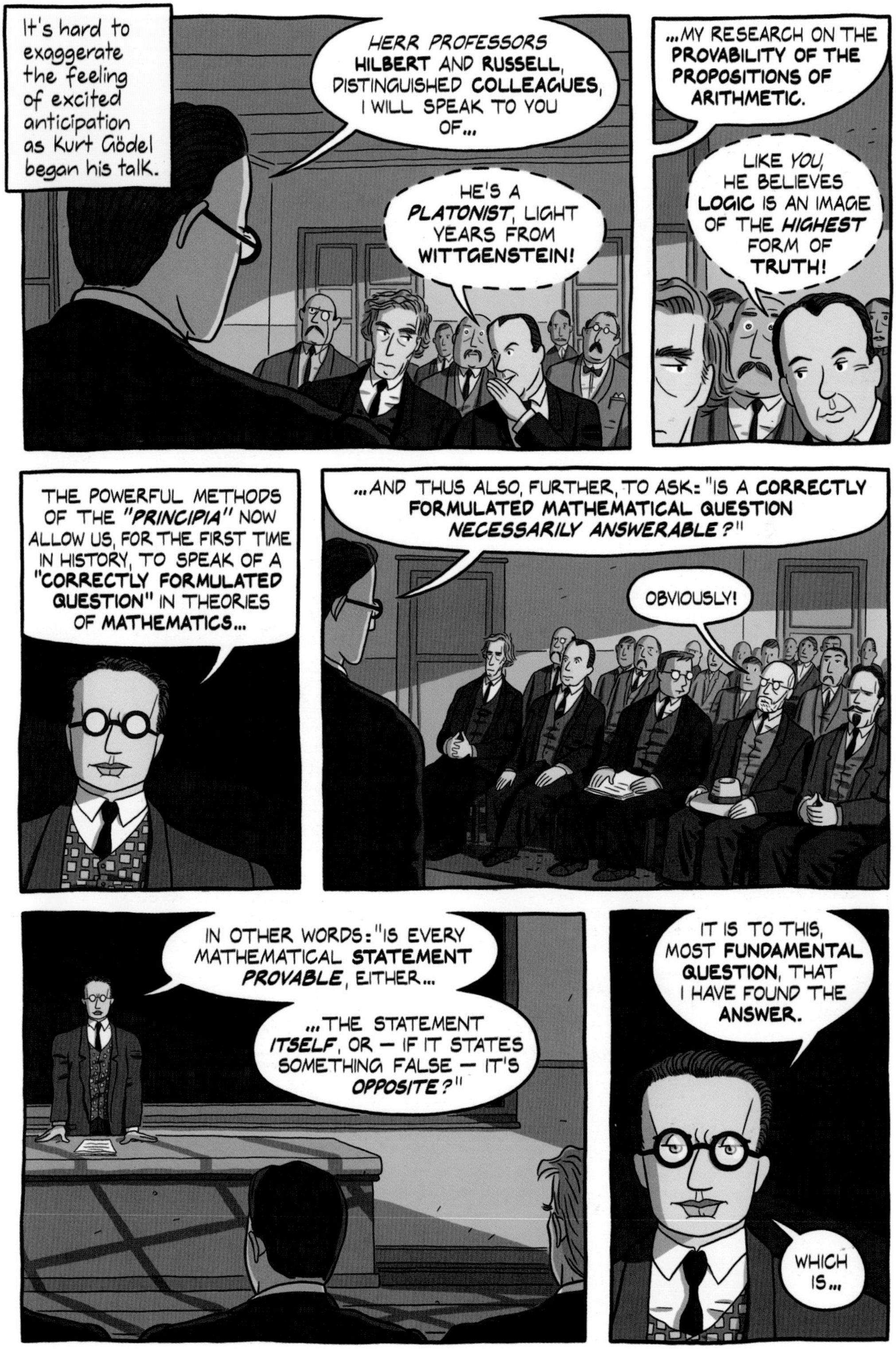
It's hard to exaggerate the feeling of excited anticipation as Kurt Gödel began his talk.
HERR PROFESSORS HILBERT AND RUSSELL, DISTINGUISHED COLLEAGUES, I WILL SPEAK TO YOU OF...
HE'S A PLATONIST, LIGHT YEARS FROM WITTGENSTEIN!
...MY RESEARCH ON THE PROVABILITY OF THE PROPOSITIONS OF ARITHMETIC.
LIKE YOU, HE BELIEVES LOGIC IS AN IMAGE OF THE HIGHEST FORM OF TRUTH!
THE POWERFUL METHODS OF THE "PRINCIPIA" NOW ALLOW US, FOR THE FIRST TIME IN HISTORY, TO SPEAK OF A "CORRECTLY FORMULATED QUESTION" IN THEORIES OF MATHEMATICS...
...AND THUS ALSO, FURTHER, TO ASK: "IS A CORRECTLY FORMULATED MATHEMATICAL QUESTION NECESSARILY ANSWERABLE?"
OBVIOUSLY!
IN OTHER WORDS: "IS EVERY MATHEMATICAL STATEMENT PROVABLE, EITHER...
...THE STATEMENT ITSELF, OR — IF IT STATES SOMETHING FALSE — IT'S OPPOSITE?"
IT IS TO THIS, MOST FUNDAMENTAL QUESTION, THAT I HAVE FOUND THE ANSWER.
WHICH IS...

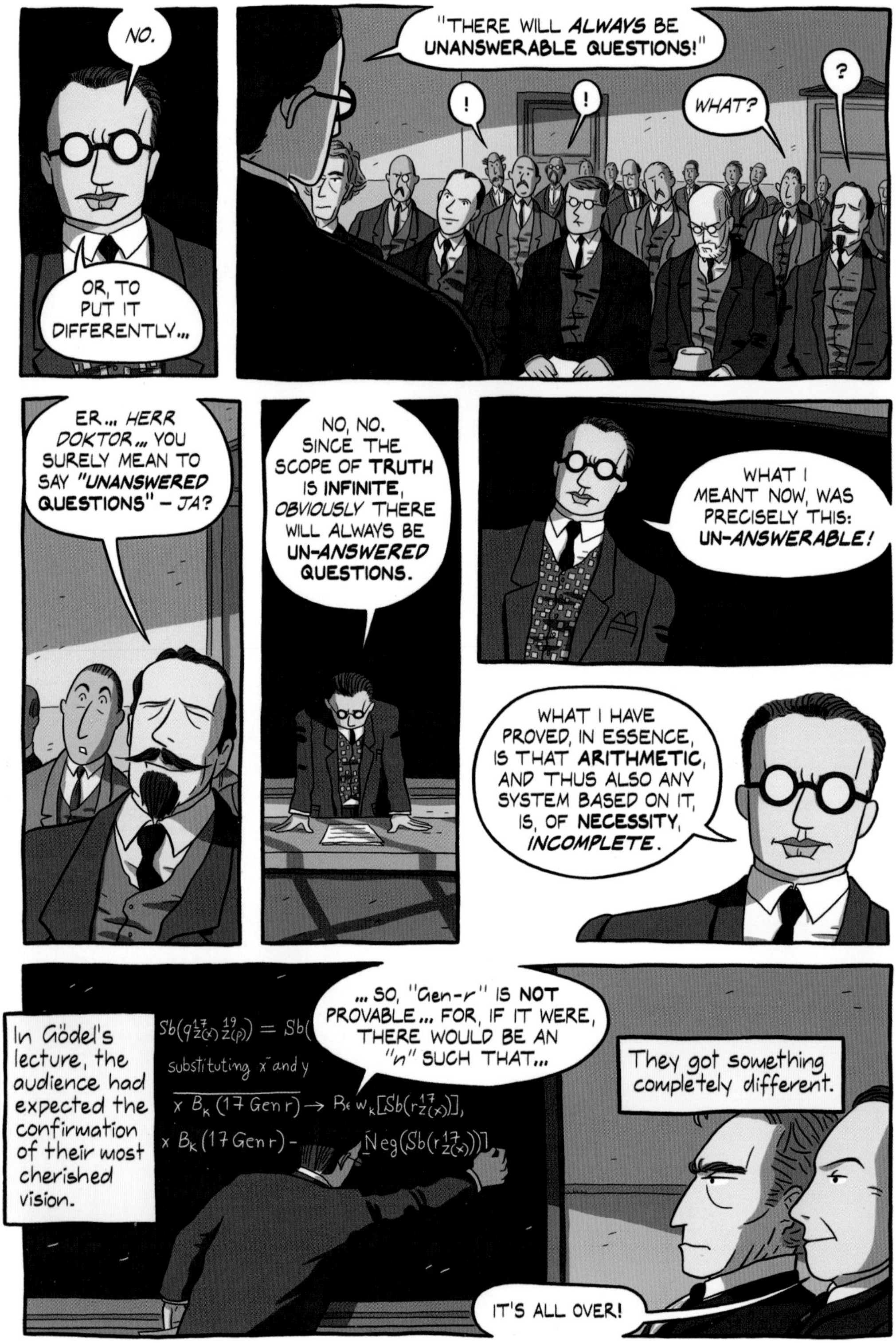
NO.
OR, TO PUT IT DIFFERENTLY...
"THERE WILL *ALWAYS* BE UNANSWERABLE QUESTIONS!"
!
!
WHAT?
?
ER... *HERR DOKTOR*... YOU SURELY MEAN TO SAY **"UNANSWERED QUESTIONS"** – *JA*?
NO, NO. SINCE THE SCOPE OF **TRUTH** IS **INFINITE**, *OBVIOUSLY* THERE WILL ALWAYS BE **UN-*ANSWERED* QUESTIONS.**
WHAT I MEANT NOW, WAS PRECISELY THIS: **UN-*ANSWERABLE!***
WHAT I HAVE PROVED, IN ESSENCE, IS THAT **ARITHMETIC**, AND THUS ALSO ANY SYSTEM BASED ON IT, IS, OF **NECESSITY**, ***INCOMPLETE***.
In Gödel's lecture, the audience had expected the confirmation of their most cherished vision.
... SO, "Gen-*r*" IS **NOT** PROVABLE... FOR, IF IT WERE, THERE WOULD BE AN "*n*" SUCH THAT...
They got something completely different.
IT'S ALL OVER!

..."All over!" Von Neumann's comment perfectly sums-up the essence of Gödel's proof.

I know it may be hard for laypersons to understand...

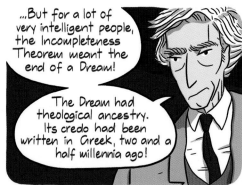

...But for a lot of very intelligent people, the Incompleteness Theorem meant the end of a Dream!

The Dream had theological ancestry. Its credo had been written in Greek, two and a half millennia ago!

And now, suddenly, the rug had been pulled from under the feet of the dreamers.

HERR PROFESSOR... WOULD YOU LIKE US TO TAKE YOU TO YOUR HOTEL?

That is the beauty, that is the terror of Mathematics...

LET HIM BE...

There's no getting round a proof...

...Even if it proves that something is unprovable!

SO, THEN... "IT'S OVER", EH?

WELL, THAT WAS VON NEUMANN'S *FIRST REACTION...* BUT THINGS LATER TURNED OUT *VERY* DIFFERENTLY!

THE JOURNEY THROUGH **ABSTRACT THOUGHT**, FROM **ARISTOTLE**, VIA **BOOLE**, ALL THE WAY TO **GÖDEL'S THEOREM**, IN EFFECT LED TO A *NEW BEGINNING*, WHICH—

LOOK, BEFORE WE EVEN *THINK* OF MENTIONING "NEW BEGINNINGS", WE HAVE OUR SERIES OF *OLD ENDINGS* TO COMPLETE!

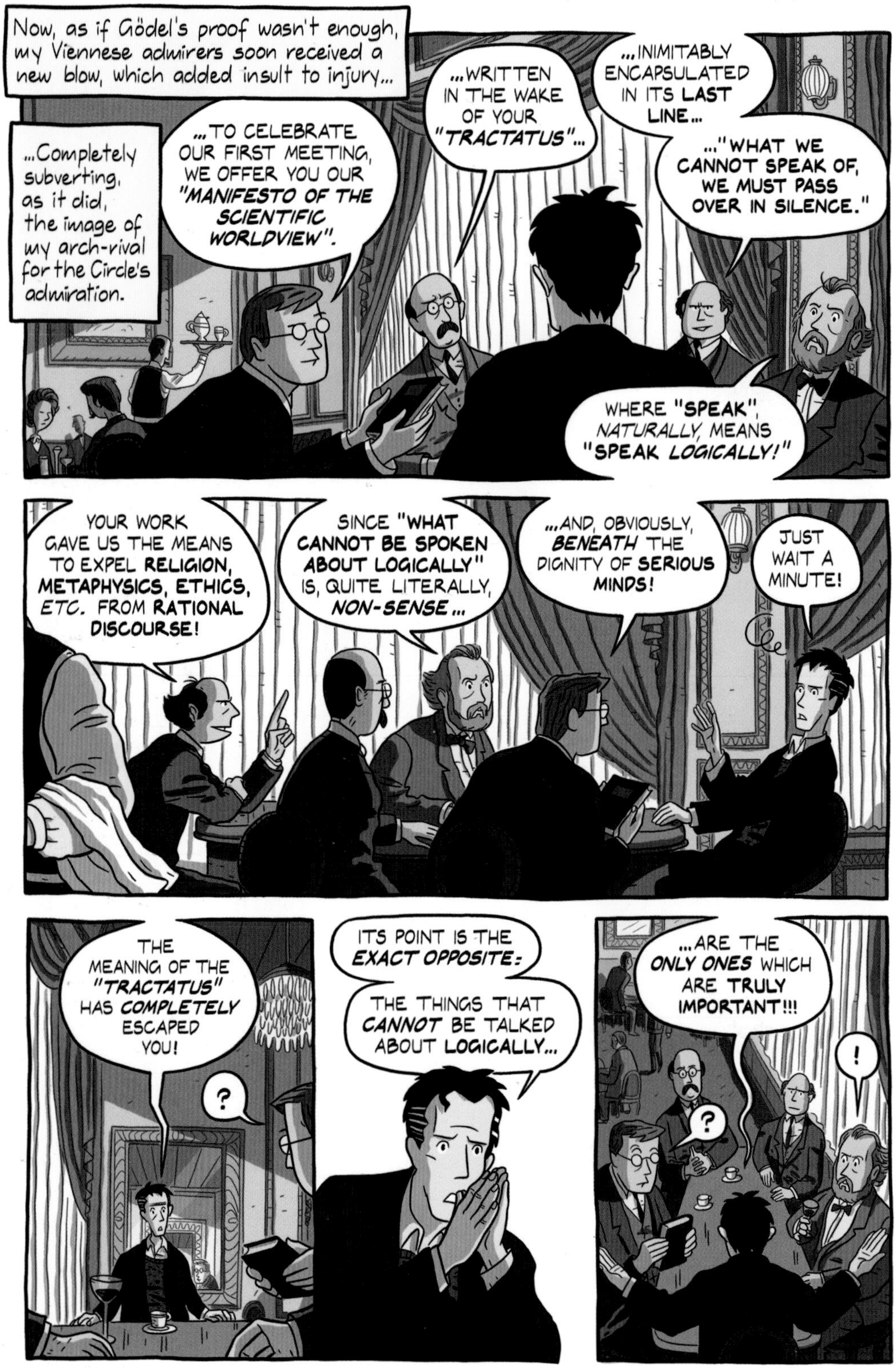

Now, as if Gödel's proof wasn't enough, my Viennese admirers soon received a new blow, which added insult to injury...
...Completely subverting, as it did, the image of my arch-rival for the Circle's admiration.
...TO CELEBRATE OUR FIRST MEETING, WE OFFER YOU OUR "MANIFESTO OF THE SCIENTIFIC WORLDVIEW".
...WRITTEN IN THE WAKE OF YOUR "TRACTATUS"...
...INIMITABLY ENCAPSULATED IN ITS LAST LINE...
..."WHAT WE CANNOT SPEAK OF, WE MUST PASS OVER IN SILENCE."
WHERE "SPEAK", NATURALLY, MEANS "SPEAK LOGICALLY!"
YOUR WORK GAVE US THE MEANS TO EXPEL RELIGION, METAPHYSICS, ETHICS, ETC. FROM RATIONAL DISCOURSE!
SINCE "WHAT CANNOT BE SPOKEN ABOUT LOGICALLY" IS, QUITE LITERALLY, NON-SENSE...
...AND, OBVIOUSLY, BENEATH THE DIGNITY OF SERIOUS MINDS!
JUST WAIT A MINUTE!
THE MEANING OF THE "TRACTATUS" HAS COMPLETELY ESCAPED YOU!
?
ITS POINT IS THE EXACT OPPOSITE:
THE THINGS THAT CANNOT BE TALKED ABOUT LOGICALLY...
...ARE THE ONLY ONES WHICH ARE TRULY IMPORTANT!!!
?
!

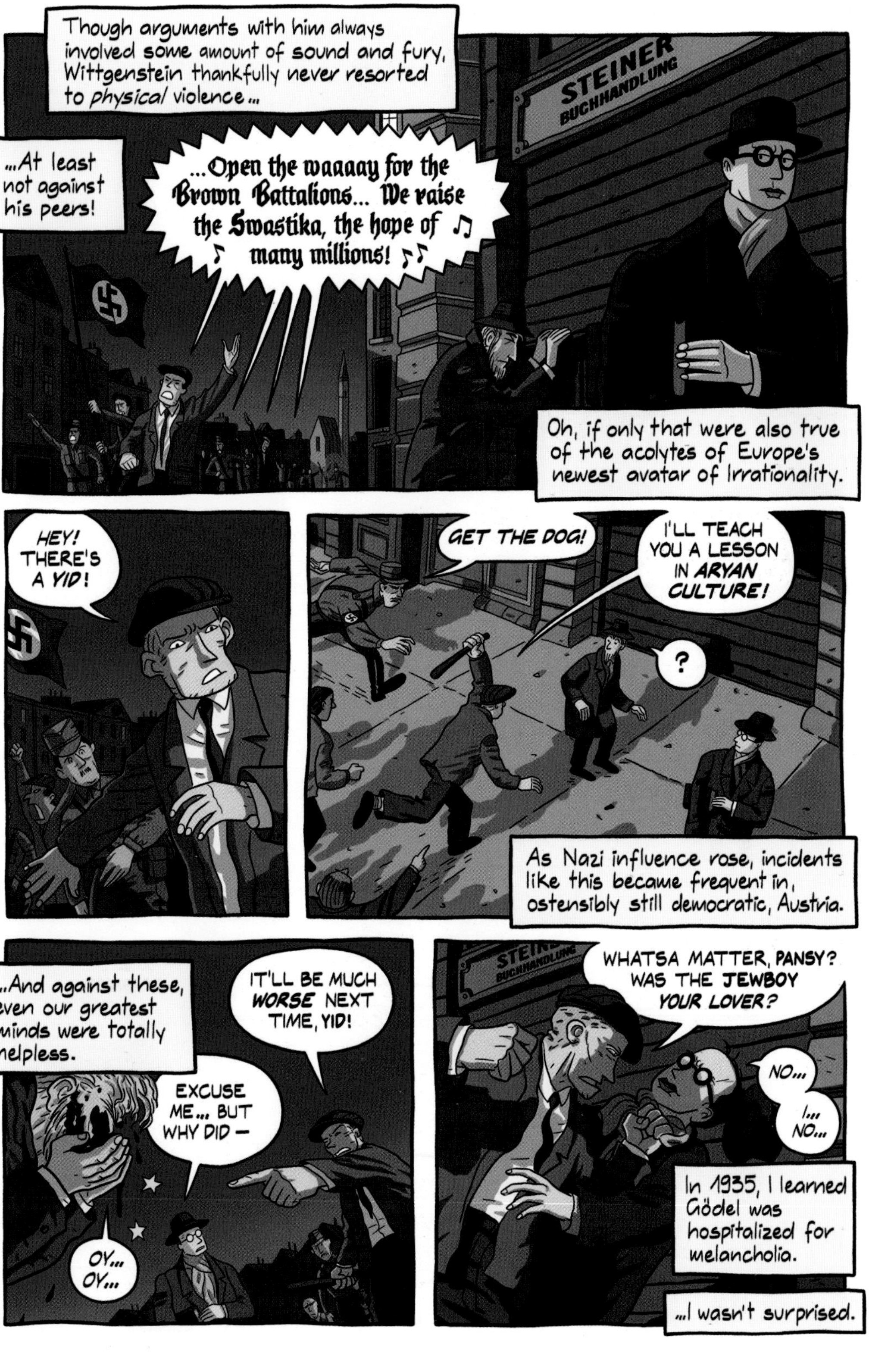

Though arguments with him always involved some amount of sound and fury, Wittgenstein thankfully never resorted to *physical* violence...
STEINER BUCHHANDLUNG
...At least not against his peers!
...Open the waaaay for the Brown Battalions... We raise the Swastika, the hope of ♫ ♪ many millions! ♪♪
Oh, if only that were also true of the acolytes of Europe's newest avatar of Irrationality.
HEY! THERE'S A YID!
GET THE DOG!
I'LL TEACH YOU A LESSON IN ARYAN CULTURE!
?
As Nazi influence rose, incidents like this became frequent in, ostensibly still democratic, Austria.
...And against these, even our greatest minds were totally helpless.
IT'LL BE MUCH WORSE NEXT TIME, YID!
EXCUSE ME... BUT WHY DID —
OY... OY...
STEINER BUCHHANDLUNG
WHATSA MATTER, PANSY? WAS THE JEWBOY YOUR LOVER?
NO...
I... NO...
In 1935, I learned Gödel was hospitalized for melancholia.
...I wasn't surprised.

Nor was I, I'm sorry to say, when the third and worst blow came.

HERR PROFESSOR SCHLICK?

The tragic final act in the short life of the Vienna Circle was played out on June 22, 1936.

I HAVE A **PRESENT** FOR YOU!

WHA—

BANG BANG BANG

The pro-Nazi newspapers declared that the Circle's rationalistic world-view had "desecrated hallowed Germanic values..."

...And thus Schlick deserved what he got.

The killer was a devotee of the new Teutonic god...

...Herr Adolf Hitler.

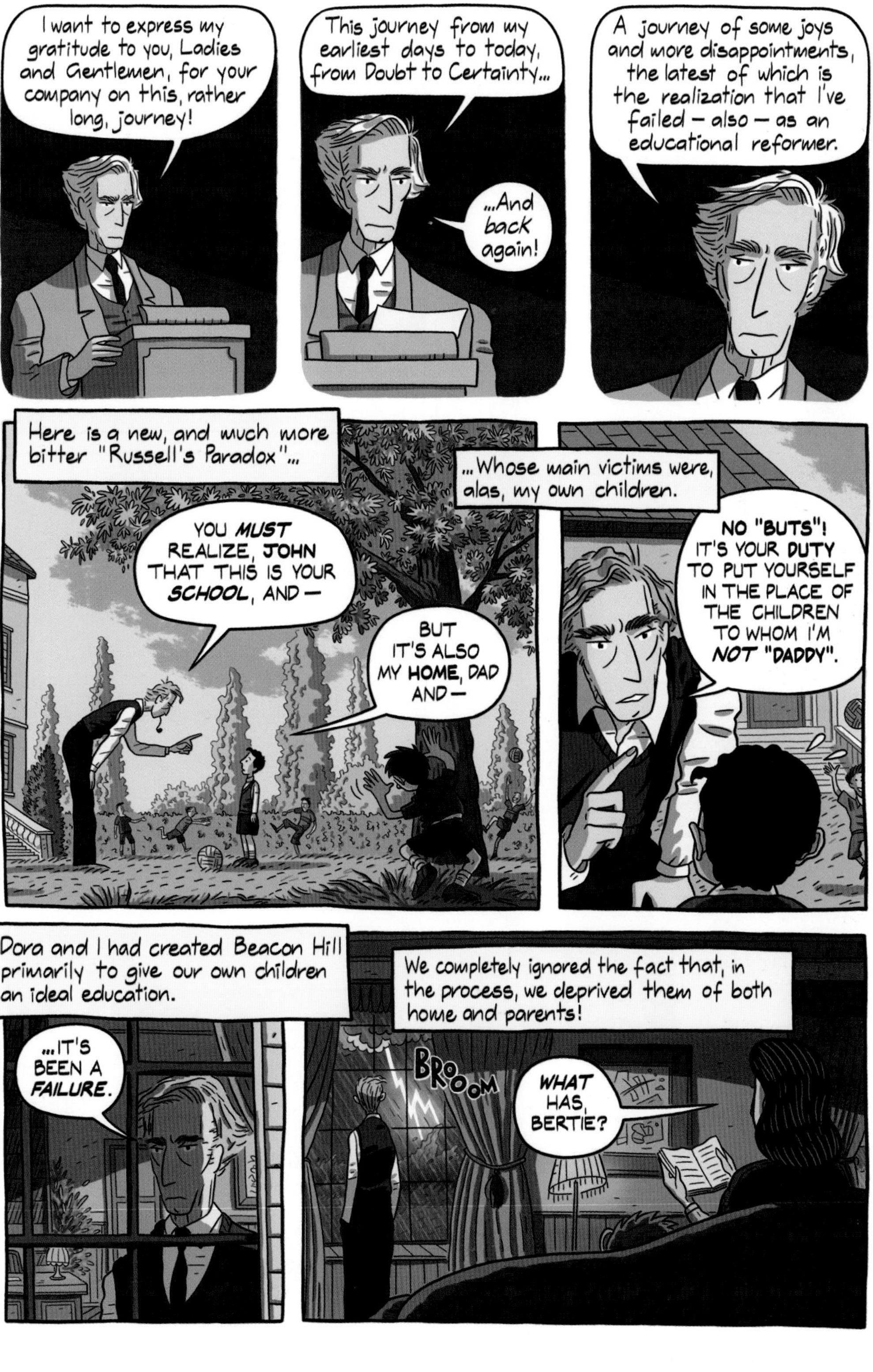
I want to express my gratitude to you, Ladies and Gentlemen, for your company on this, rather long, journey!
This journey from my earliest days to today, from Doubt to Certainty...
...And back again!
A journey of some joys and more disappointments, the latest of which is the realization that I've failed – also – as an educational reformer.
Here is a new, and much more bitter "Russell's Paradox"...
YOU MUST REALIZE, JOHN THAT THIS IS YOUR SCHOOL, AND –
BUT IT'S ALSO MY HOME, DAD AND –
...Whose main victims were, alas, my own children.
NO "BUTS"! IT'S YOUR DUTY TO PUT YOURSELF IN THE PLACE OF THE CHILDREN TO WHOM I'M NOT "DADDY".
Dora and I had created Beacon Hill primarily to give our own children an ideal education.
...IT'S BEEN A FAILURE.
We completely ignored the fact that, in the process, we deprived them of both home and parents!
BROOOM
WHAT HAS, BERTIE?

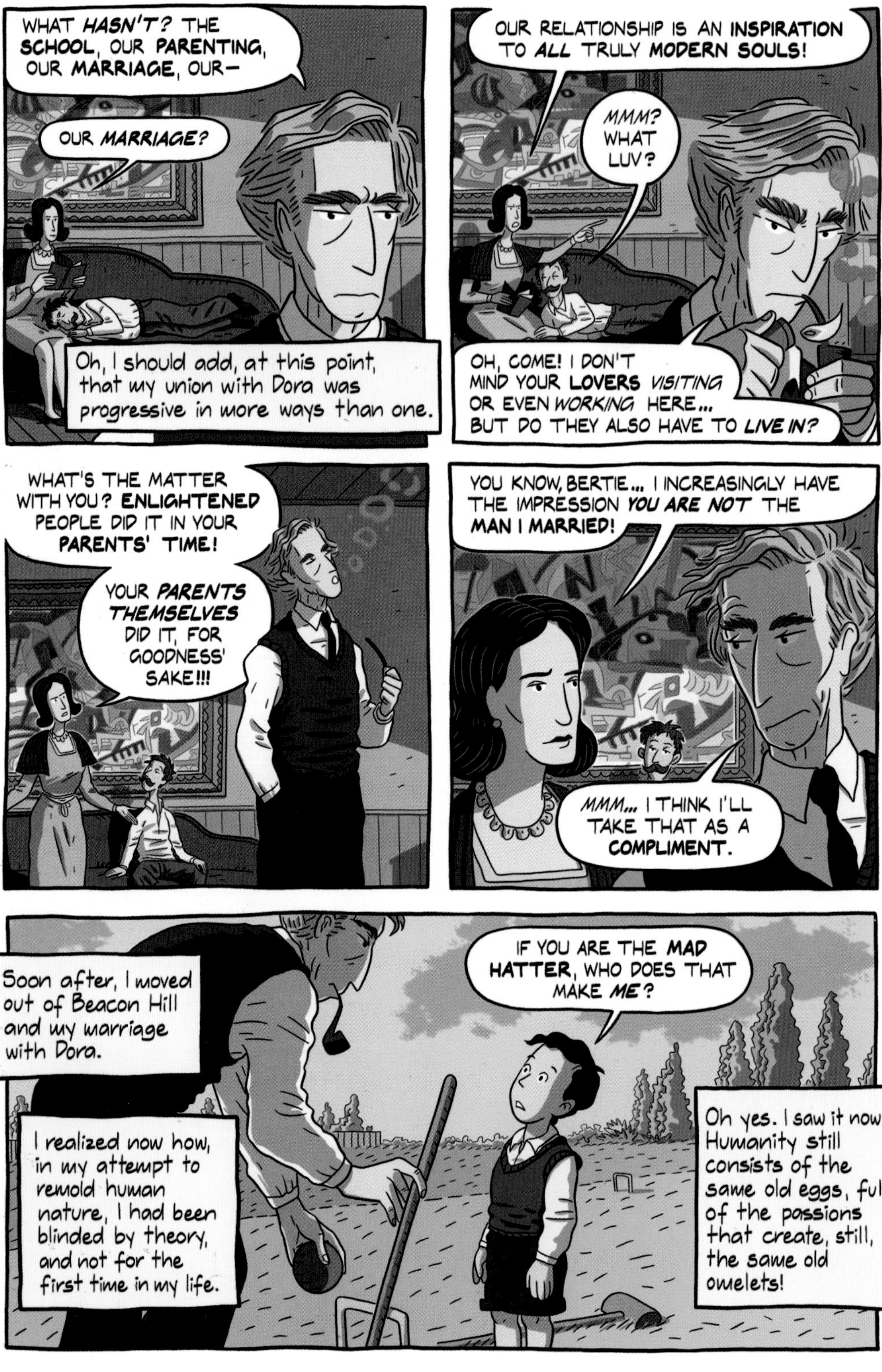
WHAT *HASN'T*? THE **SCHOOL**, OUR **PARENTING**, OUR **MARRIAGE**, OUR—
OUR *MARRIAGE*?
Oh, I should add, at this point, that my union with Dora was progressive in more ways than one.
OUR RELATIONSHIP IS AN **INSPIRATION** TO ***ALL*** TRULY **MODERN SOULS**!
MMM? WHAT LUV?
OH, COME! I DON'T MIND YOUR **LOVERS** *VISITING* OR EVEN *WORKING* HERE... BUT DO THEY ALSO HAVE TO ***LIVE IN***?
WHAT'S THE MATTER WITH YOU? **ENLIGHTENED** PEOPLE DID IT IN YOUR **PARENTS' TIME**!
YOUR ***PARENTS THEMSELVES*** DID IT, FOR GOODNESS' SAKE!!!
YOU KNOW, BERTIE... I INCREASINGLY HAVE THE IMPRESSION ***YOU ARE NOT*** THE **MAN I MARRIED**!
MMM... I THINK I'LL TAKE THAT AS A **COMPLIMENT**.
Soon after, I moved out of Beacon Hill and my marriage with Dora.
I realized now how, in my attempt to remold human nature, I had been blinded by theory, and not for the first time in my life.
IF YOU ARE THE **MAD HATTER**, WHO DOES THAT MAKE ***ME***?
Oh yes. I saw it now. Humanity still consists of the same old eggs, full of the passions that create, still, the same old omelets!

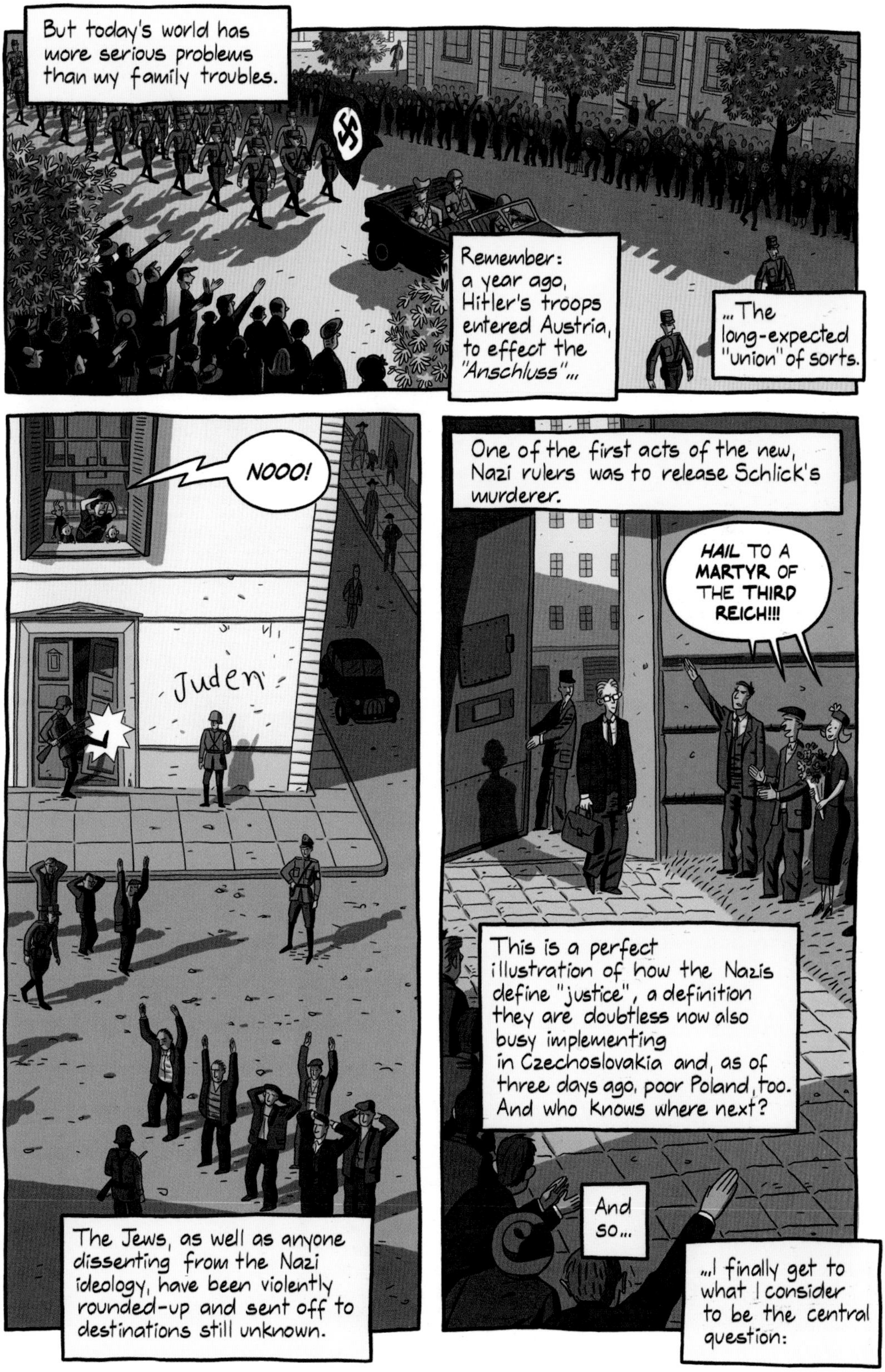
But today's world has more serious problems than my family troubles.
Remember: a year ago, Hitler's troops entered Austria, to effect the "Anschluss"...
...The long-expected "union" of sorts.
NOOO!
Juden
The Jews, as well as anyone dissenting from the Nazi ideology, have been violently rounded-up and sent off to destinations still unknown.
One of the first acts of the new, Nazi rulers was to release Schlick's murderer.
HAIL TO A MARTYR OF THE THIRD REICH!!!
This is a perfect illustration of how the Nazis define "justice", a definition they are doubtless now also busy implementing in Czechoslovakia and, as of three days ago, poor Poland, too. And who knows where next?
And so...
...I finally get to what I consider to be the central question:

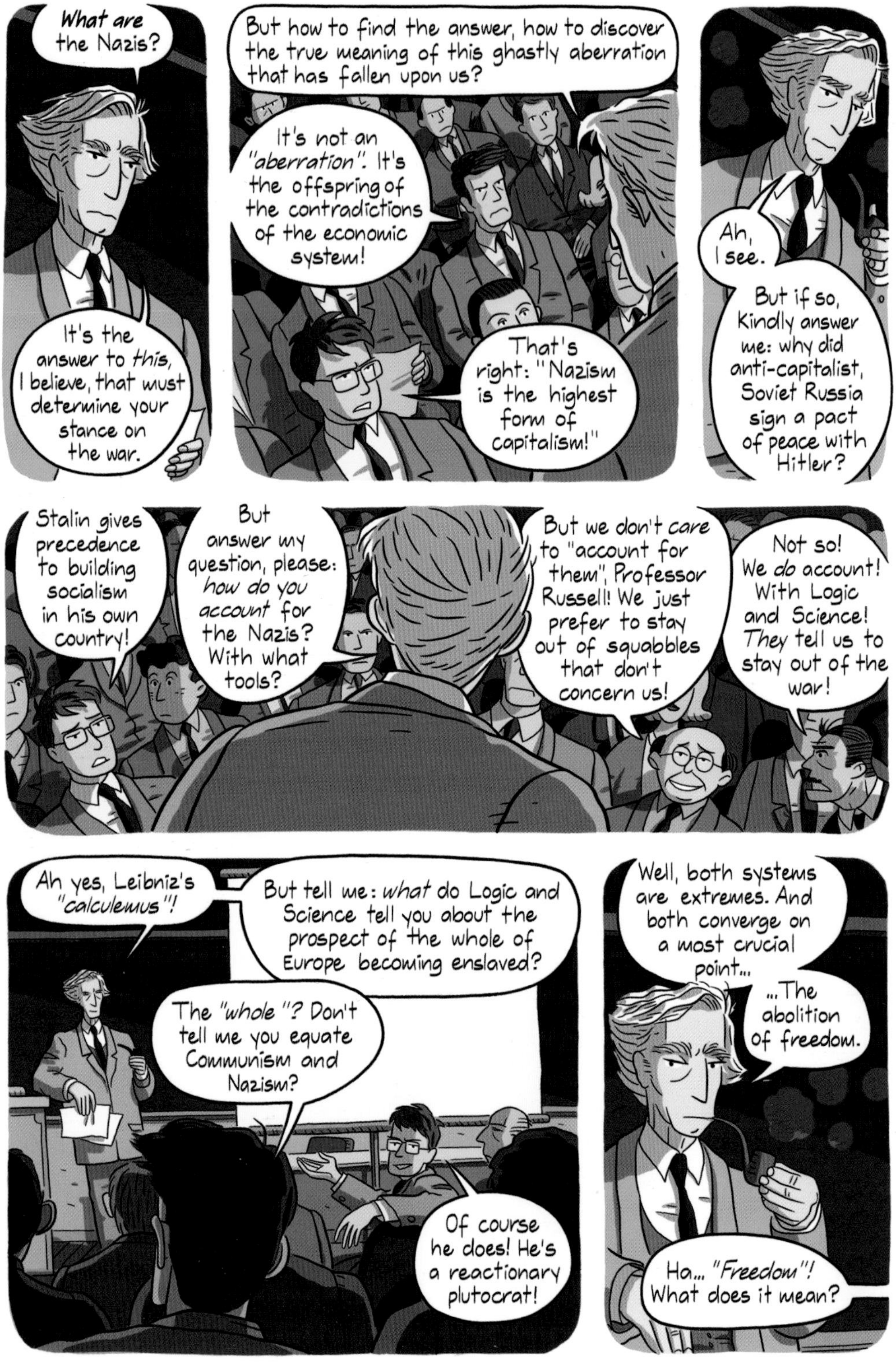
What are the Nazis?
It's the answer to *this*, I believe, that must determine your stance on the war.
But how to find the answer, how to discover the true meaning of this ghastly aberration that has fallen upon us?
It's not an "*aberration*". It's the offspring of the contradictions of the economic system!
That's right: "Nazism is the highest form of capitalism!"
Ah, I see.
But if so, Kindly answer me: why did anti-capitalist, Soviet Russia sign a pact of peace with Hitler?
Stalin gives precedence to building socialism in his own country!
But answer my question, please: *how do you account* for the Nazis? With what tools?
But we don't *care* to "account for them", Professor Russell! We just prefer to stay out of squabbles that don't concern us!
Not so! We *do* account! With Logic and Science! *They* tell us to stay out of the war!
Ah yes, Leibniz's "*calculemus*"!
But tell me: *what* do Logic and Science tell you about the prospect of the whole of Europe becoming enslaved?
The "*whole*"? Don't tell me you equate Communism and Nazism?
Of course he does! He's a reactionary plutocrat!
Well, both systems are extremes. And both converge on a most crucial point...
...The abolition of freedom.
Ha... "*Freedom*"! What does it mean?

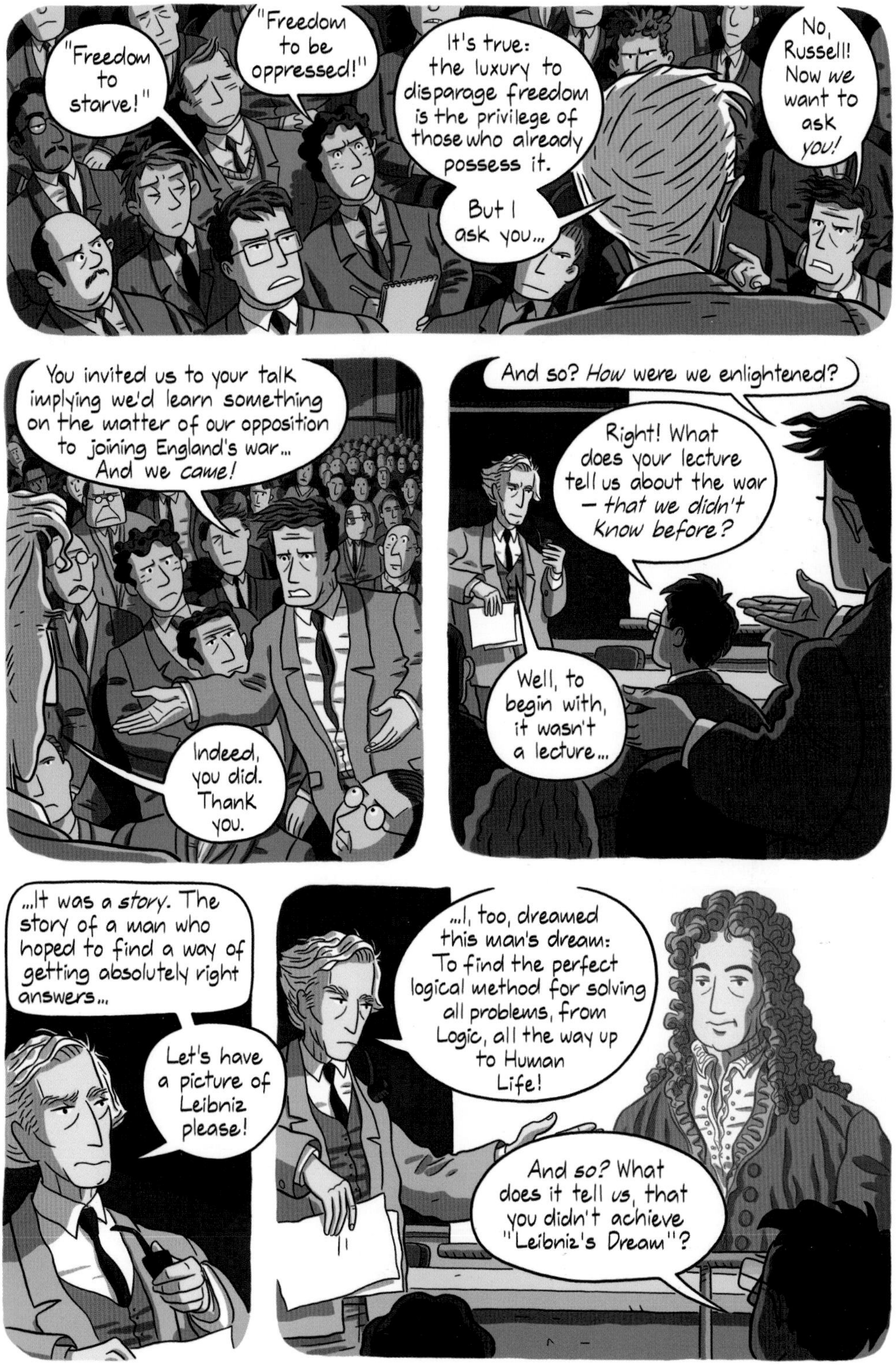
"Freedom to starve!"
"Freedom to be oppressed!"
It's true: the luxury to disparage freedom is the privilege of those who already possess it.
But I ask you...
No, Russell! Now *we* want to ask *you!*
You invited us to your talk implying we'd learn something on the matter of our opposition to joining England's war... And we *came!*
Indeed, you did. Thank you.
And so? *How* were we enlightened?
Right! What does your lecture tell us about the war — *that we didn't know before?*
Well, to begin with, it wasn't a lecture...
...It was a *story*. The story of a man who hoped to find a way of getting absolutely right answers...
Let's have a picture of Leibniz please!
...I, too, dreamed this man's dream: To find the perfect logical method for solving all problems, from Logic, all the way up to Human Life!
And *so?* What does it tell *us*, that you didn't achieve "Leibniz's Dream"?

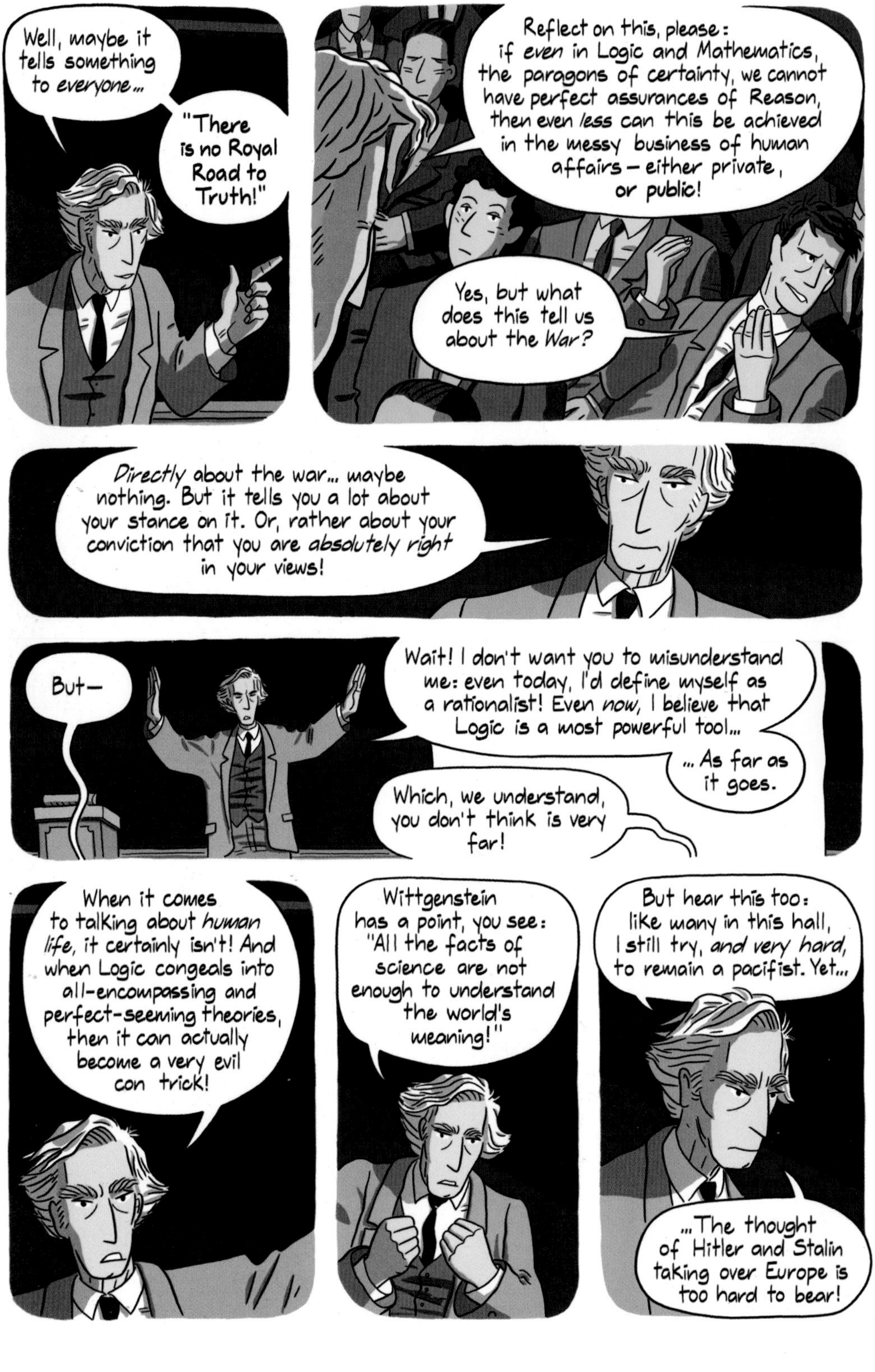
Well, maybe it tells something to *everyone*...
"There is no Royal Road to Truth!"
Reflect on this, please: if *even* in Logic and Mathematics, the paragons of certainty, we cannot have perfect assurances of Reason, then even *less* can this be achieved in the messy business of human affairs – either private, or public!
Yes, but what does this tell us about the *War*?
Directly about the war... maybe nothing. But it tells you a lot about your stance on it. Or, rather about your conviction that you are *absolutely right* in your views!
But—
Wait! I don't want you to misunderstand me: even today, I'd define myself as a rationalist! Even *now*, I believe that Logic is a most powerful tool...
... As far as it goes.
Which, we understand, you don't think is very far!
When it comes to talking about *human life*, it certainly isn't! And when Logic congeals into all-encompassing and perfect-seeming theories, then it can actually become a very evil con trick!
Wittgenstein has a point, you see: "All the facts of science are not enough to understand the world's meaning!"
But hear this too: like many in this hall, I still try, *and very hard*, to remain a pacifist. Yet...
...The thought of Hitler and Stalin taking over Europe is too hard to bear!

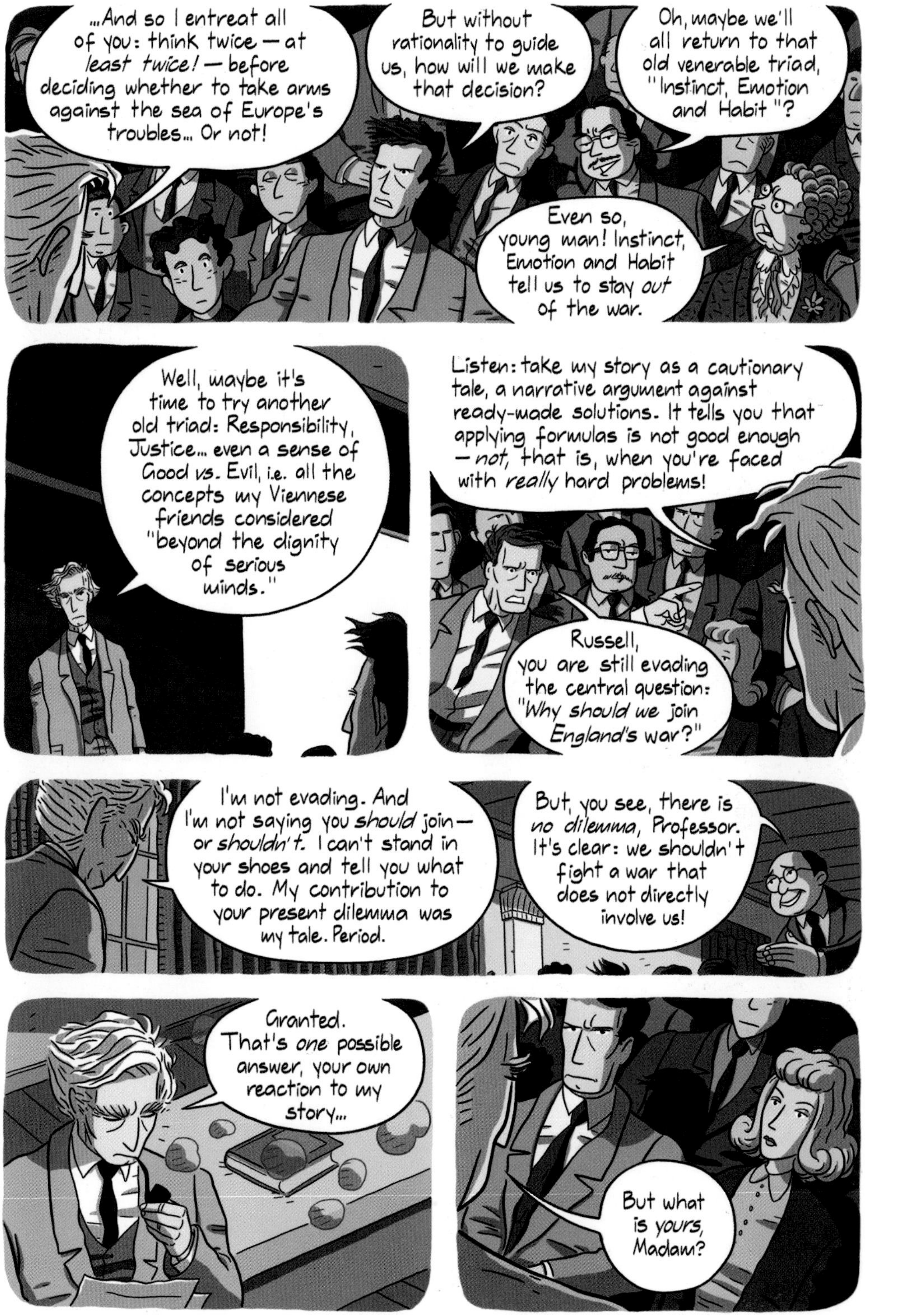
...And so I entreat all of you: think twice — at *least twice!* — before deciding whether to take arms against the sea of Europe's troubles... Or not!
But without rationality to guide us, how will we make that decision?
Oh, maybe we'll all return to that old venerable triad, "Instinct, Emotion and Habit"?
Even so, young man! Instinct, Emotion and Habit tell us to stay *out* of the war.
Well, maybe it's time to try another old triad: Responsibility, Justice... even a sense of Good *vs.* Evil, i.e. all the concepts my Viennese friends considered "beyond the dignity of serious minds."
Listen: take my story as a cautionary tale, a narrative argument against ready-made solutions. It tells you that applying formulas is not good enough — *not,* that is, when you're faced with *really* hard problems!
Russell, you are still evading the central question: "*Why should we join England's* war?"
I'm not evading. And I'm not saying you *should* join — or *shouldn't.* I can't stand in your shoes and tell you what to do. My contribution to your present dilemma was my tale. Period.
But, you see, there is *no dilemma,* Professor. It's clear: we shouldn't fight a war that does not directly involve us!
Granted. That's *one* possible answer, your own reaction to my story...
But what is *yours,* Madam?

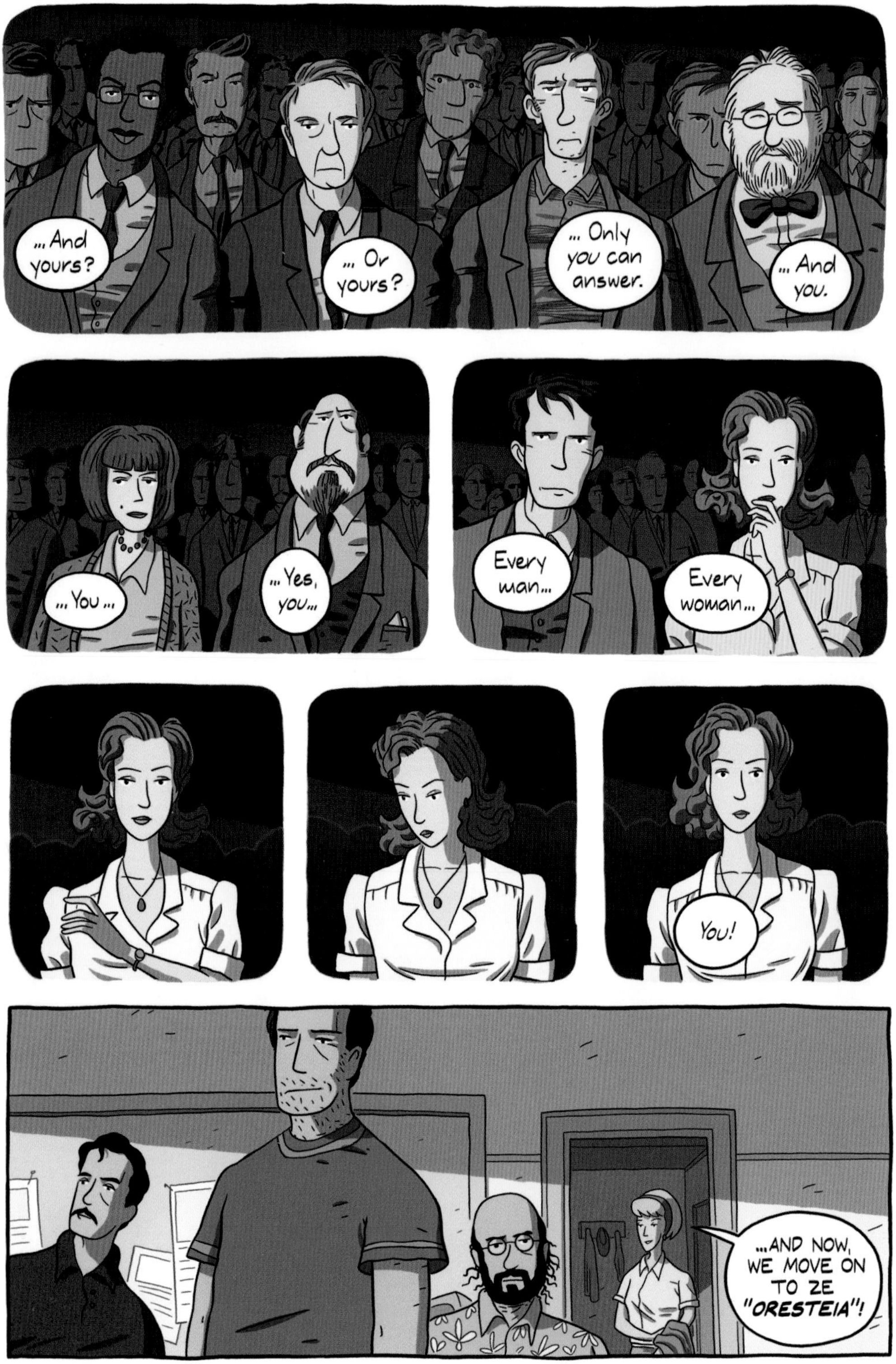
...And yours?
... Or yours?
... Only you can answer.
... And you.
... You ...
... Yes, you...
Every man...
Every woman...
You!
...AND NOW, WE MOVE ON TO ZE "ORESTEIA"!

FINALE

...THE REVENGE CYCLE BEGINS BEFORE THE *"ORESTEIA"*'S ACTION. **ORESTES'** GRANDFATHER **ATREUS** IS THE **KING OF ARGOS**, BUT HIS BROTHER, **THYESTES**, SLEEPS WITH HIS WIFE, SO—

AH, YES. CHERCHEZ LA FEMME!

...**ATREUS** GETS *SOOO* MAD, THAT HE SLAUGHTERS **THYESTES'** CHILDREN, AND SERVES THEM TO HIM FOR A MEAL!

I'LL PUKE...

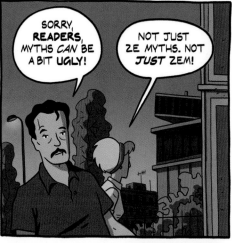

SORRY, **READERS**, MYTHS *CAN* BE A BIT **UGLY!**

NOT JUST ZE MYTHS. NOT *JUST* ZEM!

HOW NICE ZESE OLD BUILDINGS!

THEN, **THYESTES** HAS A NEW SON, **AEGISTHUS**, WHO WILL BECOME THE VEHICLE OF HIS **REVENGE**.

NOW **ATREUS'S** SON, **AGAMEMNON**, AS LEADER OF THE EXPEDITION TO **TROY**, SACRIFICES HIS OWN YOUNG **DAUGHTER**...

AH, YES, IPANEMA!

..."*IPHIGENIA*", ACTUALLY! SO, HIS WIFE **CLYTEMNESTRA**, PLOTS...

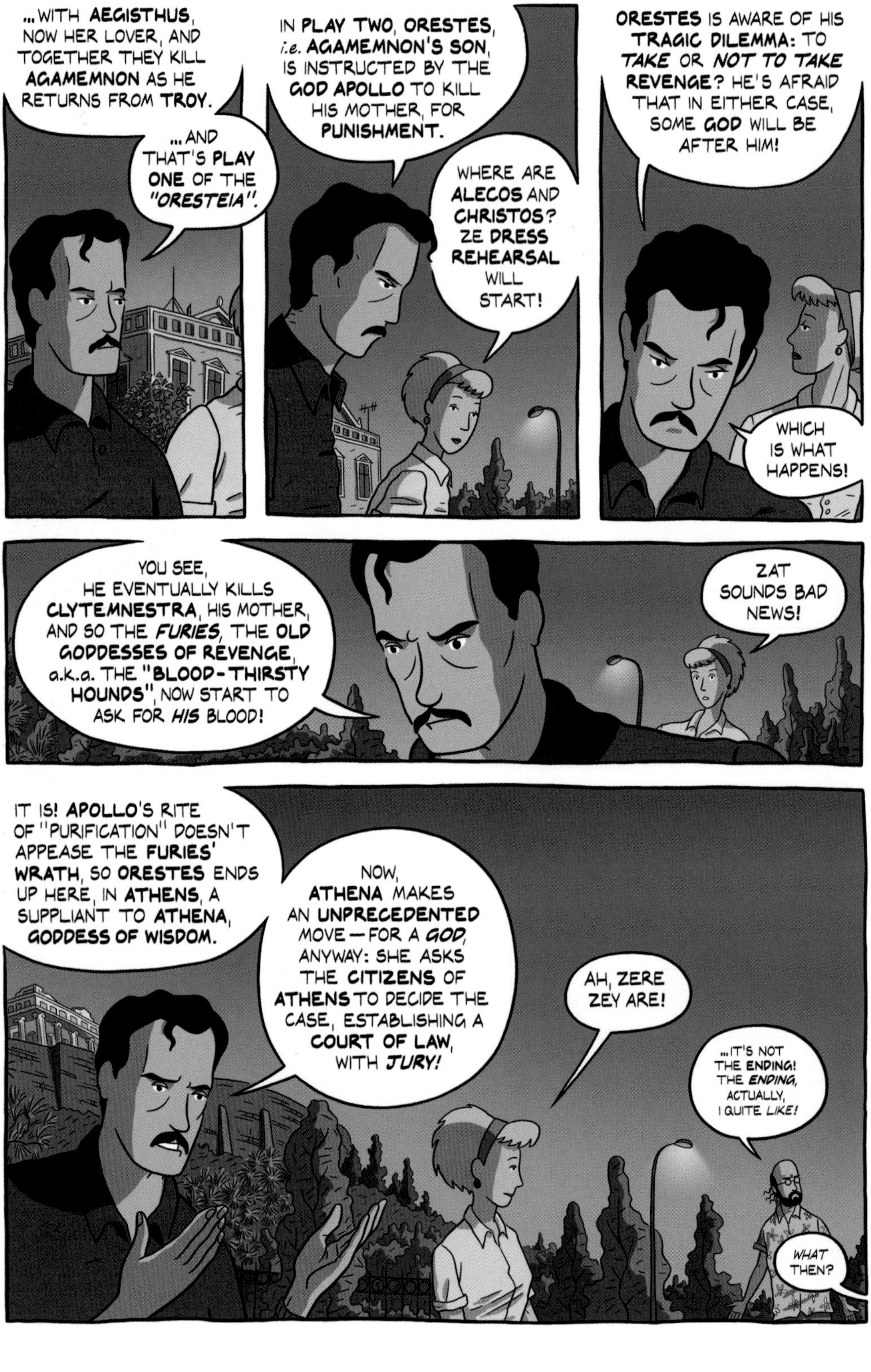
...WITH AEGISTHUS, NOW HER LOVER, AND TOGETHER THEY KILL AGAMEMNON AS HE RETURNS FROM TROY.
...AND THAT'S PLAY ONE OF THE "ORESTEIA".
IN PLAY TWO, ORESTES, i.e. AGAMEMNON'S SON, IS INSTRUCTED BY THE GOD APOLLO TO KILL HIS MOTHER, FOR PUNISHMENT.
WHERE ARE ALECOS AND CHRISTOS? ZE DRESS REHEARSAL WILL START!
ORESTES IS AWARE OF HIS TRAGIC DILEMMA: TO TAKE OR NOT TO TAKE REVENGE? HE'S AFRAID THAT IN EITHER CASE, SOME GOD WILL BE AFTER HIM!
WHICH IS WHAT HAPPENS!
YOU SEE, HE EVENTUALLY KILLS CLYTEMNESTRA, HIS MOTHER, AND SO THE FURIES, THE OLD GODDESSES OF REVENGE, a.k.a. THE "BLOOD-THIRSTY HOUNDS", NOW START TO ASK FOR HIS BLOOD!
ZAT SOUNDS BAD NEWS!
IT IS! APOLLO'S RITE OF "PURIFICATION" DOESN'T APPEASE THE FURIES' WRATH, SO ORESTES ENDS UP HERE, IN ATHENS, A SUPPLIANT TO ATHENA, GODDESS OF WISDOM.
NOW, ATHENA MAKES AN UNPRECEDENTED MOVE—FOR A GOD, ANYWAY: SHE ASKS THE CITIZENS OF ATHENS TO DECIDE THE CASE, ESTABLISHING A COURT OF LAW, WITH JURY!
AH, ZERE ZEY ARE!
...IT'S NOT THE ENDING! THE ENDING, ACTUALLY, I QUITE LIKE!
WHAT THEN?

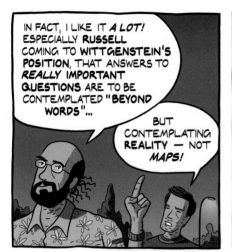

IN FACT, I LIKE IT *A LOT!* ESPECIALLY *RUSSELL* COMING TO *WITTGENSTEIN'S* POSITION, THAT ANSWERS TO *REALLY* IMPORTANT QUESTIONS ARE TO BE CONTEMPLATED "BEYOND WORDS"...

BUT CONTEMPLATING REALITY — NOT *MAPS!*

WHAT'S UP?

CHRISTOS HAS A PROBLEM WITH THE "FOUNDATIONAL QUEST"!

WRONG! I HAVE *TWO* PROBLEMS, WITH YOUR *VERSION* OF IT!

ONE, IT *DIDN'T* FAIL... AND, TWO, IT *WASN'T* A TRAGEDY!

IT ISN'T EXACTLY A COMEDY, EH?

GRANTED, THERE ARE *TRAGIC PARTS!* BUT THE ENDING IS *HAPPY*, AS IN... IN...

...THE *"ORESTEIA"!*

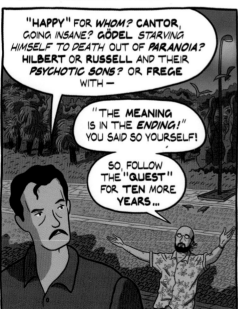

"HAPPY" FOR *WHOM?* CANTOR, GOING *INSANE?* GÖDEL *STARVING HIMSELF TO DEATH* OUT OF *PARANOIA?* HILBERT OR RUSSELL AND THEIR *PSYCHOTIC SONS?* OR FREGE WITH —

"THE MEANING IS IN THE *ENDING!*" YOU SAID SO YOURSELF!

SO, FOLLOW THE "QUEST" FOR *TEN MORE YEARS* ...

...AND YOU GET A BRAND-NEW, *TRIUMPHANT* FINALE...

...WITH THE *CREATION* OF THE *COMPUTER*, WHICH IS THE "QUEST"'S *REAL* HERO!

YOUR *PROBLEM* IS, SIMPLY, THAT YOU SEE IT AS A *STORY* OF *PEOPLE!*

WELL, STORIES *DO* TEND TO BE "ABOUT PEOPLE"!

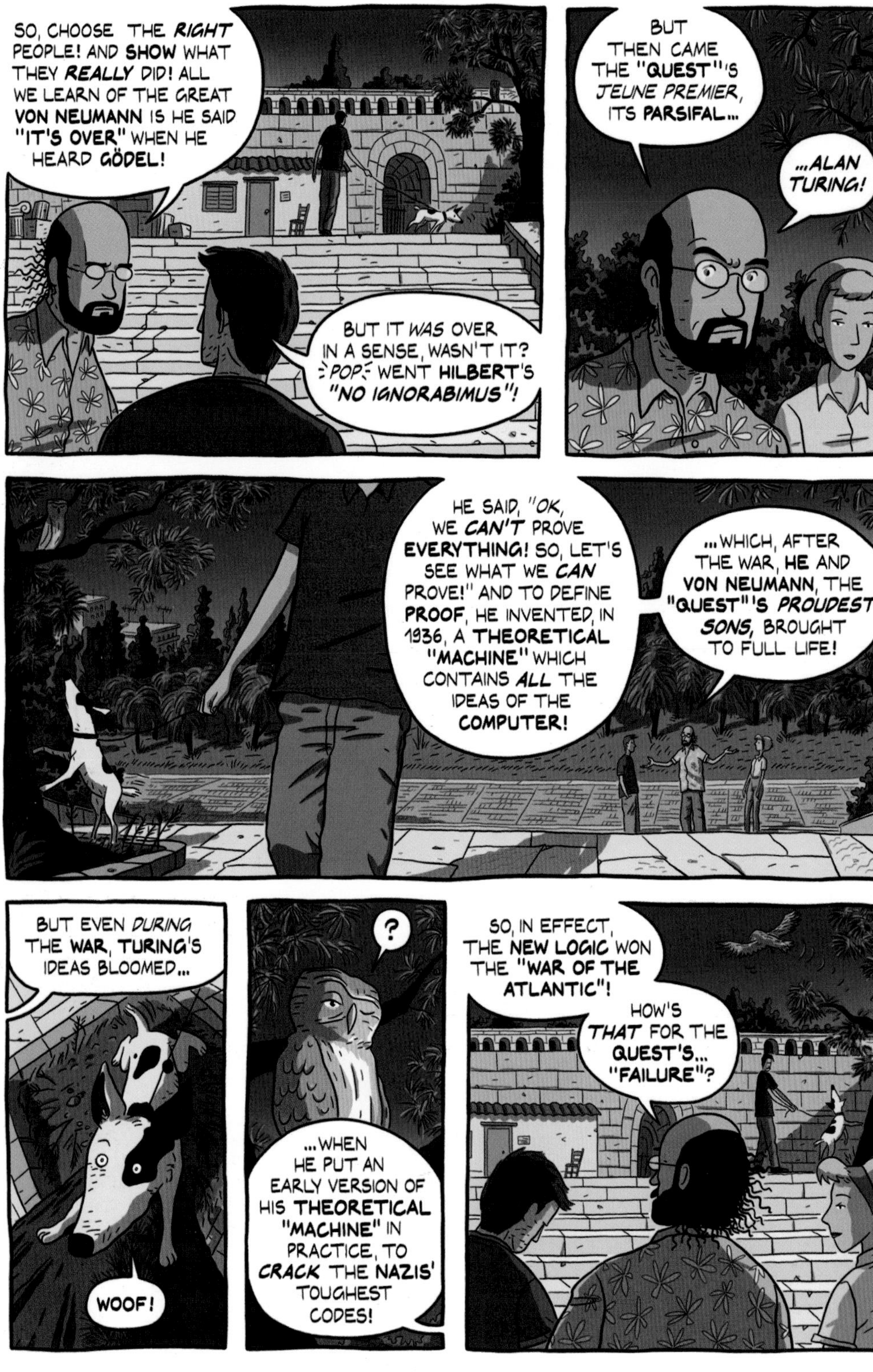

SO, CHOOSE THE RIGHT PEOPLE! AND SHOW WHAT THEY REALLY DID! ALL WE LEARN OF THE GREAT VON NEUMANN IS HE SAID "IT'S OVER" WHEN HE HEARD GÖDEL!
BUT IT WAS OVER IN A SENSE, WASN'T IT? POP WENT HILBERT'S "NO IGNORABIMUS"!
BUT THEN CAME THE "QUEST"'S JEUNE PREMIER, ITS PARSIFAL...
...ALAN TURING!
HE SAID, "OK, WE CAN'T PROVE EVERYTHING! SO, LET'S SEE WHAT WE CAN PROVE!" AND TO DEFINE PROOF, HE INVENTED, IN 1936, A THEORETICAL "MACHINE" WHICH CONTAINS ALL THE IDEAS OF THE COMPUTER!
...WHICH, AFTER THE WAR, HE AND VON NEUMANN, THE "QUEST"'S PROUDEST SONS, BROUGHT TO FULL LIFE!
BUT EVEN DURING THE WAR, TURING'S IDEAS BLOOMED...
WOOF!
?
...WHEN HE PUT AN EARLY VERSION OF HIS THEORETICAL "MACHINE" IN PRACTICE, TO CRACK THE NAZIS' TOUGHEST CODES!
SO, IN EFFECT, THE NEW LOGIC WON THE "WAR OF THE ATLANTIC"!
HOW'S THAT FOR THE QUEST'S... "FAILURE"?

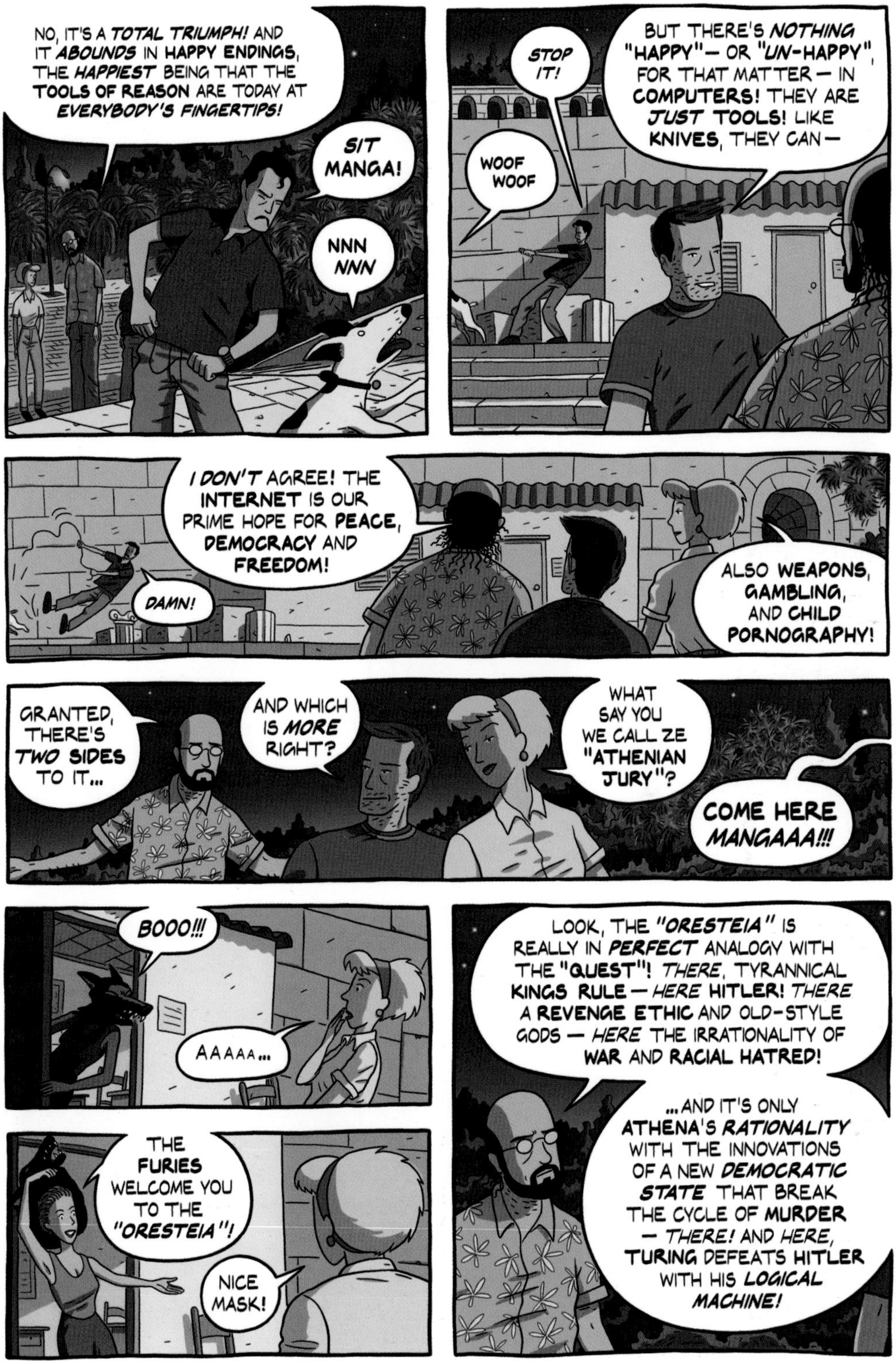
NO, IT'S A TOTAL TRIUMPH! AND IT ABOUNDS IN HAPPY ENDINGS, THE HAPPIEST BEING THAT THE TOOLS OF REASON ARE TODAY AT EVERYBODY'S FINGERTIPS!
SIT MANGA!
NNN NNN
STOP IT!
BUT THERE'S NOTHING "HAPPY"— OR "UN-HAPPY", FOR THAT MATTER— IN COMPUTERS! THEY ARE JUST TOOLS! LIKE KNIVES, THEY CAN—
WOOF WOOF
I DON'T AGREE! THE INTERNET IS OUR PRIME HOPE FOR PEACE, DEMOCRACY AND FREEDOM!
DAMN!
ALSO WEAPONS, GAMBLING, AND CHILD PORNOGRAPHY!
GRANTED, THERE'S TWO SIDES TO IT...
AND WHICH IS MORE RIGHT?
WHAT SAY YOU WE CALL ZE "ATHENIAN JURY"?
COME HERE MANGAAA!!!
BOOO!!!
AAAAA...
THE FURIES WELCOME YOU TO THE "ORESTEIA"!
NICE MASK!
LOOK, THE "ORESTEIA" IS REALLY IN PERFECT ANALOGY WITH THE "QUEST"! THERE, TYRANNICAL KINGS RULE— HERE HITLER! THERE A REVENGE ETHIC AND OLD-STYLE GODS— HERE THE IRRATIONALITY OF WAR AND RACIAL HATRED!
...AND IT'S ONLY ATHENA'S RATIONALITY WITH THE INNOVATIONS OF A NEW DEMOCRATIC STATE THAT BREAK THE CYCLE OF MURDER— THERE! AND HERE, TURING DEFEATS HITLER WITH HIS LOGICAL MACHINE!

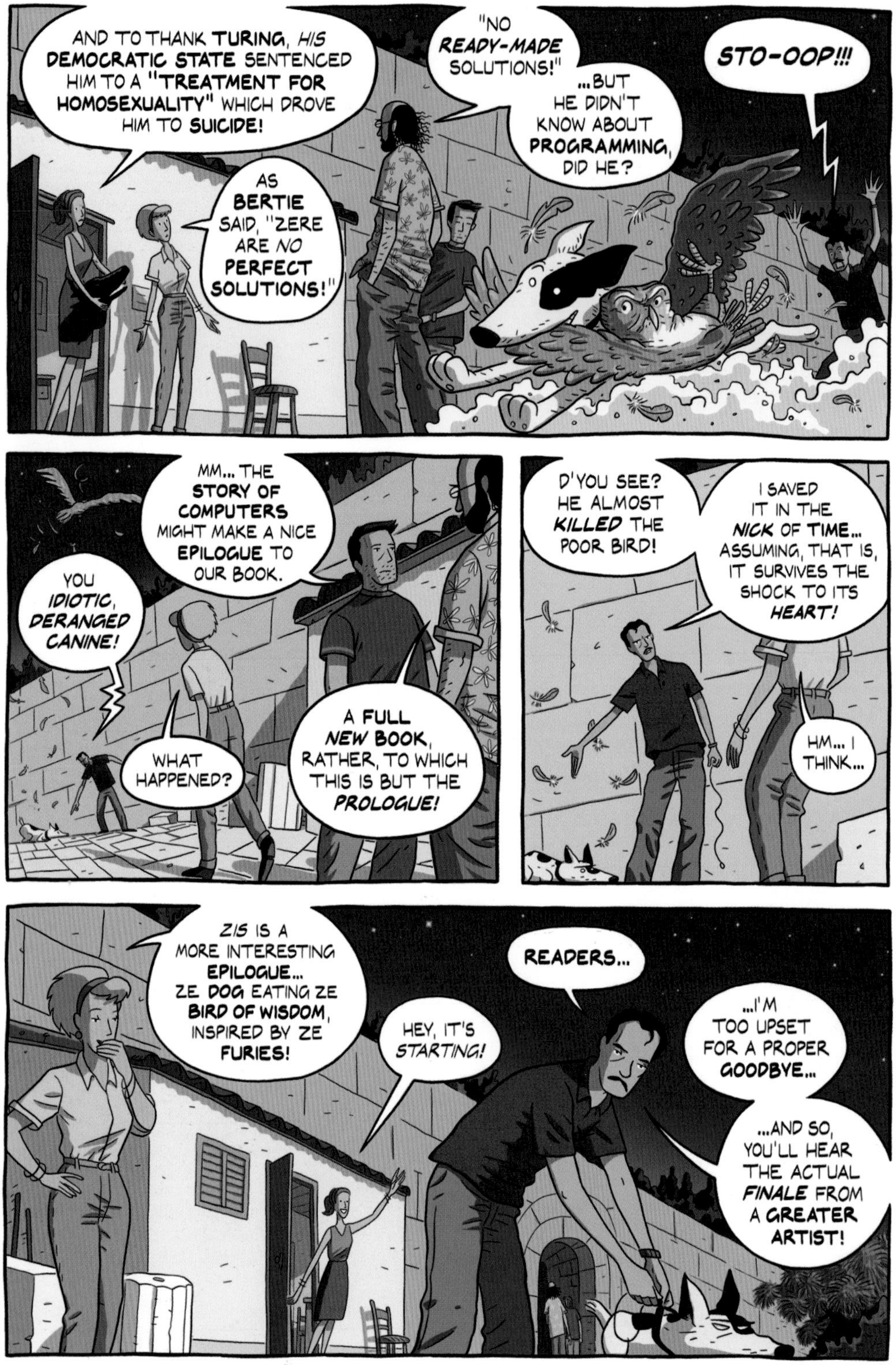
AND TO THANK **TURING**, *HIS* **DEMOCRATIC STATE** SENTENCED HIM TO A "**TREATMENT FOR HOMOSEXUALITY**" WHICH DROVE HIM TO **SUICIDE!**
AS **BERTIE** SAID, "ZERE ARE *NO* **PERFECT SOLUTIONS!**"
"NO ***READY-MADE*** SOLUTIONS!"
...BUT HE DIDN'T KNOW ABOUT **PROGRAMMING**, DID HE?
STO-OOP!!!
MM... THE **STORY OF COMPUTERS** MIGHT MAKE A NICE **EPILOGUE** TO OUR BOOK.
YOU ***IDIOTIC, DERANGED CANINE!***
WHAT HAPPENED?
A **FULL *NEW BOOK***, RATHER, TO WHICH THIS IS BUT THE ***PROLOGUE!***
D'YOU SEE? HE ALMOST ***KILLED*** THE POOR BIRD!
I SAVED IT IN THE ***NICK*** OF **TIME...** ASSUMING, THAT IS, IT SURVIVES THE SHOCK TO ITS ***HEART!***
HM... I THINK...
ZIS IS A MORE INTERESTING **EPILOGUE...** ZE **DOG** EATING ZE **BIRD OF WISDOM**, INSPIRED BY ZE **FURIES!**
HEY, IT'S *STARTING!*
READERS...
...I'M TOO UPSET FOR A PROPER **GOODBYE...**
...AND SO, YOU'LL HEAR THE ACTUAL ***FINALE*** FROM A **GREATER ARTIST!**

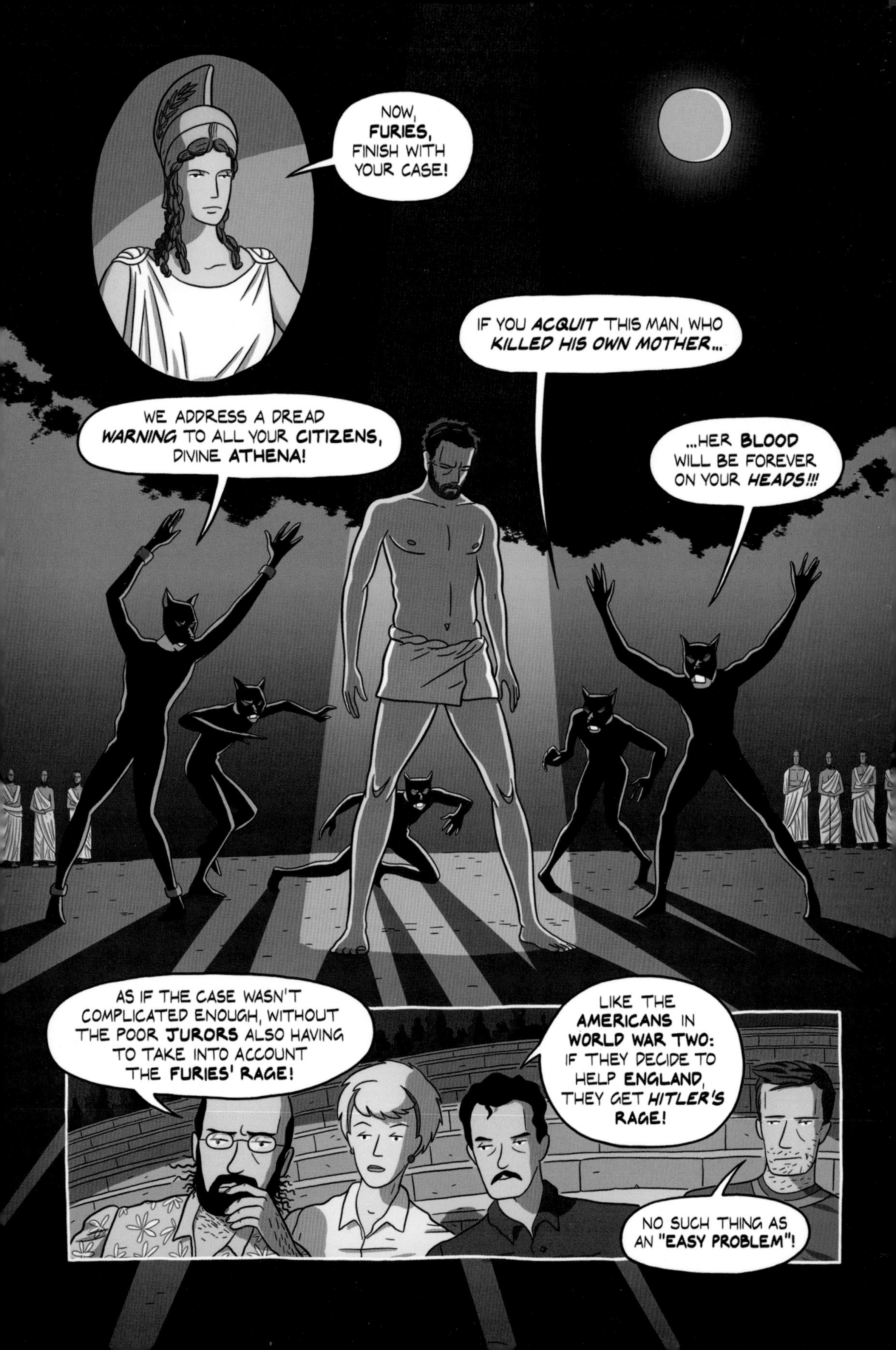
NOW, FURIES, FINISH WITH YOUR CASE!
IF YOU ACQUIT THIS MAN, WHO KILLED HIS OWN MOTHER...
WE ADDRESS A DREAD WARNING TO ALL YOUR CITIZENS, DIVINE ATHENA!
...HER BLOOD WILL BE FOREVER ON YOUR HEADS!!!
AS IF THE CASE WASN'T COMPLICATED ENOUGH, WITHOUT THE POOR JURORS ALSO HAVING TO TAKE INTO ACCOUNT THE FURIES' RAGE!
LIKE THE AMERICANS IN WORLD WAR TWO: IF THEY DECIDE TO HELP ENGLAND, THEY GET HITLER'S RAGE!
NO SUCH THING AS AN "EASY PROBLEM"!

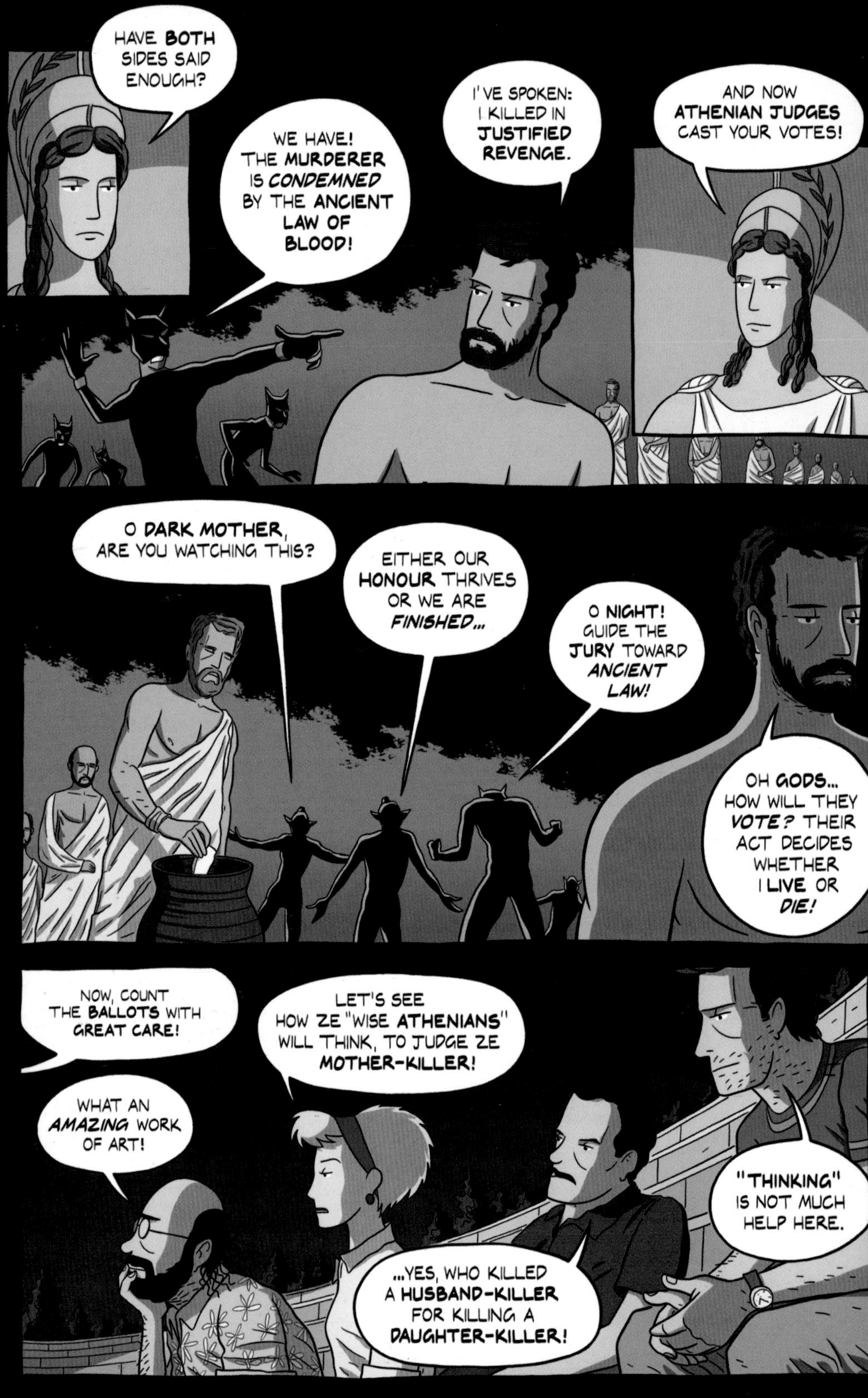
HAVE BOTH SIDES SAID ENOUGH?
WE HAVE! THE MURDERER IS CONDEMNED BY THE ANCIENT LAW OF BLOOD!
I'VE SPOKEN: I KILLED IN JUSTIFIED REVENGE.
AND NOW ATHENIAN JUDGES CAST YOUR VOTES!
O DARK MOTHER, ARE YOU WATCHING THIS?
EITHER OUR HONOUR THRIVES OR WE ARE FINISHED...
O NIGHT! GUIDE THE JURY TOWARD ANCIENT LAW!
OH GODS... HOW WILL THEY VOTE? THEIR ACT DECIDES WHETHER I LIVE OR DIE!
NOW, COUNT THE BALLOTS WITH GREAT CARE!
WHAT AN AMAZING WORK OF ART!
LET'S SEE HOW ZE "WISE ATHENIANS" WILL THINK, TO JUDGE ZE MOTHER-KILLER!
...YES, WHO KILLED A HUSBAND-KILLER FOR KILLING A DAUGHTER-KILLER!
"THINKING" IS NOT MUCH HELP HERE.

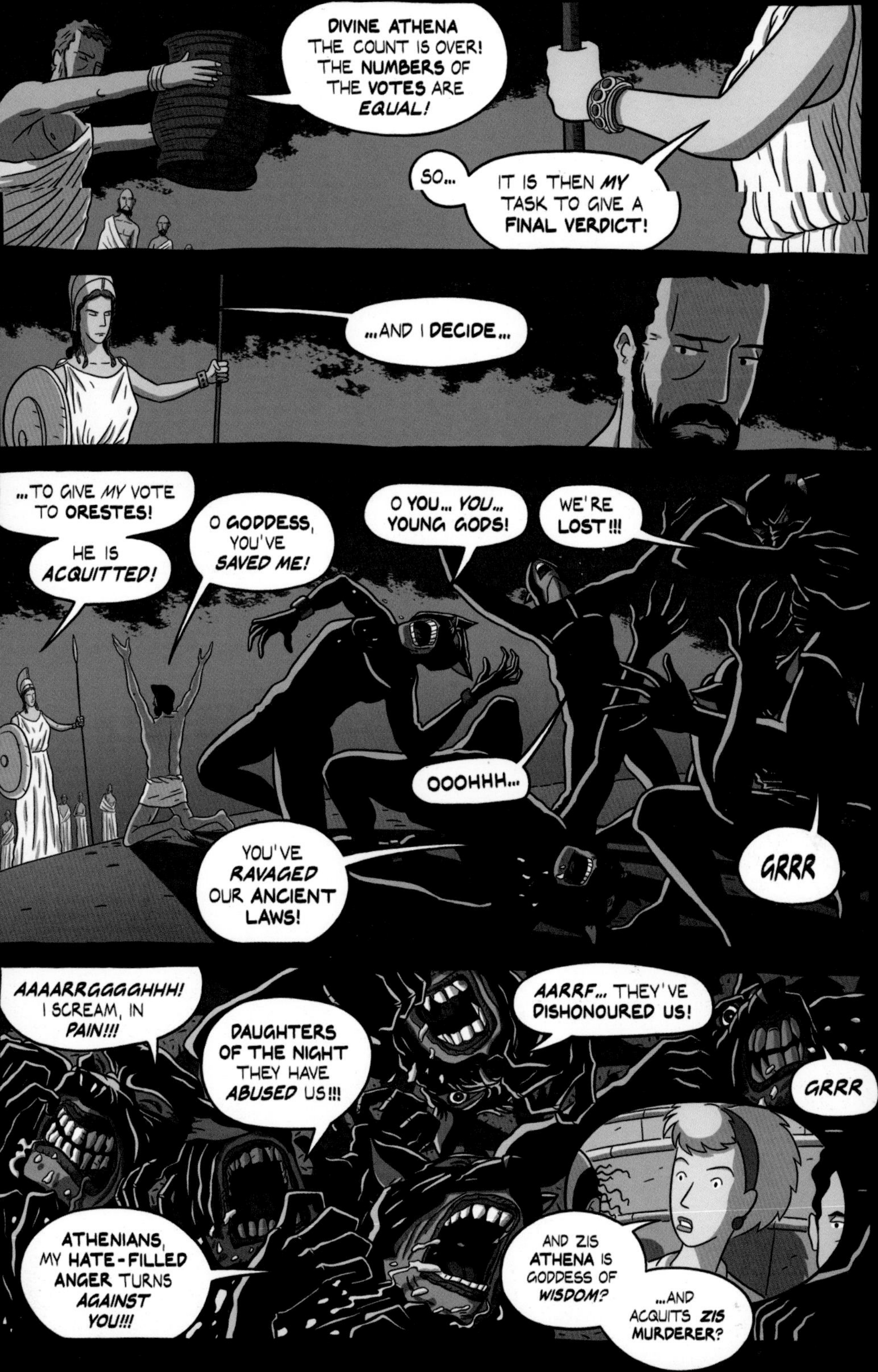

DIVINE ATHENA THE COUNT IS OVER! THE NUMBERS OF THE VOTES ARE EQUAL!
SO...
IT IS THEN MY TASK TO GIVE A FINAL VERDICT!
...AND I DECIDE...
...TO GIVE MY VOTE TO ORESTES!
HE IS ACQUITTED!
O GODDESS, YOU'VE SAVED ME!
O YOU... YOU... YOUNG GODS!
WE'RE LOST!!!
OOOHHH...
YOU'VE RAVAGED OUR ANCIENT LAWS!
GRRR
AAAARRGGGGHHH! I SCREAM, IN PAIN!!!
DAUGHTERS OF THE NIGHT THEY HAVE ABUSED US!!!
AARRF... THEY'VE DISHONOURED US!
GRRR
ATHENIANS, MY HATE-FILLED ANGER TURNS AGAINST YOU!!!
AND ZIS ATHENA IS GODDESS OF WISDOM?
...AND ACQUITS ZIS MURDERER?

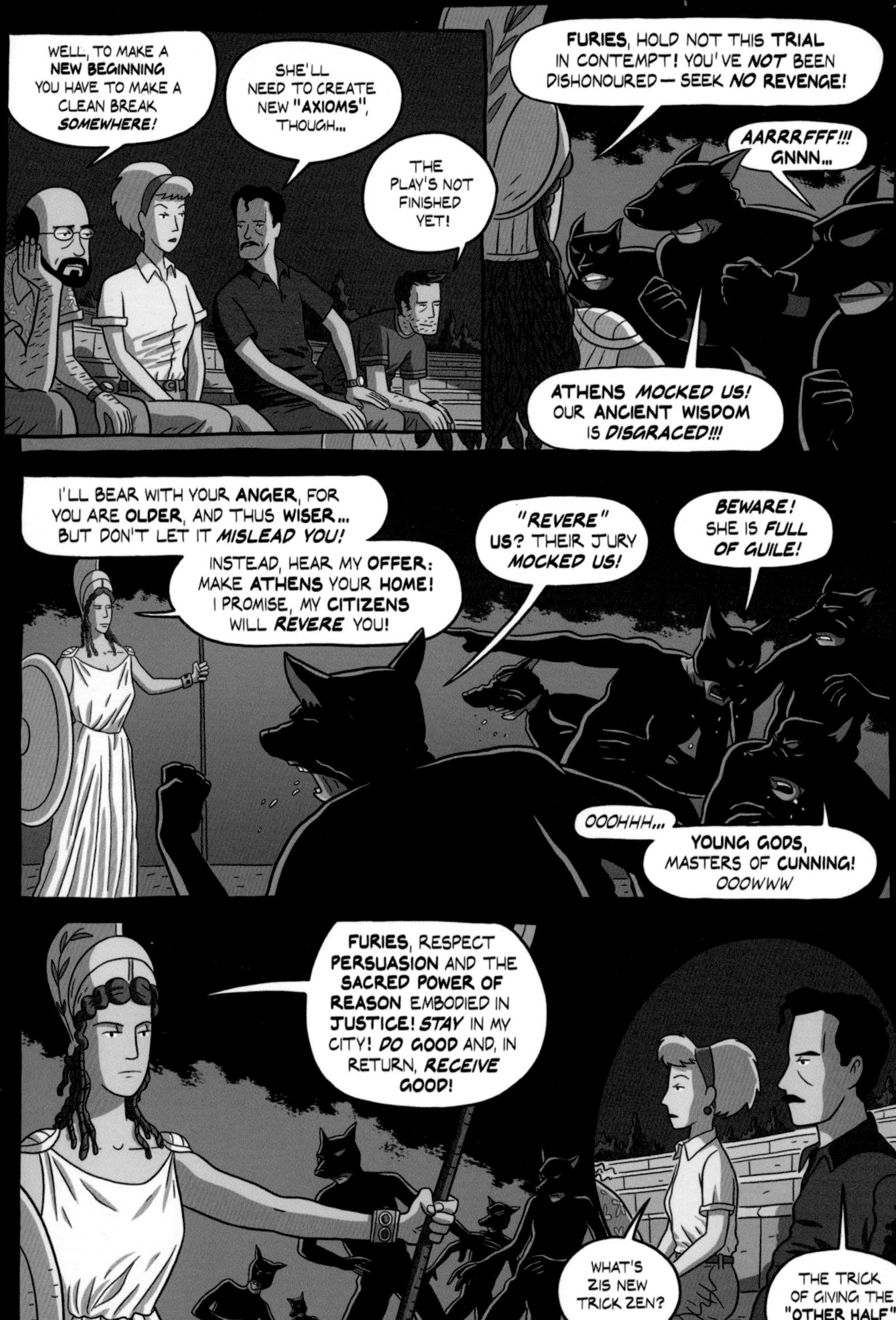
WELL, TO MAKE A NEW BEGINNING YOU HAVE TO MAKE A CLEAN BREAK SOMEWHERE!
SHE'LL NEED TO CREATE NEW "AXIOMS", THOUGH...
THE PLAY'S NOT FINISHED YET!
FURIES, HOLD NOT THIS TRIAL IN CONTEMPT! YOU'VE NOT BEEN DISHONOURED— SEEK NO REVENGE!
AARRRFFF!!! GNNN...
ATHENS MOCKED US! OUR ANCIENT WISDOM IS DISGRACED!!!
I'LL BEAR WITH YOUR ANGER, FOR YOU ARE OLDER, AND THUS WISER... BUT DON'T LET IT MISLEAD YOU!
INSTEAD, HEAR MY OFFER: MAKE ATHENS YOUR HOME! I PROMISE, MY CITIZENS WILL REVERE YOU!
"REVERE" US? THEIR JURY MOCKED US!
BEWARE! SHE IS FULL OF GUILE!
OOOHHH...
YOUNG GODS, MASTERS OF CUNNING! OOOWWW
FURIES, RESPECT PERSUASION AND THE SACRED POWER OF REASON EMBODIED IN JUSTICE! STAY IN MY CITY! DO GOOD AND, IN RETURN, RECEIVE GOOD!
WHAT'S ZIS NEW TRICK ZEN?
THE TRICK OF GIVING THE "OTHER HALF" A VOICE.

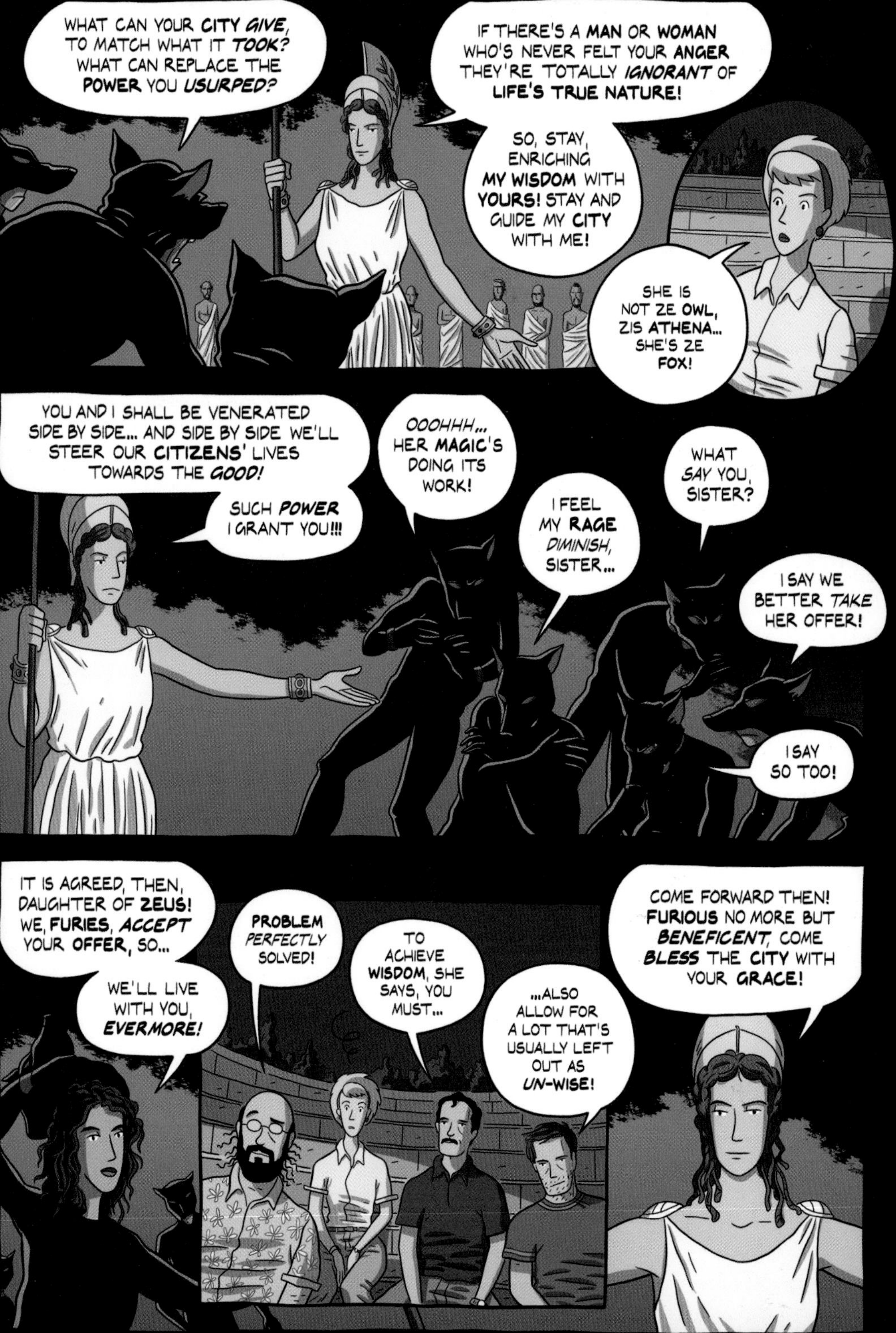
WHAT CAN YOUR **CITY** ***GIVE***, TO MATCH WHAT IT ***TOOK?*** WHAT CAN REPLACE THE **POWER** YOU ***USURPED?***
IF THERE'S A **MAN** OR **WOMAN** WHO'S NEVER FELT YOUR **ANGER** THEY'RE TOTALLY ***IGNORANT*** OF **LIFE'S TRUE NATURE!**
SO, STAY, ENRICHING **MY WISDOM** WITH **YOURS!** STAY AND GUIDE MY **CITY** WITH ME!
SHE IS NOT ZE **OWL**, ZIS **ATHENA**... SHE'S ZE **FOX!**
YOU AND I SHALL BE VENERATED SIDE BY SIDE... AND SIDE BY SIDE WE'LL STEER OUR **CITIZENS'** LIVES TOWARDS THE ***GOOD!***
SUCH ***POWER*** I GRANT YOU!!!
OOOHHH... HER **MAGIC**'S DOING ITS WORK!
I FEEL MY **RAGE** *DIMINISH*, SISTER...
WHAT *SAY* YOU, SISTER?
I SAY WE BETTER *TAKE* HER OFFER!
I SAY SO TOO!
IT IS AGREED, THEN, DAUGHTER OF **ZEUS!** WE, **FURIES**, ***ACCEPT*** YOUR **OFFER**, SO...
WE'LL LIVE WITH YOU, ***EVERMORE!***
PROBLEM *PERFECTLY* SOLVED!
TO ACHIEVE **WISDOM**, SHE SAYS, YOU MUST...
...ALSO ALLOW FOR A LOT THAT'S USUALLY LEFT OUT AS ***UN-WISE!***
COME FORWARD THEN! **FURIOUS** NO MORE BUT ***BENEFICENT***, COME ***BLESS*** THE **CITY** WITH YOUR **GRACE!**

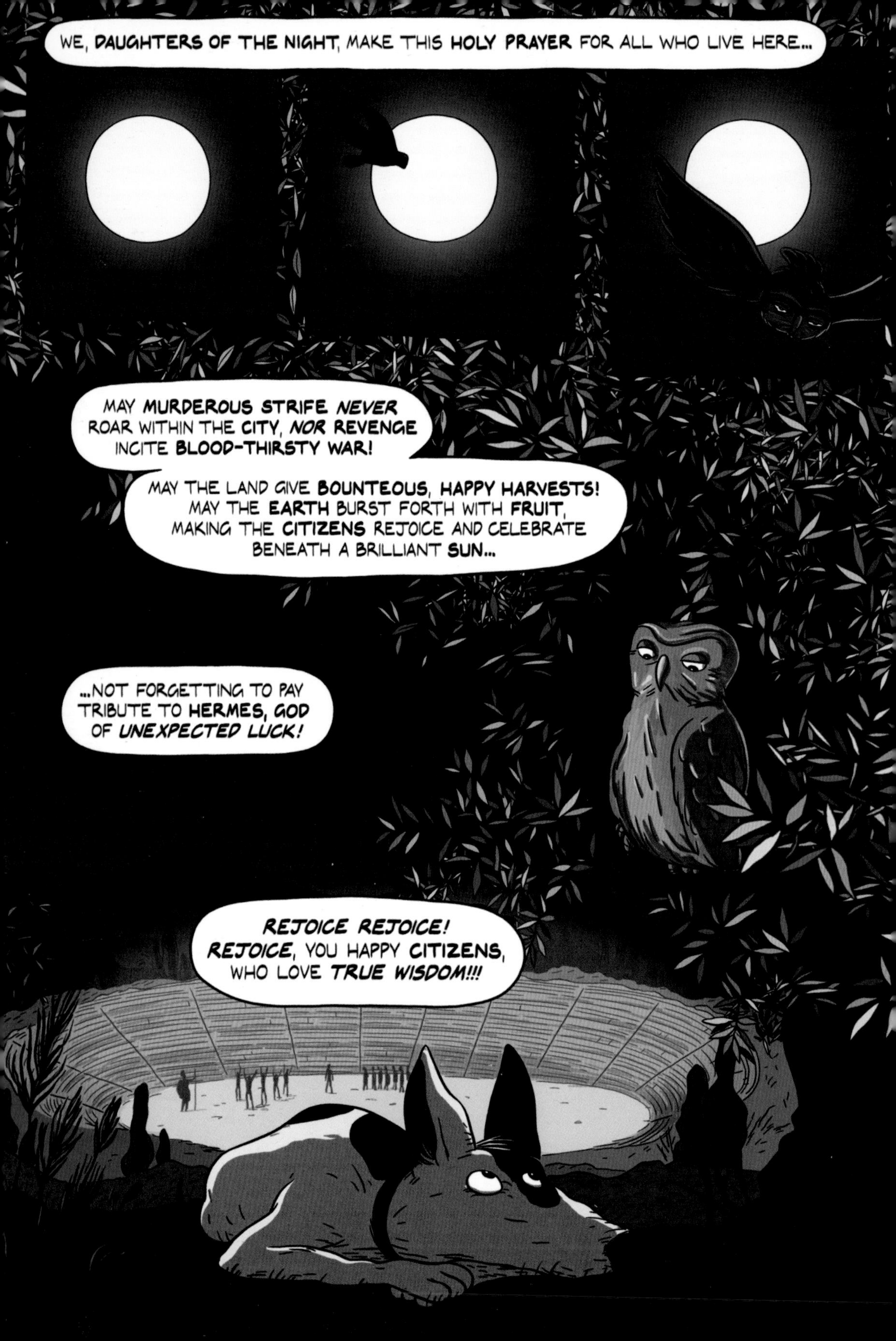
WE, DAUGHTERS OF THE NIGHT, MAKE THIS HOLY PRAYER FOR ALL WHO LIVE HERE...
MAY MURDEROUS STRIFE NEVER ROAR WITHIN THE CITY, NOR REVENGE INCITE BLOOD-THIRSTY WAR!
MAY THE LAND GIVE BOUNTEOUS, HAPPY HARVESTS! MAY THE EARTH BURST FORTH WITH FRUIT, MAKING THE CITIZENS REJOICE AND CELEBRATE BENEATH A BRILLIANT SUN...
...NOT FORGETTING TO PAY TRIBUTE TO HERMES, GOD OF UNEXPECTED LUCK!
REJOICE REJOICE! REJOICE, YOU HAPPY CITIZENS, WHO LOVE TRUE WISDOM!!!

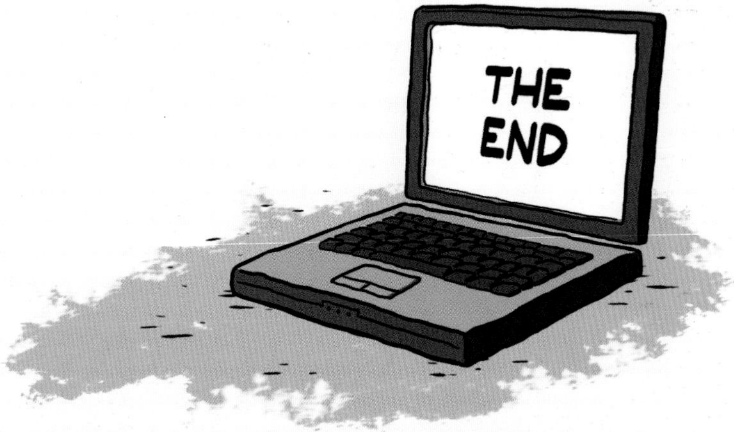
THE
END

Many thanks for their help to our friends
Aliki Chapple, Doukas Kapantais, Avraam Kawa,
Margaret Metzger, Apostolia Papadamaki, Dimitris Sivrikozis,
Chloe Theodoropoulou, Panagiotis Yiannopoulos

Logicomix and reality

Logicomix was inspired by the story of the quest for the foundations of mathematics, whose most intense phase lasted from the last decades of the 19th century to the eruption of the Second World War. Yet, despite the fact that its characters are mostly real persons, our book is definitely not — nor does it want to be — a work of history. It is — *and* wants to be — a graphic novel.

Particularly in our reconstruction of Bertrand Russell's life, we've had to wander through an immense amount of material, to select, reduce, simplify, interpret and, very often, invent. Also, though our major characters are based as closely as possible on their real-life counterparts, we have on more than one occasion departed from factual detail, in order to give our narrative greater coherence and depth. Most of these deviations consist in inventing meetings for which there is no historical evidence — or even, in some cases, where there is evidence that they did *not* occur. But these imagined meetings are always based on the actual intellectual interaction of the thinkers involved, conducted in reality either through correspondence or publications.

A few examples of such deviations from fact: from the existing evidence, or lack thereof, it is safe to assume that Russell never met Frege or Cantor in the flesh; there are no indications that he was present in Hilbert's seminal 1900 lecture on the "Problems of Mathematics", although he was certainly in Paris a few days earlier, attending the Congress of Philosophy, where he met Peano; there is no evidence whatsoever that he was in the audience during Gödel's "incompleteness" talk — he probably wasn't and Hilbert certainly wasn't, though Von Neumann certainly was, and did say "it's all over" right after. Furthermore, Russell couldn't have visited Frege right after this talk, as the latter had been dead for six years. And although the timing of Frege's rabid anti-Semitic diatribes is incongruous in our book, it is totally true that he wrote them a few years earlier.

Historically keen readers can have fun locating many more such deviations from fact. For our part, we take comfort in the words of the painter Dominikos Theotokópoulos (better known as "El Greco") explaining the freedoms he took in his painting "Storm over Toledo":

I found it necessary to reduce the size of the hospital of Don Juan
Tavera, not just because it covered the gate of Bisagra, but also
because its dome came up too high, passing the city's skyline.
And so, since I've made it smaller and moved it, I think it is better
to show its façade, rather than its other sides. As for its actual
position in the city, you can see it in the map.

Still, we must add this: apart from the simplification that was necessary to
accommodate it into a narrative work of this kind, we have not taken any
liberties with the content of the great adventure of ideas which forms our
main plot, neither with its central vision, its concepts, nor — even more
importantly — with the philosophical, existential and emotional struggles
which are inextricably bound with it.

Notebook

The following notes are by no means necessary for the enjoyment of *Logicomix*, but may give additional information on persons and ideas. A name or term in blue indicates that it also has its own entry, while *italics*, when not used for emphasis, indicate technical terms.

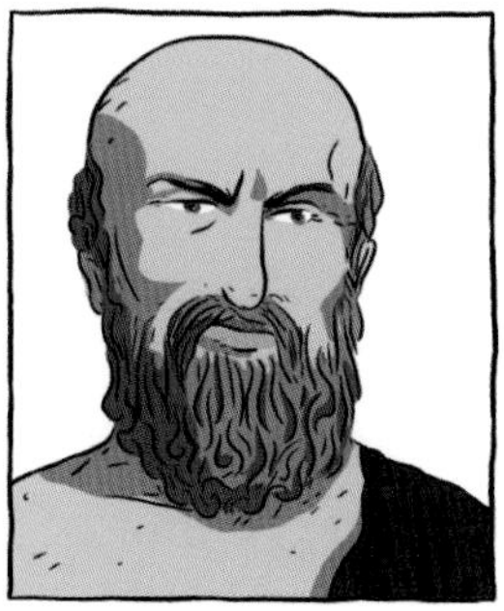

Aeschylus One of the three great Greek tragedians, the precursor of Sophocles and Euripides, Aeschylus is the creator of tragedy as we know it. He introduced a second actor into the earlier dramatic form, which only used a protagonist and the chorus, thus also inventing the technique of dramatic dialogue. Born in 525 BCE in Eleusis, near Athens, he fought against the invading armies of Darius at Marathon (490 BCE) and Xerxes at Salamis (480 BCE), the latter battle also providing the subject matter of his earliest extant play, the *Persians* (first produced in 472 BCE). The titles of seventy-nine of his plays are known to us, but only seven of these have survived in their totality, three of which constitute the *Oresteia* trilogy.

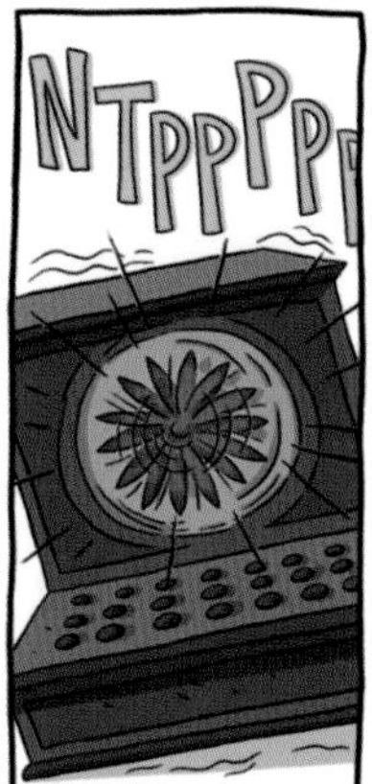
NTPPPP

Algorithm A methodical, step-by-step procedure described in terms of totally unambiguous instructions, which starts at a specified initial condition and eventually terminates with the desired outcome. Though there is no reason why a well-written cooking recipe, or the instructions for finding a certain geographical location or address cannot be called algorithms, the term originated in mathematics, where it is still mostly used. The word "algorithm" comes from a European transcription of the name of the 9th century astronomer and mathematician Al Khwarizmi of Baghdad, who catalogued and championed these methods, having invented many of them. His compendium of algorithms, the *Hisab al-jabr w'al-muqabala*, is generally considered to be the first algebraic treatise, the very words *al-jabr* in it also providing the root for our word "algebra". An example of a simple mathematical algorithm is the method we learn in elementary school for adding two integers: "write the two numbers one under the other with their rightmost digits justified to the right; add their last digits; if the sum is less than 10, write that number right under the other two; if it is greater than 10, write the second digit of the sum right under the other two, and add the first digit to the sum of the digits immediately to the left ..." and so on. Probably the earliest sophisticated Western algorithm is the one given in Euclid's *Elements* for computing the greatest common divisor of two non-negative numbers. Algorithms gained prominence in the West in the 15th century with the introduction of the decimal system, which, in stark contrast with the Roman numerical system, was amenable to fast calculations, such as the one described above. Numerical algorithms played a central part in the scientific and technological revolutions. Today, algorithms are usually coded in advanced notations called *programming languages*. They are often transmitted over the Internet, and constitute the *software* that is the workhorse, platform, and backbone of computers and the Internet.

Aristotle Born in 384 BCE, in Stageira, Chalcidice, Aristotle is, with Plato, the most influential of Greek philosophers. After he left Plato's Academy, Aristotle developed his own philosophy, which departed from his teacher's in its emphasis on the systematic observation of reality and the attempt to shape general, inductive laws. Perhaps his most lasting contribution is the systematization and exposition of logic in a series of works which later commentators edited collectively as the *Organon* ("instrument" or "tool"). The books comprising the *Organon*, i.e. *The Categories, On Interpretation, The Prior Analytics, The Posterior Analytics, The Topics* and the *Sophistical Refutations* formed the core of the canon of the study of logic until the 19th century. At the heart of Aristotle's logic is the combination of non-ambiguous statements in *syllogisms* to create new statements, different from the original but following necessarily from them. Aristotle also had a huge and lasting influence on mathematics, mainly through his emphasis on the notion of *first principles* from which any logical investigation must begin. It was this notion that found its mathematical incarnation in Euclid's concept of the axioms from which every theory has to begin. Aristotle died in 322 BCE.

Athena The ancient Greek goddess of wisdom, as well as of the arts and the city. Athena sprang in full armour from the head of Zeus, father of the Gods, whose favourite child she became. Athena was the patron goddess of ancient Athens and greatly beloved of the Athenians, to whom, according to legend, she gave the gift of the olive tree. The Parthenon, in the centre of the Acropolis, is a temple to her — the word comes from *parthenos*, meaning "virgin". Athena's role in Aeschylus' trilogy, the *Oresteia*, gives her a central role in the origin myth for the Athenian democratic invention of trial by jury, a system based on reason, as opposed to the older ones, where juridical authority emanated from a ruler's absolute power.

Axiom Since the time of Euclid, who was working in the wake of Aristotle's philosophy of logic, mathematicians agree that a workable theory must rest on some (few) agreed-upon *first principles* that don't require proof. This is a logical necessity if one wants to avoid, on the one hand, *infinite regression* (endlessly having to base something on something else) and, on the other, *circuitous* thinking (constructing proofs for statements which, however indirectly, assume the original statement to be true in the first place). Up to the 19th century, axioms were generally considered to be *self-evident* truths about the world, a view more or less still valid in Frege's idea of axioms as the reflection of an ulterior logical reality. After Hilbert, however, and under the influence of the mathematico-philosophical school of *formalism,* which developed from his ideas, axioms came to be seen as existing independently of any outside reality, the only requirements of an axiomatic system being: for the individual axioms their *grammatical correctness* (in other words, their being well-formed according to the rules of the logical language in which they are expressed), and *independence* (their not being derivable from the other axioms of the particular theory); and, for the whole set of axioms, its internal *consistency* (not containing axioms which contradict one another).

Boole, George Born in 1815, Boole was a largely self-taught mathematician who later became a professor of mathematics and logic at Queen's College in Cork, Ireland. His great contribution to mathematics is in the field of logic. In his book *An Investigation of the Laws of Thought,* Boole developed the idea that logical propositions can be expressed in a purely symbolic language which allows them to be manipulated by operations, similar to the operations of elementary arithmetic. At the heart of Boole's work is the idea of a *propositional calculus,* constructed somewhat as Leibniz imagined it. The "Boolean search" on the Internet, involving use of the logical *connectives* "and", "or" and "not", can be traced directly back to his ideas. Yet, despite the great value of his work in mathematizing logical arguments, Boole did not offer any great insights into the study of logic itself, having worked wholly within Aristotle's classical model. In Boole's system, *symbols* such as X and Y (essentially they are variables that can take only the two values 0 and 1) are joined via the three connectives mentioned above, as well as the "implies" connective envisaged by Aristotle. (Interestingly, the Stoic Chrysippus had already identified these connectives in the 3rd century BCE.) The application of algebraic identities, such as the three below, allow a logician to simplify logical expressions and deduce useful conclusions from them:

$$(X \text{ or } Y) = (Y \text{ or } X)$$
$$\text{not } (\text{not } X) = X$$
$$\text{not } (X \text{ and } Y) = (\text{not } X) \text{ or } (\text{not } Y)$$

What this logical formalism is lacking is the ability to express semantic connections between propositions. So, for example, there is no way to denote in the above that X and Y may stand for the two propositions "Plato is older than Socrates" and "Socrates is older than Plato." This weakness is remedied in the predicate calculus. Boole died in 1864.

Cantor, Georg Born in 1845, Cantor studied under some of the greatest mathematicians of his time, including Richard Dedekind and Karl Weierstrass. He spent the greatest part of his career teaching at the University of Halle, where he wrote his seminal papers demonstrating the great power of the ideas of set theory. His most famous theorem is that the set of so-called *real* numbers (all the numbers on the *number line*, i.e. the *natural* numbers 1, 2, 3... etc., together with the decimals, including 0 and the negatives) is *uncountable*, in other words cannot be put into a one-to-one correspondence with the whole numbers 1, 2, 3,... etc. On the contrary, as Cantor had already proved, the set of all *rational numbers*, i.e. all fractions of natural numbers, such as 2/3 or 11/476, is *countable* and *can* be put in such a correspondence. As both countable and non-countable sets have an infinity of elements, Cantor's results essentially proved that there are various, mutually exclusive kinds of infinity. As his theorems were extremely counter-intuitive and thus totally unexpected, they created much skepticism about set theory in the mathematical community.

One of Cantor's teachers, the great mathematician Leopold Kronecker, as well as the mathematical giant Henri Poincaré were strongly critical of sets, though the other mathematical giant of that time, David Hilbert, was one of Cantor's greatest supporters. The identification of two distinct 'sizes' of infinity in the set of real numbers, a smaller and a larger one, ushered in the question of whether there could exist a third kind: could there be a subset of the real numbers that is neither countable nor can be put in one-to-one correspondence with the reals? Cantor conjectured that none exists, a guess ever since called "the Continuum Hypothesis" — the *Continuum* being another name for the number line. Cantor worked towards a proof of the Continuum Hypothesis for many years, but never achieved it. In 1940, Kurt Gödel proved that the Continuum Hypothesis is *consistent* with the prevailing axiomatic system of set theory (which does *not* amount to a proof of it). In 1963, the young American mathematician Paul Cohen proved that it is *independent* of it, i.e. that no real proof of the Hypothesis can be established from it, or, alternatively, that the axioms of set theory are consistent with the Hypothesis being either true *or* false. This discovery earned Cohen a Fields Medal, a distinction often called "the Nobel Prize of mathematics". Cantor suffered from severe emotional problems and was repeatedly hospitalized with a diagnosis of melancholia, which certain historians of mathematics have ascribed to the hostile reactions of some mathematicians to set theory, and others to the constant anxiety resulting from his fruitless attempt to prove the Continuum Hypothesis. In the last decades of his life Cantor did no mathematical work, but wrote extensively trying to substantiate two strange theories: a) that the plays of Shakespeare were in fact written by the Elizabethan philosopher Sir Francis Bacon, and b) that Christ was the natural son of Joseph of Arimathea. The second of these is a basic component of many variations of the Holy Grail legend, and a standard part of esoteric lore. Cantor died in a mental asylum, where he had been interned against his will, in 1918.

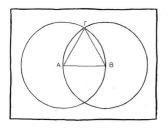

Euclid Born around 325 BCE, Euclid is the earliest Greek mathematician whose work is extant in the form in which he actually gave it — theorems of earlier mathematicians survive only as transcribed by others. He lived and worked in Alexandria, where he was associated with the Great Library. His opus magnum, the *Elements*, has been a best-seller for twenty-three centuries, and is the book with the most editions in the Western world, after the Bible. Though many of the theorems appearing in it are probably not Euclid's own discoveries, the work of compilation, classification and presentation of the existing mathematics of his day is totally his own. The *Elements* is a majestic conceptual edifice which, inspired by Aristotle and his work on logic, starts from definitions and first principles, the axioms (*aitêmata* – literally "requests" – in Euclid's original Greek) and then proceeds to arrive at all the theorems through rigorous proof. Though later students of logic, especially at the time of the quest for the foundations of mathematics and after, have criticized Euclid for relying too much on geometric insight or taking many more things for granted than his axioms, the influence of the *Elements* has been colossal, and it is rightly considered to be the fountainhead of the mathematical method. Euclid died around 265 BCE.

Foundations of Mathematics

Since the time of Pythagoras, mathematicians have wondered about the nature of mathematical truth, the ontology of mathematical entities and the reasons for the validity of proof and, more generally, mathematical knowledge. From the Enlightenment until the middle of the 19th century, the prevailing scientific ideology saw mathematics as the only way of reaching a truth that is final, absolute and totally independent of the human mind's capacity to understand it. The basic notions of mathematics were thought to reflect essential properties of the cosmos and the theorems to be the truths of a higher reality. This absolute faith in mathematics is reflected in the crowning of the discipline as the "Queen of the Sciences", a title whose previous holder, significantly, was theology. This view is usually termed *mathematical Platonism*, having its roots in the views of Plato – and, at least partly, Pythagoras before him – on the transcendent Ideas (*eidê*). Yet, in the 19th century this traditional belief was undermined in the minds of some people and eventually led to a serious *foundational crisis* in mathematics. The first of the discoveries which caused this loss of faith, dating from the time of the Renaissance, was that of the *imaginary numbers* (i.e. those involving the square root of minus one). But in the 19th century the appearance of *non-Euclidean geometries* strengthened the arguments against the "self-evident" truth of the axioms. The most troublesome of all mathematical concepts, though, was that of infinity. Problems concerning the mathematical handling of the infinite had first been alluded to by Zeno, in his paradoxes, resurfaced with the invention of the calculus in the 18th century and the counterintuitive and ill-defined concept of an *infinitesimal*, and peaked in the last two decades of the 19th century, most especially with set theory and Georg Cantor's results on infinite sets. The problems that came to the surface via set theory — chief among them Russell's Paradox — culminated in severe doubts about "self-evident" truths and thus, indirectly, about the value of all mathematical knowledge. It was principally the wish to overcome these doubts that fuelled the quest for secure foundations. The "Program" proclaimed by David Hilbert in the early 1920's bearing his name, expresses the most optimistic version of the foundational dream: the creation of a formal system for all mathematics, also containing a proof that this axiomatization is *consistent* (i.e. can lead to no contradictions), *complete* (leaves no unprovable truths) and *decidable* (one is able to decide in every occasion whether a formula follows from the axioms or not, through the application of a set of algorithms.)

Frege, Gottlob Born in 1848, Frege spent the greatest part of his mathematical life as a profesor at the University of Jena. He is generally considered to be the father of modern logic, whose notation and method he expounded first in his *Begriffsschrift* (which literally translates from the German as "concept script"), published in 1879. In it, Frege departed from the earlier logicians working in the wake of Aristotle, by explicitly introducing the notion of *variable* in logical statements. In the place of the older type of statements like "Socrates is a man", he introduced propositions like "x is a man", propositions that can be true or false according to the value given to x — this particular one, for example, is true if x is equal to "Alecos" but false if it's "Manga". Frege also invented the notion of *quantifiers*, the *universal* (written ∀) which makes a statement true "for every x"; and the *existential* (written ∃) which says that "there exists an x" which makes a statement true. He later applied his new logical system to the quest for the foundations of mathematics. His *Grundgesetze der Arithmetik* (*The Basic Laws of Arithmetic*) is the first great work of the school of *logicism,* whose central tenet is that mathematics is essentially a branch of logic. The first volume of the *Grundgesetze* was published in 1893 and the second, containing the addendum on Russell's Paradox, in 1903. Though Frege's logical symbolism has been abandoned as particularly cumbersome, most of the basic concepts and methods he invented still form the backbone of logic. After the *Grundgesetze*, Frege didn't do any important foundational work. In the last decades of his life he became increasingly paranoid, writing a series of rabid treatises attacking parliamentary democracy, labour unions, foreigners and, especially, the Jews, even suggesting "final solutions" to the "Jewish problem". He died in 1925.

Gödel, Kurt He was born in 1906 in the town of Brünn, Moravia, then a part of the Austro-Hungarian Empire (the city now called Brno, in the Czech Republic). Gödel studied mathematics in Vienna, where he became fascinated with mathematical logic and the question of the foundations of mathematics. In his doctoral dissertation, he advanced Hilbert's Program by proving his *Completeness Theorem*, a result establishing that all valid statements in Frege's *first-order logic* can be proved from a set of simple axioms. In 1931, however, he proved the Incompleteness Theorem for *second-order logic*, i.e. for a logic powerful enough to support arithmetic and equally or more complex mathematical theories. Gödel became one of the youngest members of the Vienna Circle, though his deeply-ingrained, idealist belief in the independent, Platonic existence of mathematical reality eventually alienated him from the other members, who embraced a materialist-empirical worldview. During the late thirties, Gödel was hospitalized twice for severe melancholia. In 1940, after the *Anschluss*, i.e. the annexation of Austria to Nazi Germany, he managed to escape the country with his wife and took the trans-Siberian route to the United States. He became one of the first members of the Institute for Advanced Study at Princeton, where he spent the rest of his life. His most important mathematical result from this period is the proof that Cantor's Continuum Hypothesis is *consistent* with the axioms of set theory (i.e. that it would not be in contradiction with them, if true). At Princeton, Gödel developed a close friendship with Albert Einstein and worked for a while on the theory of relativity, establishing the mathematical possibility of a non-expanding, rotating universe, in which time travel can be a physical reality. In later life, Gödel became increasingly paranoid. He died in January 1978, at the Princeton hospital, where he had been admitted for the treatment of a non-life-threatening urinary tract problem. The cause of his death was malnutrition: he refused to eat for fear that the hospital staff was attempting to poison him.

Hilbert, David Hilbert was born in 1862 in Königsberg, Prussia (now Kaliningrad, Russia) and spent the greatest part of his life at the University of Göttingen, the world's most renowned mathematical centre at that time. He is one of the greatest mathematicians in history and, with Henri Poincaré, the greatest of his era. He made important contributions to many branches of mathematics including invariant theory, algebraic number theory, functional analysis, the calculus of variations, the theory of differential equations and more, also pioneering new methods of proof. In 1899 he published *Grundlagen der Geometrie* (*Foundations of Geometry*), a book which gave geometry a firm basis, with new axioms, therein improving on the work of Euclid. In his famous 1900 talk at the International Congress of Mathematicians, in Paris, he attempted to give a bird's-eye view of the mathematics of the twentieth century, by way of twenty-three great open questions. Of these, now renowned "Hilbert's problems", eleven have been fully solved, seven partly, while the rest — the Eighth, also known as "the Riemann Hypothesis" is the most famous of these — are still unsolved. The Second Problem is the one demanding a proof of the *consistency* (the *completeness* was considered more or less obvious) of arithmetic — and it was this that spurred on a lot of the work on the foundations and logical structure of arithmetic, including Gödel's. In the 1920s, his ideas of the previous decades related to the foundations of mathematics culminated in what became known as "Hilbert's Program", i.e. a project to formalize all mathematics on an axiomatic basis, including a proof that this axiomatization is consistent. Hilbert's battle cries of "in mathematics there is no *ignorabimus*" (i.e. no "we shall not know") and "we must know, we shall know" — the latter spoken only a few days before Gödel's first announcement of his Incompleteness Theorem — encapsulate the quintessence of foundational optimism. Though the results of Gödel, Alan Turing and Alonzo Church put an end to Hilbert's grand ambition, the Program continued to exert a great influence on logic and foundational matters, and especially the development of proof theory. Though in outward appearance and behaviour Hilbert gave the impression of a paragon of normality and mental health, the way he treated his only son, Franz, raises questions. When the boy was diagnosed with schizophrenia, at age 15, his father sent him off to an asylum, where he spent the rest of his life. Hilbert never visited his son. He died in 1943.

Incompleteness Theorem In 1931, the 25 year-old Kurt Gödel proved two theorems that are sometimes referred to as "the" Incompleteness Theorem — though occasionally this form is used to denote the first of these. The *completeness* of a logical system is the property that every *well-formed* (i.e. grammatically correct by the rules of the system) proposition in it can be proved *or* disproved from the system's axioms. Gödel's earlier *Completeness Theorem* shows that there is a simple such axiomatic system for *first-order logic*. However, the holy grail of Hilbert's Program was to create a *complete* and *consistent* axiomatic system that can support *arithmetic*, i.e. the mathematics of whole numbers. Such a system would require *second-order logic*, i.e. a system that is also able to accept sets as values of variables. Gödel shocked the mathematical world by proving, in his famous paper "On Undecidable Propositions in the *Principia Mathematica* and Related Systems", that any consistent axiomatic system for arithmetic, in the form developed in the *Principia*, must of necessity be *incomplete*. More precisely, the first of the two Incompleteness Theorems establishes that in a logical axiomatic system rich enough to describe properties of the whole numbers and ordinary arithmetic operations, there will always be propositions that are grammatically correct by the rules of the system, and moreover *true*, but cannot be proven within the system. The second Incompleteness Theorem states that if such a system were to prove its own consistency it would be inconsistent. This was a new, devastating blow to Hilbert's Program, with its goal that a strong axiomatic system should be equipped with a proof of its own consistency.

Intuitionism This is the philosophy of mathematics created by the great Dutch mathematician Luitzen Egbertus Jan Brouwer (1881-1966), though some consider Henri Poincaré, with his strong belief in the role of intuition in mathematics, a clear precursor. Intuitionism is based on the belief that intuition and time are fundamental to mathematics, which cannot be made a-temporal or *formal* in the sense of Hilbert. Contrary to what *logicists* like Frege and Russell thought, Brouwer was convinced that logic is founded upon mathematics rather than the other way round. Also, he was totally against the theorems of Georg Cantor in the theory of sets, considering them ill-formed. Time-hallowed logical laws, such as that of the *excluded middle*, and mathematical techniques in use since the time of the ancient Greeks, such as the *reductio ad absurdum*, were put on trial and their use condemned. In fact, Brouwer believed that all the theorems making use of these in their proofs — where infinite sets of mathematical objects were concerned — should be excised from the body of mathematics, a view that made the brilliant British logician and mathematician Frank Ramsey call intuitionism "mathematical Bolshevism". Although his logic and mathematics were formalized by his student Arend Heyting, Brouwer remained skeptical towards any such attempt to the end of his life.

Leibniz, Gottfried This great German philosopher, mathematician, scientist and student of logic was born in 1646. He served in the courts of several German rulers as diplomat, political advisor and historian, all the while pursuing his theoretical studies. He invented the *infinitesimal calculus* concurrently with, but independently from, Isaac Newton, also proposing the notation for its operations that is still in use today. He was a strong proponent of *philosophical optimism*, with his theory that our world is the "best of all possible worlds", having been created by a God who is both loving and almighty. He is considered the most important logician after Aristotle and before Boole, having envisioned the *calculus ratiocinator*. This was a kind of computational propositional logic that would enable completely rigorous and rational decision-making which could eliminate all disagreement among rational (as Leibniz thought them) human beings. Sadly, Leibniz did not manage to realize this most coveted of his many projects. He died in 1716.

Logic The term covers a broad spectrum of disciplines — not unexpectedly, as it derives from one of the semantically richest Greek words, *logos*, some of whose meanings are *word, speech, thought, reason, ratio, rationality,* and/or *concept* — but can perhaps be best described as the study of methodical thinking, deduction and demonstration. The books of Aristotle's *Organon* present an extensive study of the deductive patterns called *syllogisms*, which for over two millennia were considered practically synonymous with logical thinking. Until the middle of the 19th century, logic was considered a branch of philosophy. But with the advent of Boole and his algebra of propositions and, more importantly, Frege and his "concept script" which led to a predicate calculus, it increasingly came within the province of mathematics. The new logic revealed both the underlying mathematical nature of the subject and its potential role in the creation of solid foundations of mathematics. The basic claim of the school in the philosophy of mathematics known as *logicism* — the school founded by Frege, of which Bertrand Russell was one of the primary exponents — was that all of mathematics can be reduced to logic or, in other words, that mathematics is essentially a *branch* of logic. After the years of the foundational quest, however, and especially after Gödel's results, logic became a well-developed, diversified field in the interface between philosophy and mathematics. In the second half of the 20th century it also found unexpected applications in computer science, where it provides solid foundations for the design and verification of software and hardware, as well as for databases and artificial intelligence.

Oresteia Written by Aeschylus and first performed in the theatre of Dionysus, in Athens, two years before the poet's death, in 458 BCE, it is the only extant trilogy of Greek dramas — although the satirical play *Proteus*, intended to be performed after the trilogy, is missing. In the trilogy's first play, the *Agamemnon*, the eponymous hero and leader of the Greek forces in Troy returns a victor to his hometown of Argos, with the captive prophetess, Cassandra. Though his wife, Clytemnestra, at first appears to rejoice at his return, she has other plans. She and her lover, Agamemnon's cousin Aegisthus, murder Agamemnon and become the new sovereigns of Argos. In the *Libation Bearers*, the second play, the chorus of women accompanies Agamemnon's daughter Electra to her father's tomb. The forlorn Electra is hoping for revenge, which she can only carry out with the help of her brother, Orestes, who is in exile. When Orestes clandestinely returns to Argos, he and Electra plan and execute the murder of Aegisthus and then, in a highly dramatic scene in which Clytemnestra bares her breasts before his naked sword, Orestes also kills her, his own mother. The third play, the *Eumenides*, or "beneficent ones", is one of the most unusual in the history of drama: all its speaking parts, apart from that of Orestes himself, are taken up by gods or other supernatural entities. The chorus consists of the *Erinyes* or Furies, archaic goddesses of revenge, who chase Orestes from the temple at Delphi, where he has been ritually purified by the god Apollo, to Athens.

In a totally unprecedented move — for a god anyway — Athena, the patron god of Athens, decides to let the citizens of Athens judge Orestes' case, thus giving a mythological origin-story for the democratic innovation of a court of law, with citizen jury. The trial and its aftermath develop as shown in our book's finale, though our text is only an approximate translation, slightly adapted, of Aeschylus' original words.

Peano, Giuseppe Born in 1858, this great Italian mathematician and logician spent the greatest part of his creative life as a professor at the University of Turin. Though his ideas were not as influential as Frege's in the search for the foundations of mathematics, Peano, like Frege, created a notation for *first-order logic* and a system of axioms for arithmetic, that is still in use — in fact, our arithmetic is formally called *Peano arithmetic*. He influenced Bertrand Russell greatly, especially with his logical notation, which was much more user-friendly than Frege's. Peano believed that all mathematics could be formalized and expressed in a common, minimal language that originates from just a few axioms. But when he tried to present his own version of this universal mathematics in textbook form and use it for teaching, his students revolted, eventually causing the book's withdrawal. Inspired by his attempts to unify all mathematics by use of a common logical language, Peano later created an international auxiliary natural language, for use among people of different linguistic backgrounds, based on a simplified form of Latin which he called *Latino sine flexione*. However, like so many other artificial international languages, such as *Esperanto, Volapük, Ido* — all of them the offspring of an overoptimistic age — Peano's brainchild proved to be a mere pipe dream. Peano died in 1932.

Poincaré, Henri Born in 1854 in Nancy, France. Although he studied engineering at the École Polytechnique and the École de Mines, Poincaré was to become, with David Hilbert, the greatest mathematician of his time. He has been called the "last universal mathematician", i.e. the last one to have profound knowledge of all the mathematics of his time. He made important contributions to many diverse fields of mathematics, among them differential equations, automorphic functions, the theory of several complex variables, probability and statistics. With his *Analysis situs* he essentially created the major 20th field of *algebraic topology*, and his work on the 3-body problem laid the groundwork for what is now called *chaos theory*. Despite his many great innovations, Poincaré was an extremely practical man, involved to the end of his life — alongside of his mathematical research — with the most down-to-earth of affairs, as for example the inspection of mines and an engineering project to make the Eiffel tower function as a huge antenna broadcasting time signals to navigators. He was probably the last of the great mathematicians to adhere to an older conception of mathematics, which championed a romantic faith in intuition over rigour and formalism. This stance was made famous by his reaction to the set theory of Georg Cantor as a "disease, from which mathematics will eventually be cured." His views on mathematical creation, encapsulated in his saying that "logic is barren, unless fertilized by intuition," are seen by many as the precursor of Luitzen Brouwer's school of intuitionism, a theory at the antipodes of Hilbert's strict *formalism*. Poincaré died in 1912.

Predicate calculus Often used synonymously with *predicate logic* and *first-order logic*, the predicate calculus is Frege's extension of the propositional logic developed by Boole. In the predicate calculus, elementary propositions (or *predicates*) are composite objects of the form $P(a, b, c, ...)$, where P is a *symbol* in the language, and a, b, c, etc. are *constants* or *variables*. For example, if "older" is a propositional symbol, "Plato" is a constant and "x" is a variable, then "older (Plato, x)" is a well-formed proposition, describing that Plato is older than x. Propositions of this type can then be combined by Boole's connectives "and", "or", "not" and "implies" and prefixed by Frege's quantifiers, such as "for all x" (written $\forall$) and "there exists y" (written $\exists$). Thus, "there exists x older (x, Plato)" means that there is (at least) one individual who is older than Plato. Evidently, this is a much more ambitious attempt at creating Leibniz's *calculus ratiocinator* than Boole's simpler formal logic. By employing symbols from various fields of mathematics (such as "<", "+", and so on) one can create predicates expressing mathematical statements in this formal, logically exact language. For example, the theorem in arithmetic stating that every integer is either odd or even can be written thus:

$$\forall x \, \exists y \, (x=y+y \text{ or } x=y+y+1)$$

Rigorously defined, the version of the predicate calculus called *first-order logic* employs simple mathematical objects as variables, whereas in *second-order logic* variables can also be sets, making possible statements like "there is a set S". This, more powerful language, can express all known mathematics. Whether a sentence in the predicate calculus, first- or second-order, is true or false depends on the *model* whereby the sentence is interpreted. Thus, for example, the simple arithmetical theorem given above is true of the whole numbers in the ordinary interpretation of "+", but becomes false if we interpret the symbol "+" as multiplication. However, some sentences — called *valid* — are true independently of interpretation, because they embody basic properties of Boolean connectives and quantifiers. Kurt Gödel's *Completeness Theorem* provides a simple, complete axiomatic system for proving validity in *first-order logic*.

Principia Mathematica The extremely influential, but highly controversial, essentially unfinished work in which Alfred North Whitehead and Bertrand Russell attempted to rescue Frege's grand project to create foundations of mathematics built on logic, in the wake of the crisis brought on by Russell's Paradox. The title *Principia Mathematica* (i.e. "Principles of Mathematics") in itself provoked controversy, as it is the exact same as that of Newton's greatest work; many in the British mathematical community thought this choice to be in bad taste, if not actually blasphemous. The three volumes of the *Principia*, published in 1910, 1912 and 1913, were based on a developed version of Russell's *theory of types*, the so-called "ramified", which imposed a hierarchical structure on the objects of set theory. This could not be made to yield the required results, however, without the addition of what Russell called an *axiom of reducibility*, which eventually became one of the main reasons for negative criticism of the whole work. Logicians found this axiom extremely counter-intuitive, a far-fetched and basically artificial method to sweep the very problem it was trying to solve under the rug. Despite the fact that the *Principia* fell short of its authors' immense ambition, it had a huge influence on the shaping of modern logic, its greatest effect possibly being the inspiration and context it provided Kurt Gödel for his groundbreaking discovery, the Incompleteness Theorem.

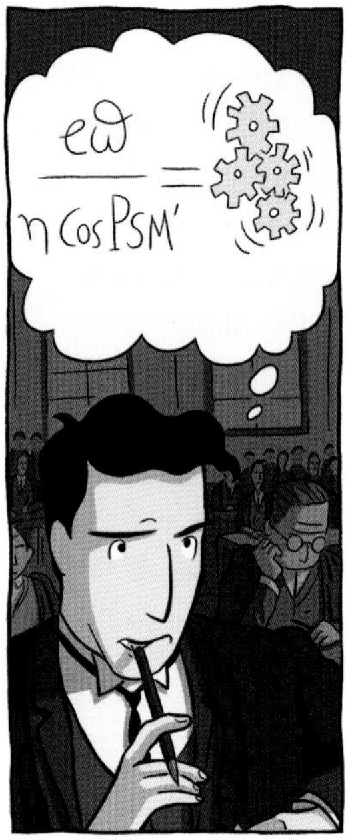
eω
η Cos PSM'

Proof The process of arriving at the logical verification of a mathematical or logical statement, starting from a set of agreed-upon first principles (these could be either axioms or already proven statements, deriving from these axioms), and proceeding by totally unambiguous and unabridged logical steps or *rules of inference*. The demonstrations of geometric propositions in Euclid's *Elements* were considered for over two millennia to set the standard of excellence to which mathematical proof should aspire. Yet, towards the end of the 19th century his method came under logical and philosophical scrutiny and was found to lack, principally, in two directions: a) in its sense of the logical "obviousness" of the axioms, and b) in its logical gaps, where intuition – which, in Euclid's case was mostly visual-geometric – took over from strict application of a formal system of rules. In a sense, Frege's and Russell and Whitehead's *logicist* project was developed as a reaction to the imperfections found in Euclid's proofs, as well as all those developed in his wake. The logicists, as well as the *formalists* working on the foundations of mathematics, aimed at a fully developed theory and practice of rigorous proof, by which arithmetic (as the basis of all mathematics) would begin from a small number of consistent axioms, and eventually lead, via proof, to the full range of truth. Hilbert's seminal question, which he called the *Entscheidungsproblem* ("decision problem"), posed in 1928 and answered seven years later by Alan Turing, is equivalent to the demand for a totally powerful apparatus of proof, which can provide a *provable* or *unprovable* response to any mathematical statement by virtue of a rigorous algorithm.

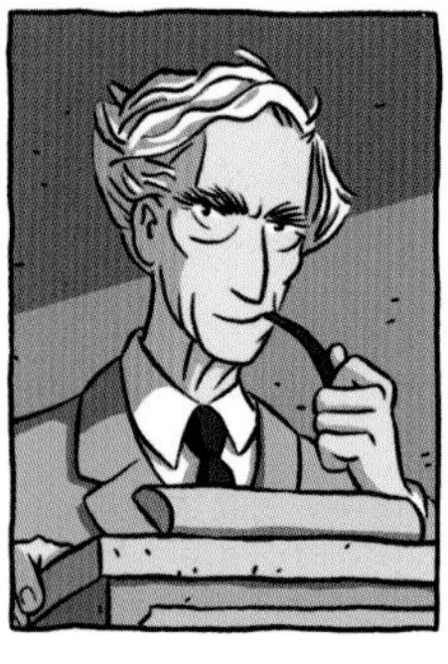

Russell, Bertrand Born in Wales, in 1872, Bertrand Arthur William, the Third Earl Russell – this is his full name, by virtue of his noble descent – was the grandson of the important politician Lord John Russell, whose title he eventually inherited. An orphan at the age of four, Russell was raised by his paternal grandparents, and after his grandfather's death two years later, exclusively by his grandmother, Lady Russell. He grew up at the family home of

Pembroke Lodge, in Richmond Park, to the west of London. Russell is now perhaps best known to a wider public for his work in philosophical exposition. His *History of Western Philosophy*, published in 1945, remains to this day a classic of idiosyncratic, yet intelligent and highly readable exposition of complex ideas. And while his later work as a pro-peace and anti-nuclear activist also earned him international fame, Russell's greatest contribution is in mathematical logic, ranking him, along with Aristotle, Boole, Frege and Gödel, with history's greatest logicians. Despite the momentous importance of his work in the establishment of a scientific logic, its direct influence on Gödel's great discoveries, and the indirect on the Vienna Circle's "scientific worldview" and the philosophies of *logical positivism* and *logical empiricism*, Russell's work in logic essentially ends with the *Principia Mathematica*, the book he co-authored with Alfred North Whitehead, completed just before he turned forty. Russell considered the *Principia* essentially a failure, as it fell short of his —and the other logicists'— grand ambition, of founding mathematics securely on logic. Russell married four times and fathered three children. His first son, John, as well as John's daughter, were diagnosed as schizophrenics, and the latter committed suicide. This pathology was very possibly another instance of the streak of mental illness running in the family, manifest both in Russell's uncle William and his aunt Agatha. During the last decades of his life, Russell gave all his energy to the struggle for nuclear disarmament, becoming an emblematic figure of pacifism. He died in 1970.

Russell's Paradox Discovered in 1901, as Russell was working on his first book on the foundations of mathematics, the *Principles of Mathematics* (published in 1903), the Paradox, in the form originally expressed, shows an essential flaw in Cantor's set theory, developed from Bolzano's simple concept of a "collection of elements with a common property".
By the generality of this definition, which Frege extended to the realm of logic, one can speak of a "sets of sets" and thus, eventually, of the "set of all sets".
Of the elements of this all-encompassing set one defines the property of "self-inclusiveness", i.e. of a set containing itself as an element. Thus, for example, the set of all sets is a set (and thus contained in itself), as is the set of all entries in a list (it can appear as an entry in a list),

but the set of all numbers is not a number and thus not contained in itself. By virtue of this property, we can define the "set of all sets which don't contain themselves", and ask, with the young Russell, the question: "Does this set contain itself or not?" See what happens: if it *does* contain itself, it follows that it is one of the sets which don't contain themselves (as this is the property that characterizes elements of this set) and thus cannot contain itself. But if it *doesn't* contain itself, then it does not have the property of not containing itself, and thus does contain itself. This situation, in which assuming something implies its negation, and vice versa, is called a *paradox*. When a paradox, such as Russell's, arises in a theory, it is a sign that one of its basic premises, definitions or axioms is faulty. Though historically developed within the context of the theory of sets, Russell himself later viewed his paradox as essentially having to do with self-reference, i.e. with statements referring to themselves, such as Euboulides' "I am now lying to you."

Self-reference Literally, the quality of a statement of referring to itself. However, it is also used more generally in logic to characterize statements which *include* themselves within their scope of reference, as in the "barber" story used to explain Russell's Paradox. The barber lives in a town wherein a law decrees that "all residents of the town must either shave themselves or be shaved by the barber." This law is *self-referential* as the barber, apart from being "the barber" referred to, is also one of the "residents of the town". Self-reference has played a seminal role in logic and mathematics, already from the time of the Greeks. From Euboulides' self-referential statements, to Cantor, whose proof of the *non-denumerability* of the *real numbers* relies heavily on a numerical variant of self-reference, to Russell and his paradox, and to Gödel. In fact, Gödel proved his Incompleteness Theorem by creating, in the context of modern logic, a statement that is quite similar in spirit to that of Euboulides, with one crucial difference: while Euboulides states "this statement is false", Gödel's ingenious variant essentially says, in the language of arithmetic, "this statement is *unprovable*." Any consistent axiomatic theory in which one can formulate such a statement must be necessarily incomplete: for either this statement is *false*, in which case it is both false *and* provable, contradicting the *consistency* of the axiomatic system, or *true*, in which case it is both true *and* unprovable, establishing its *incompleteness*.

Set theory   The study of collections of objects united by a common property — in some cases this property can be nothing more than the fact that they are defined to be members of the same set, as for example in the arbitrarily defined set whose elements are the numbers 2, 3, 8, 134, 579. Sets were first studied by the Czech mathematician Bernard Bolzano (1781-1848), who also introduced the term *Menge* ('set') and defined the notion of a set's *cardinality*, i.e. of its "size" in a way not directly involving measurement. Thus, one can speak of two sets having the same cardinality if their elements can be put in a one-to-one correspondence - without ever needing to know via the precise number of these elements. This has the great advantage that it also works for infinite sets, where the notion of number does not apply: mathematicians don't think of "infinity" as a number. However, some seeming paradoxes, such as the fact that the *whole* and the *even* numbers can be put into a one-to-one correspondence (just by multiplying each whole number by 2, or dividing each even number by 2), thus showing a subset to have the same cardinality as the containing set, prevented Bolzano from developing the theory further. The advanced mathematical discipline of set theory was arguably born on December 7, 1873, when Georg Cantor wrote to his teacher, Richard Dedekind describing his proof of the *non-denumerability* of the *real* numbers (the set of the whole numbers, decimals, zero and the negative numbers), as opposed to the *denumerability* of the *rationals* (all fractions), which Cantor also proved — denumerability is defined as a one-to-one correspondence with the *natural* numbers 1, 2, 3... etc. The concept of a set is almost too primitive to merit a mathematical definition, and is practically impossible to define informally without the use of some synonym (here we used the word "collection"). It is precisely this "naturalness" of the concept in Bolzano's and Cantor's work that led to Russell's Paradox. To overcome it, and to rule out the flawed concept of "the set of all sets" it allowed for, one has to come up with bottom-up constructions and axioms for sets, as in the *Principia Mathematica* and, later, the system called "ZFC", from the names of its two creators, Ernst Zermelo and Abraham Fraenkel, and the Axiom of Choice, a necessary additional axiom that allows the theory to deal with infinite sets. Set theory is considered by some the most basic branch of mathematics, as all others can be defined in terms of it. This was the gist of an over-ambitious project undertaken, from the 1930s onwards, by the group of brilliant French mathematicians writing under the pen name of "Nicolas Bourbaki".

Tractatus Logico-Philosophicus Ludwig Wittgenstein wrote his seminal philosophical work during WWI, building on his pre-war notebooks and ideas on logic. It contains his solution of (in his own words) "all the problems of philosophy", dealing with the world, representation, and language. Originally called *Logische-Philosophische Abhandlung* ("Logical-Philosophical Treatise"), it was renamed for its English publication under the influence of G. E. Moore, with his predilection for Latin titles. In the *Tractatus*, Wittgenstein uses many techniques and ideas from logic, especially those of Frege and Russell, as well as insights from totally different philosophical positions, mostly that of Arthur Schopenhauer. Though publication by the then totally unknown Wittgenstein was only made possible when Russell accepted to write an introduction, calling the book "an important event in the philosophical world", the *Tractatus* was the cause of the two men's falling-out. Wittgenstein considered Russell's — not altogether appreciative — introduction to his work to be fraught with misunderstandings and philosophical errors, while Russell saw in the *Tractatus* the first signs of Wittgenstein's decline — as he saw it — into mysticism. The tight structure of the book proceeds with seven main propositions, each developed in a chapter, which are further developed in propositions arranged by a rather pedantic, and often somewhat confusing, system of numbering. The first two propositions (1 and 2) expand the positions that "the world is all that is the case", and that "what is the case" are *facts*, and *combinations* of facts. This is a departure from classical philosophy and the metaphysics of Aristotle in particular, according to which the world consists of *objects*. In the logical language of the *Tractatus*, objects do figure within states of affairs, but in complex combinations and relationships with each other, and not as elementary units. The next two propositions (3 and 4) develop mostly what has been called the *picture* theory of language, whereby a "thought is a proposition with sense." Passing here to representation and language, Wittgenstein delimits thoughts to logical propositions, but within a context and in reference to the world. This is perhaps the most subtle part of the book, and also the one which relates to Wittgenstein's idea of mathematics and logic as machines for producing *tautologies*. Propositions 5 and 6 develop the idea that "propositions are truth functions of elementary propositions", in which mathematical-symbolic notation is used to explain precisely what a truth function is. Here Wittgenstein uses logic to define propositions (and thus language and thought) as the combinations of *atomic*, or elementary propositions, combined through Boole's laws of composition. This part of the book actually contains the first mention of what is now known as the "truth table method" for dealing with Boolean functions. The book's final clause, proposition 7, is: "What we cannot speak of, we must pass over in silence." (This and other quotes are from the D.F. Pears and B.F. McGuinness translation.) This last proposition was given two highly divergent interpretations, the extreme *positivist* one of the Vienna Circle, by which what one "cannot speak of" (logically) is, quite literally, *non-sense*, and the one that Wittgenstein and others himself later gave, which Russell termed "mystical", according to which what "one cannot speak of" is the truly important. The *Tractatus* is one of the most influential and closely-studied books in Western philosophy. Its influences are legion and it may have also influenced — and certainly was vindicated by — the way in which computers and databases model the world today.

Turing, Alan Born in London in 1912, this great British mathematician is generally considered to be the father of computer science. Turing contributed to many areas of mathematics, but is mostly remembered for one of his earliest results in logic. While a student at Cambridge, he became fascinated by the foundations of mathematics and especially the Incompleteness Theorem of Kurt Gödel, which inspired him to study Hilbert's *Entscheidungsproblem* ("decision problem"), a question that had survived Gödel's analysis. The *Entscheidungsproblem* asks whether, given a logical system, there is an algorithm for deciding whether a proposition is provable within the system or not. Turing's answer was a devastating "no". To reach this, he first had to define rigorously the notion of algorithm. His ingenious definition in terms of a theoretical "machine" with a *central control* and a tape for *memory*, *input* and *output*, anticipated in important ways the digital computer and has had, since then, an enormous influence on computational practice and thought. *Turing machines* — as they are now called — share with today's computers the key property of *universality*, in that a machine can carry out *any* computational task, provided it is supplied with an appropriate *program* for it. Two other mathematicians, Alonzo Church (later Turing's thesis advisor at Princeton) and Emil Post, came up independently, and at about the same time, with algorithm formalisms that were ultimately shown equivalent to Turing's. Yet his formalization had the greatest impact, mainly because of the extreme simplicity of its basic construction, which can, nevertheless, achieve extremely complex results. The work of Turing — as well as that of the others mentioned — on algorithms and methods for the general solvability of problems, is an obvious outgrowth of the foundational quest and thus, in a sense, its culmination. During World War II, Turing presided over the design and construction of two series of electronic computers, the "Bombe" and the "Colossus". These were used successfully — and crucially for the war effort — for breaking several German cryptographic codes, including the notoriously hard "Enigma" of the German navy. After the war, Turing worked in the fledgling British computer industry, did important work in biology and founded the field *of artificial intelligence* by proposing what became known as the *Turing test*, a method for determining whether an artifact "can think". Always interested in sports and games — he was an accomplished long-distance runner — Turing was the first to develop ideas for a chess-playing program, making mastery in the game one of the goals towards which the designers of intelligent machines should strive. In 1952 he was prosecuted on account of his homosexuality, then a punishable offense in Britain. As an alternative to a jail sentence, he agreed to undergo an experimental "treatment" with estrogens, which probably caused the severe depression which led him to take his own life, in 1954.

Vienna Circle A group of philosophers and philosophically-minded scientists, who met in Vienna between 1924 and 1936. Their main aim was two-fold: to build a strong empiricist philosophy using the insights into scientific methodology garnered from recent advances in logic, mathematics and physics, and to apply the methodology of the physical sciences to the social. The scientifically-trained philosopher of science Moritz Schlick is generally recognized to be the group's leader. Some of the most prominent members were: the mathematicians Hans Hahn, Olga Hahn-Neurath, Gustav Bergmann, Karl Menger, and Kurt Gödel for a short period of time; the physicist Philipp Frank; the social scientist Otto Neurath and the philosophers Viktor Kraft and Rudolf Carnap. The group met informally Thursday evenings at Vienna's "Café Central", but was later constituted as a society with public meetings. Despite the group's informal nature, the members had a common core of philosophical beliefs, expressed in a sort of manifesto, titled "The Scientific Conception of the World". The members of the Circle declared that the work of Frege, Russell and Einstein provided their first inspiration, while the *Tractatus Logico-Philosophicus* of Ludwig Wittgenstein functioned as their direct model. The philosophies of *logical positivism* and *logical empiricism*, expressing the worldview of the members of the Circle, state that knowledge comes from experience — and thus, basically, from scientific observation and experiment — developed into theory through logical analysis and synthesis. Still, following the *Tractatus*, members of the Vienna Circle held that logic and mathematics only deal in *tautologies*, and thus do not provide knowledge as such, but only one of the tools for the elaboration of empirical knowledge. According to the worldview of the Circle, statements that cannot be reduced to experience (such as theological or ethical pronouncements) cannot be right or wrong, as they are — quite literally — *non-sense*, having no meaning. The most extreme version of this tenet, due to Carnap, actually required that for a statement to be meaningful, its truth or falsity must be verifiable by an algorithm reducing it to observable truths — a new incarnation of Leibniz's *"calculemus"*. Carnap later tried to reconcile this view with the Incompleteness Theorem. Though the Vienna Circle, in its original form, was dissolved in 1936, after Schlick's murder by a paranoid ex-student and Nazi sympathizer, its spirit continued to live on. Most of its members managed to flee Austria and emigrate to England and the United States, where they had a major influence on the development of post-war philosophy.

Von Neumann, John Born in Budapest in 1903 ("John" is the anglicized form of the Hungarian "Janos"), von Neumann showed very early signs of unusual intellectual prowess, being able to do mental division of 8-digit numbers and converse in ancient Greek by the age of six. He studied mathematics in Budapest, obtaining a PhD at 22, meanwhile also working towards a degree in chemical engineering at the renowned Technical University of Zürich, to please his father.

He rapidly became the star mathematician of his generation, legendary for his penetrating and rapid-fire mathematical genius. Upon attending the lecture where Gödel announced the first Incompleteness Theorem, von Neumann was the first to realize the result's import, and did indeed proclaim "it's all over" after the talk. But he made crucial suggestions to Gödel right after it, and went on to prove the second Incompleteness Theorem — which however Gödel himself had also proven independently in the meantime. Von Neumann never worked on the foundations of mathematics again. Possessing a wide-ranging mathematical genius, he made contributions to many different branches, he has been called "the last of the great mathematicians", having made great contributions to many different branches of mathematics, among them set theory, operator algebras, ergodic theory and statistics. He also did important work in quantum theory, fluid mechanics and mathematical economics, being the co-founder (with economist Oscar Morgenstern) of the field of *game theory*. During WWII, he was one of the brains behind the atomic bomb, and after it headed the U.S. government committee in charge of the construction of the hydrogen bomb. Perhaps most important of all his work, however, was his contribution to the creation of computers. While he was working as a consultant in the design of one of the first electronic computers, in 1946, and influenced by Alan Turing's ideas, von Neumann developed an array of fundamental design principles, postulating, among others, a *central processing unit* and separate *memory* devices where both *data* and *programs* are both stored. Practically all subsequent computer designs have been based on this basic model, now known as the *von Neumann architecture*. Von Neumann went on to become one of the first great computer scientists, especially excelling in what now would be called *scientific computing*, i.e. the use of computers for scientific research. He died of cancer — possibly the result of his attendance of thermonuclear tests — in 1957.

Whitehead, Alfred North English mathematician and philosopher. Born in 1861, he studied mathematics at Cambridge, where he also taught for many decades. In 1891 he married Evelyn Wade, an Irish woman much younger than himself. Before his intense, decade-long collaboration with Bertrand Russell on the *Principia Mathematica*, Whitehead published his book *Universal Algebra*, an attempt to study the types of symbolic reasoning in various algebraic systems from a very modern — for its time — formal viewpoint. After Russell's abandonment of the *Principia*, in 1913, Whitehead tried to write a fourth volume, on geometry, but never completed it. The two men had very little interaction after the publication of the *Principia*, and Whitehead did not contribute to the 1925, second edition of the book, having moved on to mathematical physics and later philosophy. He died in 1947.

Wittgenstein, Ludwig Wittgenstein is considered by many to be the greatest philosopher of the 20th century. He was one of the eight children of industrialist Karl Wittgenstein, one of Austria's wealthiest and most powerful men, and a great patron of the arts. Of his four brothers, three committed suicide in early manhood, while the fourth, Paul, went on to become a renowned concert pianist. After two years of engineering studies, Wittgenstein developed a strong interest in logic and the foundations of mathematics. He went to see Frege, who suggested that he go to Cambridge to study with Russell, a piece of advice Wittgenstein followed. The association deeply influenced both men, but probably the teacher more than the student. During his service with the Austro-Hungarian army in WWI, Wittgenstein won several medals for his valour, his citations underlining his "*sang-froid under fire*". He was eventually captured by the enemy and completed his magnum opus, the *Tractatus Logico-Philosophicus*, in an Italian prisoners' camp. After the war he donated the huge fortune left to him by his father to his three sisters and, having, as he believed and declared, "solved all the problems of philosophy" with the *Tractatus*, he worked as a gardener, architect, and eventually as a teacher in a small village in Lower Austria. In 1929, possibly inspired by interactions with members of the Vienna Circle, as well as attending a lecture on the philosophy of mathematics by Luitzen Brouwer, on intuitionism, Wittgenstein returned to Cambridge and philosophy. He retracted his earlier work as dogmatic and went on to create a new, extremely influential philosophical stance often referred to as "the late Wittgenstein". Unlike the ideas in the *Tractatus*, Wittgenstein did not attempt to put his later philosophy in a systematic treatise, but presented them in a series of more or less independent remarks. Many of these he saw as forming a book, which was posthumously published as *Philosophical Investigations* — this, as well as a few books based on his notebooks, or transcripts of lectures or discussions, are all that we have of his later thought. This is a philosophical position of an extreme anti-dogmatic nature, focusing on *language* and *psychology* (what we now call *cognitive psychology*), instead of logic and objective truth, and on fuzzy concepts such as "family resemblance" and "language games" instead of clear definitions and propositions. In this later phase, Wittgenstein's thinking is characterized by a vicious criticism of philosophy as it had been practiced until then, by others but also himself — it was for this criticism more than anything else, that Russell was dismissive of his later work, referring to Wittgenstein's decision to "become a mystic". Most of his negative criticism of mathematics — which he increasingly came to view as a purely practical activity, a craft legitimized only by its use in application — is contained in transcriptions of his lecture notes at Cambridge. Of particular interest is the dialogue with one of the attendees at these lectures, Alan Turing, who strongly disagreed with his ideas on mathematics. Wittgenstein died in 1951.

Bibliography

In preparation for *Logicomix* we read many books — in addition to those we had read earlier, before the idea for the project was even born — and consulted many more, and even more articles. Of all these, we mention here very few, chosen either for the wealth of the information they contain, for their astuteness, profundity and/or synthetic ability. Clearly, this list represents a personal choice, and nothing more: these are the books that we most liked and found most useful.

Andersson, Stefan. *In Quest of Certainty: Bertrand Russell's Search for Certainty in Religion and Mathematics Up to the* Principles of Mathematics *(1903).* Stockholm: Almqvist & Wiksell International, 1994.

Davis, Martin. *The Universal Computer: The Road from Leibniz to Turing.* New York: W. W. Norton & Company, 2000.

Gray, Jeremy J. *The Hilbert Challenge.* Oxford: Oxford University Press, 2000.

Janik, Allan, and Stephen Toulmin. *Wittgenstein's Vienna.* New York: Simon and Schuster, 1973.

Monk, Ray. *Ludwig Wittgenstein: the Duty of Genius.* London: Jonathan Cape, 1990.

— *Bertrand Russell: the Spirit of Solitude.* London: Jonathan Cape, 1996.

— *Bertrand Russell: the Ghost of Madness, 1921-1970.* London: Jonathan Cape, 2000.

Reid, Constance. *Hilbert.* Berlin: Springer-Verlag, 1970.

Rota, Gian-Carlo. 1997. "Fine Hall in its Golden Age". In *Indiscrete Thoughts,* ed. Fabrizio Palombi, 4-20. Boston: Birkhauser Verlag AG.

Russell, Bertrand. *My Philosophical Development.* London: George Allen & Unwin, 1959.

— *The Autobiography of Bertrand Russell,* 3 vols. London: George Allen & Unwin, 1967-1969.

— Griffin, Nicholas, ed. *The Selected Letters of Bertrand Russell.* London: Routledge, 2002.

Scharfstein, Ben-Ami. *The Philosophers.* Oxford: Oxford University Press, 1980.

Stadler, Friedrich. *The Vienna Circle, Studies in the Origins, Development, and Influence of Logical Empiricism.* English translation by Camilla Nielsen. Vienna: Springer-Verlag, 2001.

Van Heijenoort, Jean. *From Frege to Gödel.* Cambridge: Harvard University Press, 1967.

Wittgenstein, Ludwig. *Tractatus Logico-Philosophicus.* (English translation: D. F. Pears and B. F. McGuinness. London: Routledge and Kegan Paul, 1961.)